Textbook of Molecular Biology

Textbook of Molecular Biology

Edited by Gildroy Swan

SYRAWOOD
PUBLISHING HOUSE
New York

Published by Syrawood Publishing House,
750 Third Avenue, 9th Floor,
New York, NY 10017, USA
www.syrawoodpublishinghouse.com

Textbook of Molecular Biology
Edited by Gildroy Swan

International Standard Book Number: 978-1-68286-411-1 (Hardback)

Cataloging-in-publication Data

Textbook of molecular biology / edited by Gildroy Swan.
        p. cm.
Includes bibliographical references and index.
ISBN 978-1-68286-411-1
1. Molecular biology. 2. Biomolecules. 3. Recombinant molecules. I. Swan, Gildroy.
QH506 .T49 2017
572.8--dc23

Printed in the United States of America.

TABLE OF CONTENTS

PREFACE

This book covers in detail some existent theories and innovative concepts revolving around molecular biology. The ever growing need of advanced technology is the reason that has fuelled the research in this field in recent times. Molecular biology refers to the study of molecular activity at the biological level. It encompasses the elements of biochemistry, biology, genetics and chemistry. It aims at examining the processes taking place in living organisms and at determining the roles and structure of biomolecules. This book explores all the important aspects of molecular biology in the present day scenario. Different approaches, evaluations, methodologies and advanced studies have been included in it. The text is appropriate for students seeking detailed information in this area as well as for experts.

This book unites the global concepts and researches in an organized manner for a comprehensive understanding of the subject. It is a ripe text for all researchers, students, scientists or anyone else who is interested in acquiring a better knowledge of this dynamic field.

I extend my sincere thanks to the contributors for such eloquent research chapters. Finally, I thank my family for being a source of support and help.

Editor

Cis-Regulatory Control of the Nuclear Receptor *Coup-TF* Gene in the Sea Urchin *Paracentrotus lividus* Embryo

Lamprini G. Kalampoki, Constantin N. Flytzanis*

Department of Biology, University of Patras, Patras 26504, Greece

Abstract

Coup-TF, an orphan member of the nuclear receptor super family, has a fundamental role in the development of metazoan embryos. The study of the gene's regulatory circuit in the sea urchin embryo will facilitate the placement of this transcription factor in the well-studied embryonic Gene Regulatory Network (GRN). The *Paracentrotus lividus Coup-TF* gene (*PlCoup-TF*) is expressed throughout embryonic development preferentially in the oral ectoderm of the gastrula and the ciliary band of the pluteus stage. Two overlapping λ genomic clones, containing three exons and upstream sequences of *PlCoup-TF*, were isolated from a genomic library. The transcription initiation site was determined and 5′ deletions and individual segments of a 1930 bp upstream region were placed ahead of a GFP reporter cassette and injected into fertilized *P.lividus* eggs. Module *a* (−532 to −232), was necessary and sufficient to confer ciliary band expression to the reporter. Comparison of *P.lividus* and *Strongylocentrotus purpuratus* upstream *Coup-TF* sequences, revealed considerable conservation, but none within module *a*. 5′ and internal deletions into module *a*, defined a smaller region that confers ciliary band specific expression. Putative regulatory *cis*-acting elements (RE1, RE2 and RE3) within module *a*, were specifically bound by proteins in sea urchin embryonic nuclear extracts. Site-specific mutagenesis of these elements resulted in loss of reporter activity (RE1) or ectopic expression (RE2, RE3). It is proposed that sea urchin transcription factors, which bind these three regulatory sites, are necessary for spatial and quantitative regulation of the *PlCoup-TF* gene at pluteus stage sea urchin embryos. These findings lead to the future identification of these factors and to the hierarchical positioning of PlCoup-TF within the embryonic GRN.

Editor: Efthimios M. C. Skoulakis, Alexander Fleming Biomedical Sciences Research Center, Greece

Funding: LGK was supported by a graduate fellowship from 'Alexander S. Onassis' Foundation. Support for this work was obtained by a 'Karatheodoris' grant from the University of Patras to CNF. The EU program ASSEMBLE supported CNF for a visit to Dr. Arnone's laboratory at Stazione Zoologica Napoli. The funders had no role in study design, data collection and analysis, decision to publish, or preparation of the manuscript.

Competing Interests: The authors have declared that no competing interests exist.

* Email: kostas@bcm.edu

Introduction

Coup-TFs (Chicken ovalbumin upstream promoter-Transcription Factors) are orphan members of the steroid-thyroid-retinoic acid super family of hormone receptors [1,2]. The structure of Coup-TFs, as for all these receptors, follows a common motif divided into several domains based on specific functional properties [3]. They regulate a plethora of target genes through activation or repression and their ability to heterodimerize with other nuclear receptors to suppress transcriptional activation has been well documented [4,5].

Coup-TFs are present in all metazoans [6,7] and show extensive protein sequence conservation across species, which suggests functional similarity [8]. They play a crucial role in homeostasis, organogenesis, neurogenesis and cellular differentiation throughout embryonic development [8]. In vertebrate embryos, Coup-TFs are expressed in the neural ectoderm and mesoderm [9–14]. In *D.melanogaster* the Coup-TF ortholog, *svp*, is expressed in the central nervous system and is involved in the differentiation of the photoreceptor cells [2]. In *C.elegans*, *Coup-TF* is a member of the uncoordinated group of genes (*unc-55*) and is involved in motor neuron function and copulation [15,16]. Knockout and loss or gain of function experiments in vertebrate and invertebrate animals lead to severe abnormalities, mainly in the developing nervous system, and lethality, emphasizing

the embryonic significance of Coup-TFs [17–19]. Knockdown of PlCoup-TF in sea urchin embryos, via egg injection of morpholino antisense oligonucleotides, results in developmental arrest and inhibition of later stage morphogenesis (unpublished results from this laboratory).

The *S.purpuratus SpCoup-TF* gene [20] is expressed throughout embryonic development [21]. At the gastrula stage, the *SpCoup-TF* gene is expressed in the presumptive oral ectoderm, and at the pluteus stage, mainly in the ciliary band. In the embryonic oral territory, it acts as a repressor of the aboral ectoderm CyIIIb actin gene [22]. The SpCoup-TF protein is maternal and oscillates between the condensed chromatin in mitosis and the nuclear periphery in interface of the early blastomeres, in a cell cycle dependent manner [23]. During larval development the *SpCoup-TF* gene is expressed specifically in neuronal cells (unpublished data from this laboratory). Although Coup-TF's embryonic role has been extensively studied, there is little information about the regulation of this gene in various organisms. To date, there are three known factors involved in the regulation of the *Coup-TFI and Coup-TFII* genes in mice. One of these is Sonic Hedgehog. This protein is secreted from the notochord and is found to play a crucial role in the induction of the chicken and mouse *Coup-TFII* in motor neurons [14,24]. The transcription factor Ets1 also acts as a positive regulator for the *mCoup-TFI* gene [25]. Retinoids also induce expression of *mCoup-TFI* and *mCoup-TFII* genes *in vitro* [26] and *in vivo* [10], suggesting that Coup-TFs might be direct targets of RA and RX receptors. In higher than physiologic concentrations, retinoic acid has also been shown to act as a ligand of Coup-TFII, which alters its LBD conformation from a repressive to an activating state. These data suggest that Coup-TF may be a ligand regulated nuclear receptor [27].

The herein experiments aim to elucidate the regulation of the *PlCoup-TF* gene in the sea urchin *Paracentrotus lividus* embryo and specifically to determine the regulatory elements that direct *PlCoup-TF* expression in the oral embryonic territory. To this end, we analyzed the *in vivo* function of *PlCoup-TF's* upstream region using the *GFP* gene in a reporter cassette [28,29]. GFP constructs harboring different deletions and specific mutations of the upstream *PlCoup-TF* region were introduced into sea urchin fertilized eggs via microinjection [30,31]. Spatial embryonic expression of GFP was monitored by fluorescent microscopy at the pluteus stage, where the definite oral ectoderm and ciliary band, the embryonic territories of endogenous *PLCoup-TF's* expression, are easily discernible. These experiments unveiled an upstream regulatory region (module *a*), which is necessary and sufficient for correct spatial expression of the reporter gene. EMSA experiments, using nuclear extracts from *P.lividus* embryos and three *in silico* identified response elements within module *a*, revealed the presence of transcription factors that specifically bind these elements. Site-specific mutagenesis indicates that correct ciliary band expression of the *PlCoup-TF* gene is mediated by the transcription factors, which bind to the three elements, RE1, RE2 and RE3, within module *a*. Identification of these response elements facilitates the recognition of the corresponding transcription factors, following isolation from embryonic nuclear extracts. This will make feasible the placement of PlCoup-TF downstream of other regulators in the oral ectoderm and ciliary band GRN of the sea urchin embryo. Furthermore, inhibition of PlCoup-TF's expression in the embryo should reveal its downstream gene targets.

Materials and Methods

PlCoup-TF gene cloning

A pair of primers (*up2: 5'atgttgtggtgcgcaggtcagc* and *do1: 5'gtccggtgtccgcataatgatccgt*) was synthesized based on the known *SpCoup-TF* sequence [20] and used in a PCR reaction with genomic DNA as substrate, isolated from the sperm of an individual *P.lividus* male. The PCR product is a 251 bp fragment within the first exon flanking the AUG codon of the *PlCoup-TF* gene. This fragment was cloned into the pCRII vector (Invitrogen) and sequenced. A *P.lividus* genomic library (a gift from Dr. Valeria Matranga, Palermo, Italy), prepared in the lambda FIXII vector (Invitrogen), was screened using the 251 bp fragment as a hybridization probe. The first round of screening involved 5×10^5 genomic clones. Isolated positives were subjected to two more rounds of screening from which, two positive overlapping clones were isolated and named "A" and "Φ". The first exon and the upstream region of the gene, contained in the λ clone "Φ", were subcloned and sequenced to facilitate the design of primers and the construction of the GFP expression cassettes.

Determination of the transcription initiation site

A *PlCoup-TF* gene fragment encompassing 1930 bp upstream and 543 bp 5'UTR region was amplified with PCR, using as substrate the genomic clone "Φ" and a pair of primers designed to be complementary to the arm of the phage (λ: 5'tctaga-

gagctcgcggcc) and the 5'UTR (*c: 5'gatggcgttgagggaatcg*) of the *Pl-Coup-TF* gene. The isolated PCR product was then cloned into the pCRII vector and sequenced. 5'RACE-PCR was performed with the FirstChoice RLM-RACE Kit (Ambion), using 2 μg of polyA$^+$ RNA isolated from the ovary of an individual *P.lividus* female. The pair of gene specific primers used for the nested PCR was *tin1 5'cagttctccacgaattgacggc* (in the *PlCoup-TF's* coding region) and *lup3 5'agcggagtaatcgcagctaa* (in the *PlCoup-TF's* 5'UTR). The PCR products were analyzed in a 1,2% agarose gel, purified and cloned into the PCRII vector and sequenced.

Upstream *PlCoup-TF/GFP* expression constructs

GFP expression constructs included 1930 bp of the upstream region, as well as deletions thereof, and were based on the EpGFPII vector previously used for expression in sea urchin embryos [28]. The first series of constructs use the 1930 bp of the upstream region (from the upstream end of the insert in clone "Φ" to +1) fused to *Endo16* gene's basal promoter carried by the vector. The EpGFPII vector also carries part of the actin gene *CyIIa* 5'UTR and its ATG codon. This 1930 bp fragment was amplified with PCR using the primers *LabUpRI*: 5'gacgatgctat-cataatagtcatgg<u>gaattc</u> and *LabUpBamH1*: 5'c<u>taggatcc</u>gactgatgttta-gatggaaag and the cloned 2.5 kb *PlCoup-TF* fragment as substrate. The *LabUpRI* primer includes an internal EcoRI site (underlined) of the substrate sequence at −1930 and the *LabUpBamH1* primer includes the most upstream transcription initiation site and has a prosthetic BamHI site (underlined) at its 5' end. The PCR product was digested with the restriction enzymes EcoRI and BamHI and ligated to the EcoRI/BglII digested EpGFPII vector. This construct was designated −1930.

Using the construct −1930 as substrate, six consecutive upstream deletions were amplified with PCR and designated **−1639**, **−1398**, **−781**, **−532**, **−232**, **−19**, respectively. The following pairs of primers were used for each PCR reaction to produce the respective deletions. **−1639**: *Rup7 5'tccttgctgtgag-caatttt/GFPpolyAright 5'gtaaaacctctacaaatgtggt*; **−1398**: *Rup6 5'gttttggtatgaagtacgaaacat/GFPpolyAright*; **−781**: *Rup2 5'gaccc-gagtaatcccaacaa/GFPpolyAright*; **−532**: *Rup1 5'agcaggacgaag-gatttgag/GFPright 5'actgggttgaaggctctcaa*; **−232**: *Rup4 5'ttccggtcttcagaaagttca/GFPright*; **−19**: *Rup3 5'ccgtca-gaaaaagctttcca/GFPright*. Thus, each PCR product contains the respective upstream *PlCoup-TF* fragment fused to the *Endo16* gene's basal promoter and the *GFP* gene. The PCR products were purified, diluted appropriately, mixed with carrier genomic *P.lividus* DNA and used for microinjections.

Expression GFP constructs with isolated upstream *PlCoup-TF* segments

The 1930 bp upstream region was divided into six sub-regions (**a–f**), which were amplified with PCR and sub-cloned into the EpGFPII vector without the basal *PlCoup-TF* promoter. The recognition sequences for the restriction enzymes EcoRI and BamH1 were added to the 5' ends of the pairs of primers used for the PCR reactions, respectively. The numbers in parentheses designate the limits of each tested segment. **a** (−532/−212): *Rup1 5'agcaggacgaaggatttgag/Pur4Bam 5'ctaggatcctgaactttct-gaagaccggaa*; **b** (−781/−513): *Rup2 5'gacccgagtaatcccaacaa/ Pur1Bam 5'ctaggatccactcaaatccttcgtcctgct*; **c** (−1079/−762): *Rup5 5'tcgaatcacacaccgaaaaa/Pur2Bam 5'ctaggatccttgttgggat-tactcgggtct*; **d** (−1398/−1121): *Rup6 5'gttttggtatgaagtacgaaa-cat/Pur5Bam 5'ctaggatccttttcggtgtgtgattcgatg*; **e** (−1639/ −1375): *Rup7 5'tccttgctgtgagcaattttt/Pur6Bam 5'ctaggatccttttcg-tacttcataccaaac*; **f** (−1930/−1681): *LabUpRI 5'gacgatgctatca-taatagtcatggaattc/Pur7Bam 5'ctaggatccaaaaattgctcacagcaagga*.

The PCR products were purified and cloned into the PCRII vector. The inserts were then recovered by double digestion with EcoRI/BamH1 from the PCRII vector and ligated to the EcoRI/BglII double digested EpGFPII vector. Plasmid DNAs were linearized with EcoRI, purified, diluted appropriately, mixed with carrier genomic *P.lividus* DNA and used for microinjections.

Upstream deletions of module *a*

The GFP fusion construct containing module *a*, was used as template to produce with PCR various upstream and internal deletions. For the upstream deletions the following pairs of primers were used (the numbers in parentheses designate the limits of each tested sub-segment). **D-a1** ($-473/-212$): *a1* 5′ctaagatcttttccatagaagcctaatccg/BglGFPright 5′ctaagatctactggggttgaaggctctcaa; **D-a2** ($-387/-212$): *a2* 5′ctaagatctatagattaatccagaagttgc/BglGFPright; **D-a3** ($-334/-212$): *a3* 5′ctaagatctcggagaataacttgtgatgtt/BglGFPright; **D-a4** ($-298/-212$): *a4* 5′ctaagatctgtttgagctccgaataccagt/BglGFPright; **D-a5** ($-232/-212$): *a5*: 5′ctaagatcttccggtcttcagaaagttca/BglGFPright. All primers carry a prosthetic BglII recognition site at their 5′ end. PCR fragments were purified, diluted appropriately, mixed with carrier genomic *P.lividus* DNA and used for microinjections.

Internal deletions into module *a*

Internal deletions into module *a*, were performed with PCR using as substrate the circular plasmid carrying module *a*, and pairs of primers flanking the respective region to be deleted (the numbers in parentheses designate the limits of each deleted internal sub-region). **D1** ($-452/-386$): *a1R* 5′ctaagatctcgcattaggcttctatggaaa/a2; **D2** ($-386/-333$): *a2R* 5′ctaagatctttcaacaaaatgacacacaat/a3; **D3** ($-312/-297$): *a3R* 5′ctaagatctaacatcacaagttattctccg/a4; **D4** ($-276/-231$): *a4R* 5′ctaagatctactggtattcggagctcaaac/a5. All primers carry a prosthetic BglII recognition site at their 5′ end. PCR fragments were purified, digested with the restriction enzyme BglII, ligated with T4 DNA ligase and used to transform DH10B bacterial cells. Selected clones were sequenced and used as templates for PCR reactions with the pair of primers BglRup1 and BglGFPright. All PCR products were purified, diluted to the appropriate concentrations, mixed with carrier *P.lividus* genomic DNA and used for microinjections.

Site specific mutations into module *a*

Site specific mutagenesis was performed by deleting or changing the selected elements with PCR using as substrate the circular plasmid carrying module *a* and pairs of primers flanking the respective site to be deleted. Mutation **−453**: EtsR 5′ctaagatctattaggcttctatggaaatt and EtsF 5′ctaagatcttatacacattaatgactcta; mutation **−432**: ApR 5′ctaagatctattaatgtgtatacttcc and ApF 5′ctaagatctacaatcaatagaaattataaa; mutation **−377**: OtxR 5′ctaagatcttctatttcaacaaaatgacac and OtxF 5′ctaagatctagaagttgcatgatattgtt; double mutation **−453/−432**: EtsR and ApF. All primers contain the BglII restriction site at their 5′ end, which substitutes for the mutated sequence, to enable religation of the amplified plasmids and cloning. Selected clones were sequenced and used as templates for PCR reactions with the pair of primers BglRup1 and BglGFPright. PCR products were purified, diluted to the appropriate concentrations, mixed with carrier *P.lividus* genomic DNA and used for microinjections.

Collection of animals and embryonic cultures

P.lividus individuals were collected from the rocky shores of the Corinth gulf at a depth of 3–5 meters. A specific permission from the Greek authorities for the collection of limited numbers of sea urchins for research purposes was not required. The collection coastal area is not part of a national park or other protected area of the sea. The animals were returned alive to the collection site following the shedding of gametes. The specific location where the animals were collected is: N38.311, E21.783.

Gametes of mature adults were obtained by injection of 0.5 ml 0.5 M KCL into the coelomic cavity. Embryonic cultures for RNA and nuclear protein extract preparations were set up at a maximum concentration of 5×10^6 embryos per liter of filtered seawater and embryos were let develop at 18°C.

Microinjection of fertilized eggs

The eggs to be used for microinjection were collected into filtered seawater, de-jellied at pH: 5.5 with the addition of citric acid for 1 min, then the pH was titrated back to 8.3 with the addition of 0.5 M Tris-HCl and filtered through a 65 µm nylon mesh. The eggs were attached onto the covers of 35 mm plastic Petri dishes, pretreated for 2 min with a 2% protamine sulphate solution. Injections were performed using the techniques previously developed [30,31] with fertilized *P.lividus* eggs. For the 3,317 pluteus stage embryos scored for this study, a total of over 20,000 eggs were microinjected in a period spanning about four years. The duration of the study was due to the short spawning season of *Paracentrotus lividus* in the shallow waters, where the collection of animals took place. In addition, not all batches of eggs give viable or well developing embryos and furthermore, a percentage of embryos do not develop properly due to the microinjection injury. Thus, only well developed plutei from microinjected eggs, were collected and observed for GFP expression. Linear DNAs for microinjection were diluted in 30% glycerol at a concentration of $3-10\times10^3$ molecules/pl, plus a five-fold excess of linearized genomic *P.lividus* DNA as carrier. An approximate volume of 3–5 pls was injected into each fertilized egg and embryos were let develop for two days to pluteus stage at 18°C. Epifluorescence observations of developed embryos were carried out with a Zeiss Axioplan microscope using the GFP filter.

In silico upstream promoter analysis

Consensus binding sites for transcription factors within module *a*, were identified using the Transfac MatInspector package (Genomatix). Homologies between upstream *PlCoup-TF* (up to -1930) and *SpCoup-TF* (up to -2500) sequences, were identified using BLAST (NCBI) and the Family Relations II software using a 20 bp window [32].

Electrophoretic mobility shift assay

Nuclear protein extracts from *P.lividus* post-hatching blastula stage embryos were prepared as described [33]. Double stranded oligonucleotides used as probes, were prepared by annealing the complementary strands of each sequence corresponding to the respective response elements. RE1: 5′cctaatccgggaagtatacaca; RE2: 5′tatacacattaatgactctacaatca and RE3: 5′gttgaaatagattaatccagaa. Binding reactions (15 µl) contained 3×10^4 cpm of ^{32}P end-labeled double stranded oligonucleotide, 1 µg of poly-dA/dT, 1 µg of poly-dI/dC, 20 mM Hepes pH: 7.9, 3 mM MgCl$_2$, 1 mM DTT, 50 mM KCl, 8% Glycerol, and 2 µl of nuclear extract. Specific competitors of unlabeled double stranded oligonucleotides, when applicable, were used at 200fold excess to the labeled one. The samples were incubated on ice for 20min and electrophoresed on 6% poly-acrylamide gels in 0,5xTBE buffer at 6°C. Dried gels were exposed to X-ray films.

WMISH

In situ hybridizations to fixed embryos of different developmental stages were performed according to previously published protocols [34,35]. Digoxygenin labeled sense and antisense hybridization probes (3 ng/ml in hybridization buffer) were prepared by *in vitro* transcription of a PlCoup-TF cDNA clone carrying the entire coding sequence, using T7 and Sp6 RNA polymerases respectively and the *in vitro* transcription kit from Roche.

Results

Embryonic expression of the *PlCoup-TF* gene

The expression of the *PlCoup-TF* gene, throughout *P.lividus* embryonic development, was determined by *in situ* hybridization (Figure 1,a-l). The maternal PlCoup-TF mRNA is detected evenly distributed in the egg (Fig. 1a) and the 16-cell stage (Fig. 1b) embryo. Presumptive zygotic transcripts are detected from the hatching blastula stage (Fig. 1c) onwards. At this, as well as the gastrula stage (Fig. 1d), the *PlCoup-TF* gene is expressed in the cells of the presumptive oral ectoderm, whereas at prism (Fig. 1e) and pluteus (Fig. 1f) stages its transcripts are predominantly detected in the ciliary band. The ciliated cells at the anal side of the band, including the anal arms, seem to be more enriched in PlCoup-TF transcripts than the oral side of the ciliary band. Expression, but to a lesser extent, is also detected at the oral face and the supra-anal ectoderm (Fig. 1e). Sense probe hybridizations result in an even, faint background throughout the embryo and not specific spatial staining (Fig. 1g-l). The presence of PlCoup-TF transcripts in the egg and throughout embryonic development was also determined by quantitative PCR experiments, using total RNAs isolated from Paracentrotus lividus eggs and embryos (not shown).

Isolation of the *PlCoup-TF* gene

5×10^5 clones of a *P.lividus* genomic library were screened with the use of a 251 bp PlCoup-TF specific probe, resulting in the isolation of two positive genomic clones. The two overlapping *P.lividus* λ clones "*A*" and "*Φ*", cover about 33.5 kb of genomic region, which encompasses approximately 17 kb of upstream sequences and 16.5 kb of the *PlCoup-TF* gene (Figure 2). The

genomic clone "*Φ*", used in this study, contains 1930 bp of 5′ upstream sequence, 543 bp of 5′UTR sequence and 3 exons, the precise positions of which are not known. The fourth and last exon of the gene, bearing the C-terminal domain of the protein and the 3′UTR are not included in the cloned region contained in the phage "*Φ*".

The *PlCoup-TF* gene has multiple transcription initiation sites

Electrophoresis of the nested PCR products indicated two major DNA bands with approximate length of 200 bp (Figure 3A). Sequencing of cloned PCR fragments yielded thirteen isolated clones identifying at least five different transcription initiation sites. The two most upstream sites (at +1 and +16) were represented more frequently by four and five clones respectively (bold As in figure 3B) and correspond to the two major 5′-RACE products observed after electrophoresis. Four additional clones revealed initiation sites at +22 (two clones), +54 (one clone) and +58 (one clone), marked with lower case bold characters in figure 3B. Considering these results the most upstream of the transcription initiation sites was designated as +1 for the herein conducted experiments. An obvious 'TATA' element is missing in the proximal upstream sequence, although a putative 'CCAAT' box is underlined at −65.

Upstream deletion analysis of the *PlCoup-TF* gene

A series of deletions into the 1930 bp upstream region was prepared by PCR using gene specific primers (the border of each deletion is marked with a vertical bar in figure 4A) and a downstream GFP primer. Linear DNAs were injected into fertilized *P.lividus* eggs and developed embryos were scored for fluorescence 2 days post fertilization at pluteus stage (figure 4B). Since embryos developing from injected eggs integrate randomly the exogenous DNA, at various early embryonic cleavage stages, ensuing embryonic lineages are mosaic. Thus, rarely all cells of a certain lineage would express the transgene. Embryos exhibiting only one or two fluorescent cells were not scored as expression positive.

A large percentage of the developed embryos exhibit fluorescence except for deletion −19. As this result is expected, since only 19 bp of *PlCoup-TF's* upstream sequence are present in this

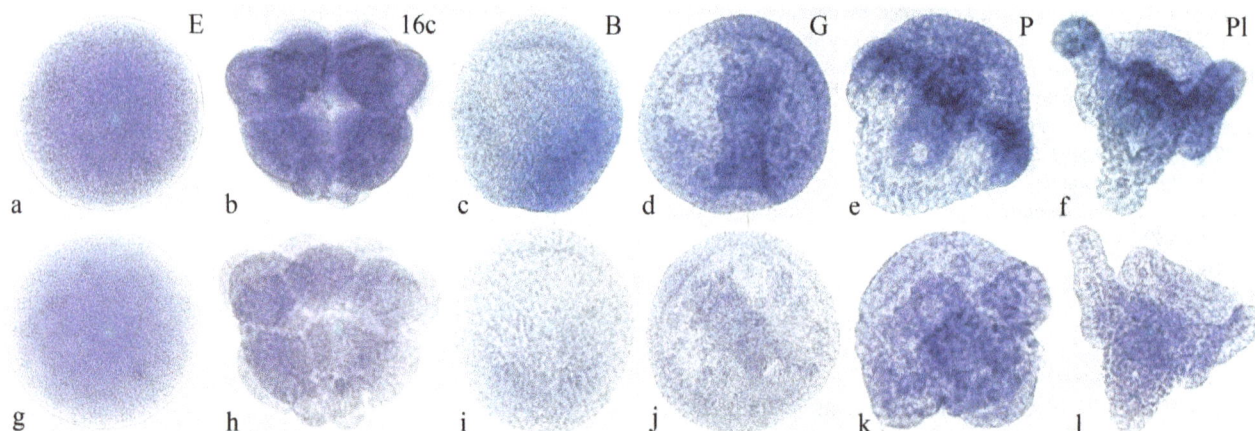

Figure 1. Spatial expression pattern of the *PlCoup-TF* gene. In situ hybridization of *P.lividus* embryos with antisense and sense digoxygenin labeled PlCoup-TF probes. **a–f**: antisense; **g–l**: sense probe hybridization. The maternal PlCoup-TF mRNA seems evenly distributed in the egg and at the 16-cell stage embryo. Zygotic transcripts are expressed in the presumptive oral ectoderm at blastula and gastrula stages and in the ciliary band at prism and pluteus stages. E: Unfertilized egg; 16c: 16-cell stage embryo; B: Hatching blastula; G: Gastrula; P: Prism and Pl: Pluteus.

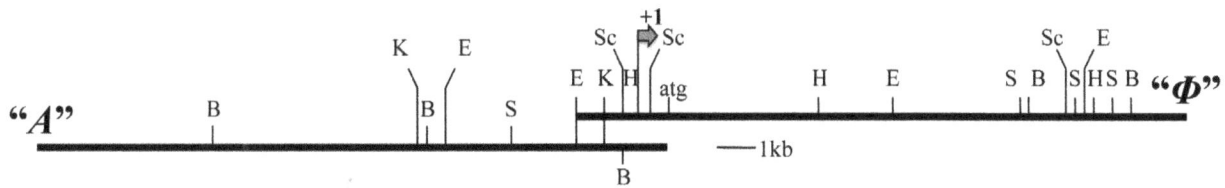

Figure 2. Restriction digest map of the overlapping inserts of genomic clones "A" and "Φ". The arrow marked by +1 symbolizes the initiation site and the direction of transcription. 'atg', marks the site of translation initiation. The scale (1 kb) is shown by a small bar. The enzymes used for mapping are: B: BamHI; E: EcoRI; H: HindIII; K: KpnI; S: SalI; Sc: SacI.

deletion, it is also indicative of *SpEndo16* basal promoter's lack of activity in the absence of additional *cis* regulatory inputs [36]. All other tested upstream sequences result in high percentage of fluorescent embryos. A significant percentage of injected embryos show fluorescence in a small number of secondary mesenchyme cells (2–3 cells) in addition to other tissues. Such fluorescence, seen at similar percentage levels in all injected groups (Me+, Fig. 4B and Table 1), is not considered to be specific expression of the various deletions, but rather background or "leakiness" of the assay.

The deletion constructs exhibiting both the highest percentage of embryos expressing GFP in the ciliary band and the lowest percentage of embryos expressing the reference gene in other embryonic territories, i.e. aboral ectoderm and endoderm, are −781 and −532. Thus, in this analysis, the smallest upstream segment that fulfills the above stated criteria is from −532 to −232 (Fig. 4B and Table 1).

Six upstream segments (a–f, Fig. 5A), cloned into the EpGFPII vector, were amplified by PCR as cassettes fused to GFP and injected into fertilized *P.lividus* eggs. Segments d, e and f do not result in any detectable GFP expression. Segment a directs GFP expression mostly in the ciliary band, while b and c show considerably less ciliary band specific expression and much higher non-specific expression in mesenchymal and endodermal cells

(Fig. 5B and Table 1). These results agree with the upstream deletion data (Fig. 4), which indicate that the minimal region conferring ciliary band specific expression extends between −532 and −232. Thus, it is evident that segment a, refer to as 'module a' from hereon, is necessary and sufficient to direct *PlCoup-TF's* specific expression in the ciliary band of the pluteus embryo.

Comparison of *Coup-TF* upstream regions between *P.lividus* and *S.purpuratus*

Using the programs BLAST and Family Relations II, we compared the upstream and the 5'UTR sequences of the orthologous *Coup-TF* genes of *P.lividus* and *S.purpuratus*. We found that the 5'UTR regions (+1 to +543) are extremely conserved (89%) and that the upstream regions of the two genes show some scattered nucleotide homology, which in places is significant. Thus, from −217 to −1 the homology is 76%; from −952 to −745, 77%; from −1201 to −953, 58% and from −1377 to −1270 the homology is 61%. To our surprise, module a (underlined in figure 6A), is not conserved between the two species, as shown by both analyses (Fig. 6A and B), although *SpCoup-TF* is also predominantly expressed in the ciliary band at pluteus stage [21].

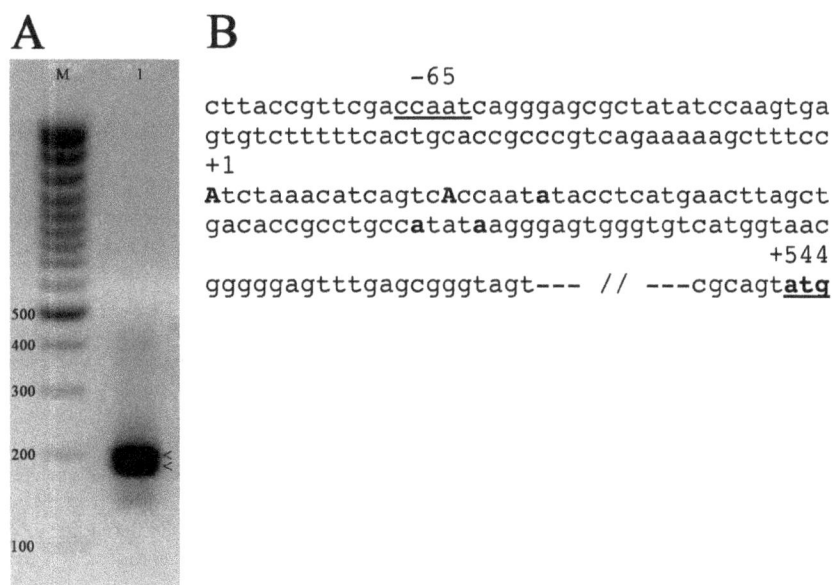

Figure 3. Identification of *PlCoup-TF* transcription initiation sites. A: Electrophoretic analysis of the 5'-RACE products. The two arrows point to the major DNA bands produced by the nested PCR (lane 1). The 100 bp ladder (NEB) was used as DNA length reference (lane M). **B**: Sequence of the proximal PlCoup-TF promoter and the positions of the multiple transcription initiation sites (capital and lower case bold characters). The most upstream site was designated as +1. A putative CCAAT box is underlined at position −65. The translation initiation site is located at +544.

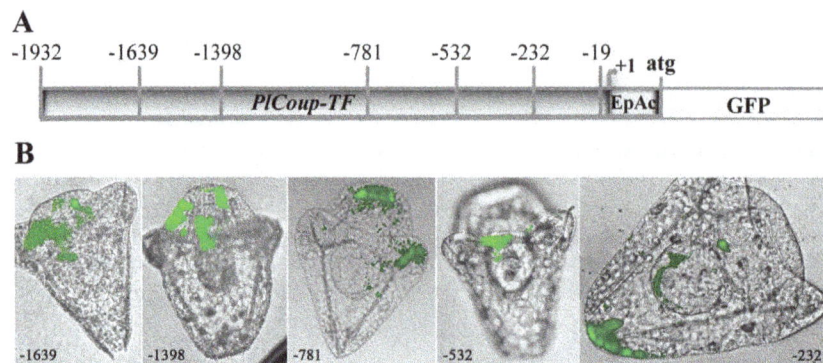

Figure 4. Spatial expression patterns generated by the upstream deletions of the GFP cassette. A: Map of *PlCoup-TF's* upstream sequence (1930 bp) fused to the *EpGFPII* reference gene. The bended arrow marks the transcription initiation site. EpAc refers to *Endo16's* basal promoter and *Cylla's* kozak sequence and ATG. The numbers above the map indicate the starting point of each upstream deletion. **B**: Composite pictures (GFP epi-fluorescence over bright field image) of embryos resulting from injection of the corresponding deletions −1639 to −232. Constructs −1639, −1398 and −781 show GFP expression in ciliary band and a few mesenchyme cells. Construct −532 shows GFP expression only in ciliary band and construct −232, in the aboral ectoderm. All embryos were photographed at pluteus stage. A picture of an embryo injected with the construct −19 is not shown, since these embryos never exhibited any GFP expression.

Upstream and internal deletions into module *a*

To explore the regulative capacity of sub-regions within module *a*, we injected progressive upstream deletions (D-a1 to D-a5), as diagrammed in figure 7A, fused to the reporter gene. Injected sequences differ only in the extent of the upstream *PlCoup-TF* region. By comparison to the entire module *a*, individual deletions resulted in considerable loss of ciliary band specific expression and a parallel increase in ectopic expression (Table 1). Thus, D-a1 exhibits increased expression in the endoderm, suggesting the existence of a negative response element, from −532 to −473, conferring suppression of endoderm expression to the *PlCoup-TF* gene. Deletion D-a2, in addition to higher endoderm expression, results in a 5 fold decreased expression in the ciliary band and an almost 4 fold increased expression in the aboral ectoderm. These results indicate the presence of a positive regulatory element from −473 to −387 regarding ciliary band specific expression, and negative regulatory elements, which suppress expression in aboral ectoderm and to some degree in the endoderm.

A further deletion into module *a*, from −387 to −334 (D-a3), shows a complete loss of spatial specificity (an increase in non specific mesenchyme cell expression), suggesting the existence of perhaps an additional regulatory element within the deleted sequence. Deletions D-a4 and D-a5 show also a high percentage of embryos expressing GFP non-specifically, in mesenchyme cells, and very low levels of other lineage specificity, indicating the absence of regulatory sites from −334 to −212 (Fig. 7B). These results imply that cis-regulatory elements driving PlCoup-TF's expression in the ciliary band, lie in a smaller 200 bp region, between −532 and −334, within module *a*.

To complement the upstream deletion experiments we performed a series of injections using internal deletions into module *a*, fused to the GFP reporter cassette. Thus, with minor nucleotide differences based on the positions of the primers used, D1 (−452 to −386) corresponds to the region between D-a1 and D-a2, D2 (−386 to −333) to D-a2 and D-a3, D3 (−312 to −297) to D-a3 and D-a4 and D4 (−276 to −232) to deletions D-a4 and D-a5 respectively (figures 8A and 7A). Deletion D1 results in ectopic expression of the reporter, i.e. we observe a three fold decrease in the percentage of embryos exhibiting expression only in the ciliary band and in addition a two and a half fold increase in the endoderm and a five fold increase in the aboral ectoderm (Table 1). Thus, compared to the expression profile of the entire

module *a*, deletion D1 exhibits similar ectopic expression as deletions D-a1 and D-a2. It is evident that the sub-region from −452 to −386 contains negative regulatory elements, suppressing the expression of *PlCoup-TF* in the endoderm and the aboral ectoderm territories of the embryo and positive regulatory elements enhancing its expression in the ciliary band (Fig. 8B and Table 1). On the contrary, deletion D2 does not influence ciliary band or endoderm expression, but leads only to ectopic expression in the aboral ectoderm. This result suggests the existence of additional negative regulatory elements between −386 and −333 that suppress PlCoup-TF expression specifically in the aboral ectoderm. The internal deletions D3 and D4, do not exhibit any remarkable differences compared to the expression profile of module *a* (Table 1). Considering the results of the entire deletion analysis of the *PlCoup-TF* upstream sequences, it is apparent that the 120 bp region from −452 to −333 includes the minimal positive and negative regulatory information for the correct spatial expression of the gene in the ciliary band of the pluteus embryo.

Nuclear factors binding to response elements RE1, RE2 and RE3

The nucleotide sequence of module *a*, was searched for transcription factor binding sites using MatInspector. This analysis produced a plethora of putative elements and we set out to test some of them within the sub-region from −452 to −333, which seems to contain significant cis-acting elements as our in vivo expression data suggest. Thus, three such elements, RE1 at −453, RE2 at −432 and RE3 at −377, were used as radioactive labeled probes to detect DNA binding proteins in embryonic (blastula stage) nuclear extracts. The EMSA experiment presented in figure 9 indicates that all three elements are specifically recognized by sea urchin transcription factors included into the nuclear extract, judged by the addition of unlabeled probes as specific competitors. RE1 and RE3 binding results in the formation of more than one specific complex, whereas RE2 seems to form a single complex.

Site-specific mutations into module *a*

Based on the EMSA results we set out to investigate the role of the three cis-acting elements, using inverse PCR with substrate module *a* fused to the GFP cassette and substituting the presumed

Table 1. Embryonic cell lineage specific expression of the GFP reference gene for each of the upstream deletions, individual upstream segments and mutations.

	# of embryos	%Fluor.embryos	% Cb	% Cb+	% En+	% Ae+	% Me+
Deletions							
−1639	151	77	19	39	29	23	51
−1398	148	75	40	75	21	17	37
−781	107	72	56	74	1	3	44
−532	102	77	51	82	6	10	37
−232	90	82	32	58	24	22	43
−19	58	0	0	0	0	0	0
Segments							
a	237	87	32	83	25	8	47
b	117	16	0	53	0	0	100
c	98	34	21	33	12	0	67
d	30	0	0	0	0	0	0
e	40	0	0	0	0	0	0
f	66	0	0	0	0	0	0
Deletions							
D-a1	199	59	15	68	48	18	44
D-a2	127	84	6	69	30	28	68
D-a3	171	83	8	31	17	8	81
D-a4	129	78	13	40	28	8	69
D-a5	130	42	0	22	15	9	83
Internal Deletions							
D1	141	96	10	76	61	41	54
D2	241	79	38	81	36	39	10
D3	193	76	21	66	35	16	57
D4	47	68	28	56	9	25	53
Mutants							
−453	204	35	72	92	14	14	3
−432	202	87	22	83	61	46	5
−377	184	66	43	79	26	36	11
−453/−432	105	18	37	47	42	37	0

Cb: Ciliary band; Cb+: Ciliary band and other cell types; En+: Endoderm and other cell types; Ae+: Aboral ectoderm and other cell types; Me+: Mesenchyme and other cell types.

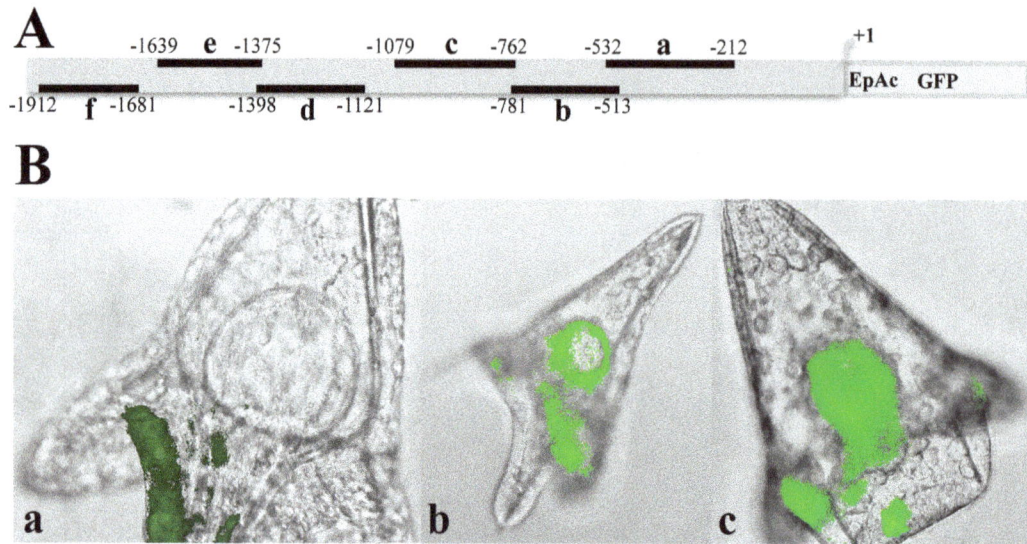

Figure 5. Spatial expression patterns generated by the individual segments a–f fused to the GFP cassette. A: Graphical positioning of the upstream *PlCoup-TF* segments (a–f) within the 1932 bp upstream sequence, which were individually fused to the *EpGFPII* reference gene. The numbers surrounding each black bar correspond to the nucleotide borders of each segment. Other designations are as in Figure 4A. **B**: Composite pictures of embryos resulting from injection of segments a–c. Segment a results in GFP expression specifically in the ciliary band, while segments b and c show GFP expression in the ciliary band, endoderm and mesenchyme cells. All embryos were photographed at pluteus stage. Embryos injected with segments d, e and f did not exhibit GFP expression.

core nucleotides of each binding site with the recognition sequence of the restriction site BglII (5′agatct). We created also the double mutant ($-453/-432$), which in addition to the two elements RE1 and RE2 lacks the 20 bp of the intervening sequence (Fig. 10A). Mutation of the RE1 site (-453) shows no ectopic expression (Fig. 10B), but a considerable decrease in the number of fluorescent embryos (Table 1). Furthermore, the fluorescence exhibited by these embryos was barely noticeable over the detection limit of the microscope. Thus, we assign to the RE1 element, a positive role in the regulation of the *PlCoup-TF* gene.

On the other hand, mutation of the RE2 site (-432) results in a two and a half fold increase of endodermal and a six fold increase in aboral ectoderm expression, with only a small decrease in ciliary band expression (Table 1). In addition, RE2 injected embryos exhibit normal levels of fluorescence. The expression of the mutant RE2 site is in agreement with the D1 and D-a2 deletions (Table 1), suggesting that RE2 is a negative regulatory element that is essential for the suppression of *PlCoup-TF* in the endodermal and aboral ectoderm territories of the pluteus embryo. Mutation of the RE3 site (-377) has no effect on ciliary band or endoderm

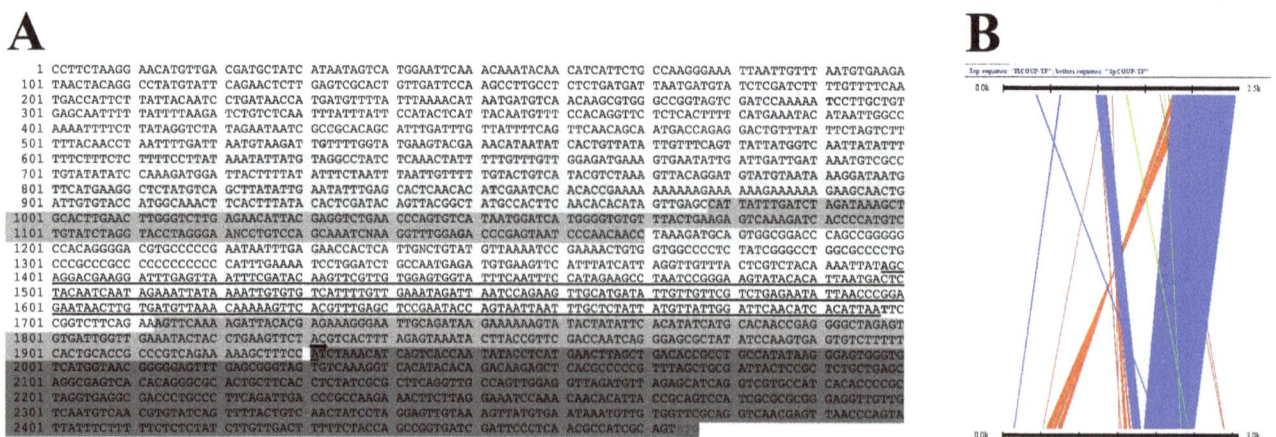

Figure 6. Comparison of *Coup-TF's* upstream and 5′UTR regions between *P.lividus* and *S.purpuratus*. A: The 5′UTR and the upstream sequence of the PlCoup-TF gene, numbered from the -1930 position. The data were obtained by subcloning and sequencing a 2.5 kb fragment of the λ clone "Φ" insert (Fig. 2). The shaded areas correspond to various degrees of homology with the corresponding sequence of *SpCoup-TF* as revealed by comparisons using the program BLAST. The lighter the shade, the lesser the homology is between the ortholog genes of the two species. Thus, the darkest shade corresponds to the 5′UTR sequence that exhibits the highest homology. The small black arrow denotes the transcription initiation site. Underlined is the upstream sequence of module *a*, which is not homologous between the two species. **B**: Graphic comparison of the 5′UTR and 5′ upstream sequences of the two orthologous genes, *PlCoup-TF* (top) and *SpCoup-TF* (bottom) using the Family Relations program (32). Crossbars joining the two sequences indicate homology, the thickest of which corresponds to the 5′UTR region.

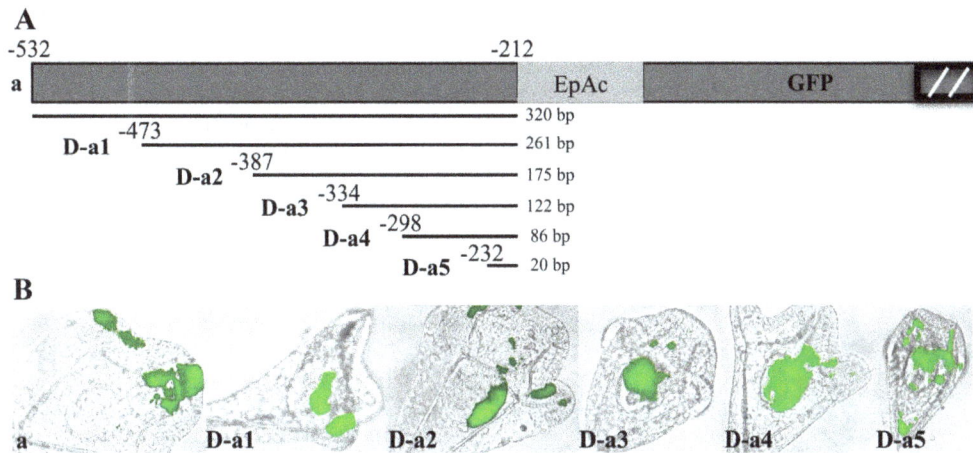

Figure 7. Spatial expression patterns generated by upstream deletions D-a1 to D-a5 of module *a*. A: Graphic presentation of the upstream deletions into module a (−532 to −212). Horizontal bars underneath module *a*, represent the size of each deletion (D-a1 to D-a5) and the numbers above them the corresponding upstream border. The numbers at the right of each bar correspond to the size of each fragment tested. EpAc refers to *Endo16's* promoter and *Cylla's* kozak sequences as in figure 4A. The two parallel bars within the black box at the right end of the graph denote that the *GFP* gene is not depicted on scale. **B**: Composite pictures of embryos resulting from injection of upstream deletions into module *a*. The entire module *a* shows GFP expression specifically in the ciliary band, while deletions D-a1, D-a2 and D-a3 show expression both in ciliary band and endoderm. Deletions D-a4 and D-a5 show expression in endoderm and mesenchyme cells. D-a5 shows GFP expression in skeletogenic mesenchyme cells (see text for non-specific expression caused by random integration of the GFP cassette). All embryos were photographed at pluteus stage.

expression (Fig. 10B), but results in a four and a half fold increase in aboral ectoderm expression, in accordance to the D2 deletion (Table 1). These results suggest that RE3 is a negative regulatory element, which suppresses *PlCoup-TF* expression specifically in the aboral ectoderm. Mutation of both RE1 and RE2 elements, results in a small percentage of fluorescent embryos as well as faint levels of fluorescence in all expressing territories. Moreover, the double mutation leads to a complete loss of spatial expression preference of the reporter gene. It is obvious that the double mutation shows the collective effects of the individually mutated RE1 and RE2 elements (Table 1). From the site-specific mutagenesis results, we conclude that the combined effects of these three upstream elements reproduce the regulatory capacity of the

entire module *a*, and suffice for the correct quantitative and spatial expression of *PlCoup-TF* in the pluteus larva.

Discussion

The herein study focuses on the regulation of the *PlCoup-TF* gene at later embryonic stages, where its expression is mainly restricted to the ciliary band of the pluteus as demonstrated by in situ hybridization (figure 1). This restriction follows a broader zygotic activation of the gene at earlier stages (blastula through gastrula), subsequent to the turnover of the maternal RNA. Thus, our study does not take into account plausible additional regulatory elements responsible for its early activation, but rather

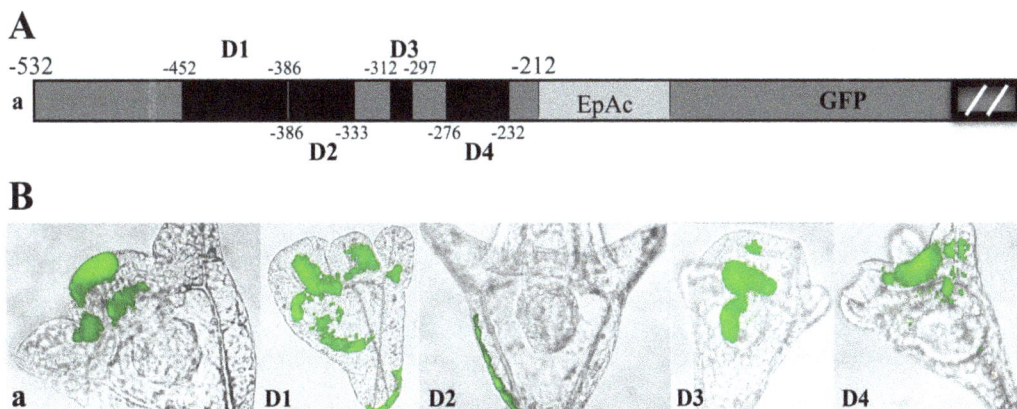

Figure 8. Spatial expression patterns generated internal deletions D1–D4 into module *a*. A: Map of the internal deletions. Black boxes correspond to the deleted regions D1–D4. The numbers surrounding each box mark the borders of each deletion. EpAc refers to *Endo16's* promoter and *Cylla's* kozak sequences as in figure 4A. The two parallel bars within the black box at the right end of the graph denote that the *GFP* gene is not depicted on scale. **B**: Composite pictures of embryos resulting from injection of the internal deletions D1–D4. The entire module *a* shows GFP expression specifically in the ciliary band, deletion D1 in ciliary band, endoderm and aboral ectoderm and D2 in aboral ectoderm. D3 shows expression in ciliary band and endoderm while D4 shows expression in mesenchyme cells. All embryos were photographed at pluteus stage.

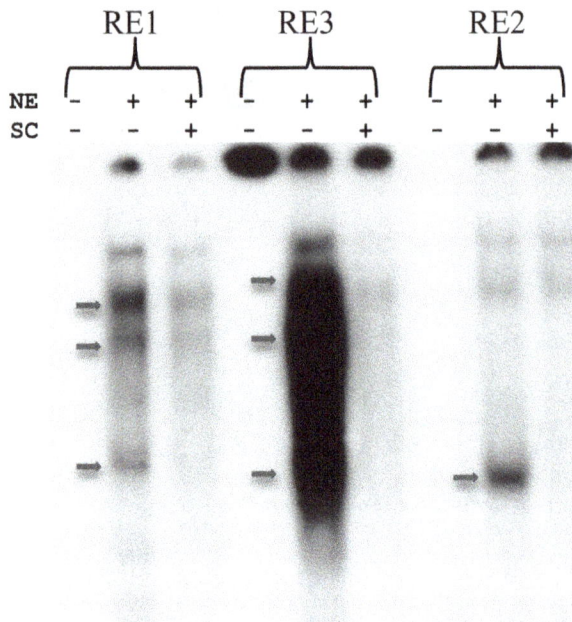

Figure 9. Specific binding of embryonic nuclear proteins to elements RE1, RE2 and RE3. The DNA binding specificity of transcription factors to the elements RE1, RE2 and RE3 was determined by electrophoretic mobility shift assays. NE: Nuclear Extract; SC: Specific Competitor. (−) and (+) denote omission and addition of nuclear extract or specific competitor to each reaction respectively. The arrows point to protein: DNA complexes, which are not formed in the presence of specific competitor.

lack gut formation and show diminished spiculogenesis (unpublished data from this laboratory). Thus, the store of maternal RNA and protein seems to be adequate for the very early embryonic regulatory functions, but later on, when maternal Coup-TF transcripts and protein are turned over, the embryo relies greatly on newly synthesized zygotic transcripts. *Coup-TF's* embryonic spatial expression profile (oral ectoderm, ciliary band) and its restricted expression in neuronal cell types of the feeding larvae (unpublished data from this laboratory) are indicative of its role in sea urchin neurogenesis, similar to a variety of other organisms [13–15]. The present study elucidates the mechanism of *PlCoup-TF's* late embryonic regulation, as a first step in placing this transcription factor within the Gene Regulatory Network of the sea urchin oral ectoderm and ciliary band.

The *PlCoup-TF* gene structure

Although the isolated overlapping P.lividus genomic clones span 33.5 kb of genomic sequences, they do not include the entire gene, since the last exon is not present. Presumably, the last intron of the gene is extremely long, as also seems to be the case for the *SpCoup-TF* gene. The scaffold containing the *SpCoup-TF* gene, assembled by the genome project of *Strongylocentrotus purpuratus*, also lacks the fourth exon.

We focused on the study of the 1930 bp upstream fragment contained in the cloned insert of phage "*Φ*", mindful that additional regulatory regions may lie further upstream or downstream of the chosen area. Two of the multiple transcription initiation sites that we found, spaced 15 bp apart, seem to account for the majority of the 5′-ends of the mature transcripts. The nucleotide sequence in the upstream vicinity of the +1 implies that the promoter of the gene is TATA-less. In addition, no other obvious promoter elements are recognized with the exception of a putative CCAAT element at position −65 (figure 3). While other unidentified promoter elements possibly exist, the proximal upstream promoter of *PlCoup-TF*, as revealed by our deletion analysis, is not capable of spatial regulation of the gene. This is evident from embryos expressing deletion −232 (figure 4), which exhibit GFP expression to the same extent in all observed embryonic territories. Thus, considering these 232 bp as the proximal promoter, they provide positive inputs to the *Endo16*

relates to the mechanism that sustains its quantitative and spatial mode of expression in the pluteus ciliary band.

Coup-TF is an essential transcription factor in early development. Knockouts of ortholog genes in various animals, result in lethal phenotypes. In *Paracentrotus lividus*, injections of PlCoup-TF morpholino antisense oligonucleotides into fertilized eggs lead to developmental arrest at the blastula stage, with embryos that

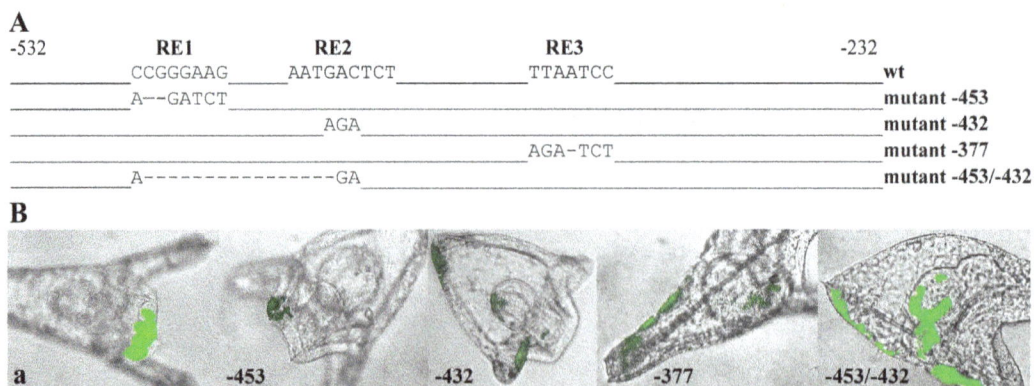

Figure 10. Spatial expression patterns of site-specific mutations into module *a*. A: Graphic presentation of the wt and mutant sequences of the three elements within module *a*. The top line depicts the 320 bp region of the wt module *a*, and the sequences that correspond to the sites RE1 (−453), RE2 (−432) and RE3 (−377) and their respective position. Each additional line shows the nucleotides that substitute for the wt sequence at each site. The bottom line depicts the double mutation, which deletes also the intervening sequences between RE1 and RE2. **B**: Composite pictures of embryos expressing GFP resulting from injection of wt and mutant module *a*. The wt module *a* and mutation −453 show GFP expression specifically in the ciliary band, while mutation −432 shows expression in ciliary band, endoderm and aboral ectoderm. Mutation −377 shows GFP expression in aboral ectoderm and the double mutation −453/−432 in aboral ectoderm, endoderm and ciliary band. All embryos were photographed at pluteus stage.

basal promoter without any territorial restriction. Further proximal promoter inputs downstream of the +1 site are also possible, but they are not included in the herein analysis.

Identification of a single regulatory module

Since the endogenous *PlCoup-TF* gene is primarily expressed in the ciliary band of the pluteus stage, groups of embryos injected by a given construct and exhibiting high percentage of fluorescence primarily in the ciliary band, would indicate that the construct contains the necessary *cis* acting elements for correct embryonic regulation of *PlCoup-TF* at this late developmental stage. It is also expected that, in this type of analysis, any given tested sequence will be expressed in more cell types than in those in which its regulatory milieu is capable of driving it. The latter, results from the fact that the injected DNA is integrated randomly into the genome [31], becoming thus influenced by surrounding *cis* acting elements.

Stepwise 5′ deletions of the 1930 bp upstream region and internal segments thereof, driving the GFP expression cassette, demonstrate that a single 320 bp module (module *a*, −532 to −232) is necessary and sufficient to recapitulate the endogenous *PlCoup-TF* expression in the pluteus larva (figures 4 and 5). Segments b and c, upstream of −532, exhibit reduced percentage of fluorescent embryos and loss of ciliary band specificity, whereas further upstream segments (d–f) are unable to drive the GFP expression cassette and thus, considered devoid of relevant cis-acting elements. Downstream though of module *a*, within the region from −232 to +1, positive acting element(s) must account for the observed expression of GFP in all embryonic territories (Table 1). Such elements, comprising perhaps the *PlCoup-TF*'s basal promoter, do not confer any restriction to spatial expression. It is evident therefore, as the −232 and −532 deletion data suggest, that module *a* must be comprised of negative regulatory elements that restrict expression of the gene to the ciliary band. Module *a* must also contain positive acting elements, recognized by ciliary band expressed factors, since it is able by itself to drive the expression of the GFP cassette (Fig. 5B) in this territory. Module *a* thus, is necessary and sufficient to provide proper spatial and quantitative regulatory inputs to the *PlCoup-TF* gene at pluteus stage.

Comparison of *PlCoup-TF*'s upstream sequence with the upstream sequence of the ortholog *SpCoup-TF* gene from *Strongylocentrotus purpuratus* should permit the identification of evolutionary conserved regulatory sites [37]. Such homologous segments are generally considered to contain important regulatory elements, conserved because of their functionality. Interestingly though, none of the identified conserved segments are found within module *a*. It is conceivable either that the conserved upstream sequences are not important for the regulation of the *PlCoup-TF* gene, or that they may be necessary for some other aspect of the gene's regulation, which is not revealed by the herein analysis. It is possible that similar sequences to module *a* maybe found in further upstream or downstream regions of the *SpCoup-TF* gene, not included in this comparison, which would suggest similar regulatory inputs for temporal and spatial embryonic regulation of the two genes. On the other hand, our results may indicate that the two genes use different regulatory factors to attain the same expression pattern in the pluteus embryo.

Regulatory sites comprising module *a*

A minimal 120 bp segment within module *a*, was shown to confer ciliary band specific expression to the *PlCoup-TF* gene. An *in silico* analysis of this segment (not shown) revealed a wealth of putative transcription factor binding consensus sequences. The position of three elements, RE1, RE2 and RE3, identified through electrophoretic mobility shift assays correlates with the functional deletion analysis of small segments within module *a*, suggesting that the three elements may be involved in functional DNA-protein interactions. This argument does not exclude the possibility that other functional cis- acting elements might exist within the minimal 120 bp segment. The assumption we took is that these three elements, to be real regulatory sites, should fulfill the following criteria, stemming from the functional deletion analysis, i.e. they should provide both positive and negative inputs to the *PlCoup-TF* gene. Thus, site-specific mutagenesis of the three discrete elements was undertaken to investigate their specific role in the context of the entire module *a*, rather than the minimal 120 bp segment, to provide a more significant functional consequence of each mutation.

Functional analysis of the cis- acting elements RE1, RE2 and RE3

Mutation of the RE1 site implies that the corresponding binding factor(s) provides a positive input into the *PlCoup-TF* gene. The observed weak fluorescent signal was mostly associated with cells of the ciliary band, suggesting that the RE1 site does not provide any spatial regulatory clues to the gene. Thus, it is through RE1 that *PlCoup-TF* receives the major positive input at pluteus stage embryos. The weak fluorescence observed in some embryos, vs. total lack of expression, is perhaps indicative of additional minor positive inputs within module *a*. The mutated sequence of RE1 (GGAAG, Fig. 10A) corresponds to the consensus binding sequence of Ets transcription factors. An identical element is found in the CyIIa gene promoter, responding also to Ets factors [38]. Two members of the sea urchin *S.purpuratus* Ets gene family, *SpElk* and *SpPea*, are expressed at the pluteus stage embryo and predominantly in the oral ectoderm [39]. As in *S.purpuratus*, the *P.lividus* ortholog genes *PlElk* and *PlPea* were shown by *in situ* hybridization to be expressed in the ciliary band of the pluteus stage embryo (unpublished results from this laboratory). Thus, one or both of these Ets factors could account for the positive regulation of the *PlCoup-TF* gene. It is interesting to note that the mouse *Coup-TFI* gene, NR2F1, is also positively regulated by Ets1, a member of the family [25]. This is not surprising, since the Ets family of transcription factors is very conserved in the animal kingdom and involved in a plethora of developmental processes.

The analysis of the RE2 and RE3 mutants suggests that these cis-acting elements are recognized by repressors, which act at embryonic territories other than the ciliary band and oral ectoderm to suppress the expression of *PlCoup-TF*. The unknown repressor that binds to the RE2 element seems to extend its influence in all such embryonic territories, i.e. endoderm, mesoderm and aboral ectoderm, since mutation of RE2 results in total loss of spatial restriction of the GFP cassette expression (Table 1). The double RE1/RE2 mutant shows an additive effect of the individual mutations, i.e. very few fluorescent embryos, weak GFP expression and total loss of spatial restriction. Thus the combinatorial effect of the RE1 and RE2 elements seems to account for most of the regulatory inputs for the proper quantitative and spatial regulation of the *PlCoup-TF* gene at pluteus stage.

Additional negative inputs are exerted to the gene through the RE3 element. These inputs though, in contrast to the RE2 element, seem to be specific to the embryonic aboral ectoderm of the pluteus, since the RE3 mutant GFP cassette looses spatial restriction only to this territory. Thus, the corresponding RE3 binding factor(s) should be a repressor expressed specifically in the aboral ectoderm of the embryo, where, in addition to the negative function of the RE2 binding factor, it silences further the expression of the PlCoup-TF gene.

The RE2 and RE3 sequences and the corresponding nuclear protein binding assays presented herein, do not give us any clues as to the identity of the two repressors. Future experiments will be designed to identify these factors, taking advantage of the significant progress made so far in the characterization of the gene regulatory networks that control the specification of the late embryonic territories [40–42].

Acknowledgments

The authors would like to acknowledge the assistance of previous laboratory members Christina Petropoulou and Sophia Papadimitriou. We are grateful to Drs. Maria Ina Arnone and Valeria Matranga for their gifts of the EpGFPII vector and the *Paracentrotus lividus* genomic library respectively and Nicholas C. Flytzanis for critical reading of the manuscript.

Author Contributions

Conceived and designed the experiments: CNF. Performed the experiments: LGK CNF. Analyzed the data: LGK CNF. Contributed reagents/materials/analysis tools: CNF. Wrote the paper: CNF.

References

1. Wang L, Tsai SY, Cook R, Beattie W, Tsai MJ, et al. (1989) COUP transcription factor is a member of the steroid receptor superfamily. Nature 340: 163–166.
2. Mlodzik M, Hiromi Y, Weber U, Goodman C, Rubin G (1990) The Drosophila seven-up gene, a member of the steroid receptor gene superfamily, controls photoreceptor cell fates. Cell 60: 211–224.
3. Achatz G, Holzl B, Speckmayer R, Hauser C, Sandhofer F, et al. (1997) Functional domains of the human orphan receptor ARP-1/COUP-TFII involved in active repression and transrepression. Mol Cell Biol 17: 4914–4932.
4. Ladias J, Karathanasis S (1991) Regulation of the apolipoprotein AI gene by ARP-1, a novel member of the steroid receptor superfamily. Science 251: 561–565.
5. Cooney A, Leng X, Tsai SY, O'Malley BW, Tsai MJ (1993) Multiple mechanisms of chicken ovalbumin upstream promoter transcription factor-dependent repression of transactivation, by the vitamin D, thyroid hormone and retinoic acid receptors. J Biol Chem 268: 4152–4160.
6. Laudet V, Hanni C, Coll J, Catzeflis F, Stehelin D (1992) Evolution of the nuclear receptor gene superfamily. EMBO J 11: 1003–1013.
7. Laudet V (1997) Evolution of the nuclear receptor superfamily: early diversification from an ancestral orphan receptor. J Mol Endocrinol 19: 207–226.
8. Tsai SY, Tsai MJ (1997) Chick ovalbumin upstream promoter-transcription factors (COUP-TFs): coming of age. Endo Rev 18: 229–240.
9. Fjose A, Nornes S, Weber U, Mlodzik M (1993) Functional conservation of vertebrate seven-up related genes in neurogenesis and eye development. EMBO J 12: 1403–1414.
10. Fjose A, Weber U, Mlodzik M (1995) A novel vertebrate svp-related nuclear receptor is expressed as a step gradient in developing rhombdomeres and is affected by retinoic acid. Mech Dev 52: 233–246.
11. van der Wees J, Matharu P, de Roos K, Destree O, Godsave S, et al. (1996) Developmental expression and differential regulation by retinoic acid of Xenopus COUP-TF-A and COUP-TF-B. Mech Dev 54: 173–184.
12. Pereira F, Qiu Y, Tsai MJ, Tsai SY (1995) Chicken ovalbumin upstream promoter transcription factor (COUPTF): Expression during mouse embryogenesis. J Steroid Biochem Mol Biol 53: 503–508.
13. Qiu Y, Cooney A, Kuratani S, DeMayo F, Tsai SY, et al. (1994) Spatiotemporal expression patterns of chicken ovalbumin upstream promoter-transcription factors in the developing mouse central nervous system: evidence for a role in segmental patterning of the diencephalon. Proc Natl Acad Sci USA 91: 4451–4455.
14. Lutz B, Kuratani S, Cooney A, Wawersik S, Tsai SY, et al. (1994) Developmental regulation of the orphan receptor COUP-TF II gene in spinal motor neurons. Development 120: 25–36.
15. Zhou M, Walthall W (1998) UNC-55, an Orphan Nuclear Hormone Receptor, Orchestrates Synaptic Specificity among Two Classes of Motor Neurons in *Caenorhabditis elegans*. J Neurosc 18: 10438–10444.
16. Shan G, Walthall W (2008) Copulation in C. elegans males requires a nuclear hormone receptor. Dev Biol 322: 11–20.
17. Qiu Y, Pereira F, DeMayo F, Lydon J, Tsai SY, et al. (1997) Null mutation of mCOUP-TFI results in defects in morphogenesis of the glossopharyngeal ganglion, axonal projection, and arborization. Genes Dev 11: 1925–1937.
18. Pereira F, Qiu Y, Zhou G, Tsai MJ, Tsai SY (1999) The orphan nuclear receptor COUP-TFII is required for angiogenesis and heart development. Genes Dev 13: 1037–1049.
19. Hiromi Y, Mlodzik M, West S, Rubin G, Goodman C (1993) Ectopic expression of seven-up causes cell fate changes during ommatidial assembly. Development 118: 1123–1135.
20. Chan SM, Xu N, Niemeyer C, Bone J, Flytzanis CN (1992) SpCOUP-TF: A sea urchin member of the steroid/thyroid hormone receptor family. Proc Natl Acad Sci USA 89: 10568–10572.
21. Vlahou A, Gonzalez-Rimbau M, Flytzanis CN (1996) Maternal mRNA encoding the orphan steroid receptor SpCOUP-TF is localized in sea urchin eggs. Development 122: 521–526.
22. Xu N, Niemeyer C, Gonzalez-Rimbau M, Bogosian E, Flytzanis CN (1996) Distal cis-acting elements restrict expression of the CyIIIb actin gene in the aboral ectoderm of the sea urchin embryo. Mech Dev 60: 151–162.
23. Vlahou A, Flytzanis CN (2000) Subcellular trafficking of the nuclear receptor COUP-TF in the early embryonic cell cycle. Dev Biol 218: 284–298.
24. Krishnan V, Pereira F, Qiu Y, Chen C, Beachy P, et al. (1997) Mediation of Sonic hedgehog-induced expression of COUP-TFII by a protein phosphatase. Science 278: 1947–1950.
25. Salas R, Petit G, Pipaon C, Tsai MJ, Tsai SY (2002) Induction of chicken ovalbumin upstream promoter-transcription factor I (COUP-TFI) gene expression is mediated by ETS factor binding sites. Eur J Biochem 269: 317–325.
26. Jonk L, deJong E, Pals C, Wissink S, Vervaart J, et al. (1994) Cloning and expression during development of three murine members of the COUP family of nuclear orphan receptors. Mech Dev 47: 81–97.
27. Kruse S, Suino-Powell K, Zhou E, Kretschman J, Reynolds R, et al. (2008) Identification of COUP-TFII Orphan Nuclear Receptor as a Retinoic Acid–Activated Receptor. PLoS Biol 6: 2002–2015.
28. Arnone MI, Bogarad L, Collazo A, Kirchhamer C, Cameron A, et al. (1997) Green Fluorescent Protein in the sea urchin: new experimental approaches to transcriptional regulatory analysis in embryos and larvae. Development 124: 4649–4655.
29. Arnone MI, Martin E, Davidson EH (1998) Cis-regulation downstream of cell type specification: a single compact element controls the complex expression of the CyIIa gene in sea urchin embryos. Development 125: 1381–1395.
30. MacMahon A, Flytzanis CN, Hough-Evans B, Katula K, Britten R, et al. (1985) Introduction of cloned DNA into sea urchin egg cytoplasm: replication and persistence during embryogenesis. Dev Biol 108: 420–430.
31. Flytzanis CN, McMahon A, Hough-Evans B, Katula K, Britten R, et al. (1985) Persistence and integration of cloned DNA in postembryonic sea urchins. Dev Biol 108: 431–442.
32. Brown T, Xie Y, Davidson E, Cameron A (2005) Paircomp, FamilyRelationsII and Cartwheel: tools for interspecific sequence comparison. BMC Bioinformatics 6: 70
33. Calzone F, Thézé N, Thiebaud P, Hill R, Britten R, et al. (1988) Developmental appearance of factors that bind specifically to cis-regulatory sequences of a gene expressed in the sea urchin embryo. Genes Dev 2: 1074–1088.
34. Arenas-Mena C, Cameron A, Davidson EH (2000) Spatial expression of Hox cluster genes in the ontogeny of a sea urchin. Development 127: 4631–4643.
35. Walton K, Croce J, Glenn T, Wu S, McClay D (2006) Genomics and expression profiles of the Hedgehog and Notch signaling pathways in sea urchin development. Dev Biol 300: 153–164.
36. Cameron A, Oliveri P, Wyllie J, Davidson EH (2004) cis-Regulatory activity of randomly chosen genomic fragments from the sea urchin. Gene Expression Patterns 4: 205–213.
37. Brown T, Rust A, Clarke P, Pan Z, Schilstra M, et al. (2002) New computational approaches for analysis of cis-regulatory networks. Dev Biol 246: 86–102.
38. Consales C, Arnone MI (2002) Functional characterization of Ets-binding sites in the sea urchin embryo: three base pair conversions redirect expression from mesoderm to ectoderm and endoderm. Gene 287: 75–81.
39. Rizzo F, Fernandez-Serra M, Squarzoni P, Archimandritis A, Arnone MI (2006) Identification and developmental expression of the ets gene family in the sea urchin *Strongylocentrotus purpuratus*. Dev Biol 300: 35–48.
40. Oliveri P, Davidson EH (2004) Gene regulatory network controlling embryonic specification in the sea urchin. Curr Opin in Gen & Dev 14: 351–360.
41. Su Y, Li E, Geiss G, Longabaugh W, Krämer A, et al. (2009) A perturbation model of the gene regulatory network for oral and aboral ectoderm specification in the sea urchin embryo. Dev Biol 329: 410–421.
42. Saudemont A, Haillot E, Mekpoh F, Bessodes N, Quirin M, et al. (2010) Ancestral Regulatory Circuits Governing Ectoderm Patterning Downstream of Nodal and BMP2/4 Revealed by Gene Regulatory Network Analysis in an Echinoderm. PLoS Genet 6(12): e1001259. doi:10.1371/journal.pgen.1001259.

2

High Efficiency *Ex Vivo* Cloning of Antigen-Specific Human Effector T Cells

Michelle A. Neller[1,3]**, Michael H.-L. Lai**[1]**, Catherine M. Lanagan**[1]**, Linda E. O'Connor**[1]**, Antonia L. Pritchard**[2]**, Nathan R. Martinez**[1¤]**, Christopher W. Schmidt**[1]*

1 Cancer Immunotherapy Laboratory, QIMR Berghofer Medical Research Institute, Brisbane, Queensland, Australia, 2 Oncogenomics Laboratory, QIMR Berghofer Medical Research Institute, Brisbane, Queensland, Australia, 3 School of Medicine, The University of Queensland Mayne Medical School, Brisbane, Queensland, Australia

Abstract

While cloned T cells are valuable tools for the exploration of immune responses against viruses and tumours, current cloning methods do not allow inferences to be made about the function and phenotype of a clone's *in vivo* precursor, nor can precise cloning efficiencies be calculated. Additionally, there is currently no general method for cloning antigen-specific effector T cells directly from peripheral blood mononuclear cells, without the need for prior expansion *in vitro*. Here we describe an efficient method for cloning effector T cells *ex vivo*. Functional T cells are detected using optimised interferon gamma capture following stimulation with viral or tumour cell-derived antigen. In combination with multiple phenotypic markers, single effector T cells are sorted using a flow cytometer directly into multi-well plates, and cloned using standard, non antigen-specific expansion methods. We provide examples of this novel technology to generate antigen-reactive clones from healthy donors using Epstein-Barr virus and cytomegalovirus as representative viral antigen sources, and from two melanoma patients using autologous melanoma cells. Cloning efficiency, clonality, and retention/loss of function are described. *Ex vivo* effector cell cloning provides a rapid and effective method of deriving antigen-specific T cells clones with traceable *in vivo* precursor function and phenotype.

Editor: Derya Unutmaz, New York University, United States of America

Funding: This work was supported by the National Health and Medical Research Council of Australia (www.nhmrc.gov.au) (Grant 290358 to CWS); Rio Tinto Ride to Conquer Cancer (www.conquercancer.org.au) QIMR Berghofer Medical Research Institute Flagship grant to ALP and CWS; Cure Cancer Australia Foundation (www.cure.org.au) fellowship to ALP; and Australian Postgraduate Award to MAN (www.uq.edu.au/grad-school/apa). The funders had no role in study design, data collection and analysis, decision to publish, or preparation of the manuscript.

Competing Interests: The authors have declared that no competing interests exist.

* Email: Chris.Schmidt@qimrberghofer.edu.au

¤ Current address: Novartis Australia Pty Ltd, North Ryde, New South Wales, Australia

Introduction

In addition to their frequent use for *in vitro* studies of immune function, antigen-specific T cell clones are important tools for identifying viral and tumour antigens. They have also been expanded to large numbers for use in adoptive immunotherapy trials [1,2]. The majority of T cell cloning methods involve stimulating unselected precursors for one or more rounds prior to limiting dilution cloning, to expand small populations of antigen-specific T cells [3,4]. Whilst this procedure facilitates the isolation of rare T cells of interest, prior *in vitro* culture can have a number of undesirable effects. For example, the choice of cytokine combination, source of antigen and antigen dose can promote selective out-growth of particular T cell subpopulations [5–8] and affect the phenotype and function of the subsequently expanded cells. The effects of extended *in vitro* culture on T cell phenotype and function therefore preclude the correlation of many T cell clonal attributes with typical *in vivo* characteristics. Alternatively, T cells may be cloned directly *ex vivo*, by sorting individual T cells based on peptide-major histocompatibility complex multimer labelling [9,10]; however, this requires knowledge of the epitope target, which prevents the use of such methods in antigen discovery or for generating clones against diverse virus or tumour antigens. A specialised technology for *ex vivo* cloning of HIV-Gag peptide-reactive CD8+ T cells, from arrays of sub-nanolitre wells that capture secreted cytokines, has also been described [11]. While this is at odds with the notion that effector cells have a limited potential for expansion in culture, as they are likely to be highly differentiated and possess short telomeres [12,13], it suggests that *ex vivo* function could provide a basis for prior selection of T cells for efficient cloning.

We here describe a novel method for cloning effector T cells based on single-cell, fluorescence activated sorting of cytokine-secreting cells *ex vivo*. Direct cloning of antigen-specific effector T cells following a brief period of antigen stimulation, enables the acquisition of information on the characteristics of individual T cell clone precursors, prior to the influences of long-term culture, and repeated rounds of cell division. The method generates effectors with diverse specificities and HLA-restrictions by stimulating with complex antigen sources, such as whole tumour cells and whole protein, and enables the selection of T cells with known precursor *ex vivo* function and phenotype. By allowing the correlation of *ex vivo* T cell characteristics with more stable attributes (such as T cell receptor usage) identified for clones *in vitro*, this method adds a new dimension to the study of T cell responses to tumours and infections.

Material and Methods

Ethics Statement

The Queensland Institute of Medical Research - Human Research Ethics Committee approved this research under protocols P962 (approval H0609-044, cancer patients) and P598 (H0306-044, healthy donors), and all patients and donors gave written, informed consent prior to enrolment.

Mononuclear cell isolation and cryopreservation

Peripheral blood from two Stage IV melanoma patients (both clinical trial participants, coded A02 and D14), and from healthy donors, was collected in tubes containing sodium heparin (BD Diagnostics, Franklin Lakes, NJ, USA). Peripheral blood mononuclear cells (PBMC) were isolated by density gradient centrifugation over Ficoll-Hypaque (GE Healthcare, Little Chalfont, UK) and cryopreserved in autologous plasma or Albumex 4 (Australian Red Cross Blood Service, Brisbane, Queensland) containing 10% dimethyl sulfoxide (Sigma-Aldrich Pty Ltd, St Louis, MO, USA). Cells were frozen in cryotubes (Thermo Fisher Scientific, Roskilde, Denmark), using Cryo 1°C Freezing Containers (Thermo Fisher), and cryopreservation was completed within 4 h of blood collection. Cells were thawed by incubating cryotubes in a 37°C waterbath, then cells were washed once in either RPMI 1640 (Life Technologies, Grand Island, NY, USA) containing 10 µg/ml DNase I (Sigma-Aldrich), for PBMC, or RPMI 1640 alone, for all other cell types.

Cell culture media

Established cell lines were grown in complete medium: RPMI 1640 containing L-glutamine (Life Technologies) and supplemented with 10% heat-inactivated foetal bovine serum (Sigma-Aldrich) and 40 µg/ml gentamicin sulfate (Pfizer, Bentley, WA, Australia).

T cells were cloned and cultured in clone medium: RPMI 1640 containing L-glutamine, supplemented with 40 µg/ml gentamicin sulfate, 1 mM HEPES (4-(2-hydroxyethyl)-1-piperazineethanesulfonic acid), 100 IU/ml IL-2 (Roche Diagnostics GmbH, Mannheim, Germany) and 10% heat-inactivated pooled human serum (Australian Red Cross Blood Service). For initial sorting and restimulation, clone medium contained 1 µg/ml Phytohaemagglutinin-L (Sigma-Aldrich) and feeder cells as detailed below.

For functional assays, T cell clones and stimulator/target cells were suspended in assay medium – RPMI 1640 containing L-glutamine, 40 µg/ml gentamicin sulfate and 5% heat-inactivated pooled human serum or foetal bovine serum. All 37°C incubations were undertaken in 5% CO_2.

Cell lines

Melanoma cell lines (denoted A02-M and D14-M) were established from metastases from trial participants A02 and D14 [14,15]. In brief, cells from mechanically disaggregated tumours were cultured in complete medium to establish cell lines, which were confirmed to be of melanoma origin by expression profiling [15], and authenticated via short tandem repeat profiling according to the manufacturer's instructions (AmpF'STR Profiler Plus ID kit; Applied Biosystems, Foster City, CA, USA). B-lymphoblastoid cell lines (LCL) were established by exogenous transformation of peripheral B cells with Epstein Barr virus (EBV), derived from the supernatant of the B95.8 cell line, and were maintained in complete medium. The K562 cell line was obtained from the American Type Culture Collection (USA). All cell lines were tested negative for mycoplasma contamination using the Venor GeM Mycoplasma Detection Kit (Minerva Biolabs GmbH, Berlin, Germany) or the MycoAlert Detection Kit (Lonza Group Ltd, Basel, Switzerland), prior to use in experiments.

Interferon (IFN)-γ capture assay and antibody labelling

The human IFN-γ secretion assay (phycoerythrin (PE) label; Miltenyi Biotec, Bergisch Gladbach, Germany) was used to detect antigen-specific T cells from patient PBMC samples. In each assay, $4\text{-}5 \times 10^6$ PBMC were tested and a modification of standard protocols [16] was used. Donor/patient PBMC were thawed rapidly, washed, and resuspended in assay medium at 10^6 cells per 200 µl well in U-bottom 96-well plates with 5×10^5 irradiated (30 Gy) autologous LCL or melanoma cells. Alternatively, PBMC were stimulated with 1/100 (vol/vol) recombinant human cytomegalovirus (CMV) phosphoprotein 65 (pp65; Miltenyi Biotec). Cells were incubated at 37°C for 14 h (or as indicated in preliminary experiments), replicates were pooled, then washed with 0.5% bovine serum albumin/phosphate buffered saline (PBS) (FACS buffer), resuspended, transferred to capped 10 ml tubes and labelled with IFN-γ catch reagent, a CD45-specific monoclonal antibody (mAb) conjugated to an anti-IFN-γ mAb. The IFN-γ catch reagent was incubated with cells at a 1/10 dilution in a 50 µl total volume, for 15 min at 4°C. Cells were then resuspended at $1\text{-}2 \times 10^4$ PBMC/ml in complete medium and incubated at 37°C for 1 h under slow rotation. For each stimulus, the optimal cell concentration for this step was determined empirically from the expected number of IFN-γ-secreting cells. Cells were subsequently washed twice with FACS buffer, and then labelled for 30 min at 4°C with pre-titred volumes of IFN-γ PE detection mAb (Miltenyi Biotec), CD8 allophycocyanin (APC; clone RPA-T8), CD4 Alexa Fluor 700 (RPA-T4), CD16 fluorescein (FITC; NKP15), CD19 FITC (HIB19), and CD14 FITC (MøP9) (BD Biosciences, Franklin Lakes, NJ, USA). Following a single wash with FACS buffer, cells were resuspended in 1 ml PBS containing 1 µg/ml propidium iodide (PI; Sigma-Aldrich).

Sorting and cloning

Cell suspensions were filtered through sterile 37 µm nylon mesh immediately prior to purification sorting of CD4+ IFN-γ+ and CD8+ IFN-γ+ populations using a MoFlo cell sorter running Summit software (Beckman Coulter, Fullerton, CA, USA). Sorting gates were determined by the bimodal expression of phenotypic markers (CD4, CD8, CD14, CD16, CD19) and IFN-γ, and in most cases were confirmed using negative controls. Subsequently a FACSVantage SE cell sorter running CellQuest and ClonCyt software (BD Biosciences) and equipped with a single cell deposition unit was used to sort single CD4+ or CD8+, IFN-γ+, CD14- CD16- CD19- cells into wells of U-bottom 96-well plates containing clone medium and feeder cells consisting of 2×10^4 irradiated allogeneic LCL (a mixture of three different lines) and 1×10^5 irradiated allogeneic PBMC per well. The overall cloning procedure is summarised in Fig. 1.

Every seven days, half the volume of medium in T cell clone plates (100 µl/well) was replaced with fresh clone medium. Proliferating clones could be visualised by microscopy after 10 days. Clones were re-plated in U-bottom 96-well plates at 5×10^4 cells per well every four weeks and re-stimulated with clone medium containing phytohaemagglutinin-L and feeder cells.

Flow cytometric T-cell receptor β variable (TRBV) chain analysis

To determine TRBV usage of T cell clones, aliquots of selected clones were labelled with a panel of 23 TRBV-specific mAb (currently available from Beckman Coulter). The T cell clones

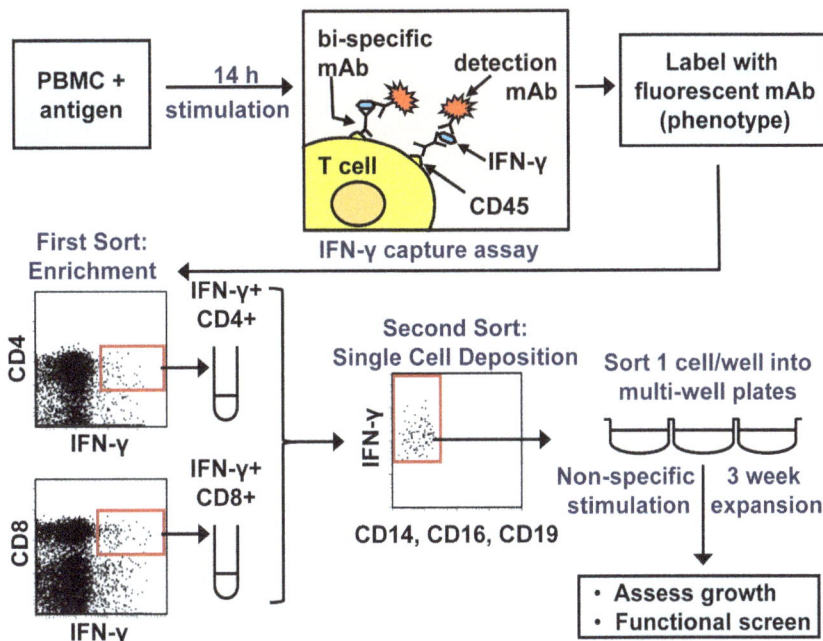

Figure 1. Overview of T cell clone generation. The procedure for cell stimulation, enrichment, single cell sorting, clone maintenance and characterisation is outlined. After PBMC were stimulated with antigen for 14 h, the IFN-γ capture assay was used to label functional cells. Cell phenotype and viability were revealed by the addition of fluorescent mAb against surface markers and propidium iodide. A MoFlo cell sorter enriched for viable, CD14– CD16– CD19– cells that were either CD4+ IFN-γ+ or CD8+ IFN-γ+. A FACS Vantage cell sorter then confirmed the phenotype of purified cell populations, and deposited functional cells into 96 well plates at one cell per well. Plates contained medium with feeder cells and phytohaemagglutinin-L, to non-specifically stimulate T cell clones. The growth and function of clones were assessed following expansion for three weeks, with weekly medium changes.

were incubated for 30 min at 4°C with CD8 APC, CD3 PE or FITC and one of the FITC- or PE-labelled TRBV-specific mAb. Cells were washed and analysed on a FACSCanto flow cytometer using FACSDiva software (BD Biosciences). Data were analysed and presented using FlowJo Software (Tree Star Inc., San Carlos, CA, USA).

Cytotoxicity assay

The cytolytic activity of T cell clones was assessed in 4 h ^{51}Cr-release assays. Autologous or allogeneic melanoma cells or LCL received fresh medium two days prior to use as targets. Target cells suspended in 50 μl assay medium were loaded with 50 μl Na$_2$ ^{51}CrO$_4$ (PerkinElmer Inc., Waltham, MA, USA) for 1 h. ^{51}Cr-labeled targets were then washed three times with RPMI 1640 and combined with effector cells at ratios ranging between 1:10 and 1:3 (10^3 targets per well) in duplicate or triplicate in a total volume of 150 μl assay medium containing a 50× target excess of unlabelled K562 cells. Maximum ^{51}Cr release was defined by incubation of targets with assay medium containing 0.1% sodium dodecyl sulfate (Sigma-Aldrich), and spontaneous ^{51}Cr release was determined by incubation of targets with assay medium alone. Plates were centrifuged at 50 g for 2 min, and then incubated for 4 h at 37°C. After incubation, 25 μl of supernatant was removed from each well, transferred to LumaPlates (Packard Bioscience, Lenexa, KS, USA) and allowed to dry overnight, prior to measurement as counts per minute (cpm) ^{51}Cr release using a TopCount Microplate Scintillation Counter (Packard). Percent specific lysis was calculated using the standard equation:

$$\% \text{ specific lysis} = 100 \times \frac{\text{mean test cpm} - \text{mean spontaneous cpm}}{\text{mean maximum cpm} - \text{mean spontaneous cpm}}$$

Clones were considered to be specific if they lysed autologous targets at >15% of maximum lysis, and lysed allogeneic targets < 5%.

IFN-γ ELISA

Antigenic stimulation was provided by autologous or allogeneic LCL or melanoma cells, or irradiated autologous LCL that had been pre-incubated with recombinant CMV pp65 (1 μl protein/ 100 μl assay medium/10^6 cells) for 1–2 h. T cell clones were combined with stimulator cells at ratios of 50:1 or 10:1 (as detailed in figure legends). After 20 h, 100 μl supernatant was removed from each well to measure IFN-γ secretion by ELISA, using standard methods (Mabtech AB, Stockholm, Sweden) in PBS. Standard curves and cytokine concentrations were calculated using SOFTmax PRO software (Molecular Devices, Sunnyvale, CA). Clones were considered to be specific if they produced > 100 pg/ml IFN-γ above controls (allogeneic LCL, allogeneic melanoma cells, or autologous LCL without CMV pp65).

Proliferation assay

CD4+ clones derived from EBV antigen stimulation were combined in assay medium with gamma-irradiated (150 Gy) autologous LCL at a stimulator: responder ratio of 1:20 in duplicate wells of a U-bottom 96 well plate. Proliferation was assessed by addition of 1μCi/well ^{3}H-thymidine (Amersham

Pharmacia Biotech Pty Ltd, Australia) for the last 6 h of the 3 day culture period. Cells were harvested onto glass fibre filter mats and ^{3}H incorporation measured as cpm on a MicroBeta scintillation counter (Wallac, Finland).

One-step TRBV gene sequence analysis

TRBV sequence analysis of T cell clones was undertaken using an adaptation of a previously described method [17]. RNA was extracted from 10^5 cells, using PureZOL RNA isolation reagent (Bio-Rad Laboratories, Hercules, CA, USA). RNA pellets were suspended in 20 μl of diethyl pyrocarbonate-treated deionised water (MP Biomedicals Inc., Solon, OH, USA). RNA was reverse transcribed and amplified in a one-step reverse transcription-PCR (RT-PCR) system (Life Technologies). Primers used in one-step reactions and subsequent sequencing reactions are as follows: Degenerate forward primers VP1 (5′-GCIITKTIYTGGTAYM-GACA-3′) and VP2 (5′-CTITKTWTTGGTAYCIKCAG-3′), specific forward primers VP3 (5′-ATCCTTTATTGGTATC-GACGT-3′) and VP4 (5′-ATGTTTACTGGTATCATAAG-3′), specific reverse primer CP1 (5′-GCACCTTCCTTCCCATT-CAC-3′). Each initial 25 μl reaction contained 12.5 μl of 2× reaction mix, 0.5 μl RT/Platinum Taq mix, 1 μl RNA, 200 nM CP1 and 2 μM VP1. If no product was detected in the PCR product by agarose gel electrophoresis, the one-step reaction was repeated using CP1 (200 nM), VP2 (2 μM), VP3 and VP4 (200 nM). RT-PCR cycling was performed at 50°C for 30 min and 94°C for 2 min. PCR cycling was then performed at 94°C for 20 s, 50°C for 40 s then 72°C for 40s for 40 cycles, with a final 10 min extension at 72°C. International ImMunoGeneTics information system (IMGT) nomenclature is used throughout this report [18].

Sequencing of PCR products

PCR products were sequenced with BigDye Terminator v3.1 Cycle Sequencing Kit (Applied Biosystems, Foster City, CA, USA), according to the manufacturer's protocol, using a VP1 or VP2 primer. Sequences were then analysed on an ABI Prism 310 Genetic Analyzer (Applied Biosystems Inc., Foster City, CA, USA). Product sequences were assessed using Chromas software (Technelysium Pty Ltd, Tewantin, Qld, Australia). *TRBV(D)J* usage was determined using the IMGT V-QUEST online analysis tool [19].

Statistics

Standard statistical tests (99% confidence intervals, standard errors of the mean, Kolmogorov-Smirnov test for differences in data distribution) were performed and graphs created in Prism v6.02 (GraphPad Software, San Diego, CA, USA).

Results

Optimisation of T cell stimulation

The human IFN-γ secretion assay (Miltenyi Biotec) uses cytokine capture to identify viable, functional T cells responding to antigenic stimulation. This kit has been used to purify IFN-γ-secreting cells prior to cloning by limiting dilution [20], but we wanted to determine if it could be combined with single-cell sorting to clone effector T cells *ex vivo*. Non-specific staining of bystander T cells in close proximity to cytokine-secreting effectors ("cross-feeding") is a well recognised problem with the cytokine capture technique. In preliminary tests of the IFN-γ detection kit, a shift in the IFN-γ negative population was occasionally observed, despite following the manufacturer's recommendations (Fig. 2A). By decreasing the concentration of cells during the one hour IFN-

γ secretion period to $1-2\times10^4$ PBMC/ml, the negative population consistently remained in the left quadrants defined by stained, unstimulated PBMC (Fig. 2B).

The period of stimulation prior to IFN-γ capture was also varied, to identify a timepoint at which IFN-γ-secreting T cells could be detected prior to significant antigen-specific proliferation. Although responses were detected as early as 45 min in some analyses, extensive longitudinal experiments consistently showed that 10-23 h maximised responses (Fig. 2C). In subsequent experiments, PBMC were stimulated for 14 h, a convenient period minimising the possibility of *in vitro* division prior to sorting, and maximising the number of cells detected.

Identification and sorting of antigen-specific effector T cells

Current T cell cloning methods frequently require prior, antigen-driven cell expansion, to overcome low precursor

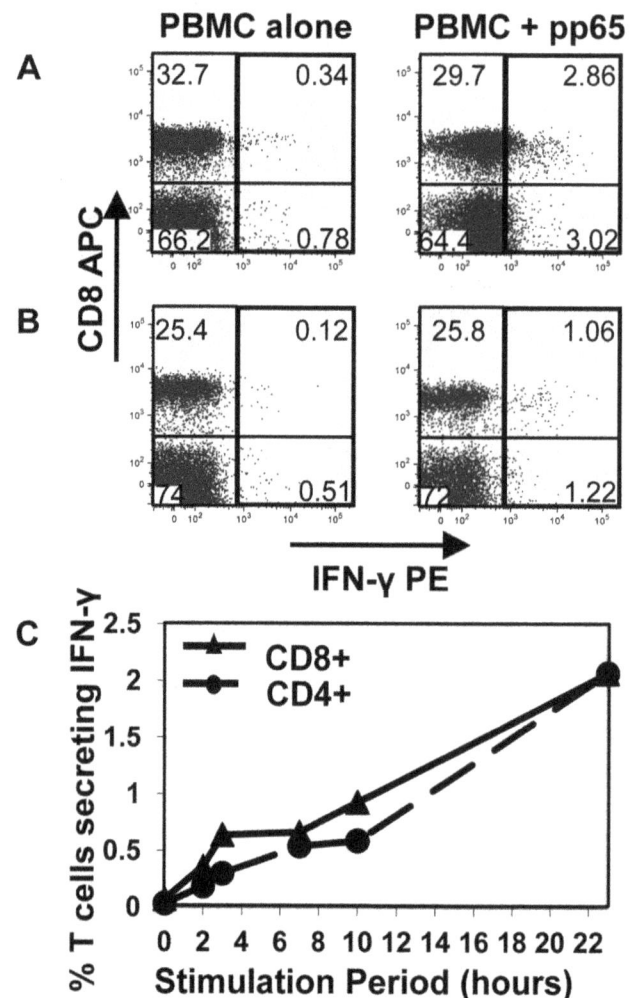

Figure 2. Optimisation of IFN-γ capture assay. (A, B) PBMC from healthy donors were stimulated for 14 h with recombinant CMV pp65, or (C) for increasing periods of time with the autologous lymphoblastoid cell line (LCL), then assessed for reactivity by IFN-γ capture assay. Cells were maintained at (A) too high a concentration during IFN-γ secretion, or (B) an optimal concentration. Cells are gated on viable, CD14– CD16– CD19– subsets. (C) The percentage of CD8+ and CD4+ T cells secreting IFN-γ, gated on viable CD3+ cells. Results are representative of (A, B) two, or (C) three independent experiments.

frequencies. However, in chronic infections, antigen-specific effector cell frequencies are elevated. Initial experiments indicated that IFN-γ production by T cells stimulated with autologous cell lines presenting cancer or viral antigens could be detected using flow cytometry, suggesting that *ex vivo* sorting based on function could enrich for antigen-specific cloning. In four separate experiments, different antigen sources were used to stimulate antigen-specific T cells: autologous melanoma cells (for two different patients), autologous LCL, or recombinant CMV pp65. PBMC from each donor were thawed, stimulated for 14 h with a single antigen source, and then IFN-γ-secreting T cells were identified using the cell surface capture IFN-γ detection kit. As expected, the magnitude of the responses to each antigen source varied, with IFN-γ-secreting T cells ranging from 0.1% of all T cells from melanoma patient A02 to 2.8% in the EBV model (Fig. 3). Although *ex vivo* T cell responses were very low for melanoma cell-stimulated PBMC, they were at least 2 fold higher than backgrounds detected in unstimulated PBMC from each donor. In summary, in each case precursor frequencies were sufficient to detect and sort antigen-reactive effector T cells based on phenotype and IFN-γ secretion.

Cloning virus- and tumour-specific CD8+ and CD4+ T cells

Using the method outlined in Fig. 1, T cells specific for viral and tumour antigens were identified and sorted in two steps: following a preliminary enrichment for CD4+ IFN-γ+ and CD8+ IFN-γ+ T cells using a MoFlo cell sorter, single IFN-γ+ cells were then deposited into individual wells of a 96 well plate using a FACS Vantage SE cell sorter. This two step, tandem procedure eliminated the need to re-configure the primary sorter for single cell deposition. In four separate experiments, 26-79% of single-sorted T cells proliferated as clones after three weeks in culture (Table 1). CD4+ and CD8+ T cells from the healthy donors (stimulated with CMV pp65 or autologous LCL) and melanoma patient A02 (stimulated with the autologous melanoma cell line) cloned at comparable efficiencies (26-47% of all wells grew). However, CD8+ T cells from melanoma patient D14 had a substantially higher cloning efficiency, as all wells proliferated initially, and 79% of the sorted T cells generated long term clones. No relationship was found between cloning efficiency and precursor frequency, as PBMC from patient D14 had one of the lowest *ex vivo* responses to antigenic stimulation. The subsequent growth of a subset of wells seeded with CD8+ T cells producing IFN-γ following CMV pp65 stimulation was assessed, using the "index sorting" function available in ClonCyte software. A significantly different distribution of IFN-γ production was observed between wells that grew *vs.* those that died (P<0.0001; Kolmogorov-Smirnov test), consistent with a lower cloning efficiency for cells producing low amounts of IFN-γ *ex vivo* (Fig. S1). This suggests that, in this instance, the fitness of CD8 T cells for subsequent expansion was related to their functional response to cognate antigen.

Clones were re-stimulated with mitogen every four weeks, and using this procedure, it was possible to expand clones to obtain up to 10^8 cells per clone, after repeated stimulation.

Functional assessment of T cell clones

The use of T cell clones is generally dependent on their expression of some antigen-specific function. We therefore analysed clones for functional characteristics in response to antigenic stimulation, given that effector T cells can alter their expression of cytokines following expansion [21].

CD4+ and CD8+ clones were stimulated with autologous or allogeneic melanoma cells, LCL or LCL loaded (or untreated) with CMV pp65 protein, as appropriate, and then culture supernatants were analysed for IFN-ã by ELISA (Fig. 4 A–D). The proportion of clones that retained IFN-γ production varied from 8% (CD8+ clones derived from LCL stimulation) to 77% (CD8+ clones from melanoma patient D14; Table 1). As with cloning efficiency, the ability of clones to secrete IFN-γ was not related to higher precursor frequency – indeed, the opposite may be true, as the high proportion of cells producing IFN-γ *ex vivo* in response to CMV and EBV translated into lower proportions of functional CD8+ T cells (Table 1).

To determine whether T cells sorted on the basis of IFN-γ production were capable of killing, selected LCL- and melanoma-stimulated CD8+ clones were tested for cytolytic activity against autologous and allogeneic cell lines. Strong activity was demonstrated for all tested CD8+ clones (Fig 5 A, B, C), with minimal reactivity against the allogeneic control. To obtain representative data across a broader range of clones, we screened all patient D14 CD8$^+$ T cell clones against autologous and allogeneic melanoma lines using ^{51}Cr-release assays (Fig 5D). Cytolytic activity was demonstrated for 75% of clones, of which 16% were scored negative for IFN-γ production in the parallel assay (Fig 5D). Likewise, a proportion of clones secreted IFN-γ but lacked cytolytic activity against the autologous melanoma line. Overall, only 7% of sorted, proliferating D14 CD8+ T cell clones failed to exhibit significant functional activity in either of these assays.

Selected CD4+ clones derived by LCL stimulation of healthy donor PBMC were also tested for proliferation in response to autologous antigen using a tritiated thymidine assay; a total of 14/63 clones proliferated significantly above background defined by non-stimulated controls (Figure S2).

Overall, between 12% and 93% of clones expressed some functional response to antigen stimulation. This is likely to be an underestimate of the true total for some stimuli, as (except for patient D14) only selected clones were tested for cytolytic activity or proliferation (Table 1).

Assessment of clonality by TRBV analysis

Clonality is required for many applications, for example, antigen discovery and studies of T cell receptor usage. The most common method of T cell cloning, limiting dilution, relies on the statistical probability that no more than one proliferating cell will be placed in each well during plate setup. This makes the method inefficient, as many wells will not contain T cells, and re-cloning is often required due to outgrowth of mixed populations. Cloning using single cell sorting avoids this problem, by providing a definitive method of obtaining clonal T cells, whilst employing comparatively few plates. The only confounding factor would be if cross-contamination occurred during sorting, or during extended culture, which could compromise the outcome.

To confirm that T cell cultures generated by single-cell sorting were truly clonal, we undertook TRBV expression analysis of eleven clones from the LCL sort, using a panel of 23 mAb specific for different TRBV (representative clones shown in Fig. 6A). Six clones stained uniformly with a single mAb: TRBV20, TRBV22, and TRBV3 (2 clones) and TRBV2 (2 clones). These six clones were not stained by the other 22 TRBV mAb. The clones for which a TRBV was not identified (*i.e.* they were not stained by any of the 23 mAb) likely express TRBV that were not part of the mAb set. These data definitively support the clonality of these cultures, and show that feeder cells do not interfere with the analysis. Flow cytometric data was confirmed by sequencing the TRBV regions of these clones (Fig. 6C), and both forms of analysis

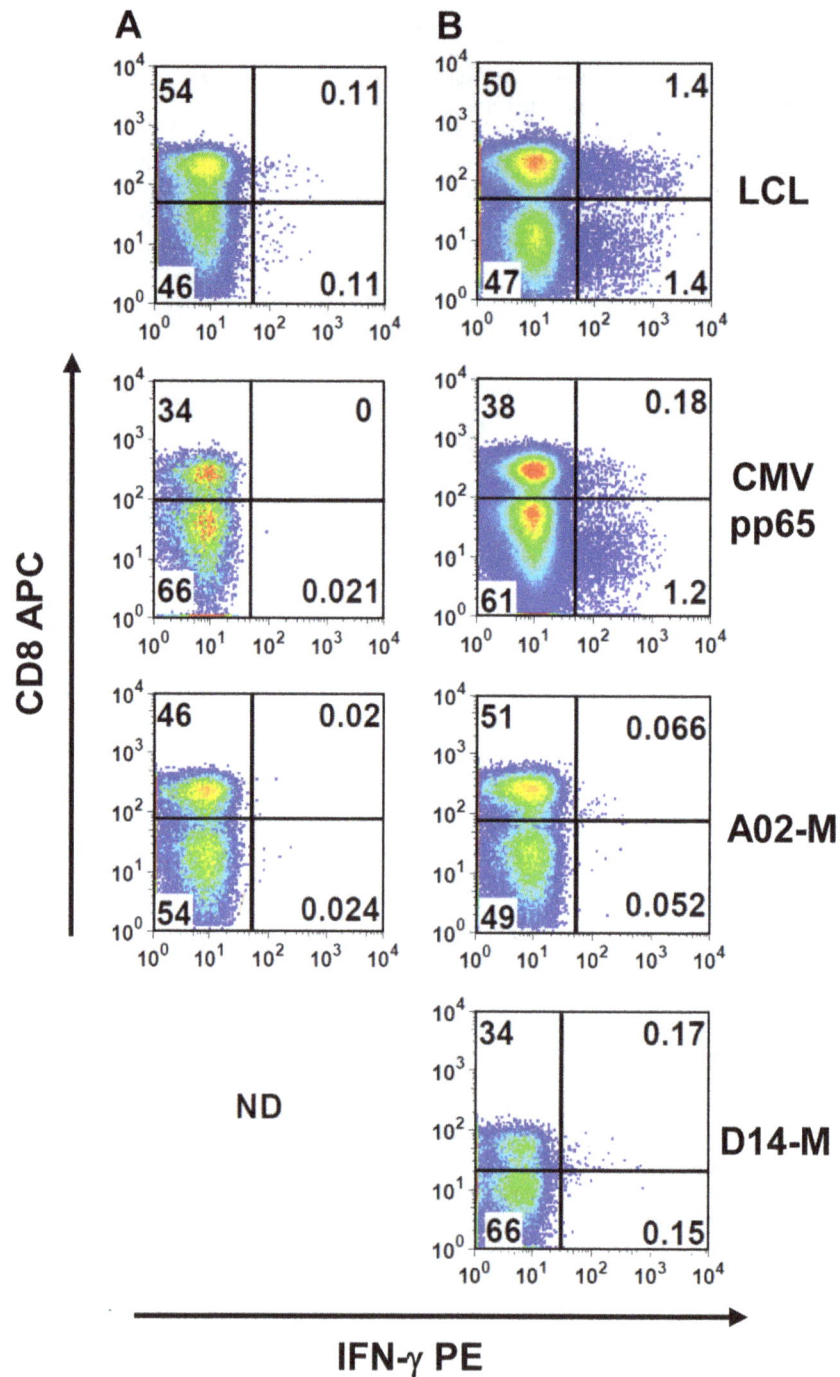

Figure 3. Detection of antigen-specific effector T cells *ex vivo*. IFN-γ production measured by the capture assay in (A) unstimulated PBMC samples and (B) PBMC from healthy donors stimulated with the autologous lymphoblastoid cell line (LCL) or CMV pp65 (top two panels), or PBMC from melanoma patients A02 or D14 stimulated with autologous melanoma cells (bottom two panels), as indicated. Cells were gated on the single cell, low scatter, viable, CD14–CD16–CD19– subpopulation. Percentages are shown in each quadrant; ND = not done.

were repeated on tumour antigen-specific T cell clones (Fig. 6B, D). All T cell clones stained positively with a maximum of one TRBV mAb and only one TRBV region was detected in T cell clone RNA, therefore we conclude that the generated T cells were clonal.

Discussion

Here we describe a rapid and highly efficient method for cloning effector CD4+ and CD8+ T cells *ex vivo*, which can use a range of antigen sources as stimuli, and generates T cell clones with diverse specificities. The method enables the selection of T cell clones with known function *ex vivo*, without the need for prior, multiple rounds of *in vitro* stimulation and cell division.

Table 1. Cloning efficiency and functional retention of T cell clones.

Stimulus	T cell Subset	Wells Seeded	Established Clones[a]	IFN-γ+ clones[b]	Functional clones[d]
LCL	CD4+	234	62 (26%)	5/5[c]	15/65 (23%)
	CD8+	246	116 (47%)	9/107 (8%)	15/126 (12%)
CMV pp65	CD4+	250	87 (35%)	41/87 (47%)	–
	CD8+	230	66 (29%)	19/66 (29%)	–
A02-M	CD4+	114	32 (28%)	19/31 (61%)	–
	CD8+	162	57 (35%)	23/63 (37%)	23/63 (37%)
D14-M	CD4+	ND	ND	ND	ND
	CD8+	288	227 (79%)	221/288 (77%)	267/288 (93%)

[a] Number of growing clones, percentage of wells seeded in brackets.
[b] Number of clones producing >100 pg/ml more IFN-γ in response to autologous/antigen loaded autologous targets than to control targets, as a fraction of total clones tested; percentage in brackets. In some cases clones that did not proliferate long term were tested, so the denominator is different to the number of established clones.
[c] CD4+ clones from LCL stimulation were pre-selected from a subset which were positive for antigen-stimulated proliferation, and thus may not be representative of the 62 proliferating clones.
[d] Clones were tested for cytolytic activity (CD8+ clones from LCL and melanoma stimulation only), IFN-γ production (CD4+ and CD8+ clones from each stimulation), or proliferation (CD4+ LCL clones only). Data represent number of clones with any significant response to autologous antigen, as a fraction of total number of clones tested.
ND, not done; – indicates no test additional to IFN-γ production was performed.

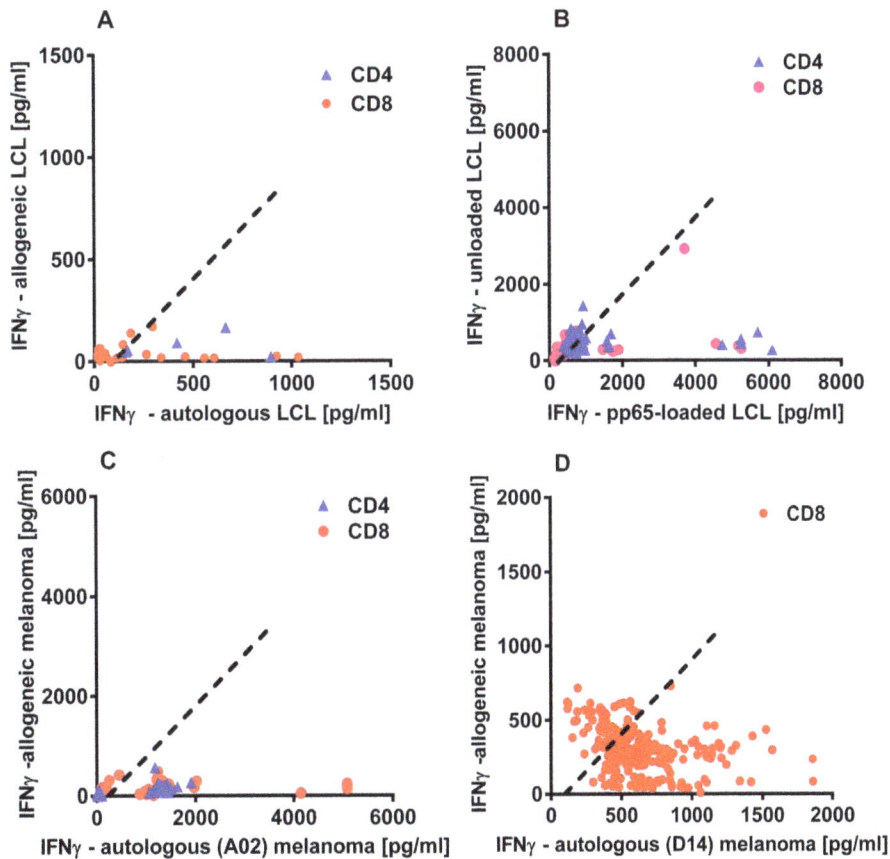

Figure 4. Retention of IFN-γ production by T cell clones. ELISA was used to measure IFN-γ in supernatants of T cell clones (▲CD4+; ● CD8+) stimulated with (A) autologous or allogeneic lymphoblastoid cell lines (LCL), (B) autologous LCL either unloaded or pre-incubated with recombinant pp65 for 1-2 h, or (C, D) autologous or allogeneic melanoma cells (as indicated in graph). The stimulator: T cell responder ratios were (A–C) 50,000:1,000 or (D, patient D14) 100,000:10,000. Dotted line indicates >100 pg/ml in excess of control, the cut-off for positivity used in Table 1, which summarises the data from this figure.

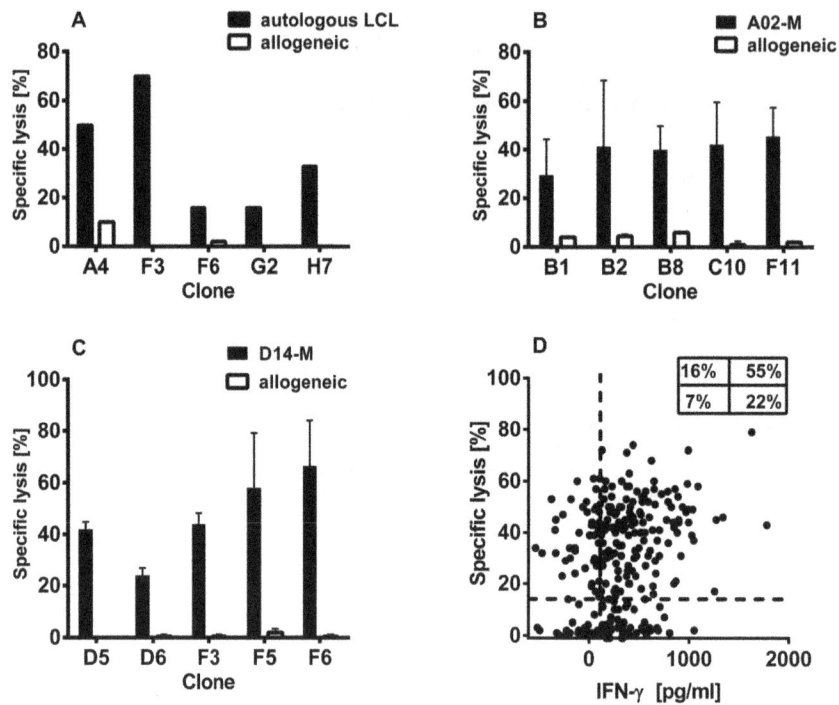

Figure 5. Cytotoxicity of CD8+ T cell clones. Cytotoxicity of selected CD8+ T cell clones from (A) healthy donor, (B) melanoma patient A02, and (C, D) melanoma patient D14, was assessed against the autologous and allogeneic lymphoblastoid cell line (LCL) or melanoma cells by chromium release assay at an effector: target ratio of 10:1. Data indicate the mean specific lysis from (A) one; (B) three; and (C) two independent experiments, and error bars represent the standard error of the mean. (D) T cell clones from patient D14 were simultaneously assessed for IFN-γ production in response to autologous and allogeneic melanoma cells (as described in Fig 4). The specific lysis of autologous melanoma cells (vertical axis) is plotted against IFN-γ production (data corrected by subtracting allogeneic release). Dotted lines indicate cut-offs for positivity (see methods); inset indicates percentages positive in each quadrant.

Subpopulations of T cells <0.1% above background (Fig. 1) were cloned and expanded to large numbers, and the capability of recording *ex vivo* phenotypic characteristics enables the linkage of these attributes with data subsequently generated *in vitro*.

Current cloning methods do not allow precise cloning efficiencies to be calculated, since limiting dilution analysis frequently underestimates the precursor frequency [22]. In the case of known epitopes, multimer technology does allow precursor frequencies to be estimated. Dunbar *et al.* sorted single HLA-A2-restricted, melanoma-epitope tetramer+ CD8+ T cells and cloned them at an average efficiency of 6.5% [9]. In contrast, the method described in this paper generates CD4+ and CD8+ clones without prior knowledge of epitope specificity or limitation of particular HLA types, and is thus suitable for goals such as antigen discovery and the analysis of broad responses to antigen. Since flow cytometric and sequence analysis of TRBV regions of the clones indicated that each clone had indeed originated from a single precursor, the method provides a direct estimate of cloning efficiency. Clonal expansion was successful for an average of 30% (CD4+) and 48% (CD8+) of effector T cells that were sorted from PBMC in response to autologous melanoma cells, LCL or a CMV protein. The proportion of clones that retained antigen-specific functionality varied widely among donors, and between CD4+ and CD8+ subsets. Although this could reflect the sorting and clonal expansion of IFN-γ false-positive (e.g., cross feeding) contaminants, the high functional capacity of CD8+ T cells cloned from melanoma patient D14 (despite a low precursor frequency) indicates that instability of functional characteristics likely contributes to this phenomenon. Importantly, the method described herein could be used to investigate this further. For

example, index sorting data reveal that effector CD8+ T cells producing low amounts of INF-γ in response to CMV pp65 have a significantly lower cloning efficiency, conceivably due to a lessened fitness of T cells to replicate or to survive. It would be of interest to investigate whether this characteristic has clinical implications.

T cells are often cloned after cultures of PBMC are stimulated and allowed to expand *in vitro* [10,23,24]. This prevents the analysis of some precursor characteristics, based on cell surface marker or cytokine expression, due to the differentiation of cells within the *in vitro* cultures. The method described in this paper enables highly efficient selection of subpopulations of interest, by avoiding this extended *in vitro* culture step. Short-term stimulation with antigen limits the opportunity for T cell replication, and phenotypic and functional modulation, and also enables direct quantitation of the T cells of interest. Index sorting allows the "history" of each T cell clone to be examined retrospectively, allowing connections to be made between *ex vivo* and *in vitro* T cell characteristics. This becomes important when assessing parameters used to define T cell characteristics *in vivo* [25,26], which may not be stable *in vitro*, e.g. CD27 and CD45RA.

By removing the need for extended stimulation with antigen and antigen-presenting cells, T cell clones can be generated rapidly. Clones are available for use within three weeks of sorting, and can be expanded to large numbers (10^7–10^8) after a further three weeks of culture. There is no need for re-cloning, which is often required following limiting dilution cloning. In addition, by avoiding the selective outgrowth of more rapidly dividing cells, direct cloning likely enhances the repertoire diversity of the resulting collection of clones.

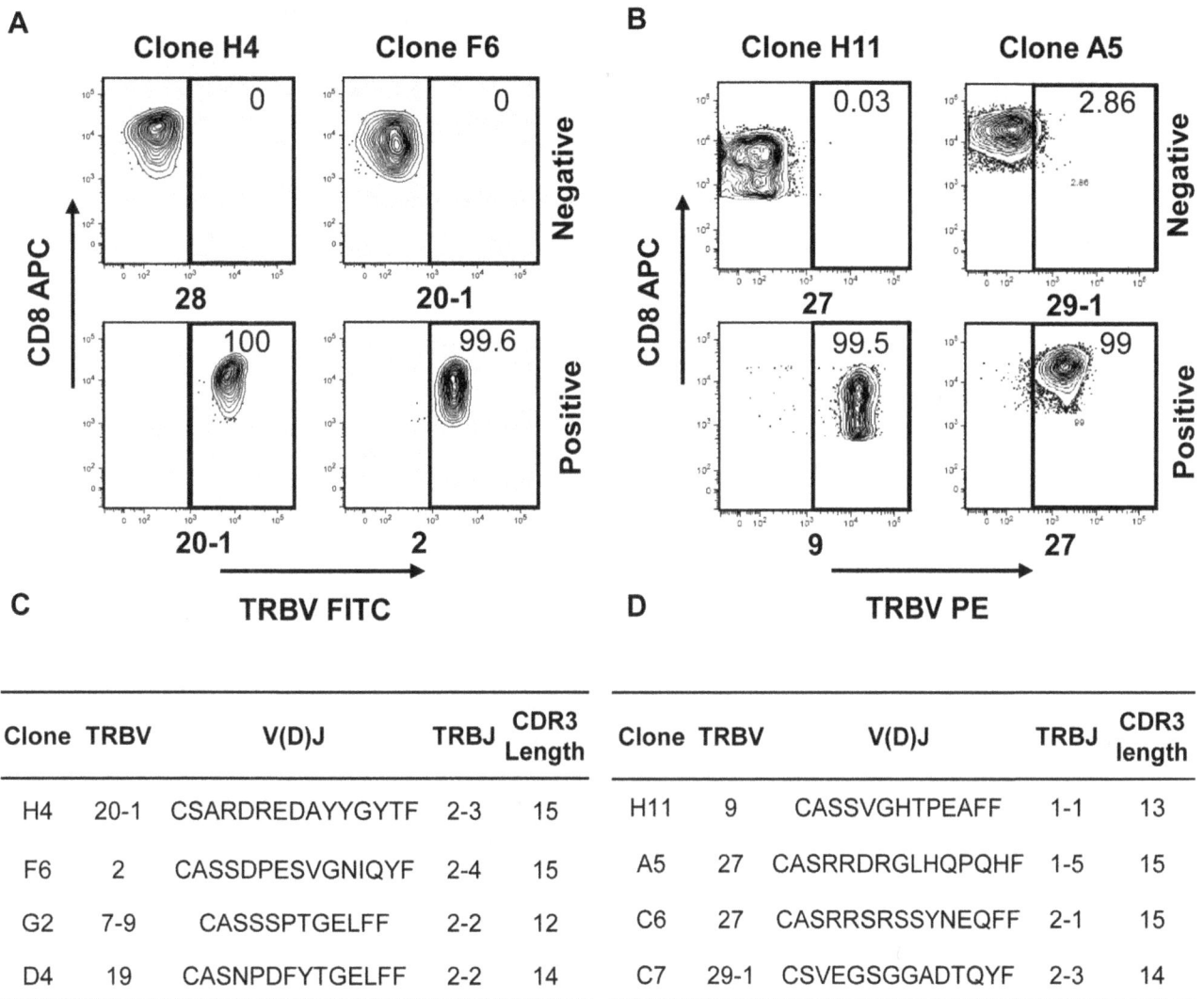

Figure 6. TRBV analysis of LCL- and tumour-reactive CD8+ T cell clones. CD8+ T cell clones specific for (A, C) LCL or (B, D) tumour antigens were assessed for clonality by flow cytometric and molecular analysis. (A, B) Clones were stained with a panel of fluorescently-labelled TRBV mAb, then analysed on a flow cytometer. Positive and negative TRBV staining is shown for representative T cell clones. (C, D) To confirm the flow cytometric findings and determine unidentified TRBV regions, RNA was extracted from T cell clones and TRBV regions reverse transcribed, amplified, then sequenced. The designations TRBV and TRBJ follow the TCR gene nomenclature specified by IMGT [18].

Clone	TRBV	V(D)J	TRBJ	CDR3 Length
H4	20-1	CSARDREDAYYGYTF	2-3	15
F6	2	CASSDPESVGNIQYF	2-4	15
G2	7-9	CASSSPTGELFF	2-2	12
D4	19	CASNPDFYTGELFF	2-2	14

Clone	TRBV	V(D)J	TRBJ	CDR3 length
H11	9	CASSVGHTPEAFF	1-1	13
A5	27	CASRRDRGLHQPQHF	1-5	15
C6	27	CASRRSRSSYNEQFF	2-1	15
C7	29-1	CSVEGSGGADTQYF	2-3	14

The addition of exogenous IL-2 or a co-stimulatory antibody against CD28 did not noticeably improve the *ex vivo* response to melanoma cell stimulation in preliminary studies, but the proportion of T cells activated might, however, be enhanced through the use of these additional signals when using other antigen sources, or by using other growth factors e.g. GM-CSF [27,28].

Although our cloning technique as described here is based on IFN-γ secretion, the method could be adapted to select T cells based on production of other cytokines for which secretion assays are available, such as TNF, IL-2 or IL-17, alone or in combination. We sorted directly from stained PBMC, but commercially available cytokine capture assays allow magnetic enrichment of extremely low precursor numbers, so the extension of our method to very rare subsets would be straight forward. Direct cloning might also be applied to other non-lethal methods of identifying antigen reactive T cells, such as membrane-bound TNF [29] and the expression of a wide variety of activation markers (reviewed in [30]). However, the reliance of direct cloning on effector function makes it unsuitable for expanding naïve precursors, which is achievable by limit dilution cloning of amplified T cells [31].

T cell clones generated using direct cloning have advantages for some specific applications that are of potential clinical importance. The method preferentially selects effector-memory T cells, with immediate functional ability, which presumably reflects the active, circulating anti-pathogen or anti-tumour population. Since the cloning efficiency is high, the application of clones to defining the antigenic repertoire or discovery of novel antigens [32] would enable establishment of an immunodominance hierarchy for identified epitopes. Furthermore, by sequencing T-cell receptors of clones recognising an epitope of interest, the level of diversity within the effector response can be determined.

In conclusion, the ex vivo effector cell cloning method described here provides a rapid, powerful and effective method of deriving antigen-specific T cells clones with traceable in vivo precursor function and phenotype, thus improving on functionality and applicability compared to traditional T cell cloning techniques.

Supporting Information

Figure S1   (A) Index sorting was used to assign *ex vivo* IFN-γ secretion levels (arbitrary fluorescence units), in response to CMV pp65, to individual CD8+ T cells seeded into wells, which were scored for subsequent growth into long-term clones ("Grew" *vs.* "Died"). (B) Percentile plot showing cumulative percentage of cells that established clones (– –) or died (—) according to IFN-γ production. The distribution of IFN-γ production differed significantly between the two groups of cells (Kolmogorov-Smirnov test; P< 0.0001).

Figure S2   CD4+ clones derived from EBV stimulation were tested for proliferation in response to the autologous LCL by tritiated thymidine incorporation; the background proliferation (Nil stimulation) of a random selection of clones was also assessed. Data indicate means of duplicate measurements in a single experiment. The geometric mean (—) and upper 99% confidence limit of the geometric mean (---) of proliferation of unstimulated clones are indicated.

Acknowledgments

The authors wish to acknowledge the Grace Chojnowski and Paula Hall from the Flow Cytometry facility for their technical support, and the Australian Red Cross Blood Service, for provision of blood products.

Author Contributions

Conceived and designed the experiments: MAN CWS NRM MHLL. Performed the experiments: MAN MHLL CML LEO NRM. Analyzed the data: MAN CWS NRM. Contributed reagents/materials/analysis tools: CWS NRM. Wrote the paper: MAN CWS NRM ALP. Data interpretation: MAN NRM ALP CWS.

References

1. Hunder NN, Wallen H, Cao J, Hendricks DW, Reilly JZ, et al. (2008) Treatment of metastatic melanoma with autologous CD4+ T cells against NY-ESO-1. N Engl J Med 358: 2698–2703.
2. Riddell SR, Watanabe KS, Goodrich JM, Li CR, Agha ME, et al. (1992) Restoration of viral immunity in immunodeficient humans by the adoptive transfer of T cell clones. Science 257: 238–241.
3. Fonteneau JF, Larsson M, Somersan S, Sanders C, Munz C, et al. (2001) Generation of high quantities of viral and tumor-specific human CD4+ and CD8+ T-cell clones using peptide pulsed mature dendritic cells. J Immunol Methods 258: 111–126.
4. Gervois N, Labarriere N, Le Guiner S, Pandolfino MC, Fonteneau JF, et al. (2000) High avidity melanoma-reactive cytotoxic T lymphocytes are efficiently induced from peripheral blood lymphocytes on stimulation by peptide-pulsed melanoma cells. Clin Cancer Res 6: 1459–1467.
5. Alexander-Miller MA, Leggatt GR, Sarin A, Berzofsky JA (1996) Role of antigen, CD8, and cytotoxic T lymphocyte (CTL) avidity in high dose antigen induction of apoptosis of effector CTL. J Exp Med 184: 485–492.
6. Geginat J, Sallusto F, Lanzavecchia A (2001) Cytokine-driven proliferation and differentiation of human naive, central memory, and effector memory CD4(+) T cells. J Exp Med 194: 1711–1719.
7. Kim M, Moon HB, Kim K, Lee KY (2006) Antigen dose governs the shaping of CTL repertoires in vitro and in vivo. Int Immunol 18: 435–444.
8. Leggatt GR, Narayan S, Fernando GJ, Frazer IH (2004) Changes to peptide structure, not concentration, contribute to expansion of the lowest avidity cytotoxic T lymphocytes. J Leukoc Biol 76: 787–795.
9. Dunbar PR, Chen JL, Chao D, Rust N, Teisserenc H, et al. (1999) Cutting edge: rapid cloning of tumor-specific CTL suitable for adoptive immunotherapy of melanoma. J Immunol 162: 6959–6962.
10. Yee C, Savage PA, Lee PP, Davis MM, Greenberg PD (1999) Isolation of high avidity melanoma-reactive CTL from heterogeneous populations using peptide-MHC tetramers. J Immunol 162: 2227–2234.
11. Varadarajan N, Kwon DS, Law KM, Ogunniyi AO, Anahtar MN, et al. (2012) Rapid, efficient functional characterization and recovery of HIV-specific human CD8+ T cells using microengraving. Proc Natl Acad Sci U S A 109: 3885–3890.
12. Fletcher JM, Vukmanovic-Stejic M, Dunne PJ, Birch KE, Cook JE, et al. (2005) Cytomegalovirus-specific CD4+ T cells in healthy carriers are continuously driven to replicative exhaustion. J Immunol 175: 8218–8225.
13. Plunkett FJ, Franzese O, Belaramani LL, Fletcher JM, Gilmour KC, et al. (2005) The impact of telomere erosion on memory CD8+ T cells in patients with X-linked lymphoproliferative syndrome. Mech Ageing Dev 126: 855–865.
14. O'Rourke MG, Johnson M, Lanagan C, See J, Yang J, et al. (2003) Durable complete clinical responses in a phase I/II trial using an autologous melanoma cell/dendritic cell vaccine. Cancer Immunol Immunother 52: 387–395.
15. Pavey S, Johansson P, Packer L, Taylor J, Stark M, et al. (2004) Microarray expression profiling in melanoma reveals a BRAF mutation signature. Oncogene 23: 4060–4067.
16. Campbell JD (2003) Detection and enrichment of antigen-specific CD4+ and CD8+ T cells based on cytokine secretion. Methods 31: 150–159.
17. Zhou D, Srivastava R, Grummel V, Cepok S, Hartung HP, et al. (2006) High throughput analysis of TCR-beta rearrangement and gene expression in single T cells. Lab Invest 86: 314–321.
18. Giudicelli V, Duroux P, Ginestoux C, Folch G, Jabado-Michaloud J, et al. (2006) IMGT/LIGM-DB, the IMGT comprehensive database of immunoglobulin and T cell receptor nucleotide sequences. Nucleic Acids Res 34: D781–784.
19. Giudicelli V, Brochet X, Lefranc MP (2011) IMGT/V-QUEST: IMGT standardized analysis of the immunoglobulin (IG) and T cell receptor (TR) nucleotide sequences. Cold Spring Harb Protoc 2011: 695–715.
20. Manley TJ, Luy L, Jones T, Boeckh M, Mutimer H, et al. (2004) Immune evasion proteins of human cytomegalovirus do not prevent a diverse CD8+ cytotoxic T-cell response in natural infection. Blood 104: 1075–1082.
21. Doyle AG, Buttigieg K, Groves P, Johnson BJ, Kelso A (1999) The activated type 1-polarized CD8(+) T cell population isolated from an effector site contains cells with flexible cytokine profiles. J Exp Med 190: 1081–1092.
22. Ogg GS, McMichael AJ (1998) HLA-peptide tetrameric complexes. Curr Opin Immunol 10: 393–396.
23. Ho WY, Nguyen HN, Wolfl M, Kuball J, Greenberg PD (2006) In vitro methods for generating CD8+ T-cell clones for immunotherapy from the naive repertoire. J Immunol Methods 310: 40–52.
24. Riddell SR, Greenberg PD (1990) The use of anti-CD3 and anti-CD28 monoclonal antibodies to clone and expand human antigen-specific T cells. J Immunol Methods 128: 189–201.
25. Hamann D, Baars PA, Rep MH, Hooibrink B, Kerkhof-Garde SR, et al. (1997) Phenotypic and functional separation of memory and effector human CD8+ T cells. J Exp Med 186: 1407–1418.
26. Sallusto F, Lenig D, Forster R, Lipp M, Lanzavecchia A (1999) Two subsets of memory T lymphocytes with distinct homing potentials and effector functions. Nature 401: 708–712.
27. Morrissey PJ, Bressler L, Park LS, Alpert A, Gillis S (1987) Granulocyte-macrophage colony-stimulating factor augments the primary antibody response by enhancing the function of antigen-presenting cells. J Immunol 139: 1113–1119.
28. Martinuzzi E, Afonso G, Gagnerault MC, Naselli G, Mittag D, et al. (2011) acDCs enhance human antigen-specific T-cell responses. Blood 118: 2128–2137.
29. Haney D, Quigley MF, Asher TE, Ambrozak DR, Gostick E, et al. (2011) Isolation of viable antigen-specific CD8+ T cells based on membrane-bound tumor necrosis factor (TNF)-alpha expression. J Immunol Methods 369: 33–41.
30. Bacher P, Scheffold A (2013) Flow-cytometric analysis of rare antigen-specific T cells. Cytometry A 83: 692–701.
31. Geiger R, Duhen T, Lanzavecchia A, Sallusto F (2009) Human naive and memory CD4+ T cell repertoires specific for naturally processed antigens analyzed using libraries of amplified T cells. J Exp Med 206: 1525–1534.
32. Lennerz V, Fatho M, Gentilini C, Frye RA, Lifke A, et al. (2005) The response of autologous T cells to a human melanoma is dominated by mutated neoantigens. Proc Natl Acad Sci U S A 102: 16013–16018.

3

Genetic Diversity of the Coat Protein of Olive Mild Mosaic Virus (OMMV) and Tobacco Necrosis Virus D (TNV-D) Isolates and Its Structural Implications

Carla M. R. Varanda[1]*, Marco Machado[1], Paulo Martel[2], Gustavo Nolasco[3], Maria I. E. Clara[1], Maria R. Félix[1]

1 Laboratório de Virologia Vegetal, Instituto de Ciências Agrárias e Ambientais Mediterrânicas Universidade de Évora, Évora, Portugal, **2** Departamento de Ciências Biológicas e Bioengenharia, Faculdade de Ciências e Tecnologia da Universidade do Algarve, Faro, Portugal, **3** Laboratório de Virologia Vegetal, Universidade do Algarve, Faro, Portugal

Abstract

The genetic variability among 13 isolates of *Olive mild mosaic virus* (OMMV) and of 11 isolates of *Tobacco necrosis virus D* (TNV-D) recovered from *Olea europaea* L. samples from various sites in Portugal, was assessed through the analysis of the coat protein (CP) gene sequences. This gene was amplified through reverse transcriptase polymerase chain reaction (RT-PCR), cloned, and 5 clone sequences of each virus isolate, were analysed and compared, including sequences from OMMV and TNV-D isolates originally recovered from different hosts and countries and available in the GenBank, totalling 131 sequences. The encoded CP sequences consisted of 269 amino acids (aa) in OMMV and 268 in TNV-D. Comparison of the CP genomic and amino acid sequences of the isolates showed a very low variability among OMMV isolates, 0.005 and 0.007, respectively, as well as among TNV-D isolates, 0.006 and 0.008. The maximum nucleotide distances of OMMV and TNV-D sequences within isolates were also low, 0.013 and 0.031, respectively, and close to that found between isolates, 0.018 and 0.034, respectively. In some cases, less variability was found in clone sequences between isolates than in clone sequences within isolates, as also shown through phylogenetic analysis. CP aa sequence identities among OMMV and TNV-D isolates ranged from 84.3% to 85.8%. Comparison between the CP genomic sequences of the two viruses, showed a relatively low variability, 0.199, and a maximum nucleotide distance between isolates of 0.411. Analysis of comparative models of OMMV and TNV-D CPs, showed that naturally occurring substitutions in their respective sequences do not seem to cause significant alterations in the virion structure. This is consistent with a high selective pressure to preserve the structure of viral capsid proteins.

Editor: Hanu Pappu, Washington State University, United States of America

Funding: Carla Marisa R. Varanda is the recipient of a Post doctoral fellowship from Fundação para a Ciência e a Tecnologia (FCT), SFRH/BPD/76194/2011. This work has been supported by FEDER and National funds, through the Programa Operacional Regional do Alentejo (InAlentejo) Operation ALENT-07-0262-FEDER-001871/Laboratório de Biotecnologia Aplicada e Tecnologias Agro-Ambientais. The funders had no role in study design, data collection and analysis, decision to publish, or preparation of the manuscript.

Competing Interests: The authors have declared that no competing interests exist.

* Email: carlavaranda@uevora.pt

Introduction

Olive mild mosaic virus (OMMV) and *Tobacco necrosis virus D* (TNV-D) originally placed in the Necrovirus genus, were recently divided and included into the new genera *Alphanecrovirus* and *Betanecrovirus*, respectively, based on the level of sequence diversity in their polymerases.

Complete genomic sequences of TNV-D were obtained from isolates recovered from bean in England [1], tobacco in Hungary [2] and, more recently, from an olive tree in Portugal [3].

In 2004, an olive isolate initially identified as a TNV-D isolate based on the sequencing of the coat protein (CP) gene [4] was later considered a distinct species, OMMV [5], following the complete sequencing of its genome. Based on the deduced amino acid (aa) CP sequence, OMMV showed 85.1% identity with that of TNV-D [3], which explains its initial diagnosis as a TNV-D isolate.

Since then, OMMV has been recorded infecting other hosts such as spinach in Greece [6] and tulip in the Netherlands, revealing that OMMV is part of the viral complex associated with the Augusta disease, previously ascribed to TNV [7].

Differentiation between OMMV and TNV-D is only possible through PCR based assays using specific primers [8] or through complete genome sequencing. Thus, TNV isolates referred in earlier literature may reveal themselves to be either TNV-D or OMMV.

OMMV and TNV-D particles are oligomers consisting of 180 copies of the CP polypeptide [5]. Both viruses have isometric particles, *ca.* 28 nm in diameter, with single-stranded positive-sense RNA and a genome of *ca.* 3.7 kb in length. OMMV genome has 5 Open Reading Frames (ORF) and TNV-D has 6. The 5′-proximal ORF1 of OMMV encodes a polypeptide of 202 aa with a molecular weight (MW) of 23 kDa (p23) and that of TNV-D has

22 KDa (p22) MW. ORF1RT results from the read-through of the amber stop codon, and encodes a 82 kDa protein predicted to be the viral RdRp. OMMV ORF2 overlaps ORF1RT by 17 nts and encodes a 8 kDa polypeptide with 73 aa (p8) and ORF 3 encodes a 56 aa polypeptide with a molecular mass of 6 kDa (p6). These two small proteins are predicted to be involved in virus movement based on the high aa sequence identity with the OLV-1 movement proteins p6 and p8. As for TNV-D, ORFs 2, 3 and 4 are predicted to encode small peptides with about 7 kDa designated $p7_1$ (62 aa), p7a (65 aa) and p7b (66 aa) respectively. The 3′–proximal ORF5 of OMMV and TNV-D encode a 269 aa polypeptide with 29 kDa (p29), identified as the virus CP.

Most CPs of plant icosahedral positive-stranded RNA viruses have four distinct structural domains: an 'R' domain involved in the interaction with RNA, a connecting arm 'a', a central shell domain 'S' and a C-terminal projecting 'P' domain. Necroviruses particles do not have a protruding domain [9,10]. The 'S' domain comprises 8 anti-parallel beta-strands, which form a twisted sheet or jelly-roll fold [11] and presents a signature pattern, consisting of 26 amino acid residues ([FYW]-x-[PSTA]-x(7)-G-x-[LIVM]-x-[LIVM]-x-[FYWIL]-x(2)-D-x(5)-P). In OMMV this pattern was detected in positions OMMV CP 134 to 159 aa [4,12]. OMMV and TNV-D CP show significant homology (≈ 45% identity) with the *Tobacco necrosis virus A* coat protein, whose quaternary structure was solved at 2.25 Å resolution [13]. The quaternary structure of the OMMV coat arrangement has been determined based on the TNV-A structure through comparative modelling approaches and the effect of two mutations in the virion structure was assessed [14].

The CP has been shown to be involved in many non-structural functions such as virus movement within the plant, genome activation and elicitation of symptoms, as well as in suppression of RNA silencing and vector transmission [15–17]. Recent studies showed that a single mutation in the CP gene of OMMV was responsible for the loss of transmission by the fungus *Olpidium brassicae* [14].

RNA viruses are expected to undergo high mutation rates potentially leading to a substantial genetic variation. Continuous generation of new mutants gives rise to a high diversity in nucleotide sequences among viruses leading to populations made up of many different yet related genomes. Conversely, as a result of factors such as natural selection, bottleneck effect occurrences and even natural host range, this genetic diversity has often showed to be lower than expected [18–21]. Recently, analysis of the genetic diversity of the CP of the alphanecrovirus *Olive latent virus* 1 (OLV-1), that shares an identity of 49.6% with OMMV CP and 48.8% with that of TNV-D [3,5], showed a very low value of genetic diversity, 0.02, among 25 isolates obtained from different hosts and locations [22].

Knowledge on the diversity in the viral CP gene will help to understand how viruses become adapted to hosts and vectors and contribute to more efficient and durable diagnostic methods. In this work we analysed the CP molecular diversity of the two viruses, OMMV and TNV-D, that share an economically important host, olive, and determined the implications of that diversity in the virions structure.

Materials and Methods

Viral isolates used in this study were obtained from olive fruits and leaves collected during the year of 1995, during two research projects of University of Évora (1995–1998 - Project PAMAF-IED N° 4057 'Valorization of Olea europaea L. cultivars "Negrinha de Freixo" and "Santulhana" in Trás-os-Montes region of Portugal',

coordinated by Estação Agronómica Nacional, Oeiras. Person in charge: Eng° Fausto António Leitão; and 1995–1998 - Project PAMAF-IED N° 2064 'Selection of clones of Olea europaea L. cultivars used in the production of Moura olive oil for the renovation of olive orchards in Serpa and Moura, south of Portugal', coordinated by Instituto de Biologia Experimental e Tecnológica, IBET, Oeiras. Person in charge: Professor Pedro Fevereiro. No protected species were sampled. Isolates have been maintained, since then, in *Chenopodium murale* plants in Laboratory of Plant Virology of University of Évora, and have been used in several studies.

Virus isolates

Thirteen OMMV isolates and eleven TNV-D isolates obtained from field olive trees growing in different regions of Portugal were used in this study. Viral isolates were collected during the year of 1995, during two research projects of University of Évora. Isolates have been maintained, since then, in *Chenopodium murale* plants in Laboratory of Plant Virology of University of Évora, and have been used in several studies.

Some of the field isolates revealed to be either OMMV (five) or TNV-D (three) whereas eight were a complex of both (Table 1). OMMV3 resulted from the inoculation in *C. murale* plants of OMMVpUFLOMMV-3 (gb|HQ651832.1|), a full length clone derived from GP isolate (NC_006939.1); OMMVL11 is an OMMV3 variant obtained after 15 serial passages of single local lesions in *C. murale* plants which has shown *O. brassicae* non transmissible [17]. V8ia1 isolate was obtained following 10 serial passages of an initial single local lesion induced in *C. murale* plants by V8i.

Additional CP gene sequences from 3 isolates of TNV-D and of 8 OMMV of various origins were retrieved from Genbank and included in the present study (Table 1).

Nucleic acid extraction

Total RNA was extracted from 100 mg of symptomatic leaves of *C. murale* plants previously inoculated with the isolates indicated in Table 1, following maceration in liquid nitrogen and use of the commercial RNeasy Plant Mini Kit (Qiagen) according to manufacturer's instructions.

RT-PCR

For cDNA synthesis, 1 μg of total RNA was used together with 1.5 μg of random hexamers (Promega) followed by denaturation at 70°C for 10 minutes and incubation on ice for 15 minutes. Reverse transcription (RT) reaction was performed in the presence of 200 U of M-MLV reverse transcriptase (Invitrogen), 50 mM Tris-HCl pH 8.3, 75 mM KCl, 3 mM $MgCl_2$, 10 mM DTT and 0.5 mM dNTPs.

For amplifications, initially, one common pair of primers was designed based on OMMV and TNV-D Genbank sequences, to amplify their CP ORFs (OMMVTNVDcoat5′: TAATCATGCC-TAAGAGAGG and OMMVTNVDcoat3′: ATCCTTCCAT-TAACGTTTA) and used. Later, two pairs of specific primers, one designed to amplify the CP ORF of OMMV (OMMVcoat5′: GACATTTCGCAACTCTCT and OMMVcoat3′: CACAAC-GATGGGTGAGTTGC) and the other designed to amplify the CP ORF of TNV-D (TNVDcoat5′: TCGGAGGATCAACCAC-TACAAA and TNVDcoat3′: CCGGAAGACGGGTCTAT-GAAA) were used. In all reactions, one μL of cDNA was used in a 50 μL reaction with 2.5 U of FideliTaqDNA Polymerase (USB corporation) performed in 10 mM Tris HCl (pH 8.6), 50 mM KCl, 1.5 mM MgCl2, 0.2 mM dNTPs, 0.3 μM of each primer. Amplifications were performed in a Thermal Cycler

Table 1. OMMV and TNV-D isolates present in single and in double infected plant samples and Genbank sequences used in coat protein variability analysis.

Plant Host and Virus origin	Virus	Viral Isolates	Accession number*
Olive, Portugal	OMMV	A4P5	KM355270 (1); KM355271 (2); KM355272 (2)
		V4	KM355275 (5)
		V8	KM355276 (4); KM355277 (1)
		OMMV3	KM355247 (3); KM355248 (2)
		OMMVL11	KM355249 (5)
		V8i _OMMV	OMMV: KM355278 (4); KM355279 (1)
		V10_OMMV	OMMV: KM355280 (1); KM355281 (1); KM355282 (1); KM355283 (2)
		gp _OMMV	OMMV: KM355250 (1); KM355251 (2); KM355252 (2)
		A1P2_OMMV	OMMV: KM355253 (1); KM355254 (1); KM355255 (1); KM355256 (1); KM355257 (1)
		A4P2_OMMV	OMMV: KM355258 (1); KM355259 (2); KM355260 (1); KM355261 (1)
		A5P2_OMMV	OMMV: KM355262 (2); KM355263 (1); KM355264 (1); KM355265 (1)
		A10P2_OMMV	OMMV: KM355266 (2); KM355267 (1); KM355268 (1); KM355269 (1)
		A6P5_OMMV	OMMV: KM355273 (2); KM355274 (3)
	TNV-D	gp_TNV-D	TNV-D: KM355307 (2); KM355308 (3)
		V4PB	KM355284; KM355285; KM355286; KM3555287; KM355288
		V6	KM355309 (1); KM355310 (1); KM355311 (1); KM355312 (2)
		V8ia1	KM355294 (1); KM355295 (2); KM355296 (1); KM355297 (1)
		V8i_TNV-D	TNV-D: KM355313 (5)
		V10_TNV-D	TNV-D: KM355289 (2); KM355290 (1); KM355291 (1); KM355292 (1)
		A1P2_TNV-D	TNV-D: KM355298 (2); KM355299 (2); KM355300 (1)
		A4P2_TNV-D	TNV-D: KM355301 (2); KM355302 (3)
		A5P2_TNV-D	TNV-D: KM355293 (5)
		A10P2_TNV-D	TNV-D: KM355305 (3); KM355306 (2)
		A6P5_TNV-D	TNV-D: KM355303 (4); KM355304 (1)
		GenBank sequences	
Olive, Portugal	OMMV	GP	NC_006939.1
		GP puFLOMMV3	HQ651832.1
		GP puFLOMMV4	HQ651833.1
		GP puFLOMMV5	HQ651834.1
Tulip, The Netherlands		TNV5	EF201607.1
		Inzell	EF201606.1
		BKD	EF201605.1
Spinach, Greece		OMMV-spinach	JQ288895.1
Tobacco, Hungary	TNV-D	Hungarian	NC_003487.1
Bean, United Kingdom		English	D00942.1
Olive, Portugal		Portuguese	FJ666328.1

* Number of clones that present the same sequence are between brackets.

(BioRad) following initial denaturation at 94°C for 1 min, 35 cycles at 94°C for 1 min, 52°C for 2 min and 68°C for 1 min and 30 seconds, and a final extension step of 68°C for 5 min. The use of the common primers OMMVTNVDcoat in RT-PCR assays originated a fragment of ≈813 bp for OMMV and ≈810 bp for TNV-D and the use of the specific pair of primers for OMMV and TNVD, originated a fragment of ≈877 bp and ≈872 bp, respectively.

Cloning and sequence analysis

RT-PCR products were purified using GFX PCR DNA Purification kit (GE Healthcare Biosciences) and cloned into pGEM-T easy vector (Promega), in accordance with the manufacturer's instructions. Plasmid DNA was extracted from *E.coli* JM109 cells using GenElute HP Plasmid Miniprep kit (Sigma) in accordance with the manufacturer's instructions, after growing cells in low salt LB medium (1% tryptone, 0.5% yeast extract, 0.5% NaCl, pH 7.5) supplemented with 100 μg/mL of ampicillin and grown overnight at 37°C at 175 rpm. DNA sequencing reactions were performed on both strands, by Macrogen (The Netherlands). Five complete CP sequences for each isolate were determined. Sequences were deposited in GenBank database. These, as well as those of the OMMV and TNV-D CP gene sequences available from the GenBank database (Table 1), totalling 131 sequences, were compared. The search for homologous sequences was done using BLAST. Multiple sequence alignment was performed with BioEdit 7.1.3.0 [23] and CLUSTAL W in MEGA 5.2 software [24]. The best fit nucleotide substitution model for these data was the Kimura 2-parameter model in the MEGA 5.2 software, showing the lowest Bayesian information criterion (BIC) score. This model was used to estimate nucleotide distance, diversity and phylogenetic relationships which were inferred using neighbour-joining (NJ) method. To validate phylogenetic tree analysis from the NJ method, trees were produced using Minimum Evolution, Maximum Parsimony and Maximum Likelihood methods in the MEGA 5.2 software. Bootstrap analyses with 1000 replicates were performed to evaluate the significance of the inner branches.

Potential recombination events in the aligned sequences were evaluated by RDP, GENECONV, Chimaera, 3Seq and SiSCAN in the RDP4.18 software, using default settings and a Bonferroni-corrected highest P value of 0.05. To identify specific amino acid sites under selective constraints, the difference between non synonymous (d_N) and synonymous (d_S) substitution rates was estimated for each position in the alignments using the FEL method (0.1 significance level) as implemented in the HYPHY server (http://www.datamonkey.org) [25].

Secondary structure prediction

An analysis of the potential effect of sequence variants in virion structure was carried using the SIFT algorithm [26], which predicts potentially deleterious mutations based on multiple sequence alignment and position-dependent scoring matrices.

Sequences containing variations marked as potentially deleterious by the SIFT algorithm were set to the JPRED secondary structure prediction server (based on the Jnet neural network algorithm [27]). The predictions for different sequence variants were aligned to highlight differences relative to the reference TNV-D and OMMV sequences.

Homology Modeling

Given the high similarity of the OMMV and TNV-D sequences ($\approx$85% identity), very similar protein 3D structures were expected. Since the structure of a close homolog ($\approx$45% identity) has been determined by X-ray crystallography (PDB code 1C8N), reliable 3D structural models of the two proteins could be built. To this task, the MODELLER v9 software was used [28]. The models were built as trimmers, since this was the oligomerization state of the asymmetric unit in the 1C8N structure, and the sequences included only the residues present in the crystal structure (78–269 in chains A and B, and 44–269 in chain C). The viral capsid has T3 type icosahedral structure with a total of 180 (60×3) polypeptide chains, and can be produced by replication of the original trimmer using the appropriate symmetry operators. In this study a "minimal" fragment with only 3 trimmers (9 chains) was used, since it contains all possible interfaces between different subunits.

Results

Amplicons sized $\approx$810 nt corresponding to the CP gene of OMMV and/or TNV-D were obtained from all isolates when the common primers OMMVTNVDcoat were used in RT-PCR.

Each amplicon was cloned and 5 clones per isolate were randomly selected and sequenced. All clone sequences, either from single OMMV or double OMMV and TNV-D infected isolates, revealed a near 100% homology to OMMV in all isolates. Clone sequences obtained from the isolates that were single infected with TNV-D isolates (V4PB, V6 and V8ia1), exhibited near 100% homology to TNV-D.

The use of those primers in amplification reactions applied to RNA obtained from double infected tissues, preferentially amplified OMMV sequences over those of TNV-D. This was overcome using specific primers for each OMMV and TNV-D, resulting in products of the expected size, $\approx$877 bp and $\approx$872 bp, respectively.

The CP sequences (5 per isolate) of thirteen OMMV isolates and eleven TNV-D isolates were compared together with the 8 OMMV and 3 TNV-D CP gene sequences available from GenBank, totalling 131 sequences.

All sequences encoding the OMMV CP possessed 269 aa and all sequences encoding the TNV-D CP possessed 268 aa, in accordance to previously published sequences. Pairwise distances of all OMMV sequences ranged from 0.000 to 0.251. The highest value was observed between isolate BKD (from tulip) and most olive isolates, and the lowest, 0.000 to 0.036, was observed among isolates infecting olive, being the highest value (0.036) observed between OMMV clone sequences within sample A1P2. Genetic distance of OMMV clone sequences within samples ranged from very low values, 0.000–0.004 (OMMVL11; V4; OMMV3; A6P5; A5P2; A4P5; V8; V8i; V10; GP), to low, 0.007–0.011 (A10P2; A4P2; A1P2), averaging 0.003, a value near to the OMMV sequence diversity between samples, 0.005, as expected by the coefficient of differentiation (0.607). OMMV sequences from samples OMMVL11 and V4 did not show any diversity.

Pairwise distances of all TNV-D sequences ranged from 0.000 to 0.218. The highest value was observed between the Hungarian and English isolate. As observed for OMMV, the lowest pairwise distance range (0.000 to 0.034) was also obtained among isolates infecting olive. The highest value (0.034) was observed between TNV-D clones of the same sample (A1P2) and between some clones of A1P2 and V8ia1. Genetic distance of TNV-D clone sequences within samples ranged from very low values, 0.000–0.008 (A5P2; V8i; A6P5; A10P2; GP; TNVD; A4P2), to low, 0.012–0.021 (V6; V4PB; V10; A1P2), averaging 0.007, near to the TNV-D sequences diversity between samples (0.006), as shown by the value of the coefficient of differentiation (0.439). TNV-D sequences from samples A5P2 and V8i did not show any diversity.

Pairwise distances of all OMMV and TNV-D sequences taken together ranged from 0.000 to 0.411. The highest value was observed between OMMV and TNV-D clones of sample A1P2 and the lowest was observed between several clone sequences of the same virus from either within or between samples. Diversity among OMMV and TNV-D isolates CP genes reached 0.199.

Along OMMV CP ORF (Figure 1, top right box), a slight increased variable region is noticed around nt 82 to 163 (aa 28 to 54) (equivalent to 0.2 position on the CP) and from nt 488 to 810 (aa 163 to 269) (0.7 position), more evident in deduced aa sequences. Along the TNV-D CP ORF (Figure 1, bottom right box) the highest diversity was observed in regions nt 1 to 81 (aa 1 to 27) (0.1 position), nt 326 to 406 (aa 109 to 135) (0.5 position) and from nt 650 to 807 (aa 217 to 268) (0.9 position). Comparisons of OMMV and of TNV-D clone CP sequences (Figure 1, left box) have shown the highest diversity (near 0.4) in region nt 82 to 163 (aa 28 to 54) (0.2 position). A high diversity value (0.3) was also observed in region nt 488 to 568 (position 0.7). In terms of amino acid sequence, diversity levels were maintained under 0.06 in region from aa 82 to 243 (position 0.2). From nt 731 to 807 (aa 244 to 268) (0.9 position), there was an increase in nt and aa sequence diversity.

The CP 'S' domain starts at aa 53 and ends at aa 265 in OMMV CP (equivalent to positions 0.29 to 0.99 of the CP in figure 1, top right box) and aa 52 to aa 264 in TNV-D CP (equivalent to positions 0.29 to 0.99 of the CP in figure 1, bottom right box) (Jones et al., 2014). The typical signature pattern for this domain ([FYW]-x-[PSTA]-x(7)-G-x-[LIVM]-x-[LIVM]-x-[FY-WIL]-x(2)-D-x(5)-P) was recognized at 134 to 159 aa in OMMV and at 131 to 156 aa in TNV-D (equivalent to position 0.6 in figure 1). It consists of highly conserved motif (YIPKCPTTTQ GSVVMAIVYDAQDTVP) in all OMMV sequences as well as in all TNV-D sequences except for a single difference in the eighth aa (in bold) where in TNV-D the 'T' is substituted by 'S'. No isolate showed variation in the four amino acids predicted to be involved in Ca^{2+} binding: two residues of aspartic acid (D), one of threonine (T) and one of asparagine (N) [13].

The CP genes of the isolates of both virus species did not show distinct recombination sites, when examined using the RDP software (data not shown) suggesting that recombination has not occurred.

The phylogenetic tree deduced from the CP alignment (Figure 2) revealed segregation of isolates under study into 2 main clusters. As expected from the matrix of sequence identity, isolates were grouped according to the virus species OMMV (I) or TNV-D (II) and each of the 2 main clusters appears divided into 2 subgroups: OMMV olive isolates (I-A) and OMMV tulip and spinach isolates (I-B) and TNV-D tobacco isolate (II-A) and TNV-D bean and olive isolates (II-B). For the phylogenetic tree all clone sequences that were different between them were used instead of using a single consensus sequence per isolate, this way it is visible that clones from different isolates are often closer than clones within an isolate.

The difference between d_N and d_S at each individual codon was statistically tested by the FEL method to determine if negative selection had some role in the low genetic variability observed. Data showed only 1 positively-selected codon in OMMV (189) and

2 in TNV-D (128, 201), whereas 5 codons were under negative selection in OMMV (44, 172, 230, 257, 264) and 8 codons in TNV-D (136, 155, 187, 199, 238, 261, 263), suggesting negative or purifying selection.

Secondary structure prediction

Analysis with the SIFT server of the OMMV and TNV-D sequence variants has provided a list of substitutions likely to cause structure alterations. These are Q20L, Q32R, I133V, A216T (for OMMV) and G5R, F13I, A23N, S129P, N227D, C230R, A244S, E261K (for TNV-D). Running secondary structure predictions with the JPRED server produced the results shown in Figures 3 and 4. In both cases the results are practically identical, with the substitutions causing little or no effect in the positions and lengths of helices and sheets. The exception is a small length of beta sheet (residues 208–212) which is not predicted when the substitutions N227D and C230R of the TNV-D sequence are not present. The overall result is not surprising, given the high degree of similarity among each sequence group (>85%), only slight changes in structure are to be expected (see below Homology Modelling). Replacements G5R, F13I, A23N (TNV-D) and Q20L, Q32R (OMMV) fall within the putative peptide signal sequence, for which no structural information is available. In this situation predictions are unreliable and no further consideration was given to such substitutions.

Homology Modelling

Given the high degree of identity (≈45%) between the template and target sequences, the virus models were expected, and here confirmed, to be very similar to the template structure. The backbone trace was almost identical in models and experimental structure, with a 0.3 Å RMSD between C-α atoms. The models were based on the crystallographic trimmer (chains A, B and C), and the symmetry operators provided in the PDB entry were used to recreate the entire icosahedral T3 capsid, with a total of 180 chains – this was required to analyze also possible interfacial contacts between monomers. The C chain is 35 residues longer in the models, because its leading sequence is better defined in the X-

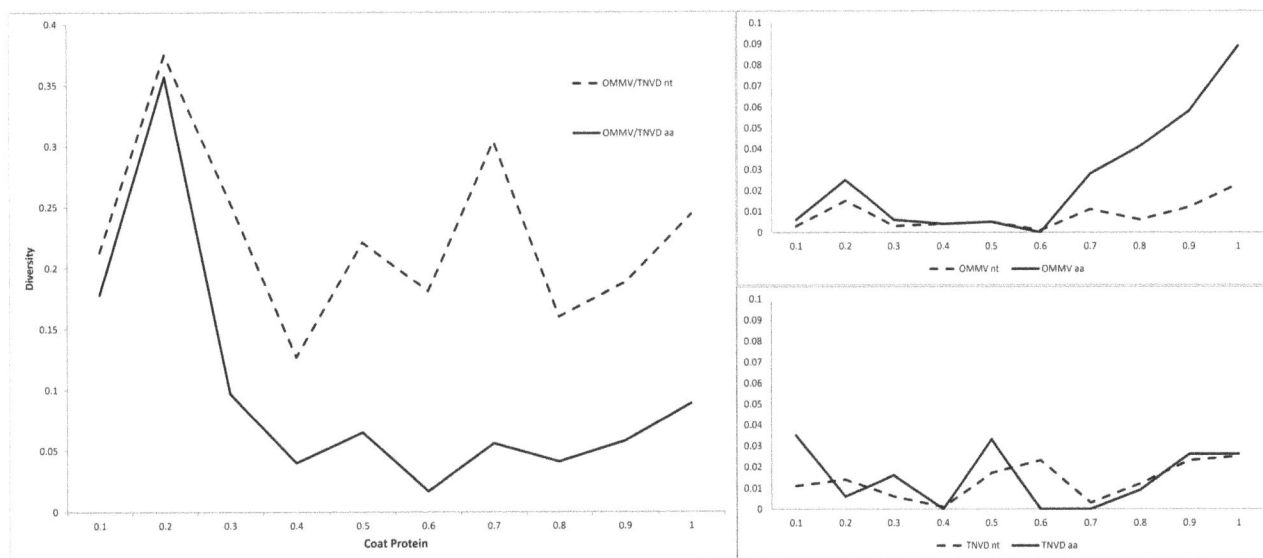

Figure 1. Nucleotide and deduced amino acid diversities along the OMMV and TNV-D CP gene. Left - OMMV and TNV-D aa and nt CP sequences; Right top – OMMV aa and nt CP sequences; Right bottom – TNV-D aa and nt CP sequences. Nucleotide diversity values were obtained in successive windows of 81 nt and amino acid diversity was obtained in successive windows of 26 aa.

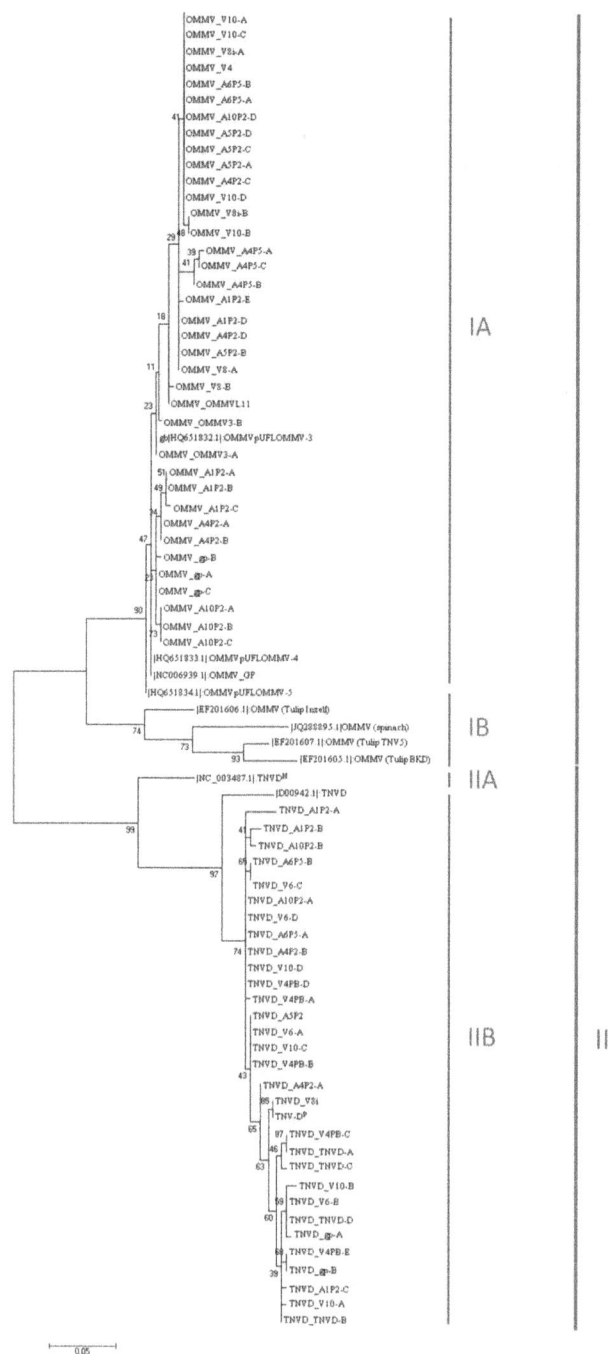

Figure 2. Phylogenetic tree analysis of OMMV and TNV-D isolates based on CP nt sequences. NJ tree was constructed from the sequence alignment of ca. 870 nts of OMMV and TNV-D CP coding region from 65 clones of OMMV and 55 TNV-D isolates and from 11 sequences retrieved from the GenBank database, totalling 131 sequences. Repeated sequences within each isolate were omitted. Each sequence variant was named with a letter (A, B, C,...) following the isolate designation to allow differentiation. Phylogenetic analysis included 78 sequences. Multiple sequence alignments were generated using MEGA 5.1, and phylogenetic tree was constructed by the NJ algorithm, based on calculations from pairwise nt sequence distances for gene nt analysis. Bootstrap analysis was done with 1000 replicates. Numbers above the lines indicate bootstrap scores out of 1000 replicates.

ray structure than the corresponding sequences of A and B chains. This difference may be related to a specific role of C chains in the assembly of viral capsids. For the analysis of all possible monomer-monomer interactions, a *minimal contact unit* consisting of 3 trimmers was used, where all possible contacts are represented (Figure 5, A).

The models were analysed to investigate the role of the sequence variants that were previously identified as having a potentially disruptive effect on the sequence. As for OMMV sequence variants, the replacement I133V corresponds to the removal of a methyl group from the side chain of I133, creating a small void that may affect interaction with close partner L186. Movement of residue 186 from a C chain could in turn affect interaction with a neighbour C chain. Given the important role of C-chain dimmers in the assembly of this type of virus, this effect cannot be neglected. Residue A216 is in a well exposed location of the outer side of the viral capsid. Replacement by a Threonine residue should have little or no consequence, given the conservative character of the replacement and its unencumbered location. As for TNV-D sequence variants, the mutation S129P, is located on the inner side of the capsid, and not close to any other replaced residue. Proline is a well-known helix breaker, but since the replacement does not occur on a helix no major effects are expected. The Serine residue is polar and hydrogen bond former, while Proline is neither, and that could influence the hydrogen bonding network of residues on the inner side of the capsid. The mutation N227D changes a neutral (but polar) into a negatively charged residue. Given the presence of the nearby (in contact distance) D153, a strong repulsion is expected to arise, that could change the structure of monomer C at its interface with A. Since this replacement is observed together with C230R, another strong attractive interaction will arise between the new Arginine and Aspartate residues at positions 230 and 227. Residue N227 is not in the interior of the capsid, so the replacement should not affect interaction with the viral genetic material. The strong interaction of C230R with residue 227D has been already discussed, but 230R could interact with K226, resulting in a repulsion between these two positively charged residues. (Note: the sequence variants N227D and C230R are the ones where the small segment of β-sheet between residues 208–212 is missing). Mutation A244S results in a change in a residue that is part of the ring structures around the 5-fold and 6-fold axis of the virus capsid. Since there is only a very slight change in size and character, the replacement is not expected to have an impact on the capsid structure. Residue E261 is buried, and lies in contact distance of a group of charged residues, both positive (Lysine, Arginine) and negative (Aspartate, Glutamate). Replacement of the negatively charged Glutamate side chain with a positive Lysine will cause several strong and very close charge-charge interactions that could locally perturb the structure and induce structural changes.

Discussion and Conclusions

The genetic diversity in OMMV and TNV-D coat protein was evaluated by sequence analysis of the CP of thirteen OMMV isolates and eleven TNV-D isolates, eight of which were obtained from double infected tissues. Additional eight OMMV and three TNV-D CP sequences were retrieved from Genbank and included in this study.

For the amplification of the CP sequences present in the total RNA tissue samples, a single pair of primers common to OMMV and TNV-D was initially used. This resulted in the amplification of TNV-D CP only on the three single TNV-D-infected samples which was confirmed after cloning. In the eight isolates containing

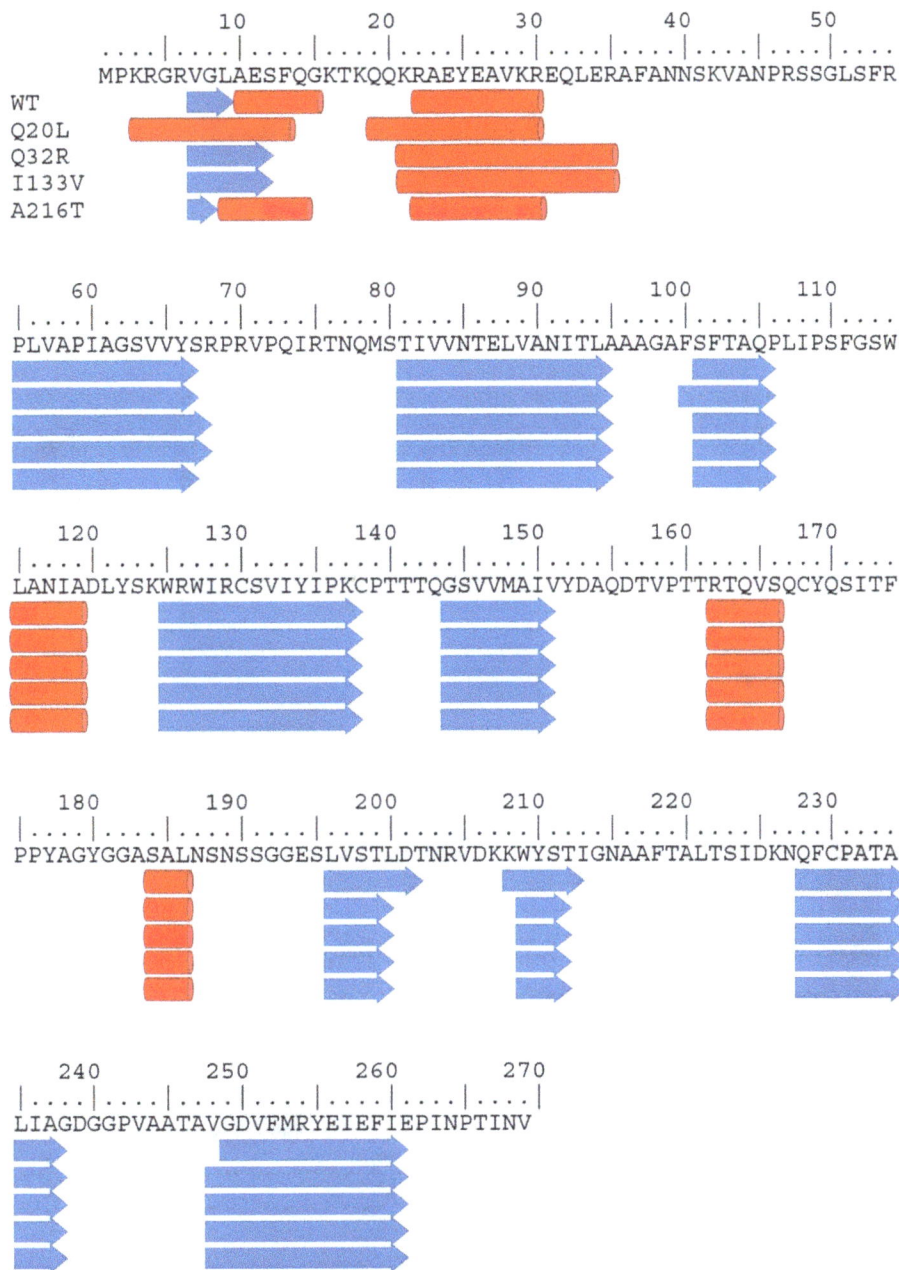

Figure 3. Multiple sequence alignment of the deduced amino acid sequences of OMMV variants analysed: WT (OMMV_GP, reference sequence, NC_006939.1), Q20L, Q32R, I133V and A216T.

mixed infections of OMMV and TNV-D (Table 1), only the presence of OMMV was detected indicating replicase preference for this virus RNA over that of TNV-D. This tallies with previous results that have demonstrated recurrent higher levels of OMMV over TNV-D in olive orchards suggesting that the former is better fitted to the host thus becoming predominant [8]. It is possible that the higher predominance of OMMV is due to a higher virulence when compared to that of TNV-D as observed with other viruses [29–32]. Mixed infections, frequent in necroviruses [8], may have important evolutionary implications since they can affect the within isolate population of variants and allow interaction and/or recombination between different viruses thus affecting pathogenicity and adaptability [33]. In fact, it has been proposed that OMMV has originated from genetic recombination between

OLV-1 (a virus with which it shares a high identity in the polymerase region) and TNV-D (a virus with which it shares a high identity in the CP region) during their simultaneous replication in the same cell [5].

The analysis of CP gene and of deduced aa sequences revealed that overall diversity of OMMV (average 0.006) and TNV-D (average 0.011) was low either between or within isolates. Diversity values between and within samples were always very near in both virus species isolates, contributing equally to total diversity. This is contrary to results obtained on the alphanecrovirus OLV-1 showing a much higher variability between, rather than within sequences from different OLV-1 isolates [22]. Similar studies on other members of *Tombusviridae* (*Carnation mottle virus* [34], *Pelargonium flower break virus* [21], *Sweet potato chlorotic stunt*

```
              10        20        30        40        50
    ....|....|....|....|....|....|....|....|....|....|....|
    MPKRGRVGLAESFQSKSKKQKEAEYNAFQREKMERALVNNATAARKGSGMSFRP
WT
G5R
F13I
S17T
A23N
S129P
N227D
C230R
A244S
E261K
```

```
      60        70        80        90       100       110
    ....|....|....|....|....|....|....|....|....|....|....|
    LTVPVAGSVIYSRPRVPQVRTNQMSTFVVNTELVANITLAAAGAFSFTTQPLIPSFGSWL
```

```
     120       130       140       150       160       170
    ....|....|....|....|....|....|....|....|....|....|....|
    ANIADLYSKWRWISCSVVYIPKCPTSTQGSVVMAIVYDAQDTVPTTRTQVSQCYQSITFP
```

```
     180       190       200       210       220       230
    ....|....|....|....|....|....|....|....|....|....|....|
    PYAGYGGASALNHKGSSGESLVSTLDTNRVDKKWYSTIGNAAFTALTSIDKNQFCPATAI
```

```
     240       250       260       270
    ....|....|....|....|....|....|....|....|
    IAGDGGPVAATAVGDIFMRYEIEFIEPVNPTINI
```

Figure 4. Multiple sequence alignment of the deduced amino acid sequences of TNV-D variants analysed: WT (TNV-D[H], NC_003487.1), G5R, F13I, S17T, A23N, S129P, N227D, C230R, A244S, E261K.

virus [35], *Olive latent virus 1* [22]) have shown diversity values for virus CP sequences of near 0.03.

Sequence conservation among the CP of OMMV isolates (nt pairwise distance <0.018 and genetic diversity 0.008) and among TNV-D isolates (nt pairwise distance <0.034 and genetic diversity 0.013) was extremely high in olive isolates. These values were, as expected, higher when sequences of the two viruses were compared (reaching nt pairwise distances and genetic diversities values of 0.411 and 0.199, respectively). However, similar values of pairwise distances and genetic diversities have been observed in other studies among different isolates from a unique virus (*Grapevine leafroll-associated virus* 4) [35]. Based on the deduced amino acid (aa) CP sequences, identities between the several OMMV and TNV-D isolates ranged from 84.3% to 85.8%, as previously observed [3] who found an identity of 85.1% between isolates OMMV_GP and TNV-D[P].

The comparison among isolates infecting other hosts has shown nt pairwise distances of <0.251 and <0.218 for OMMV and TNV-D, respectively, and genetic diversity of 0.034 and 0.021 for

OMMV and TNV-D, respectively. These results seem to indicate some host specificity as already shown in other studies [21,22,36,37].

The 'S' domain is the only conserved domain found in OMMV and TNV-D CP sequences, when examined by NCBI conserved domain search software, which may explain the low diversity found since it is the most conserved region in the CP of small plant viruses, indicating that it is where more functional or structural constraints are located [9,21]. Residues in the 'S' domain have shown to be better conserved in all OMMV and TNV-D isolates, whereas the sequences containing the N-terminal region, are slightly less conserved, as previously observed with other TNV strains [13]. The diversity between OMMV and TNV-D in the N-terminal region reaches its higher value of 0.4. Several studies have shown that viruses present lower identity values in this region than in the rest of the CP region [21,38]. TNV-D, which presents a mean identity of 0.45 in the CP protein with TNV-A has shown a sequence identity of less than 0.2 in this region [13]. Although a high number of basic residues is maintained in this region, residues

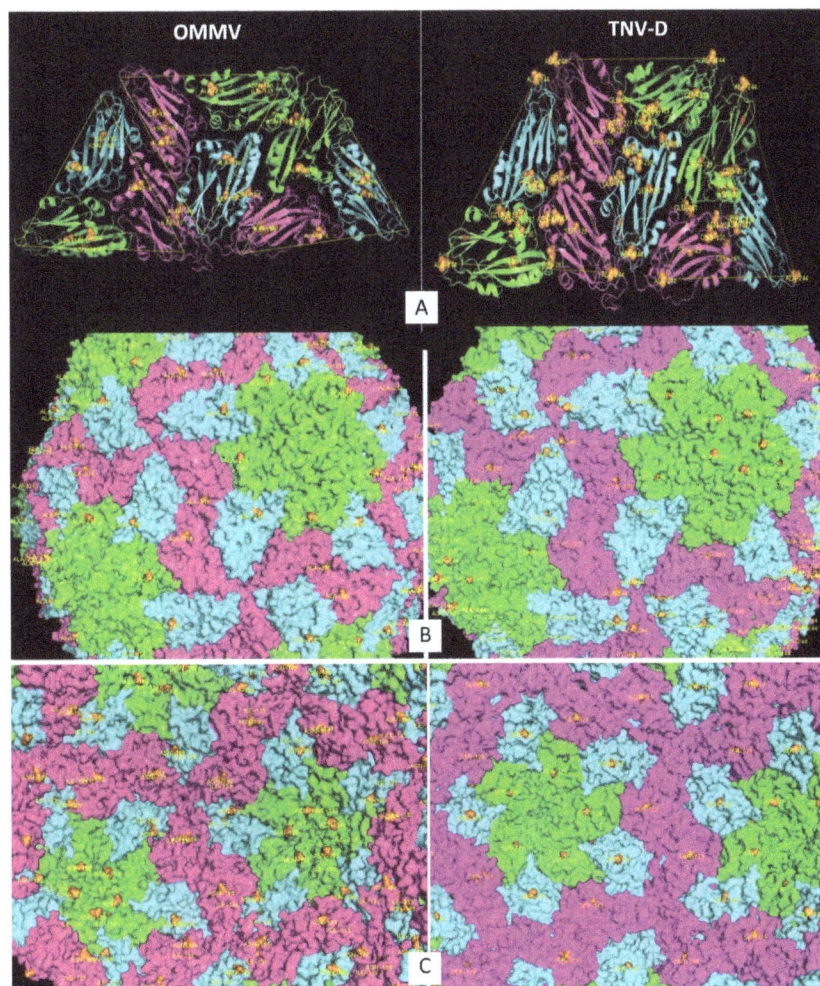

Figure 5. Three-dimensional representation of the OMMV (left) and TNV-D (right) coat protein, highlighting mutations. Mutations are rendered in spheres. Each trimmer is coloured in light green, magenta and cyan. A – Virus minimum contact unit; B – The surface exterior; C – The surface interior.

may have diverged during evolution, probably due to the lack of constraints that act in this N-terminal region, except for the predicted interaction with the RNA inside the particle. The fact that the N-terminal region is buried inside the particle [13] may explain why the surface features of the particle are little affected by this region, and why OMMV and TNV-D presenting a diversity of 40% in this region, show identical serological reactions.

Among the CP sequences compared in detail, 143 residues are completely conserved. Among these are 10 residues previously shown to be conserved among CP sequences of different genera in the *Tombusviridae* [38] and predicted to have an important structural role.

The relatively low genetic diversity found for both OMMV and TNV-D suggests that negative selection restricts the number of molecular variants. This was confirmed after evaluation of selective constraints by comparing rates of synonymous and nonsynonymous substitutions across codon sites. Only one positively-selected codon was detected in OMMV CP and 2 in TNV-D CP, whereas for OMMV, 5 codons were under negative selection, and for TNV-D there were 8, indicating that changes in aa residues would result in functional or structural disadvantages, indicative of a strong negative or purifying selection. All negatively selected sites were, for both viruses, in the 'S' domain region, with the exception of 1 codon of OMMV CP gene that was located in the N-terminal region, however it did not correspond to any of the basic residues which are likely to be critical for RNA-binding capability. This has also been observed with other viruses, suggesting that selection is acting on the preservation of the proper conformation of the RNA-binding motif [21,39].

Negative constraints that viral CPs are subjected to, may be due to multiple functions, including genome encapsidation and protection, cell to cell movement, transmission between plants, host and/or vector interactions, suppression of gene silencing. Chare and Holmes [40] analysed selection pressures in the capsid genes of plant RNA viruses and found that vector borne viruses are subjected to a greater selection than non-vectored viruses. Soil transmission of OMMV is known to be more efficient in the presence of the fungus *Olpidium brassicae* [17], however, a single mutation N189Y, located in the virus particle interior, rendered it non fungus-transmitted probably due to changes in particle conformation affecting the recognition site [14]. Interestingly, this mutation was found in all clones of 7 of the 21 OMMV isolate sequences analysed. A remarkable covariation was found between amino acid positions 189 and 216 of the OMMV CP sequence. Asparagine in position 189 correlated with Alanine in position 216, and the change N189Y correlated to the change A216T, a mutation that is located in the surface of the particle but has however not shown to influence virus transmissibility by the fungus [14]. This covariation suggests the existence of tertiary interactions between these regions of the molecule and that mutation may result from a readjustment of the viral particle towards stability. However, single directed mutants were constructed with each one of these mutations, and the virus remained stable and infectious [14]. As for TNV-D isolates, in place of the mutation A216T, an alanine residue is present (position 213) and in the place of the mutation N189Y, TNV-D isolates show a lysine residue instead of asparagine (position 186). This substitution of a polar aa to other polar aa does not suggest alterations in virus conformation.

Based on the prediction of structure/function of the CP based on the sequences, we found that only the mutations Q20L, Q32R, I133V, A216T in OMMV and G5R, F13I, A23N, S129P, N227D, C230R, A244S and E261K in TNV-D are likely to cause significant alterations in virus structure.

Analysis of the TNV-D and OMMV comparative models shows that replacements A216T of OMMV and C230R and A244S of TNV-D are located in the outer surface of the capsid and replacements I133V of OMMV and S129P of TNV-D are located in the capsid interior (figure 5), while N227D and E261K of TNV-D are buried below the surfaces. The location of the OMMV mutation A216T in the particle surface was predicted previously [14]. OMMV replacement I133V could affect interaction with L186 which in turn could affect interaction with a neighbour C chain, disturbing the capsid assembly and/or changing virion conformation. This may have several implications, namely by changing the accessibility of exposed amino acids in particle surface as was suggested by several authors as a probable reason why interior mutations render loss of vector transmission [14,41]. Replacement A216T produces a new polar residue at an exposed zone which may have an impact on the interaction of the virus with the host cell. The replacement N189Y has previously been shown to influence virus transmission [14] and it was decided to include this mutation in the analyses even though it was not marked by SIFT as potentially critical. This shows that the preservation of structure does not always correlate with preservation of functions. Residue 189 is placed around the 5- and 6-fold axis, on the inner side of the capsid. They are very well exposed at the tip of a loop thus not very likely to cause steric clashes when replaced by tyrosine. However, this residue is well placed for interaction with the viral genome, and as such their replacement could potentially affect viral properties.

As for TNV-D, replacements N227D and C230R may give rise to strong repulsions due to nearby identically charged residues (D153 and K226) and this could have a significant effect on the capsid assembly and stability, since these residues lie close to the monomer-monomer interfaces. The replacement S129P may affect interaction with the nucleic acid in the capsid interior.

In general, point aminoacid replacements are expected to cause only small effects upon the structure, which could form a strong selective pressure to preserve the coat protein structure. This notion is supported through comparison of the coat proteins of TNV and *Tobacco bushy stunt virus*, with very similar structures (2.5 Å RMSD) but very divergent sequences (14% identity).

All the substitutions under study were naturally occurring variants filtered by natural selection, and not comparable to isolated mutational events many of which would most likely result in large structural changes and malformed or non-assembling viral capsids.

Few studies have been done on the functional role of the CP in OMMV and TNV-D and these focus mainly on the ability of the virus particles to bind to fungal zoospores and act in *in vitro* transmission events [14,17]. Two single OMMV mutations (N189Y and A216T) did not alter virus infectivity nor systemic movement. However other mutations may act differently at various levels. Thus it would be of interest to find the role of all the predicted important structural mutations found in the CP to investigate the potential implications of these alterations in virus movement, elicitation of symptoms, host range and particle assembly. Some studies concerning the CP of OLV-1 [42] and *Carnation ringspot virus* [43], also of the *Tombusviridae* family, have highlighted the importance of the C-terminal domain on the systemic movement. Among the predicted important structural mutations found, A216T in OMMV and N227D, C230R, A244S and E261K in TNV-D are in the C-terminal region and may have implications at that level.

A possible explanation for the high stability found either in OMMV or TNV-D isolates, sometimes even greater between different isolates than within an isolate, may be that these isolates have evolved from a single original sequence in an original host, possibly olive. The fact that olive material is propagated through cuttings, and that these viruses occur symptomless, contributes to the rapid multiplication and dissemination of the viruses and may lead to a high degree of virus CP similarities. OMMV, as suggested, may have originated from recombination events between OLV-1 and TNV-D. Recombination may play a significant role in the evolution of necrovirus by producing new viruses, rather than point mutations. Several members of *Tombusviridae* have shown to contain putative recombination signals in the genome of some of their members, including TNV-D and OMMV. TNV-D is suggested to result from recombination between BBSV and OMMV isolates and OMMV from recombination between OLV-1 and TNV-D [5,44]. In the family *Tombusviridae*, recombination in the genera *Carmovirus* and *Tombusvirus* has been shown to repair damaged deleted 3′ends of virus associated RNAs, as well as generate new satRNAs or DIRNAs [45]. Genetic drift may also have contributed to the low variability after bottlenecks virus populations undergo such as systemic movement or transmission between plants. The major changes observed seem to be the result of successful adaptation to other hosts.

Data here presented agree with those of other authors who have reported high genetic stability for several RNA viruses, suggesting the existence of very strong selective pressures for the preservation of their biological functions [19,46–48] and the most variability exists between virus isolates due to host speciation [22]. In addition, it was shown that mere preservation of virus structural identity does not always correlate with functional conservation.

Acknowledgments

The authors are grateful to Mrs. Maria Mário Azedo for technical assistance.

Author Contributions

Conceived and designed the experiments: CMRV MIEC MRF. Performed the experiments: CMRV MM. Analyzed the data: CMRV MM PM. Contributed reagents/materials/analysis tools: CMRV PM GN MRF. Wrote the paper: CMRV MM MIEC MRF.

References

1. Coutts R, Rigden J, Slabas A, Lomonossoff G, Wise P (1991) The complete nucleotide sequence of tobacco necrosis virus strain D. J Gen Virol 72: 1521–1529.

2. Molnar A, Havelda Z, Dalmay T, Szutorisz H, Burgyan J (1997) Complete nucleotide sequence of tobacco necrosis virus strain D-H and genes required for RNA replication and virus movement. J Gen Virol 78: 1235–1239.

3. Cardoso J, Félix M, Clara M, Oliveira S (2009) Complete genome sequence of a Tobacco necrosis virus D isolate from olive trees. Arch Virol 154: 1169–1172.

4. Cardoso J, Felix M, Oliveira S, Clara M (2004) A Tobacco necrosis virus D isolate from Olea europaea L.: viral characterization and coat protein sequence analysis. Arch Virol 149: 1129–1138.

5. Cardoso J, Félix M, Clara M, Oliveira S (2005) The complete genome sequence of a new necrovirus isolated from *Olea europaea L.* Arch Virol 150: 815–823.

6. Gratsia ME, Kyriakopoulou PE, Voloudakis AE, Fasseas C, Tzanetakis IE (2012) First Report of Olive mild mosaic virus and Sowbane mosaic virus in Spinach in Greece. Plant Dis 96: 1230–1230.

7. Kassanis B (1949) A necrotic disease of forced tulips caused by tobacco necrosis viruses. Ann Appl Biol 36: 14–17.

8. Varanda C, Cardoso J, Félix M, Oliveira S, Clara M (2010) Multiplex RT-PCR for detection and identification of three necroviruses that infect olive trees. J Plant Pathol 127: 161–164.

9. Meulewaeter F, Seurinck J, Vanemmelo J (1990) Genome structure of Tobacco necrosis virus strain-A. Virology 177: 699–709.

10. King A, Adams M, Lefkowitz E, Carstens E (2012) Virus Taxonomy: Classification and Nomenclature of Viruses: Ninth Report of the International Committee on Taxonomy of Viruses. USA: Elsevier Academic Press.

11. Dolja V, Koonin EV (1991) Phylogeny of capsid proteins of small icosahedral RNA plant viruses. J Gen Virol 72: 1481–1486.

12. Felix MR, Cardoso JMS, Oliveira S, Clara MIE (2005) Viral properties, primary structure and phylogenetic analysis of the coat protein of an Olive latent virus 1 isolate from Olea europaea L. Virus Res 108: 195–198.

13. Oda Y, Saeki K, Takahashi Y, Maeda T, Naitow H, et al. (2000) Crystal structure of tobacco necrosis virus at 2.25 Å resolution. J Mol Biol 300: 153–169.

14. Varanda C, Félix M, Soares C, Oliveira S, Clara M (2011) Specific amino acids of Olive mild mosaic virus coat protein are involved on transmission by Olpidium brassicae. J Gen Virol 92: 2209–2213.

15. Callaway A, Giesman-Cookmeyer D, Gillok E, Sit T, Lommel S (2001) The multifunctional capsid proteins of plant RNA viruses. Ann Rev Phytopathol 39: 419–460.

16. Qu F, Morris T (2005) Suppressors of RNA silencing encoded by plant viruses and their role in viral infections. FEBS Lett 579: 5958–5964.

17. Varanda C, Silva M, Félix M, Clara M (2011) Evidence of Olive mild mosaic virus transmission by Olpidium brassicae. Eur J Plant Pathol 130: 165–172.

18. Garcia-Arenal F, Fraile A, Malpica J (2001) Variability and genetic structure of plant virus populations. Ann Rev Phytopathol 39: 157–186.

19. Rubio L, Ayllón M, Kong P, Fernández A, Polek M, et al. (2001) Genetic variation of Citrus tristeza virus isolates from California and Spain: evidence for mixed infections and recombination. J Virol 75: 8054–8062.

20. Manrubia S, Lázaro E, Pérez-Mercader J, Escarnis C, Domingo E (2003) Fitness distributions in exponentially growing asexual populations. Phys Rev Let 90: 188101–188104.

21. Rico P, Ivars P, Elena S, Hernández C (2006) Insights into the selective pressures restricting Pelargonium flower break virus genome variability: evidence for host adaptation. J Virol 80: 8124–8132.

22. Varanda C, Nolasco G, Clara M, Félix M (2014) Genetic diversity of the coat protein gene of Olive latent virus 1 isolates. Arch Virol 159: 1351–1357.

23. Hall T (1999) BioEdit: a user-friendly biological sequence alignment editor and analysis program for Windows 95/98/NT. Nuc Ac Symp Ser 41: 95–98.

24. Tamura K, Peterson D, Peterson N, Stecher G, Nei M, et al. (2011) Mega5: Molecular evolutionary genetics analysis using maximum likelihood, evolutionary distance, and maximum parsimony methods. Mol Biol Evol 28: 2731–2739.

25. Kosakovsky Pond S, Frost S (2005) Not So Different After All: A Comparison of Methods for Detecting Amino Acid Sites Under Selection. Mol Biol Evol 22: 1208–1222.

26. Ng P, Henikoff S (2003) SIFT: predicting amino acid changes that affect protein function. Nuc Ac Res 31: 3812–3814.

27. Cuff J, Barton G (1999) Application of enhanced multiple sequence alignment profiles to improve protein secondary structure prediction. Proteins 40: 502–511.

28. Sali A, Blundell T (1993) Comparative protein modelling by satisfaction of spatial restraints. J Mol Biol 234: 779–815.

29. Sambade A, Rubio L, Garnsey S, Costa N, Muller G, et al. (2002) Comparison of viral RNA populations of pathogenically distinct isolates of Citrus tristeza virus: application to monitoring cross-protection. Plant Pathol 51: 257–265.

30. Powell C, Pelosi R, Rundell P, Cohen M (2003) Breakdown of cross protection of grapefruit from decline-inducing isolates of Citrus tristeza virus following introduction of the brown citrus aphid. Plant Dis 87: 1116–1118.

31. Moreno P, Ambrós S, Albiach-Marti M, Guerri J, Pena L (2008) Citrus tristeza virus: a pathogen that changed the course of the citrus industry. Mol Plant Pathol 9: 251–268.

32. Matos L, Hilf M, Cayetano X, Feliz A, Harper S, et al. (2013) Dramatic change in Citrus tristeza virus populations in the Dominican Republic. Plant Dis 97: 339–345.

33. Hull R (2014) Plant Virology. USA: Elsevier Academic Press.

34. Canizares M, Marcos J, Pallás V (2001) Molecular variability of twenty-one geographically distinct isolates of Carnation mottle virus (CarMV) and phylogenetic relationships within the Tombusviridae family. Arch Virol 146: 2039–2051.

35. Rubio L, Guerri J, Moreno P (2013) Genetic variability and evolutionary dynamics of viruses of the family Closteroviridae. Front Microbiol 4: 151

36. Ayllón M, Rubio L, Sentandreu V, Moya A, Guerri J, et al. (2006) Variations in two gene sequences of Citrus tristeza virus after host passage. Virus Genes 32: 119–128.

37. Scott K, Hlela Q, Zablocki O, Read D, vanVuuren S, et al. (2013) Genotype composition of populations of grapefruit-cross-protecting *Citrus tristeza virus* strain GFMS12 in different host plants and aphid-transmitted sub-isolates. Arch Virol 158: 27–37.

38. Saeki K, Takahashi Y, Oh-Oka H, Umeoka T, Oda Y, et al. (2001) Primary structure and phylogenetic analysis of the coat protein of a Toyama Isolate of Tobacco necrosis virus. Biosco Biotechnol Biochem 65: 719–724.

39. Castano A, Ruiz L, Elena S, Hernández C (2011) Population differentiation and selective constraints in Pelargonium line pattern virus. Virus Res 155: 274–282.

40. Chare E, Holmes E (2004) Selection pressure in the capsid genes of plant RNA viruses reflect mode of transmission. J Gen Virol 85: 3149–3157.

41. Kakani K, Sgro JY, Rochon D (2001) Identification of specific cucumber necrosis virus coat protein amino acids affecting fungus transmission and zoospore attachment. J Virol 75: 5576–5583.

42. Pantaleo V, Grieco F, Di Franco A, Martelli GP (2006) The role of the C-terminal region of olive latent virus 1 coat protein in host systemic infection. Arch Virol 151: 1973–1983.

43. Sit T, Haikal P, Callaway A, Lommel S (2001) A single amino acid mutation in the Carnation ringspot virus capsid protein allows virion formation but prevents systemic infection. J Virol 75: 9538–9542.

44. Boulila M (2011) Positive selection, molecular recombination structure and phylogenetic reconstruction of members of the family Tombusviridae: Implication in virus taxonomy. Gen Mol Biol 34: 647–660.

45. Lommel S, Sit TL (2008) Tombusviruses. In: Eds. Marc Van Regenmortel and Brian Mahy. Desk Encyclopedia of Plant and Fungal Virology. USA: Elsevier.

46. Fraile A, Malpica J, Aranda M, Rodriguez-Cerezo E, Garcia-Arenal F (1996) Genetic diversity in tobacco mild green mosaic tobamovirus infecting the wild plant Nicotiana glauca. Virology 223: 148–155.

47. Vives M, Rubio L, Galipienso L, Navarro L, Moreno P, et al. (2002) Low genetic variation between isolates of Citrus leaf blotch virus from different geographical origins. J Gen Virol 83: 2587–2591.

48. Lin H, Rubio L, Smythe A, Jiminez M, Falk BW (2003) Genetic diversity and biological variation among California isolates of Cucumber mosaic virus. J Gen Virol 84: 249–258.

4

A Tyrosine-Rich Cell Surface Protein in the Diatom *Amphora coffeaeformis* Identified through Transcriptome Analysis and Genetic Transformation

Matthias T. Buhmann[1], Nicole Poulsen[1], Jennifer Klemm[1], Matthew R. Kennedy[3], C. David Sherrill[3], Nils Kröger[1,2]*

1 B CUBE Center for Molecular Bioengineering, Technische Universität Dresden, Dresden, Germany, **2** Department of Chemistry and Food Chemistry, Technische Universität Dresden, Dresden, Germany, **3** School of Chemistry and Biochemistry, Georgia Institute of Technology, Atlanta, Georgia, United States of America

Abstract

Diatoms are single-celled eukaryotic microalgae that are ubiquitously found in almost all aquatic ecosystems, and are characterized by their intricately structured SiO_2 (silica)-based cell walls. Diatoms with a benthic life style are capable of attaching to any natural or man-made submerged surface, thus contributing substantially to both microbial biofilm communities and economic losses through biofouling. Surface attachment of diatoms is mediated by a carbohydrate- and protein- based glue, yet no protein involved in diatom underwater adhesion has been identified so far. In the present work, we have generated a normalized transcriptome database from the model adhesion diatom *Amphora coffeaeformis*. Using an unconventional bioinformatics analysis we have identified five proteins that exhibit unique amino acid sequences resembling the amino acid composition of the tyrosine-rich adhesion proteins from mussel footpads. Establishing the first method for the molecular genetic transformation of *A. coffeaeformis* has enabled investigations into the function of one of these proteins, AC3362, through expression as YFP fusion protein. Biochemical analysis and imaging by fluorescence microscopy revealed that AC3362 is not involved in adhesion, but rather plays a role in biosynthesis and/or structural stability of the cell wall. The methods established in the present study have paved the way for further molecular studies on the mechanisms of underwater adhesion and biological silica formation in the diatom *A. coffeaeformis*.

Editor: Douglas Andrew Campbell, Mount Allison University, Canada

Funding: This work was supported by the US Air Force Office of Scientific Research (AFOSR grant number FA9550-12-1-0093) to NP and NK, and by the Deutsche Forschungsgemeinschaft (DFG grant number KR 1853/5-1) to NK. The funders had no role in study design, data collection and analysis, decision to publish, or preparation of the manuscript.

Competing Interests: The authors have declared that no competing interests exist.

* Email: kroeger@bcube-dresden.de

Introduction

Diatoms are a large group of single-celled microalgae that are ubiquitously present in water habitats, and are among the most prolific biological primary producers in the oceans [1]. The hallmark of diatoms is that each cell is encased by a wall made of intricately patterned SiO_2 (silica). Diatoms are widely studied due to their importance for ocean ecosystems, their physiological capabilities, their complex evolutionary history, and their ability to adhere to any natural or man-made surface underwater. The colonization of submerged surfaces ("biofouling") by bacteria, microalgae (including diatoms), and multicellular organisms (e.g., barnacles, mussels and macroalgae) can lead to the development of biofilms that are several centimeters thick [2]. Biofouling causes enormous costs world-wide due to the increase in hydrodynamic drag of ships, and damage to aquaculture equipment [3,4]. Therefore, substantial efforts are being spent on developing environmentally friendly agents that prevent biofilm formation by inhibiting the initial attachment of bacteria and diatoms [5,6]. On the other hand, the adhesive components produced by diatoms provide a paradigm for the development of underwater glues for numerous applications in technology and medicine [7]. For both, the prevention of diatom adhesion and the development of underwater glues, it is necessary to identify the adhesive biomolecules of diatoms and understand their molecular mechanism of adhesion.

While little is known about the attachment of microalgae, several extracellular proteins required for the adhesion of bacterial biofilms have been identified. These include amyloid-fibers that provide biofilm matrix cohesiveness [8], and flagella that are required for bacterial attachment to abiotic surfaces [9]. Diatoms, however, adhere with fundamentally different mechanisms that do not involve flagella but rather adhesive strands, which are secreted through a slit-like cell wall opening termed "the raphe" [10]. Recently, it has been demonstrated that diatom adhesive material is composed of both protein and carbohydrate, and their amino acid and monosaccharide compositions have been determined [11]. However, so far no sequence information from these macromolecules has been obtained.

To date, the molecular mechanisms of underwater adhesion have been studied in most detail in animals, particularly the marine mussel *Mytulis edulis*. Mussels attach to surfaces via macroscopic fibers that contain tyrosine-rich proteins at their sticky end (i. e., the mussel foot). In many mussel foot proteins tyrosine residues have been post-translationally hydroxylated to 3,4-dihydroxyphenyl-L-alanine (Dopa) [12]. The presence of Dopa seems to play an important role in both structural integrity of the filaments and underwater adhesion to surfaces by forming covalent cross-links and coordination bonds with metal ions, as well as by forming hydrogen bonds with the surface. Also the adhesive proteins of other organisms, like polycheates, invertebrates, turbellarians, hydroids and tunicats contain significant amounts of Dopa (for a review, see [13]).

The pivotal role of Dopa-rich proteins in underwater surface adhesion of lower animals has prompted the question as to whether diatoms employ similar proteins. This question has been addressed in the present work using a bioinformatics-based approach. *Amphora coffeaeformis* was chosen for these studies, because it is one of the most common biofouling diatoms, and it has been used in many studies as a model organism for underwater bioadhesion [14–16]. Previously, *A. coffeaeformis* had not been investigated on a molecular level, and thus neither genome nor transcriptome data were available for this species at the onset of this study. Through the present work we have made *A. coffeaeformis* amenable for investigations on the molecular mechanism of underwater adhesion through establishing a transcriptome database and a method for its molecular genetic transformation. These tools have then been employed to identify *A. coffeaeformis* proteins with similarities to mussel adhesion proteins, and first steps have been taken towards their functional characterization.

Materials and Methods

Strains and culture conditions

Cultures of *Amphora coffeaeformis* (C. Agardh) Kuetzing clone CCMP126 were grown on the bottom of 1 Liter Fernbach flasks in an artificial seawater medium (coined NEPC medium) according to the North East Pacific Culture Collection (http://www3. botany.ubc.ca/ccm/NEPCC/esaw.html). Cultivation conditions were 18°C and constant light at an intensity of 40–60 µmol photons $m^{-2} s^{-1}$, using cool white and warm white fluorescent tubes as light source. Bacteria-free (axenic) cultures were obtained by a 3-day treatment with antibiotics (1 mg mL^{-1} penicillin, 0.5 mg mL^{-1} streptomycin) and subsequent recovery under antibiotic-free conditions.

RNA isolation and transcriptome sequencing

Total RNA was extracted from cells that were attached to and actively moving on polystyrene petri dish surfaces using the RNAqueous Micro kit (Ambion, Carlsbad, CA, USA) by applying the lysis buffer directly to the petri dish surface after decanting the culture medium. The resulting total RNA (100 µg) was sent to Eurofins Genomics (Hunstville, AL, USA) for generation of a normalized random primed cDNA library and subsequent sequencing. Briefly, first strand cDNA was synthesized from isolated polyA$^+$ mRNA using random hexamers with the subsequent ligation of 3′ adapters. Second strand synthesis was performed using the 3′ adapter sequence. The cDNA library was then size-fractionated and normalized. Subsequently, the normalized cDNA was sequenced by Roche GS FLX technology using Titanium series chemistry (half a plate). The contig assemblies were performed by Eurofins Genomics. The transcriptome

sequence data have been deposited in the NCBI Sequence Read Archive (SRA) under accession number SRP046053.

Bioinformatics analysis

The database screening tool was written in Python, utilizing the common gateway interface (CGI) to interact with the web page at the URL http://vergil.chemistry.gatech.edu/cgi-bin/proteomics. py.

The *A. coffeaeformis* transcriptome sequence database was translated in all three forward frames and compiled in FASTA format. Once the composition and domain size are selected, the database is read in and each sequence is analyzed for composition and checked against all the composition requirements. If a domain length requirement of *n* is given, the first *n* amino acids are checked against the composition requirements. If the requirements are not met, then the first amino acid in the window being analyzed is subtracted out, the next amino acid is added in, and the requirements are checked again.

Construction of expression vector pPhaT1/YFP+fcpA/nat

A *Phaeodactylum tricornutum* expression vector was generated to allow for the single step construction of genes that encode fusion proteins carrying C-terminal YFP. In the first step the *eyfp* gene was amplified using the sense primer 5′-**GAA TTC** <u>TAC GTA</u> *GCA TGC* ***TCT AGA*** GGC GGA ATG GTG AGC AAG GGC GAG G-3′ and the antisense primer 5′-**<u>AAG CTT</u>** TTA CTT GTA CAG CTC GTC CAT-3′, which introduced a single copy of each of the following restriction sites *Eco*RI (bold), *Sna*BI (underlined), *Sph*I (italics), *Xba*I (bold, italics), and *Hind*III (bold, underlined). The resulting PCR products were introduced into pJet1.2 (Thermo Fisher Scientific; Waltham, MA, USA), sequenced, and then subcloned into the *Eco*RI and *Hind*III sites of pPhaT1/nat, which was kindly provided by Kirk Apt [17]. This generated the plasmid pPhaT1/YFP, which also contains the *ble* gene for resistance to zeocin under the control of the fcpB promoter (fcpB/*ble*). The fcpB/*ble* gene fragment was excised from pPhaT1/YFP using *Pst*I and *Xho*I (blunted using T4 DNA polymerase), and replaced by the fcpA/*nat* gene fragment derived from pPhaT1/*nat* by digestion with *Xho*I (blunted with T4 DNA polymerase) to generate pPhat1/YFP+fcpA/*nat*. This vector contains a short multiple cloning site (*Sna*BI, *Sph*I, *Xba*I) that allows for the generation of C-terminally tagged YFP fusion proteins.

Construction of fusion genes

Total RNA isolation, synthesis of a cDNA library attached to oligo(dT)$_{25}$ magnetic beads (Invitrogen, Carlsbad, CA, USA), and rapid amplification of cDNA ends (RACE PCR) was performed as described previously [18]. The resulting PCR products were cloned into pJet1.2 and sequenced by Eurofins Genomics (Ebersberg, Germany). Genomic DNA was isolated according to an established protocol [19]. For the generation of C-terminal YFP fusions genes the full length gene was amplified from genomic DNA using oligonucleotide primers that introduced either a *Sna*BI or *Eco*RV restriction site at the 5′-end and an *Xba*I restriction site at the 3′-end of the gene. The genes were amplified using Phusion DNA polymerase (Thermo Fisher Scientific), and the resulting PCR products were cloned into pJet1.2 and sequenced. The pJet1.2/AC genes were subsequently digested with the appropriate restriction enzyme and cloned into the SnaBI/XbaI site of pPhat1/YFP+fcpA/*nat*. Due to the apparent instability of the *AC4076-YFP* fusion gene it was necessary to use SURE2 *E. coli* cells (Agilent; Waldbronn, Germany) for its cloning. All other cloning procedures were performed using DH5α.

Transformation of Amphora coffeaeformis

Exponentially growing cells were harvested and concentrated by centrifugation for 5 min at 3,220 g. A total of 10^8 cells were plated in a 5 cm circle in the center of a NEPC medium agar plate (1.5% agar; Difco, Becton, Dickinson and Company, UK), and allowed to dry. Bombardment with DNA-coated tungsten particles was performed using the Biolistic PDS-1000/He particle delivery system (Bio-Rad, Hercules, CA, USA) as previously described [20]. DNA-coated tungsten particles were prepared by mixing 300 μg of tungsten particles of different diameter (i. e., 40 nm from Chempur, Karlsruhe and 400 nm, 700 nm, 1100 nm from Bio-Rad, München, Germany) with 5 μg of circular plasmid DNA using the CaCl$_2$-spermidine method [21]. For bombardment of the cells, agar plates were placed at a distance of 7 cm using either 1,500 psi rupture disks (1,400–1,600 psi) or 2,000 psi rupture disks (1,800–2,100 psi) (Bio-Rad, München, Germany). Immediately after bombardment the agar dishes were covered with liquid NEPC medium and incubated for 16 h under constant illumination. Subsequently, cells were spread on selective plates (5×10^6 cells per plate) containing 300 μg mL^{-1} of nourseothricin (clonNat; Werner Bioagents, Jena, Germany), and incubated at cultivation conditions either in continuous light or at a 14 hours light:10 hours dark rhythm.

Fluorescence and bright field microscopy

To screen for fluorescent transformant cells, clones from selective agar plates were inoculated into NEPC medium containing 300 μg mL^{-1} of nourseothricin in a 96-well glass bottom optical plate (Corning, Kaiserslautern, Germany) and incubated for 24–48 h prior to observation with an Axioplan 200 epifluorescence microscope (Zeiss, Jena, Germany) equipped with a YFP band pass filter set (EX: HQ500/20x; BS: Q515lp; EM: HQ535/30m; Chroma, Bellows Falls, VT, USA). Confocal fluorescence microscopy images were acquired with an inverted laser scanning microscope LSM 780/FLIM equipped with 32-channel GaAsP spectral detector and a multiline argon laser (458, 488, 514 nm) (Zeiss, Jena, Germany) using an alpha Plan-Apochromat 63x/1.46 Oil Korr M27 objective (Zeiss) and the Zen software (2011 version; Zeiss). YFP fluorescence was excited with the argon laser at 488 nm and detected using a 535/30 nm bandpass filter. During epifluorescence microscopy isolated cell walls were imaged using an Axioplan 200 epifluorescence microscope equipped with an eYFP 535/30 nm bandpass filter set (Zeiss).

Immunofluorescence images were acquired using an NSTORM laser microscope equipped with an Andor Ixon Ultra 897 camera (Nikon Instruments Europe B.V., Amsterdam, Netherlands) and a CFI TIRF Apochromat 100x/1.49 Oil objective (Nikon Instruments Europe B.V.). YFP fluorescence was excited at 488 nm and detected using a bandpass filter (502–548 nm, F37–521; AHF Analysentechnik, Tübingen, Germany), Alexa Fluor 647 fluorescence was excited at 647 nm and detected using a bandpass filter (663–738 nm, F47–700, AHF Analysentechnik).

Diatom trails were stained with 1 mg mL^{-1} 1-Ethyl-2-[3-(1-ethylnaphtho[1,2-d]thiazolin-2-ylidene)-2-methylpropenyl]naphtho [1,2-d]thiazolium bromide, 3,3′-Diethyl-9-methyl-4,5,4′,5′-dibenzothiacarbocyanine bromide (Stains-All; Sigma-Aldrich) in formamide for 15 min, washed excessively with H$_2$O, and then analyzed with an Axioplan 200 epifluorescence microscope (Zeiss) in the bright field mode as previously described [11].

Scanning electron microscopy

To prepare A. coffeaeformis biosilica for electron microscopy, cells were extracted twice in methanol (95°C) for 15 minutes each, and then washed three times with H$_2$O. The methanol-extracted cell walls were then treated with 1% SDS at 95°C for 30 min, washed four times with H$_2$O, and dried onto 0.5″ aluminium stubs (Agar Scientific, Stansted, UK). Electron microscopy was performed with a JSM7500F SEM (Jeol, München, Germany) at 10 kV acceleration voltage.

Cell wall isolation

A total of 3×10^9 A. coffeaeformis cells were harvested by decanting ~90% of the culture supernatant, resuspending the attached biofilm in the remaining supernatant using a cell scraper, and pelleting the cells by centrifugation for 5 min at 3220 g. The pelleted cells were thoroughly resuspended in buffer A (100 mM Tris-acetate pH 8, 50 mM EDTA, EDTA-free Pierce protease inhibitor (Thermo Fisher Scientific)), followed by extraction with 1 v/v-% Triton X-100 in buffer A at room temperature for 10 min. The cells were pelleted by centrifugation for 5 min at 3220 g and washed by resuspending the pellet in buffer B (100 mM Tris-acetate pH 8). This washing procedure was repeated three times. Finally, the cells were extracted with 1 w/v-% SDS in buffer B at room temperature for 10 min, centrifuged and washed in buffer B as described above, and extracted with 50% acetone until the material was colorless. All extractions were supported by vortexing and gentle sonication in a water bath at room temperature. Finally the cell walls were washed three times with buffer B as above.

Preparation of cell wall extracts

A total of 1.5×10^9 cell walls were resuspended in 2 mL SDS-extraction buffer (1 w/v-% SDS in buffer B) and shaken for 10 min at 55°C and 1,400 rpm in an thermomixer (Eppendorf AG, Hamburg, Germany). After pelleting by centrifugation at room temperature for 10 min at 4,000 g, the SDS extract was collected, the pellet resuspended in the same extraction buffer, and incubated under the same conditions. This procedure was performed three times in total, and the extracts were combined. The final pellet was washed twice by resuspension in buffer B and then centrifuged as above. The supernatants of the washing steps were combined with the supernatants of the extraction steps yielding the SDS cell wall extract. The SDS cell wall extract was diluted with H$_2$O to a final concentration of 0.1% SDS, and then concentrated by ultrafiltration (10 kDa MWCO; Amicon, Millipore, Darmstadt, Germany) to approximately one eightieth of the original volume. The original volume was restored by adding 200 mM ammonium acetate, and subsequently concentrated again to approximately one eightieth of the volume. The dilution-concentration cycles were repeated four times (i. e., until the SDS was largely removed) to a final volume of 300 μL.

For extraction with ammonium fluoride, 1.5×10^9 cell walls were resuspended in 6 mL 10 M NH$_4$F, adjusted to pH 4–5 by drop-wise addition of 6 M HCl, and incubated for 30 min at room temperature. After centrifugation for 15 min at 3,220 g at 4°C, the extraction procedure was repeated. The ammonium fluoride insoluble material was finally pelleted as before, washed with 200 mM ammonium acetate and pelleted again. All supernatants were unified, desalted and concentrated by ultrafiltration (10 kDa MWCO; Amicon, Millipore) as described above.

Western Blot

Protein extracts from 2.3×10^8 cell walls were separated on 10% Laemmli gels by gel electrophoresis, and transferred to nitrocellulose membranes (Protran BA 85; Whatman, GE Healthcare Europe, Freiburg, Germany) using a wet-blotting system (Bio-Rad, München, Germany) at 100 V for one hour. Towbin buffer [22] (25 mM Tris, 192 mM glycine) with 20 v/v-% methanol and 0.05% SDS was used for the transfer. Western Blots were probed with polyclonal Rabbit anti-GFP antibody (diluted 1:2,000 in Roti Block; referred to as "anti-YFP antibody", Cat. No 632592, Living Colours full-length A.v. polyclonal; Clontech, Mountain View, CA, USA) and secondary antibody peroxidase conjugate (diluted 1:10,000 in Roti Block; Anti-Rabbit IgG, A0545; Sigma-Aldrich, München, Germany). After 5 min development with chemiluminescent substrate (SuperSignal West Pico; Thermo Fisher Scientific) signals were detected on X-ray film (CL- XPosue Film; Thermo Fisher Scientific).

Immunolabeling

For immunolabeling of cell walls and ammonium fluoride insoluble material, the samples were resuspended in 1× blocking agent (Roti Immunoblock, Carl Roth, Karlsruhe, Germany) and incubated for 1 hour at room temperature with constant gentle shaking. Subsequently, anti-YFP antibody (see "Western Blot") was added to a final dilution of 1:300, and the samples were incubated with gentle shaking for an additional 1 hour. Afterwards the samples were pelleted by centrifugation (10 min at 3,220 g) and resuspended in 1× blocking agent. The centrifugation-resuspension procedure was repeated two more times, and after the last step Alexa Fluor 647-labeled anti-rabbit IgG (F(ab')₂ (H+ L) fragment developed in goat; Invitrogen, Germany) was added to a final dilution of 1:8,000 in 1× blocking agent. Before imaging, the samples were washed at room temperature by centrifugation and resuspenion three times in 1x blocking agent, and three times in 100 mM Tris-acetate pH 8.

Results

Generation of a transcriptome database for Amphora coffeaeformis

RNA was isolated from *A. coffeaeformis* cells that were adhered to polystyrene petri dishes, and a normalized *A. coffeaeformis* cDNA library was generated. Sequencing of the cDNA library yielded a total of 659,065 raw reads, with an average length of 341 bp. After trimming for the adaptors and primer sequences, 29,306 sequences were removed due to their short length resulting in 568,626 high quality (HQ) reads. The HQ reads were assembled into 41,824 contiguous sequences (contigs) ranging in size from 41 to 8,684 bp, with an average length of 799±509 bp.

Bioinformatics search for candidate adhesion proteins in A. coffeaeformis

Screening of the *A. coffeaeformis* transcriptome database for diatom proteins with sequence similarity to underwater adhesion proteins from mussels and the sandcastle worm was unsuccessful. Therefore, in an extension of our previous work [23], we have developed here a bioinformatics analysis tool that enables the in-depth screening of sequence databases for proteins based on amino acid composition rather than amino acid sequence. The tool allows for the identification of proteins that exhibit amino acid compositions of interest within a defined sequence domain. Both amino acid composition (in mol-%) and domain size (in number of amino acids) are freely selectable. Additionally, proteins can be screened simultaneously for the presence of an N-terminal signal peptide (required for most secreted proteins) according to the SignalP algorithm [24]. This new amino acid composition-based database screening tool is available at a publicly accessible website (http://vergil.chemistry.gatech.edu/cgi-bin/proteomics.py).

By applying this bioinformatics tool, the transcriptome of *A. coffeaeformis* was screened for putative adhesion proteins based on amino acid composition similarity to mussel foot proteins FP-3 and FP-5. Both proteins are present in the adhesive pads, and are believed to be directly involved in surface adhesion [25,26]. The unmodified FP-3 and FP-5 polypeptides exhibit molecular masses of 7.5–8.9 kDa and 8.9–12.2 kDa respectively, and a high content of tyrosine (Y>10 mol-%) [27]. Protein FP-1 is a coating rather than adhesive protein, yet it has also a high tyrosine content (up to 15 mol-%) and exhibits adhesive properties when the tyrosine residues are converted to Dopa [26,27]. Additional FP-1 characteristics are a high content of lysine (K>11 mol-%), and a highly repetitive sequence structure. An essential selection criterion in screening the diatom transcriptome database for candidate adhesion proteins was the presence of an N-terminal signal peptide, because the diatom adhesion proteins are expected to be assembled in and transported through the secretory pathway [28]. When screening the *A. coffeaeformis* transcriptome database for proteins with high tyrosine and high lysine content, five predicted proteins were retrieved that matched the search criteria (Table 1). The cDNA sequences of the encoding genes were validated by reverse transcriptase (RT) PCR (including 5′- and 3′- RACE PCR), except for Ac203 for which 3′-RACE PCR failed. The polypeptide sequences of the five proteins and the sequences of oligonucleotide primers used for RT PCR are shown in Table S1 and Table S2, respectively.

Interestingly, the putative adhesion proteins from *A. coffeaeformis* share additional features with mussel adhesion proteins that had not been selected for in the bioinformatics screen. These features are a high glycine content (9.2 mol-% for AC203; 23.4 mol-% for AC1077; 16.3 mol-% for AC714), and a high proline content (23.4 mol-% for AC1077; 14.3 mol-% for AC3362) (Table S3). Standard BLAST searches in the NCBI database using the complete sequences of the Y-rich proteins from *A. coffeaeformis* did not reveal homologous proteins in diatoms or other organisms. However, the Position-Specific Iterated (PSI) BLAST algorithm revealed that certain segments of proteins AC4076, AC714, and AC3362 exhibited sequence similarity to proteins from other organisms. A tyrosine- and histidine-rich stretch of AC4076 was similar to a putative adhesion protein from the fungus *Naumovozyma dairenensis* (29% identity over 103 amino acids, E-value: 3e-43; Table S4, Figure S1). A tyrosine-, lysine- and glycine-rich region of AC714 showed high similarity to a domain from a putative collagen alpha-1V chain from the insect *Danaus plexippus* (41% identity over 179 amino acids, E-value: 3e-36, Table S4, Figure S1). A repetitive proline- and lysine-rich region of AC3362 that also exhibited a high tyrosine and threonine content was highly similar to a domain from a fungal protein of unknown function (67% identity over 185 amino acids, E-value: 1e-68; Table S4, Figure S1). However, PSI-BLAST analysis did not detect sequence similarity of regions from the Y-rich proteins to any domains from the mussel foot proteins.

As we were unable to obtain information on the full length cDNA sequence of AC203 further analysis of this protein was discontinued. The amino acid sequences of the remaining four Y-rich proteins (AC1077, AC4076, AC714, AC3362) exhibit a modular structure which often contain repetitive peptide motifs (Figure 1, File S1). In most domains except for those of AC1077 charged and/or polar amino acids dominate (>25 mol-%), giving AC4076, AC714, and AC3362 a strongly hydrophilic character despite the high tyrosine content. In all four proteins, lysine rather than arginine residues account for almost all positively charged amino acid residues. There are also several proline-rich modules,

Table 1. Candidate adhesion proteins identified by an amino acid composition-based bioinformatics screen of the *A. coffeaeformis* transcriptome database. n. a. = not applicable.

Search criteria	Domain size (amino acids)	Protein ID
>10 mol-% Y	n. a.	AC1077, AC4076, AC714
>8 mol-% Y and >20 mol-% K	50–200	AC203
>10 mol-% Y and >20 mol-% K	100–300	AC3362

in which proline residues account for 16–54 mol-% of the amino acid residues.

To gain the first insight into the function of the *A. coffeaeformis* Y-rich proteins, we intended to overexpress them as fluorescent fusion proteins in *A. coffeaeformis*, and study their locations *in vivo*. To enable such experiments, it was first necessary to establish a genetic transformation method for this diatom species.

Genetic transformation of Amphora coffeaeformis

To date, all routinely used methods for the genetic transformation of diatoms rely on microparticle bombardment (termed biolistic transformation). In this method tungsten particles are coated with plasmid DNA containing both the gene of interest and an antibiotic resistance gene. In most cases endogenous diatom specific promoters are used for driving the expression of both genes, but also heterologous diatom promoters or even nondiatom promoters have been shown to work in some cases [29–32]. Promoters from *A. coffeaeformis* genes are not available due to the complete lack of sequence information from the genome of

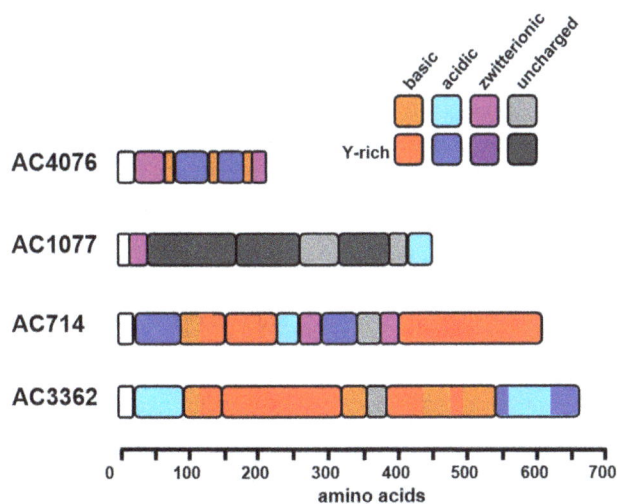

Figure 1. Schematic primary structures of the Y-rich proteins from *A. coffeaeformis*. Domains with a high abundance of charged amino acids (≥25 mol-%) are colored. Red color indicates modules with a surplus of positively charged amino acids (mol-% ratio D+E: K+R≥2), blue modules are dominated by negatively charged amino acids (mol-% ratio K+R: D+E ≥2), and purple color indicates zwitterionic modules in which negatively and positively charged amino acids are almost balanced (mol-% ratio K+R: D+E>0.8 or <1.2). Gray modules are essentially devoid of charged amino acids (<10 mol-%). In all modules dark color indicates regions with a tyrosine content ≥6 mol-%. Note that the average contents of tyrosine in all proteins in the Swiss-Prot database is 2.9%, of D+E is 12.2%, and of K+R is 11.4% [46]. White modules indicate the signal peptide.

this species. Therefore, transformation experiments had to be been performed using promoters from other diatom species (see below).

Growth of *A. coffeaeformis* wild type cells in liquid culture and on agar plates can be completely inhibited by the antibiotics zeocin (at ≥600 µg mL^{-1}) or nourseothricin (at ≥300 µg mL^{-1}). Therefore, for biolistic transformation experiments with *A. coffeaeformis* the *ble* gene (resistance to zeocin) and the *nat* gene (resistance to nourseothricin) were chosen as selection markers. Diatom specific expression vectors containing the antibiotic resistance genes under control of the *fcp* promoters from the diatoms *Phaeodactylum tricornutum*, *Cylindrotheca fusiformis*, and *Thalassiosira pseudonana* (pPhat1/*nat*, pCfcp/*ble*, Tpfcp/*nat*, Tpfcp/*ble*) [17,20,33] were coated onto tungsten microparticles, and used in biolistic transformation experiments. The promoters from all three diatom species appeared to be functional in *A. coffeaeformis* as indicated by the growth of *A. coffeaeformis* transformant clones on both on agar plates, and in liquid medium in the presence of antibiotic concentrations that were lethal to the wild type cells.

The highest number of antibiotic resistant *A. coffeaeformis* clones was obtained with the *P. tricornutum* fcp promoter (pPhat1) in combination with the *nat* resistance gene. Therefore, pPhat1/*nat* was used as the selection marker in all subsequent transformation experiments. However, even with pPhat1/*nat* the efficacy of genetic transformation was rather low, yielding a maximum of only 50 antibiotic resistant cells per 10^8 bombarded cells. Through systematic variation of experimental conditions (i. e., tungsten particle size, particle acceleration pressure, and cultivation conditions) the transformation efficacy could be drastically improved consistently yielding 800 antibiotic resistant cells per 10^8 bombarded cells (Figure 2). This yield is even higher than in currently used transformation protocols for *T. pseudonana* or *P. tricornutum*, which generate up to 400 transformant clones per 10^8 cells [17,20].

After spreading the cells on antibiotic containing agar plates, cultivation in a 10:14 hour light:dark cycle rather than continuous light appeared to be beneficial (data not shown). Tungsten particles of average diameters of 700 nm and 1,100 nm were more effective than smaller particles (Figure 2), while lower particle acceleration (pressure of 1,550 psi) proved to be more effective except for the largest tungsten particles (Figure 2).

Localization of AC3362

For four of the five candidate adhesion proteins full-length cDNA sequences could be obtained by RACE PCR (i. e., AC1077, AC714, AC3362, AC4076). We constructed expression plasmids in which each of the four *A. coffeaeformis* genes was fused to the YFP gene under the control of the *P. tricornutum* fcpA promoter (pPhat1), which were then introduced into *A. coffeaeformis* using the transformation method described above. At least 27 nourseothricin-resistant transformant clones were analyzed by fluorescence microscopy for each expression construct

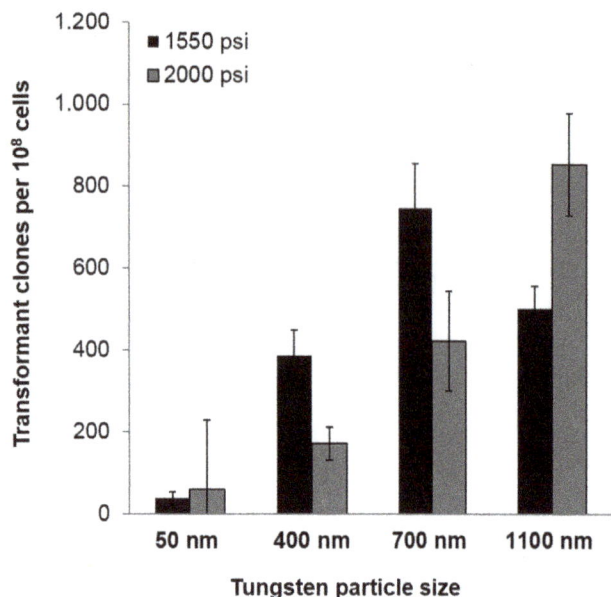

Figure 2. Influence of tungsten particle size and acceleration pressure on transformation efficacy. Tungsten particles of the indicated sizes were coated with plasmid DNA and shot at plated *A. coffeaeformis* cells using a helium pulse of 1550 psi (black bars) or 2000 psi (grey bars). Transformed clones were quantified by counting the number of colonies that grew on agar plates containing the antibiotic nourseothricin.

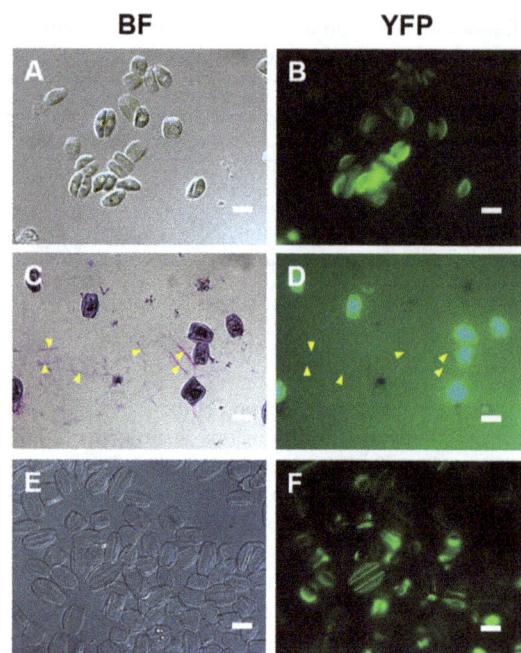

Figure 3. Localization of AC3362-YFP in *A. coffeaeformis*. (A, C, E) Bright field microscopy images, and **(B, D, F)** corresponding epifluorescence microscopy images. **(A, B)** Live cells. **(C, D)** Cells after treatment with the polycationic dye 'Stains-All'. For orientation some trails are labeled by arrowheads. Image **(D)** was deliberately overexposed to check for YFP fluorescence in the trails. **(E, F)** Isolated cell walls. Bars = 10 μm.

(Table S5). Only transformants harboring the AC3362-YFP fusion gene exhibited YFP fluorescence in the cells (Figure 3A, B), transformation with the other YFP fusion genes (i. e., AC1077, AC714, AC4076) did not result in fluorescent clones (data not shown). None of the AC3362-YFP transformant cells exhibited YFP fluorescence in the adhesive material that is deposited on the glass slides as characteristic trails (Figure 3C, D). The AC3362-YFP fusion protein appeared to be located at the periphery of the cell (Figure 3A, B) suggesting that it is a cell wall associated protein. This was confirmed by isolating the cell walls, which still exhibited strong YFP fluorescence (Figure 3E, F).

To investigate the localization of AC3362-YFP in the cell wall in more detail, epifluorescence and confocal fluorescence microscopy analysis of both intact and fragmented cell walls were performed. Like with all other diatoms *A. coffeaeformis* cell walls are composed of two silica plates, each termed valve, which are connected by circular silica strips, termed girdle bands. In most diatoms the two valves are located on the opposing poles of the cell, but the genus *Amphora* is characterized by the two valves being positioned on the same side of the cell (i. e., the ventral side; Figure 4A, B). As a consequence the girdle bands of *A. coffeaeformis* are wedge shaped, being narrower on the ventral side than on the opposing side (i. e., the dorsal side; Figure 4A-C). When isolated cell walls were treated with controlled doses of ultrasound, the girdle bands became separated from the valves (Figure 4D). Epifluorescence microscopy revealed that the AC3362-YFP fusion protein was present both in the girdle bands and the valves (Figure 4D, E). In the valves the YFP fluorescence appeared to be most abundant at the edges and significantly weaker in the central region around the so-called raphe (Figure 4D, E). The raphe contains a longitudinal slit through which the adhesive material is secreted [34]. Confocal fluorescence microscopy confirmed that the AC3362-YFP fusion protein was

present almost everywhere on the valve surface except for the region of the raphe (Figure 4F-I). The absence of AC3362-YFP in the raphe, through which the adhesive material is being secreted, is consistent with the absence of this fusion protein in the trails that contain the adhesive material (see Figure 3C, D). The dorsal side of intact isolated cell walls contains only girdle bands and exhibits a striated fluorescent pattern that is congruent with the longitudinal axes of the girdle bands (Figure 4J, K).

Characterizing the Biosilica Association of AC3362-YFP

The results presented above strongly suggest that AC3362 is a component of the *A. coffeaeformis* cell wall rather than a protein involved in surface adhesion. The amino acid sequence of AC3362 does not exhibit any similarities to previously characterized diatom cell wall proteins, which are frustulins, pleuralins, silaffins, silacidins, cingulins, and p150 [23,35-37]. However, regarding the high content of lysine (15.5 mol-%) and tyrosine (8.0 mol-%) residues, AC3362 resembles the cingulins of the diatom *T. pseudonana* [23] and the silaffin-1 peptides from *C. fusiformis* [38]. Cingulins and silaffins are tightly associated with the biosilica, and thus cannot be extracted from the cell walls even when using solutions of SDS at elevated temperature [23,35,39]. To determine the stability of the interaction between AC3362-YFP and the biosilica, isolated cell walls were treated with 1 w/w-% SDS at 55°C and YFP-fluorescence intensity of the cell walls was monitored by epifluorescence microscopy. After this treatment the YFP fluorescence in the cell walls was substantially decreased seemingly indicating that the majority of AC3363-YFP became extracted (compare Figures 5B and G). However, isolated cell walls before and after treatment with hot SDS solution exhibited comparable fluorescence intensity following indirect immunofluorescence labeling using

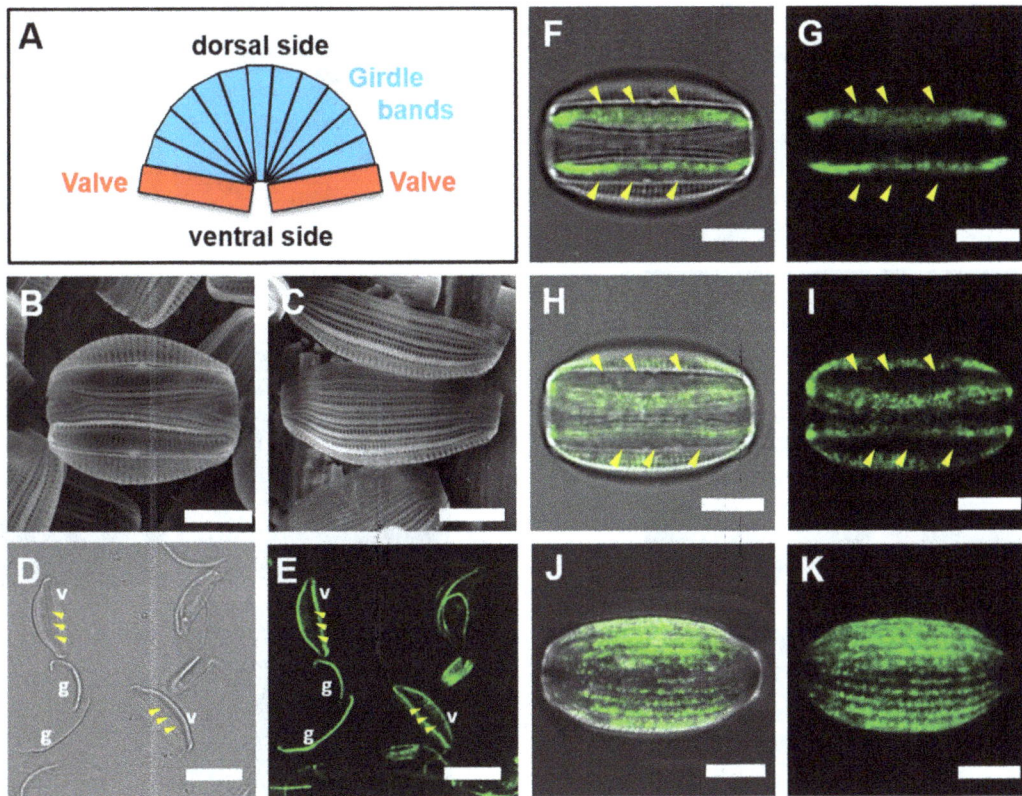

Figure 4. Localization of AC3362-YFP within the cell wall. (A) Schematic of an *A. coffeaeformis* cell wall. **(B, C)** Scanning electron micrographs of *A. coffeaeformis* cell walls in **(B)** ventral view, and **(C)** dorsal view. **(D, E)** Sonicated cell walls imaged with bright field microscopy **(D)**, and confocal fluorescence microscopy **(E)**. Valves (v) and girdle bands (g) are labeled. **(F-K)** Confocal fluorescence microscopy images of isolated cell walls in **(F-I)** ventral view, and in **(J, K)** dorsal view. The images represent 2D projections of confocal Z-stacks of the cell wall in contact with the surface-substratum **(F, G)**, around the raphe region **(H, I)**, and throughout the girdle band region **(J, K)**. **(F, H, J)** show overlays of confocal fluorescence microscopy images and the corresponding bright field images, while **(G, I, K)** show the confocal fluorescence microscopy images only. In **(D-I)** the positions of the raphes are indicated by a triplet of arrowheads. Bars: (B, C, F-K) = 5 µm, (D, E) = 10 µm.

primary antibodies against YFP and an Alexa Fluor 647-labeled fluorescent secondary antibody (compare Figures 5C and H). Wild type cells walls exhibited no fluorescence following immunolabeling (Figure 5D, E, I, K) demonstrating the specificity of the immuno-labeling procedure. Altogether, the immunolabeling data indicate that most, if not all, of the AC3362-YFP fusion protein is still present in the cell wall after extraction with hot SDS solution. The strongly reduced YFP fluorescence in cell walls after treatment with hot SDS must therefore be mainly due to a loss of fluorescence by partial denaturation of YFP. The SDS-resistant incorporation into the cell wall identifies AC3362 as a biosilica-associated protein.

Previously, two types of biosilica-associated proteins have been characterized: (i) proteins that become soluble when the silica is dissolved using an ammonium fluoride solution at pH 5 (e. g., silaffins; [35]), and (ii) proteins that are constituents of an organic matrix that remain insoluble after ammonium fluoride treatment (cingulins; [23]). To investigate the type of biosilica association of AC3362, isolated *A. coffeaeformis* cell walls bearing AC3362-YFP were resuspended in a solution of ammonium fluoride at pH 5. After dissolution of the silica, the insoluble material was recovered by centrifugation and investigated by fluorescence microscopy. The ammonium fluoride-insoluble material exhibited YFP fluorescence (Figure 6 A, C) thus indicating the presence of AC3362. This was confirmed by indirect immunolabeling using the same primary and secondary antibodies as above (Figure 6A, B). No fluorescence was observed when immunolabeling was performed

with ammonium fluoride-insoluble material from wild type cell walls (Figure 6D, E), demonstrating the specificity of the immunolabeling procedure. To investigate whether any of the AC3362-YFP fusion protein was extracted during ammonium fluoride treatment, Western Blot analyses were performed using the anti-YFP antibody. No proteins were detected in the ammonium fluoride extracts from the AC3362-YFP bearing cell walls and wild type cell walls (Figure 7). A loading control (i. e., recombinant silaffin3-GFP purified from *E. coli*) indicated that the Western Blot procedure had worked (Figure 7) leaving the options that either the ammonium fluoride extract did not contain AC3362-YFP, or an insufficient amount of ammonium fluoride extract was loaded on the Western blot. The latter possibility could be ruled out by analyzing the hot SDS extract of AC3362-YFP bearing cell walls as described in the following paragraph.

In the hot SDS-extract from AC3362-YFP bearing cell walls three bands of ~40 kDa, ~45 kDa, and ~55 kDa were recognized by the anti-YFP antibodies (Figure 7). The ~40 kDa band was also present in the hot SDS-extract from wild type cell walls and thus resulted from non-specific cross-reaction with an unidentified protein. In contrast, the ~45 kDa and ~55 kDa bands were only present in the hot SDS-extract from the transformant cell walls, and thus must be related to the AC3362-YFP fusion protein. Considering the molecular mass of YFP (27 kDa) it can be concluded that the ~45 kDa and ~55 kDa proteins in the SDS extract contain ~150 and ~250

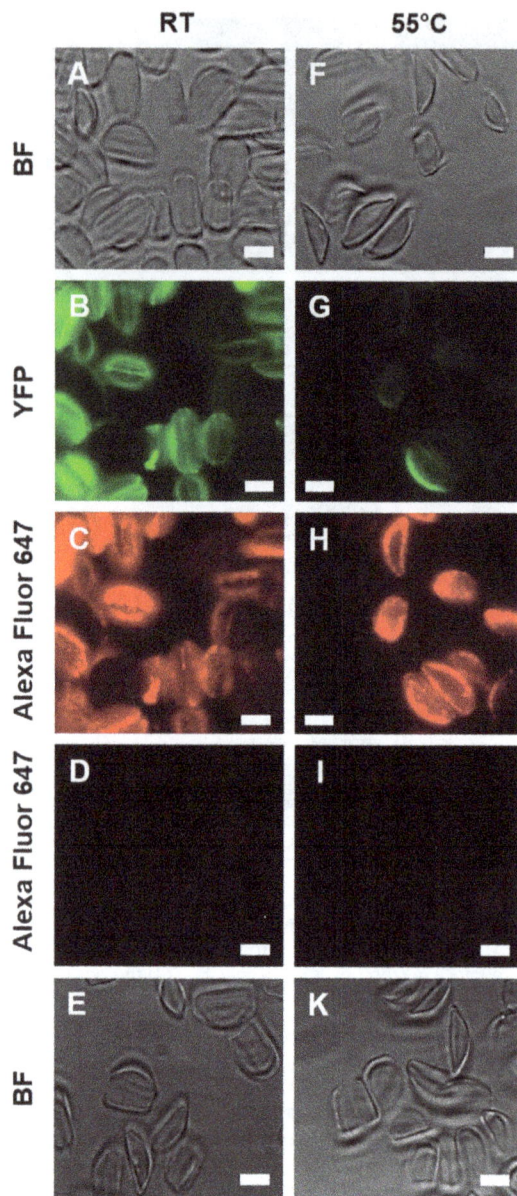

Figure 5. Biosilica association of AC3362-YFP. Isolated cell walls (RT) and isolated cell walls after hot SDS teatment (55°C) were analyzed by bright field microscopy (BF), by direct fluorescence microscopy (YFP), and by indirect immunofluorescence microscopy (Alexa Fluor 674) using an anti-YFP primary antibody and an Alexa Fluor 647-labeled secondary antibody. (**A-C**) Cell walls from a transformant clone expressing AC3362-YFP, and (**F-H**) cell walls from the same clone after extraction with a hot SDS solution. Cell walls from wild type (**D, E**) before and (**I, K**) after extraction with a hot SDS solution. Bars = 5 μm.

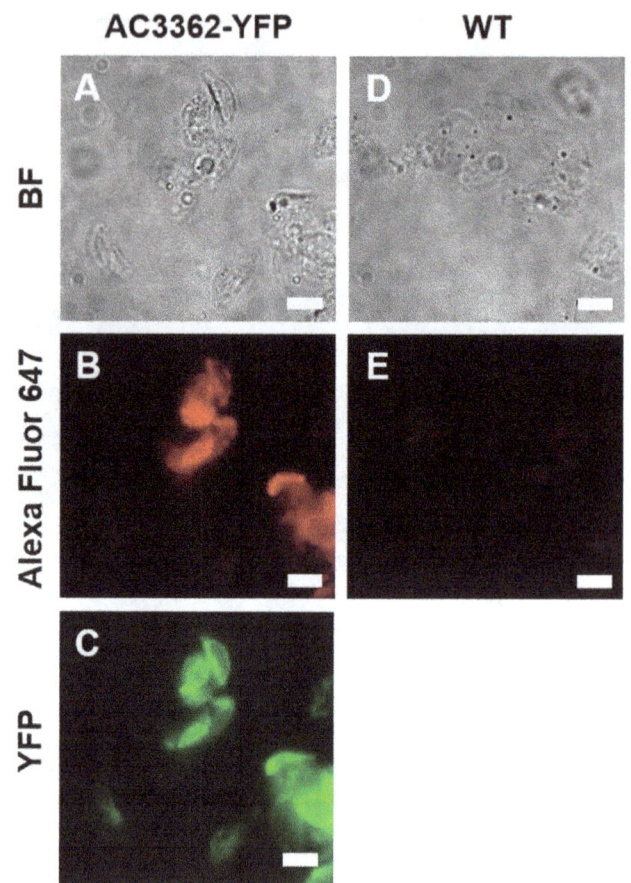

Figure 6. Ammonium fluoride extraction of cell walls. The ammonium fluoride insoluble materials from the cell walls of a AC3362-YFP expressing transfomant clone (**A-C**)) and from wild type (**D, E**) were analyzed by bright field microscopy (BF), by direct fluorescence microscopy (YFP), and by indirect immunofluorescence microscopy (Alexa Fluor 647) using an anti-YFP primary antibody and an Alexa Fluor 647-labeled secondary antibody. Bars = 5 μm.

amino acids, respectively, of the C-terminal end of AC3362 provided that posttranslational modifications are absent. The immunofluorescence data shown above (see Figure 5) have demonstrated that hot SDS treatment extracts only a very small fraction of the AC3362-YFP fusion protein from the cell walls. This small amount could be detected in the Western Blot experiment indicating the high sensitivity of the method. As no YFP fusion protein was detected in the ammonium fluoride extract that was prepared from the same amount of cell walls, it was concluded that the vast majority of AC3362-YFP molecules is present in the ammonium fluoride-insoluble material.

Altogether the data from fluorescence microscopy and Western Blot analyses (see Figures 5–7) suggest that the majority of the AC3362-YFP fusion protein is a constituent of a biosilica-associated, insoluble organic matrix. The small amount of biosilica-associated YFP fusion protein extractable by hot SDS solutions is composed of relatively short fragments of AC3362 ($\leq$ 23% and $\leq$38%) that may lack the domain(s) required for incorporation into the ammonium fluoride insoluble matrix. Determining the apparent molecular mass of the AC3362-YFP fusion protein in the ammonium fluoride insoluble organic matrix has not been possible, because attempts to solubilize this material have so far failed. Regarding itsincorporation in an insoluble organic matrix AC3362 resembles cingulins rather than silaffins.

Discussion

Here we describe the identification of the first cell surface protein, AC3362, from *A. coffeaeformis*, a model species for studying diatom underwater adhesion to surfaces [14–16]. AC3362 was identified by screening a normalized *A. coffeaeformis* transcriptome database, which has been established in the present study (note: during the course of this work additional transcrip-

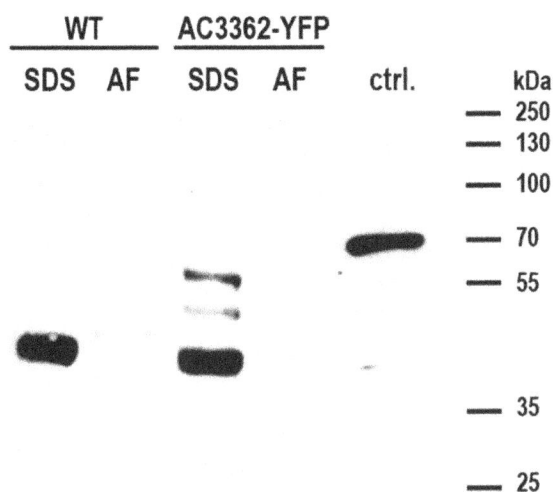

Figure 7. Western blot probed with an anti-YFP antibodies. The extracts obtained from identical amounts of either wild type (WT) or AC3362-YFP bearing cell walls by treatment with a hot SDS solution (SDS) or with an ammonium fluoride solution at pH 4–5 (AF) were loaded. In lane 'ctrl.' 100 ng recombinant Sil3-GFP (isolated from *E. coli*) was loaded.

tome databases for *A. coffeaeformis* have been published by other groups [40]). The database screen was performed using a novel amino acid composition-based bioinformatics screening software that we have developed here and made publicly available (http://vergil.chemistry.gatech.edu/cgi-bin/proteomics.py). The screening parameters were based on proteins that mediate underwater adhesion in marine mussels, which are highly enriched in both lysine and tyrosine residues [12,13]. AC3362 contains two lysine- and tyrosine-rich domains (amino acids 95–146 and 147–325), but does not exhibit sequence similarity to mussel adhesion proteins.

Studies on the functional characterization of AC3362 relied on a genetic transformation system for *A. coffeaeformis* that has also been established in the present study. This enabled the expression of an AC3362-YFP fusion protein and investigation of its location by fluorescence microscopy using both direct imaging of the YFP fusion protein and indirect immunolabeling with anti-YFP antibodies. The data clearly indicate that AC3362 is not a component of the adhesive material that is secreted by the diatom cell. Instead, AC3362 is part of an insoluble organic matrix associated with the biosilica of the cell wall, similar to the cingulin-containing microrings recently described from *T. pseudonana* [23]. Several biosilica associated proteins, have been implicated in cellular biosilica formation, and the AC3362 protein resembles the cingulins from the diatom *T. pseudonana*, which are also rich in tyrosine residues [23,35–37,41].

As the AC3362-YFP fusion protein was accessible to antibody molecules in immunolabeling it appears to be partially or fully exposed on the biosilica surface, rather than completely embedded within the biosilica. Ammonium fluoride-insoluble organic matrices that are composed of proteins and polysaccharides and exposed on the biosilica surface have recently been identified in several diatoms (not including *Amphora* species) [42]. Tesson and Hildebrand argued that insoluble organic matrices embedded within diatom biosilica may not exist as they were unable to detect ammonium fluoride-insoluble organic matrices from acid hydrolyzed biosilica [42]. However, their data were inconclusive, because their argument was based on the resistance to acid

hydrolysis of the biosilica-associated long-chain polyamines, which are devoid of acid-labile bonds [38]. Therefore, it cannot be argued that silica-embedded organic matrices have to be resistant to acid hydrolysis conditions. In contrast, as biosilica is highly porous and hydrated it can be expected that protons can easily penetrate throughout the biosilica, and thus polysaccharide- and protein-based insoluble organic-matrices should become completely degraded under acid hydrolysis conditions, regardless of whether they are embedded within or exposed on the surface of biosilica. In future research, analysis by immunofluorescence microscopy, using the method described in the present study, may be able to validate whether the cingulin-containing organic matrix of *T. pseudonana* is embedded inside the biosilica or located on the biosilica surface.

It has been discussed that biosilica-associated organic matrices may have a role in biosilica formation, mechanical support of biosilica, the stabilization of biosilica against dissolution, or combinations of these functions [35,42,43]. One possibility to further investigate the function of the AC3362 containing organic matrix would be the generation of knock-down mutants. This technique is established for the diatoms *P. tricornutum* [44] and *T. pseudonana* [45], and should be also possible for *A. coffeaeformis* by utilizing the transformation system that has been established in this study.

The bioinformatics screen for mussel-like putative adhesion proteins in *A. coffeaeformis* has yielded three additional tyrosine-rich proteins (AC714, AC1077, AC3362). However, we were unable to detect by fluorescence microscopy the production of the resulting YFP fusion proteins in *A. coffeaeformis* transformants. Expression rate of the YFP fusion proteins may have been too low, the C-terminal YFP-tag may be proteolytically cleaved from the mature protein, or the YFP domain may have interfered with folding or stability of these tyrosine-rich proteins. Whatever the case, due to the presence of the encoding mRNAs it is reasonable to assume that the three proteins are produced in *A. coffeaeformis* wild type cells. Furthermore, the presence of an N-terminal signal peptide for co-translational ER import and the absence of transmembrane helices in each of these proteins strongly suggest that they become secreted into the medium, or are incorporated into the cell wall, or are targeted to another intracellular compartment. Protein AC714 contains repeats of the dipeptide KG (amino acids 419–599) which are also present in the cell wall protein AC3362 (amino acids 95–146, 326–357, 385–537). In the silaffin family of diatom proteins, lysine-rich repeats have been shown to promote association with the biosilica *in vivo*.

During the course of our work on the tyrosine-rich *A. coffeaeformis* proteins, a different study has provided evidence for the absence of Dopa in *A. coffeaeformis* [11]. As the adhesiveness of the tyrosine-rich proteins depends on conversion of tyrosine to Dopa residues [13], the tyrosine-rich *A. coffeaeformis* proteins AC714, AC1077, and 4076, which we have identified in the present study cannot be Dopa-dependent adhesion proteins. However, recently a Dopa-independent mechanism for underwater adhesion of the sandcastle worm has been proposed, which is based on complex coacervation [47]. Complex coacervation involves the aggregation of polyelectrolyte chains (here: proteins) resulting in liquid-liquid phase separation of a polyelectrolyte-rich and a polyelectrolyte-depleted aqueous phase [48,49]. Aggregation of polyelectrolyte chains can be mediated through neutralization of oppositely charged polyelectrolytes or through the hydrophobic effect, yielding coacervate phases with low interfacial energies that are conducive to spreading on surfaces [47,48]. AC714 and AC4076 exhibit domains with high densities of negative and positive charges (see Fig. 1) and might therefore self-aggregate or aggregate with other zwitterionic biomolecules. In contrast, AC1077 is mainly composed of uncharged amino acid residues with high proportions of tyrosine (17.6 mol-%) and proline (23.4 mol-%), which might promote hydrophobically driven complex coacervation. Investigating whether these unique proteins are involved in surface adhesion of *A. coffeaeformis* will require successful expression of tagged fusion proteins or the generation of specific antibodies to enable their localization, and establishing gene knockdown mutants followed by phenotype analysis.

The value of amino acid composition-based bioinformatics screens for the identification of novel diatom cell surface proteins has previously been demonstrated through the identification of cingulins [23] and a frustule-associated protein [32]. The amino acid-based screening method established in the present study substantially extends the possible search parameters and largely simplifies the procedure by providing a simple web-based interface. Therefore, we expect that amino acid-based screening methods will become increasingly used to identify proteins whose function is independent on a particular 3D fold and requires characteristic non-complex amino acid compositions.

Supporting Information

Figure S1   Alignments of the Y-rich proteins from *A. coffeaeformis*. AC4076 (**A**), AC714 (**B**) and AC3362 (**C**) were aligned with their respective best BLAST hit using Clustal Omega (http://www.ebi.ac.uk/Tools/msa/clustalo/). Note that BLAST analyses did not yield hits for AC1077.

Table S1   Sequences of the Y-rich proteins. Predicted N-terminal signal peptide sequences are underlined. Note that for AC203 5′-RACE PCR confirmed the sequence of amino acids 1-41, whereas the 3′-RACE PCR primers failed to yield a product.

Table S2   Oligonucleotide sequences of primers used in the present study. Note that the 3′-RACE PCR primers for AC203 failed to yield a product.

Table S3   Amino acid composition of the Y-rich proteins from *A. coffeaeformis*. Amino acid frequencis are shown in mol-%. Analyses were performed with ProtParam (http://web.expasy.org/protparam/). Note that the sequence of AC203 could be not verified by RACE PCR.

Table S4   PSI-BLAST hits for the Y-rich proteins from *A. coffeaeformis*. Note that AC203 was not analyzed as its polypeptide sequence could not be verified by RACE PCR.

Table S5   Number of transformant clones analyzed by fluorescence microscopy. The clones were selected on nourseothricin containing agar plates.

File S1   Domain analysis of the tyrosine-rich proteins.

Acknowledgments

We thank Philip Gröger (B CUBE, TU Dresden, Germany) for help with laser epifluorescence microscopy, Kirk Apt (Martek Biosciences) for providing the pPhat1/nat plasmid, Anusuya Willis (Georgia Institute of Technology, Atlanta, USA, now at Griffith University, Australia) for the construction of pPhat/YFP+fcpA/*nat* plasmid, and Thomas Kurth (CRTD, TU Dresden, Germany) for the SEM service.

Author Contributions

Conceived and designed the experiments: MTB NP NK. Performed the experiments: MTB NP JK. Analyzed the data: MTB NP NK. Contributed reagents/materials/analysis tools: MRK CDS. Wrote the paper: MTB NP MRK CDS NK.

References

1. Field CB, Behrenfeld MJ, Randerson JT, Falkowski P (1998) Primary Production of the Biosphere: Integrating Terrestrial and Oceanic Components. Science 281: 237–240.
2. Smith AM, Callow JA (2006) Biological Adhesives; Smith AM, Callow JA, editors. Berlin: Springer, 302p.
3. Schultz MP, Bendick JA, Holm ER, Hertel WM (2010) Economic impact of biofouling on a naval surface ship. Biofouling 27: 87–98.
4. Fitridge I, Dempster T, Guenther J, de Nys R (2012) The impact and control of biofouling in marine aquaculture: a review. Biofouling 28: 649–669.
5. Callow JA, Callow ME (2011) Trends in the development of environmentally friendly fouling-resistant marine coatings. Nat Commun 2: 244.
6. Molino PJ, Wetherbee R (2008) The biology of biofouling diatoms and their role in the development of microbial slimes. Biofouling 24: 365–379.
7. Vournakis JN, Fischer T, Lindner HB, Demcheva M, Seth A, et al. (2012) Poly-N-acetyl Glucosamine Nanofibers Derived from a Marine Diatom: Applications in Diabetic Wound Healing and Tissue Regeneration. In: Le L-A, Hunter R, Preedy V, editors. Nanotechnology and Nanomedicine in Diabetes. St. Helier, Jersey: CRC Press Tyler & Francis Group. pp. 345.
8. Romero D, Aguilar C, Losick R, Kolter R (2010) Amyloid fibers provide structural integrity to *Bacillus subtilis* biofilms. Proc Natl Acad Sci USA 107: 2230–2234.
9. Friedlander RS, Vlamakis H, Kim P, Khan M, Kolter R, et al. (2013) Bacterial flagella explore microscale hummocks and hollows to increase adhesion. Proc Natl Acad Sci USA 110: 5624–5629.
10. Drum R, Hopkins J (1966) Diatom locomotion – an explanation. Protoplasma 62: 1–33.
11. Poulsen N, Kröger N, Harrington MJ, Brunner E, Paasch S, et al. (2014) Isolation and biochemical characterization of underwater adhesives from diatoms. Biofouling 30: 513–523.
12. Waite JH, Tanzer ML (1981) Polyphenolic substance of *Mytilus edulis*: novel adhesive containing L-dopa and hydroxyproline. Science 212: 1038–1040.
13. Nicklisch SC, Waite JH (2012) Mini-review: The role of redox in Dopa-mediated marine adhesion. Biofouling 28: 865–877.
14. Finlay JA, Callow ME, Ista LK, Lopez GP, Callow JA (2002) The influence of surface wettability on the adhesion strength of settled spores of the green alga *Enteromorpha* and the diatom *Amphora*. Integr comp biol 42: 1116–1122.
15. Hudon C, Bourget E (1981) Initial colonization of artificial substrate: community development and structure studied by scanning electron microscopy. Can J Fish Aquat Sci 38: 1371–1384.
16. Hodson OM, Monty JP, Molino PJ, Wetherbee R (2012) Novel whole cell adhesion assays of three isolates of the fouling diatom *Amphora coffeaeformis* reveal diverse responses to surfaces of different wettability. Biofouling 28: 381–393.
17. Apt KE, Grossman A, Kroth-Pancic P (1996) Stable nuclear transformation of the diatom *Phaeodactylum tricornutum*. Mol Gen Genet 252: 572–579.
18. Poulsen N, Kröger N (2004) Silica morphogenesis by alternative processing of silaffins in the diatom *Thalassiosira pseudonana*. J Biol Chem 279: 42993–42999.
19. Jacobs JD, Ludwig JR, Hildebrand M, Kukel A, Feng T-Y, et al. (1992) Characterization of two circular plasmids from the marine diatom *Cylindrotheca fusiformis*: plasmids hybridize to chloroplast and nuclear DNA. Mol Gen Genet 233: 302–310.
20. Poulsen N, Chesley PM, Kröger N (2006) Molecular genetic manipulation of the diatom *Thalassiosira pseudonana* (Bacillariophyceae). J Phycol 42: 1059–1065.
21. Dhir S, Pajeau M, Fromm M, Fry J (1994) Anthocyanin genes as visual markers for wheat transformation. Improvement of cereal quality by genetic engineering: Springer. pp. 71–75.
22. Towbin H, Staehelin T, Gordon J (1979) Electrophoretic transfer of proteins from polyacrylamide gels to nitrocellulose sheets: procedure and some applications. Proc Natl Acad Sci U S A 76: 4350–4354.
23. Scheffel A, Poulsen N, Shian S, Kröger N (2011) Nanopatterned protein microrings from a diatom that direct silica morphogenesis. Proc Natl Acad Sci USA 108: 3175–3180.
24. Petersen TN, Brunak S, von Heijne G, Nielsen H (2011) SignalP 4.0: discriminating signal peptides from transmembrane regions. Nat Meth 8: 785–786.

25. Hwang DS, Zeng H, Masic A, Harrington MJ, Israelachvili JN, et al. (2010) Protein-and metal-dependent interactions of a prominent protein in mussel adhesive plaques. J Biol Chem 285: 25850–25858.

26. Lin Q, Gourdon D, Sun C, Holten-Andersen N, Anderson TH, et al. (2007) Adhesion mechanisms of the mussel foot proteins mfp-1 and mfp-3. Proc Natl Acad Sci USA 104: 3782–3786.

27. Silverman HG, Roberto FF (2010) Byssus formation in *Mytilus*. Biological Adhesive Systems: Springer. pp. 273–283.

28. Wetherbee R, Lind JL, Burke J, Quatrano RS (1998) Minireview—The first kiss: establishment and control of initial adhesion by raphid diatoms. J Phycol 34: 9–15.

29. Miyagawa A, Okami T, Kira N, Yamaguchi H, Ohnishi K, et al. (2009) Research note: High efficiency transformation of the diatom *Phaeodactylum tricornutum* with a promoter from the diatom *Cylindrotheca fusiformis*. Phycol Res 57: 142–146.

30. Sakaue K, Harada H, Matsuda Y (2008) Development of gene expression system in a marine diatom using viral promoters of a wide variety of origin. Physiol Plant 133: 59–67.

31. Miyagawa-Yamaguchi A, Okami T, Kira N, Yamaguchi H, Ohnishi K, et al. (2011) Stable nuclear transformation of the diatom *Chaetoceros* sp. Phycol Res 59: 113–119.

32. Muto M, Fukuda Y, Nemoto M, Yoshino T, Matsunaga T, et al. (2013) Establishment of a genetic transformation system for the marine pennate diatom *Fistulifera* sp. strain JPCC DA0580—A high triglyceride producer. Mar Biotechnol (NY) 15: 48–55.

33. Poulsen N, Kröger N (2005) A new molecular tool for transgenic diatoms. FEBS J 272: 3413–3423.

34. Edgar LA, Pickett-Heaps J (1983) The mechanism of diatom locomotion. I. An ultrastructural study of the motility apparatus. Proc R Soc Lond B Biol Sci: 331–343.

35. Kröger N, Poulsen N (2008) Diatoms-from cell wall biogenesis to nanotechnology. Annu Rev Genet 42: 83–107.

36. Wenzl S, Hett R, Richthammer P, Sumper M (2008) Silacidins: highly acidic phosphopeptides from diatom shells assist in silica precipitation in vitro. Angew Chem Int Ed Engl 120: 1753–1756.

37. Davis AK, Hildebrand M, Palenik B (2005) A stress-induced protein associated with the girdle band region of the diatom *Thalassiosira pseudonana* (Bacillariophyta). J Phycol 41: 577–589.

38. Kröger N, Deutzmann R, Sumper M (1999) Polycationic peptides from diatom biosilica that direct silica nanosphere formation. Science 286: 1129–1132.

39. Poulsen N, Scheffel A, Sheppard VC, Chesley PM, Kröger N (2013) Pentalysine clusters mediate silica targeting of silaffins in *Thalassiosira pseudonana*. J Biol Chem.

40. Stief P (2011) Marine Microbial Eukaryote Transcriptome Sequencing Project. Available: http://camera.crbs.ucsd.edu/mmetsp/details.php?id=MMETSP031 6. Accessed: 2014 Sep 27.

41. Sumper M, Brunner E (2008) Silica biomineralisation in diatoms: the model organism *Thalassiosira pseudonana*. ChemBioChem 9: 1187–1194.

42. Tesson B, Hildebrand M (2013) Characterization and localization of insoluble organic matrices associated with diatom cell walls: insight into their roles during cell wall formation. PloS one 8: e61675.

43. Bidle KD, Azam F (1999) Accelerated dissolution of diatom silica by marine bacterial assemblages. Nature 397: 508–512.

44. De Riso V, Raniello R, Maumus F, Rogato A, Bowler C, et al. (2009) Gene silencing in the marine diatom *Phaeodactylum tricornutum*. Nucleic Acids Res 37: e96–e96.

45. Trentacoste EM, Shrestha RP, Smith SR, Glé C, Hartmann AC, et al. (2013) Metabolic engineering of lipid catabolism increases microalgal lipid accumulation without compromising growth. Proc Natl Acad Sci USA 110: 19748–19753.

46. Bairoch A, Apweiler R (2000) The SWISS-PROT protein sequence database and its supplement TrEMBL in 2000. Nucleic Acids Res 28: 45–48.

47. Stewart RJ, Weaver JC, Morse DE, Waite JH (2004) The tube cement of *Phragmatopoma californica*: a solid foam. J Exp Biol 207: 4727–4734.

48. Bungenberg de Jong HG (1932). Die Koazervation und ihre Bedeutung für die Biologie. Protoplasma 15: 110–73.

49. Cooper CL, Dubin PL, Kayitmazer AB, Turksen S. (2005). Polyelectrolyte-protein complexes. Curr Opin Colloid Interface Sci 10: 52–78.

5

Rapid Restriction Enzyme-Free Cloning of PCR Products: A High-Throughput Method Applicable for Library Construction

Vijay K. Chaudhary[1]*, Nimisha Shrivastava[1]◕, Vaishali Verma[1]◕, Shilpi Das[1], Charanpreet Kaur[1], Payal Grover[1], Amita Gupta[2]*

1 Department of Biochemistry, University of Delhi South Campus, Benito Juarez Road, New Delhi 110021, India, **2** Department of Microbiology, University of Delhi South Campus, Benito Juarez Road, New Delhi 110021, India

Abstract

Herein, we describe a novel cloning strategy for PCR-amplified DNA which employs the type IIs restriction endonuclease BsaI to create a linearized vector with four base-long 5'-overhangs, and T4 DNA polymerase treatment of the insert in presence of a single dNTP to create vector-compatible four base-long overhangs. Notably, the insert preparation does not require any restriction enzyme treatment. The BsaI sites in the vector are oriented in such a manner that upon digestion with BsaI, a stuffer sequence along with both BsaI recognition sequences is removed. The sequence of the four base-long overhangs produced by BsaI cleavage were designed to be non-palindromic, non-compatible to each other. Therefore, only ligation of an insert carrying compatible ends allows directional cloning of the insert to the vector to generate a recombinant without recreating the BsaI sites. We also developed rapid protocols for insert preparation and cloning, by which the entire process from PCR to transformation can be completed in 6–8 h and DNA fragments ranging in size from 200 to 2200 bp can be cloned with equal efficiencies. One protocol uses a single tube for insert preparation if amplification is performed using polymerases with low 3'-exonuclease activity. The other protocol is compatible with any thermostable polymerase, including those with high 3'-exonuclease activity, and does not significantly increase the time required for cloning. The suitability of this method for high-throughput cloning was demonstrated by cloning batches of 24 PCR products with nearly 100% efficiency. The cloning strategy is also suitable for high efficiency cloning and was used to construct large libraries comprising more than 10^8 clones/μg vector. Additionally, based on this strategy, a variety of vectors were constructed for the expression of proteins in *E. coli*, enabling large number of different clones to be rapidly generated.

Editor: Odir A. Dellagostin, Federal University of Pelotas, Brazil

Funding: This work was supported by grants from the Department of Biotechnology, Government of India to VKC and AG. VV is the recipient of a research fellowship from Council of Scientific and Industrial Research, India. The funders had no role in study design, data collection and analysis, decision to publish, or preparation of the manuscript.

Competing Interests: The authors have declared that no competing interests exist.

* Email: vkchaudhary@south.du.ac.in (VKC); amitagupta@south.du.ac.in (AG)

◕ These authors contributed equally to this work.

Introduction

The availability of genome sequences from a large number of organisms has created a wide-spread need to clone complete sets of open reading frames (ORFs) followed by the expression and purification of the encoded gene products in a high-throughput manner to examine the functions of genes and proteins in organisms. Using a variety of thermostable polymerases, the target ORFs can be efficiently amplified by polymerase chain reaction (PCR) with virtually no errors. However, high efficiency and high-throughput cloning of amplified products into a suitable expression vector still remains an arduous task.

An ideal high-throughput cloning method should include a minimal number of steps without the need for intermediate purification and preferably be performed in the same tube in which the gene-of interest was amplified. For such cloning, (i) the reagents used to prepare the vector and insert should not be proprietary and should be available from multiple sources; (ii) the vector should not self-ligate and the insert should not form concatemers, rather ligate to the vector in the desired orientation; (iii) the primers for amplification should carry only small additional sequences without the need for any modifications, and thereby only add a sequence encoding a few amino acids at the vector/insert junctions to serve as spacers, and finally; (iv) the vector design should be amenable to modifications such as incorporation of different tags to enable purification and/or enhance solubility, or changing the promoter, antibiotic selection marker and even the origin of replication.

Traditional methods such as restriction enzyme-based strategies cannot be considered suitable for the high-throughput cloning of a large set of different ORFs due to the presence of numerous restriction enzyme sites within the ORF sequences [1]. As an alternative, the commonly used TA cloning methodology is simple, but lacks directionality in cloning and requires the inserts to be amplified using enzymes that have template-independent terminal transferase activity; these polymerases have a low fidelity resulting in mutations in the amplified DNA [2]. Different formats of the ligation-independent cloning (LIC) method have been described for efficient high-throughput cloning, but rely on the annealing of complementary single-stranded DNA and capability of bacterial cells to repair the resulting gaps within double-stranded DNA. LIC involves amplification of genes of interest with primers containing an additional sequence of 15–20 bases, which is then used to create long cohesive ends on the amplified inserts by T4 DNA polymerase or exonuclease III treatment under controlled conditions *in vitro*. The long overhangs of these inserts are then annealed to the linearized vector that carries compatible long cohesive ends created by specific treatments, then without prior ligation, the vector-insert mixture is transformed into a bacterial host in which the nicks/gaps are filled [3]. The disadvantage of this method is the addition of unwanted amino acids at both ends of the expressed protein. Methods such as sequence and ligation-independent cloning (SLIC) have been described as an alternative to LIC; however, their cloning efficiency in the absence of a recombinase, RecA, is low [4]. Recently described methods like Seamless Ligation Cloning Extract, SLiCE, also require the addition of up to 42–52 bp to the insert to provide end homology in order to achieve a very high efficiency. In fact, inserts without end homology or less than 10 bp end homology did not yield any recombinant clones [5]. Recombination-based cloning methods including Gateway [6,7], Creator [8] etc., have been developed for high-throughput cloning; however, all of these methods make use of long site-specific sequences for recombination that must be incorporated on either side of the ORF to enable its insertion into the desired vector. Moreover, these methods are expensive and impose restrictions in terms of the sequences and hosts. Commercially available cloning systems such as TOPO cloning [9], Infusion [10] etc., require specialized vectors and/or proprietary reagents to create cohesive ends in the vector and inserts. Recently, simple high-throughput cloning strategies based on type IIs restriction enzymes have been described [11,12], but require the inserts to be amplified with primers carrying the restriction site of choice. However, the limitation of all type IIs restriction enzyme-based cloning is the possibility that the insert may contain the same internal restriction enzyme site, requiring the use of alternative strategies such as site-directed mutagenesis of the endogenous restriction site. Another major drawback of most of the strategies described above is the requirement to purify the PCR products prior to ligation to the vector.

In this paper, we present a highly versatile and efficient restriction enzyme-free cloning strategy for rapid and high-throughput cloning of PCR-amplified DNA fragments into the desired vector. We demonstrate the strategy is equally effective for cloning inserts of various sizes (0.5 kbp to 2.2 kbp) in single tube format suitable for high-throughput applications, and also for high efficiency cloning during the construction of genome-scale libraries (more than 10^8 clones). Based on this strategy, several expression vectors were constructed and employed for cloning hundreds of mycobacterial genes to produce proteins containing different tags. The versatility and high efficiency of this strategy has been extended by constructing vectors for a number of different applications including phage display of gene-fragments and construction of mouse antibody libraries.

Materials and Methods

Construction of vectors

The expression vector pVNLEBAP1306 [13] containing a backbone obtained from pET11a [14] was modified by oligonu-

cleotide-directed mutagenesis to delete a 1.32 kbp fragment between the *laci* and *rop* genes to derive the pVLExp vector backbone (Fig. 1 & Fig. S7). A stuffer flanked by two BsaI sites in the appropriate orientations for cloning and containing other unique restriction sites, and the sequences encoding fusion tags was introduced between the NdeI and EcoRI sites. The NdeI site harboring the initiation codon (CAT**ATG**) was followed by a sequence encoding an N-terminal tag (Tag1), NheI site, 5′ inverted BsaI site, 1.8 kbp stuffer, 3′ BsaI site followed by a Bsu36I restriction site, the sequence encoding a C-terminal tag (Tag2) and the stop codon (TAA). Digestion of the vector with BsaI linearizes the vector, creating a 5′-GCCG-3′ overhang close to the 3′ end of the promoter and 5′-GGAG-3′ overhang close to the EcoRI site, and removes the stuffer so that the BsaI recognition sites are lost upon ligation with the insert.

The vector for cloning was prepared by digesting 10 μg plasmid DNA in a total volume of 400 μl with 400 units of BsaI (New England Biolabs, Ipswich, MA, USA) added in four aliquots during the 4 h incubation at 50°C. The digested DNA was extracted with phenol: chloroform, ethanol precipitated, resuspended in 0.1X TE and then separated on 1.2% Sea Plaque GTG agarose (Lonza, Rockland, ME, USA) to purify the linearized vector from the stuffer fragment using the Qiaquick Gel Extraction kit (Qiagen, Hilden, Germany).

Insert preparation

The forward and reverse primers used for amplification of genes of interest carried 5′-CGGCAGC and 5′-CTCCACC, respectively, as seven base-long extensions in addition to the approximately 20–26 base-long gene-specific sequence (Fig. 2C). The sequences of the primers used to amplify the different mycobacterial genes for the initial optimizations are shown in Table S1. Primers used for amplification were either molecular biology grade (without any purification) or HPLC-purified (IBA GmbH, Goettingen, Germany).

For initial optimizations, PCR was performed in 100 μl reactions containing 10 ng of plasmid carrying the test genes or 100 ng of *Mycobacterium tuberculosis (Mtb)* genomic DNA as a template, as well as 200 μM dNTPs, 50 pmoles each of the forward and reverse primers and 2 U of Expand High Fidelity PCR System (Expand HF polymerase; Roche, Mannheim, Germany). The amplification steps involved initial denaturation at 95°C for 3 min followed by 25 cycles of denaturation at 95°C for 30 sec, annealing at 55°C for 30 sec, polymerization at 72°C for variable periods of time depending upon the size of the genes (as per the manufacturer's instructions) and 2 sec extension in each cycle followed by a final polymerization at 72°C for 4 min. Five microlitres of each reaction was analyzed by agarose gel electrophoresis to estimate the amount of PCR product; the remaining product was processed by the following methods to prepare the inserts for cloning.

Single tube method

Ten microlitres of PCR product was mixed with 1 μl Exonuclease I-Shrimp Alkaline Phosphatase (Exo-SAP; Affymetrix, Santa Clara, CA, USA) in a 0.2 ml PCR tube, vortexed, centrifuged and incubated in a thermocycler programmed for 60 min incubation at 37°C followed by heat inactivation at 80°C for 20 min, then the tube was held at 4°C. One microlitre of dTTP (20 mM; 1.7 mM final concentration) was added to each tube containing the Exo-SAP treated inserts, mixed well by mild vortexing followed by brief centrifugation at 4°C, and 1.5 units (0.5 μl) of T4 DNA polymerase (New England Biolabs) were added. The contents were mixed well, centrifuged at 4°C and then

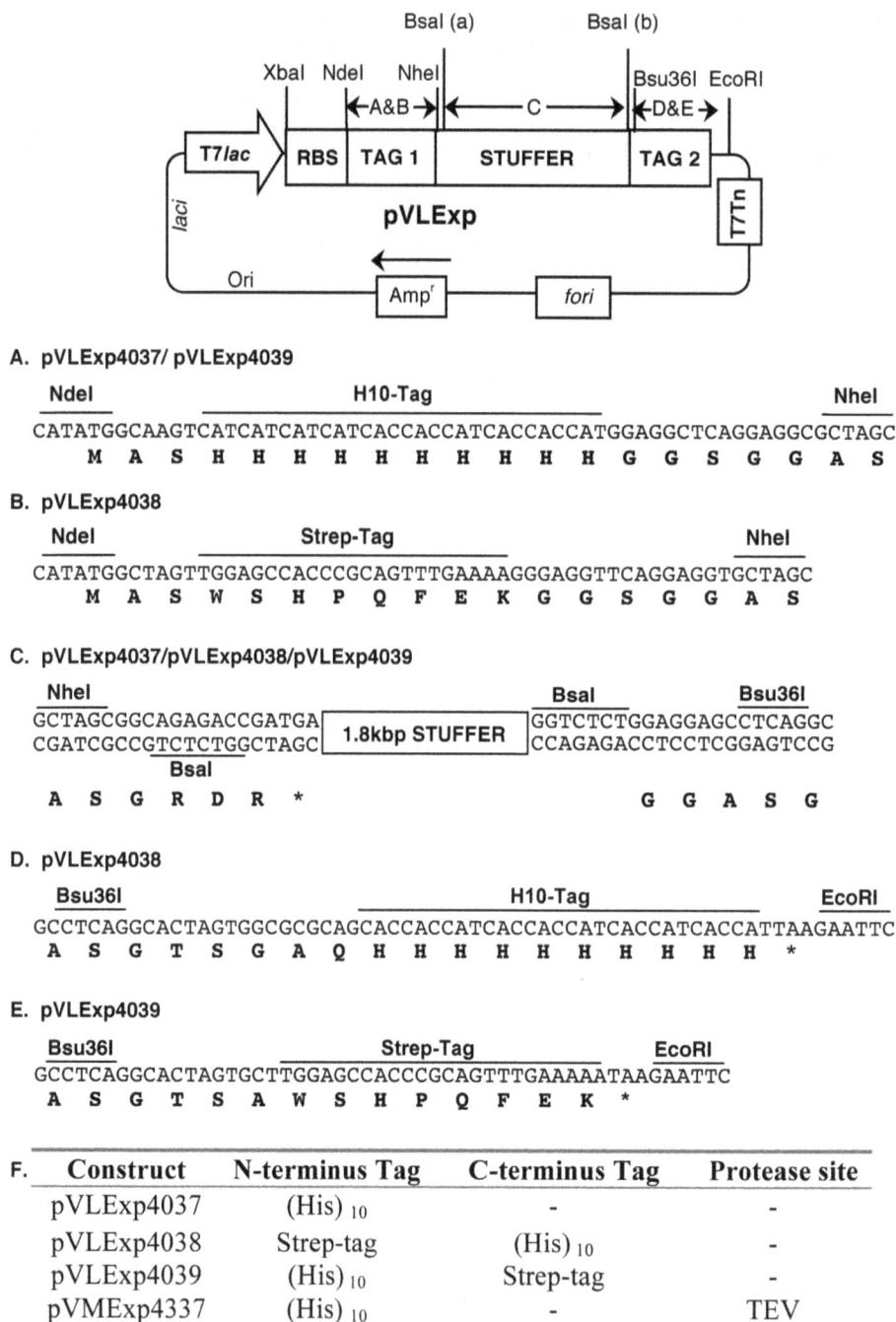

Figure 1. Diagrammatic representation of pVLExp4037/4038/4039 vectors. Only relevant genes and restriction sites are shown. The maps are not to scale. T7*lac*, T7 promoter lac operator; RBS ribosome-binding site; T7Tn, T7 transcription terminator; *fori*, origin of replication of filamentous phage; Amp[r], β lactamase gene; Ori, Col E1 origin of replication; *laci*, lac repressor encoding gene rop/rom; Stuffer, 1.8 kbp stuffer flanked by BsaI sites. The encoded amino acids are shown in single letter code (bold) below the nucleotide sequence (A–E). F, shows the summary of vectors containing different fusion tags and protease sites.

placed in a thermocycler for incubation at 15°C for 60 min followed by heat inactivation at 75°C for 20 min.

Ligation was performed in a total volume of 10 µl containing 1 µl of 10X ligation buffer (New England Biolabs), 25 ng BsaI-HF-digested vector, 1 µl of T4 DNA Polymerase-treated insert (2–5 fold molar excess relative to the vector) and 1.0 unit of T4 DNA ligase (Roche) for 16 h at 16°C followed by 1 h at 37°C and heat inactivation at 65°C for 10 min. For electroporation, the ligation mixture was diluted ten-fold in water and then 1 µl of the diluted ligation sample (containing 250 pg vector equivalent) was electroporated into 25 µl of *E. coli* competent cells [BL21 (DE3) RIL, Agilent Technologies, CA, USA; efficiency 3–5×10[8]/µg]. The resulting recombinants were analyzed by colony PCR followed by sequencing using an ABI 3730 DNA sequencer (Life Technologies Corporation, USA). To check for protein expression, individual colonies were inoculated into 150 µl MDA-

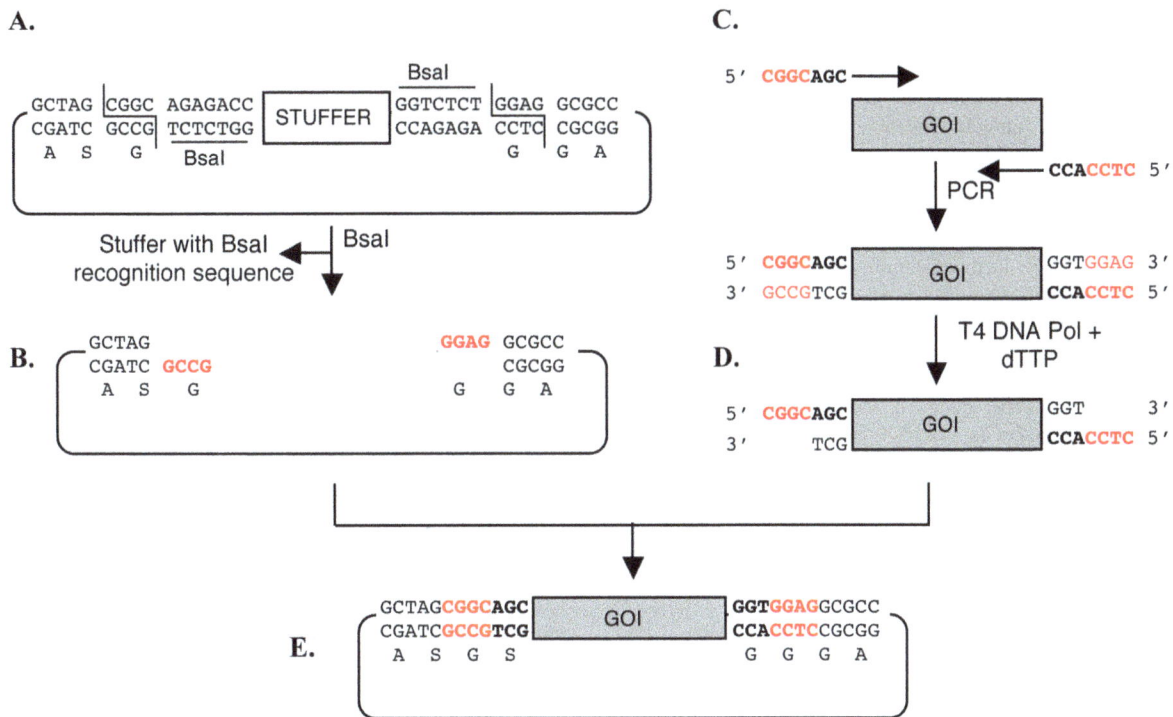

Figure 2. Cloning strategy. The vector contains two appropriately oriented BsaI sites (A) upon digestion with BsaI linearized vector is obtained with ends having 4-base 5′-overhangs (B) shown in red. The recognition sequence of restriction enzyme BsaI are underlined and the cleavage site is marked. The Gene Of Interest (GOI) is amplified using two gene-specific primers with 7-base long additional sequence at the 5′ end (C) shown in bold. Treatment of PCR product with T4 DNA polymerase and dTTP produces two different four-base overhangs that are complementary to two ends of the linearized vector shown in red (D). The ligation results in direction cloning of the insert into the vector (E).

GAmp$_{100}$ medium [15] in a 96 well microtiter plate and cultured at 37°C with shaking for 3 h, and an aliquot of each suspension was subjected to PCR. Additionally, 20 µl of the cultures were transferred to 96-well deep well plate (2 ml capacity) containing 0.6 ml ZYM5052Amp$_{100}$ auto-inducing medium [15] and incubated at 30°C with shaking at 300 rpm. After 16 h of induction, 50 µl of each culture was mixed with 50 µl of 2X Laemmli reducing dye and analyzed by SDS-PAGE using gradient gel (8–20%) electrophoresis and the bands were visualized using Coomassie Brilliant blue R-250 staining.

Rapid single tube strategy

To further optimize the single tube strategy for rapid cloning with reduced incubation times for T4 DNA polymerase treatment and ligation, three mycobacterial genes of various sizes [*Rv1886c* (Ag85B), *Rv1908c* (MPT64) and *Rv3763* (19 KDa)] were amplified using the gene-specific primers containing specific 7-base extensions (Table S1) in a total volume of 50 µl. The inserts were processed as described above for the single tube strategy, except that the time of incubation of the inserts with T4 DNA polymerase to generate 5′ overhangs in the presence of dTTP was systematically reduced from 60 min to 5 min at 15°C followed by heat inactivation at 75°C for 20 min. The timing of the ligation was optimized by testing 60, 30 or 15 min incubation at room temperature (25°C) followed by heat inactivation at 65°C for 10 min. Then, 1 µl of the ligation mixtures were directly electroporated into 25 µl of BL21 (DE3) RIL electrocompetent cells (efficiency 2×10^8/µg supercoiled plasmid DNA) and plated onto non-inducing MDAGAmp$_{100}$plates. Furthermore, five different scFv genes cloned into phage display vectors (carrying ampicillin antibiotic resistance genes) were PCR amplified and sub-cloned into

an Arabinose promoter-based expression vector pVMAAscFvclo 0001(carrying a ampicillin antibiotic resistance gene) using the rapid single tube method, and the resulting recombinants were analyzed by colony PCR and sequencing as described above.

Column method

For PCR products obtained using high fidelity polymerases [16] such as PfuUltra II Fusion HS DNA Polymerase, Herculase II Fusion DNA Polymerase etc. (error rate $\sim 4.3 \times 10^{-7}$) that have very high 3′ to 5′ exonuclease activity, the PCR-amplified product needs to be purified using a column to remove the polymerase, excess dNTPs and primers.

Therefore, the remaining $\sim$85 µl of the amplified products (out of initial 100 µl) were column purified using the Qiaquick PCR purification kit (Qiagen) as per the manufacturer's instructions and the DNA was eluted in 60 µl of 1X TE. For treatment with T4 DNA polymerase, 10 µl of purified PCR product was mixed with 1 µl of 10X NEB 2 buffer (New England Biolabs), 0.1 µl of bovine serum albumin (BSA) and 1.0 µl of 20 mM dTTP (1.7 mM final concentration), briefly centrifuged at 4°C and 0.5 µl (1.5 units) of T4 DNA polymerase was added to obtain a total volume of 12.6 µl. The tubes were incubated at 15°C for 60 min, followed by heat inactivation of the enzyme at 75°C for 20 min, ligation and electroporation were performed as described above for the single tube strategy, and the resulting clones were analyzed by colony PCR and sequencing.

Preparation of inserts for library-scale cloning

The PCR products (preparation described in Fig. S8) were purified using the Qiaquick PCR purification kit following the manufacturer's instructions and subjected to agarose gel electrophoresis to estimate the DNA concentration. In a 1.5 ml microfuge tube, 5 µg of the purified PCR product was mixed with 10 µl of 10X NEB 2 buffer, 2.5 µl of 20 mM dTTP (500 µM dTTP final concentration) and the volume was adjusted to 98.5 µl with H$_2$O. While keeping the tube on ice, 1.5 µl of T4 DNA polymerase (3 U/µl) was added, the contents were mixed by mild vortexing, centrifuged at 4°C, incubated at 15°C for 60 min and the enzyme was inactivated by adding 5 volume of Qiaquick buffer PB. The inserts were further column purified and analyzed by agarose gel electrophoresis. Then, 10 µl ligation reactions were set-up in multiple tubes, each containing 1 µl of 10X ligation buffer, 100 ng BsaI-digested vector, 25–100 ng T4 DNA polymerase-treated purified inserts (2–8 fold molar excess) and 1.0 unit of T4 DNA ligase (Roche) and incubated for 16 h at 16°C, 60 min at 37°C and heat inactivated at 65°C for 10 min. For electroporation, the contents of ligation reaction were pooled and 4 µl ligation mixture was directly electroporated into 100 µl of competent *E. coli* cells (efficiency 7–8×10^9/µg) and several such electroporations were performed to obtain a library of ~10^8 clones.

Results

Concept of the cloning strategy

The cloning strategy described here is based on a combination of techniques involving the use of the type IIs restriction endonuclease BsaI to create a linearized vector with 4 base-long 5′-overhangs, and T4 DNA polymerase treatment of the insert in the presence of a single dNTP to create vector-compatible 4 base-long overhangs. The cohesive ends produced in the vector and the insert are non-compatible and non-palindromic in *cis* to prevent vector self-ligation or insert concatemerization, but allow directional ligation of the insert to the vector (Fig. 2).

The BsaI sites in the vector are oriented in such a manner that the stuffer, along with the recognition sequences of both sites, is removed upon digestion with BsaI. Thus, digestion of the vector with BsaI produces a linearized vector with two four base-long overhangs: 5′-GCCG-3′ on one end close to the initiation codon and 5′-GGAG-3′ at the other end near to the EcoRI site (Fig. 2A & 2B). The sequence of the four base-long overhangs produced by BsaI cleavage were designed in such a way that they are non-palindromic, non-compatible to each other and most importantly, do not produce compatible ends even if one or more bases of the overhangs are excised. Therefore, ligation of any insert carrying compatible ends would generate a recombinant without recreating the BsaI sites.

Design and construction of the vector

The vectors (pVLExp) described in this work were derived from the T7 promoter-based *lac* operator (T7-Lac) expression vector pVNLEBAP1306 [13] carrying a pET11a backbone (Fig. 1). The pVNLEBAP1306 vector was modified by oligonucleotide-directed mutagenesis to delete the non-functional *tet*^{r} gene between *laci* and the plasmid origin of replication and to also destroy a BsaI restriction site located in the β-lactamase gene. A 1.8 kbp stuffer flanked by two appropriately-oriented BsaI sites was cloned between the NdeI site (CAT<u>ATG</u>) that contains the initiation codon (underlined) and the EcoRI site. This stuffer was preceded by a sequence encoding Tag1 and a NheI restriction site upstream of the 5′ BsaI restriction site, and Bsu36I and Tag2 sequences downstream of the 3′ BsaI site, just before the stop codon. A series of expression vectors was constructed (Fig. 1 & Fig. S5) that contain different N- and/or C-terminal tags that can be used for

purification purposes. The vector pVLExp4037 carries a decahistidine (H10) as Tag1 so that after cloning an insert, the genes of interest are expressed with an N-terminal H10 tag and 5-amino acid spacer (GSGGG) to allow spatial accessibility of the tag for purification. The vectors pVLExp4038 and pVLExp4039 are similar and allow cloning of genes of interest to express proteins with an N-terminal poly-His and C-terminal Strep tag or with an N-terminal Strep tag and C-terminal poly-His, respectively. Prior to cloning, each vector was prepared by digestion with BsaI to remove the 1.8 kbp stuffer.

Insert preparation and cloning strategy

Insert preparation involves PCR amplification of genes of interest using primers containing a 20–26 base-long gene-specific annealing sequence with an additional seven defined nucleotides at the 5′ end (Fig. 2C). To make the PCR product compatible with cloning, it was treated with T4 DNA polymerase in the presence of dTTP to create the four base-long overhangs 5′-CGGC and 5′-CTCC at the 5′ and 3′ ends of the insert, respectively (Fig. 2D). In this particular case, the 3′- exonuclease activity of T4 DNA polymerase would remove A, G and C bases and then stall when polymerase will encounter T because of the presence of dTTP in the reaction mix (due to equilibrium between the 3′-exonuclease and polymerase activities of the enzyme). These ends are compatible with the overhangs generated in the vector by digestion with BsaI (Fig. 2B). Thus, ligation of the linearized vector and T4 DNA polymerase-treated insert results in directional cloning of the insert to obtain a recombinant without recreating the BsaI sites (Fig. 2E).

However, before T4 DNA polymerase treatment, the insert has to be purified from the excess primers, dNTPs and polymerase used for amplification. Two rapid strategies were developed for this purpose. The first strategy (column method) involves use of silica-based columns for purification of the PCR product to remove excess primers, dNTPs and also the thermostable DNA polymerase. This method requires ~2 µg amplified product (50–100 µl of PCR mix) for purification on the column. The subsequent step involves treatment of a fraction of the purified PCR product (250–500 ng in a volume of 10 µl) with T4 DNA polymerase in the presence of the appropriate buffer and dTTP in a total volume of 12.6 µl. The second strategy is a single tube method performed in a thermocycler. For this method, 10 µl of PCR mix (containing 250–500 ng crude amplified product) is first treated with a mixture of Exonuclease I (Exo I) and shrimp alkaline phosphatase (SAP), which is commercially available as Exo-SAP. Exo I degrades the primers and SAP dephosphorylates the released nucleotides and excess dNTPs present the PCR mix after amplification. Both of these enzymes are then heat-inactivated in the same tube, followed by addition of dTTP and T4 DNA polymerase without the need for adding any buffer, as T4 DNA polymerase is compatible with most buffers used for PCR. After 1 h incubation, the contents of the tube are heated to inactivate T4 DNA polymerase. Thus, using the single tube strategy, an insert compatible with cloning is prepared in the same tube without any need for transferring the amplified product.

Both insert preparation methods were initially evaluated by cloning four mycobacterial genes (*Rv1827, Rv3029c, Rv1077* and *Rv1908c*) of sizes, approximately 0.5, 1, 1.5 and 2.2 kbp. These inserts were amplified in 100 µl PCR reactions using either HPLC-purified or machine grade (unpurified) primers containing 24–26 base-long gene-specific annealing sequences along with a seven base-long additional sequence (Table S1). Then, 3 µl of each reaction mixture was analyzed by agarose gel electrophoresis; both types of primers yielded comparable amounts of amplified products, suggesting that the purity of the primers does not affect the quality and the quantity of the amplified insert (Fig. S1). The amplified products were processed to generate cloning-compatible inserts via both the single tube method and column method, as described in the *Materials and Methods*. In both cases, ligation was performed using 1 μl of insert (~30 ng) and 25 ng of vector in a total reaction volume of 10 μl, and then 250 pg vector equivalent of the ligated sample was electroporated into competent *E. coli* cells.

Using column-purified inserts, ~1–5×10^6 transformants per μg vector DNA were obtained. The number of transformants was not influenced by the size of the insert and nearly 100% of the clones obtained were recombinants (Table 1). Furthermore, DNA sequencing of the recombinant inserts demonstrated that over 90% of the clones contained the correct nucleotide sequences. Expression analysis of selected clones revealed that nearly all of the recombinants expressed proteins of the expected sizes (Fig. S2). The non-recombinant clones obtained were undigested vector sequences. The transformation efficiency of the ligation mixture prepared with inserts via the 'single tube' strategy was ~5 times lower; however, over 95% of these transformants were recombinant. More than 80% contained the correct DNA sequences, and nearly 100% of clones expressed the desired proteins (Table 1 and data not shown). Thus, the 'single tube' strategy described in this paper is a simple, robust and rapid cloning method that requires small volumes of PCR amplified product and provides results comparable to the 'column method'.

To make the cloning strategy more rapid, such that the entire process from PCR to transformation could be completed in about 6–8 h, the timings of different treatment steps post-clean up (by either Exo-SAP treatment or column) were systematically evaluated using three PCR-amplified mycobacterial genes (*Rv1886c* [860 bp], *Rv1980c* [680 bp] and *Rv3763* [480 bp]; Fig. S3A). Eventually, the "time saver" cloning strategy was devised, in which the T4 DNA polymerase treatment time was reduced from 60 min to 5 min at 15°C and the ligation time was reduced from 2 h to 15 min at room temperature. Thus, after PCR clean up, the processing time up to transformation was reduced from 4 h to just 1.5 h. This reduction in treatment time did not affect the transformation efficiency, and 100% of the transformants obtained using the "time saver" cloning strategy were recombinants (Fig. S3B). The reproducibility of this new cloning strategy was further confirmed by cloning five different scFv genes into an arabinose promoter-based expression vector pVMAAscFvclo 0001. Analysis of the transformants obtained showed concordant results, as 100% of the clones screened were recombinants (Fig. S4). Thus, the

"time saver" cloning strategy is a highly efficient and rapid method that can be completed within few hours post-PCR (Fig. 3).

More than 300 mycobacterial genes were cloned following the 'single tube' strategy using PCR products amplified with the Expand High Fidelity PCR System (Expand HF Polymerase) or Expand Long Template PCR System (Expand LT Polymerase). These enzymes are a cocktail of Taq DNA polymerase (low fidelity and high processivity) and Pwo DNA polymerase (high fidelity, lower processivity and high 3′ exonuclease activity). Since the amount of Pwo DNA polymerase in the cocktail is expected to be low (although unknown), the presence of 3′-exonuclease activity has no effect on the insert preparation process. However, the presence of Taq DNA polymerase creates PCR-borne mutations necessitating the use of other high fidelity and high processivity thermostable DNA polymerases to obtain error-free PCR products. However, one has to be careful while using high-fidelity enzymes with single tube cloning strategy described here. We presume that in the absence of dNTPs during Exo- SAP treatment, and at higher temperature required for its heat inactivation, the thermostable high-fidelity polymerases might exhibit uncontrolled 3′- exonuclease activity, which might remove much more bases than desired (i.e. only 4 bases). It should be noted that Exo-SAP-based clean-up of inserts in the 'single tube method' takes about 80 min and the 'column method' can produce PCR-amplified product free of polymerase, excess primers and dNTPs within the same time frame. The use of columns for clean-up does not necessarily take longer and up to 24 single PCR clean-up columns can be easily handled in a single batch within 80–90 min.

The proposed cloning strategy is highly versatile with the possibility of being used efficiently in either 'single tube method' suitable for the inserts amplified using thermostable polymerases with low/no 3′ exonuclease activity or 'column method' involving prior clean-up of the inserts amplified using a high fidelity thermostable polymerases with high 3′ exonuclease activity (Fig. 3).

High-throughput application of the single tube strategy

The main objective of developing this novel cloning strategy was to clone and express the entire proteome of *Mtb*; therefore, the protocol was developed for high-throughput cloning and expression of genes. For high-throughput cloning, a batch of 24 genes was selected for ease of handling. As an example, 24 mycobacterial genes of various sizes ranging from 170 bp to 396 bp (Table 2) were amplified from *Mtb* genomic DNA using gene-specific primers and Expand HF Polymerase in the presence of 1%

Table 1. Cloning efficiency of four genes using column method and single tube method.

Gene	Size in (bp)	Transformants/μg DNA[a]	Efficiency obtained using column method		Single tube method
			Recombinants/Total amplified (%)	Number of protein Expressing clones	Recombinants/Total amplified (%)
Rv1827	496	~3×10^6	30/30 (100)	8/8	29/32 (91)
Rv3029c	1001	~1×10^6	31/32 (97)	3/4	32/32 (100)
Rv1077	1402	~5×10^6	28/29 (96.5)	8/8	29/30 (97)
Rv1908c	2230	~1×10^6	30/31 (97)	8/8	28/29 (96.5)

[a]The competent cells [BL21(DE3)RIL] used had efficiency of 1×10^8 transformants/μg supercoiled pGEM3Z DNA.

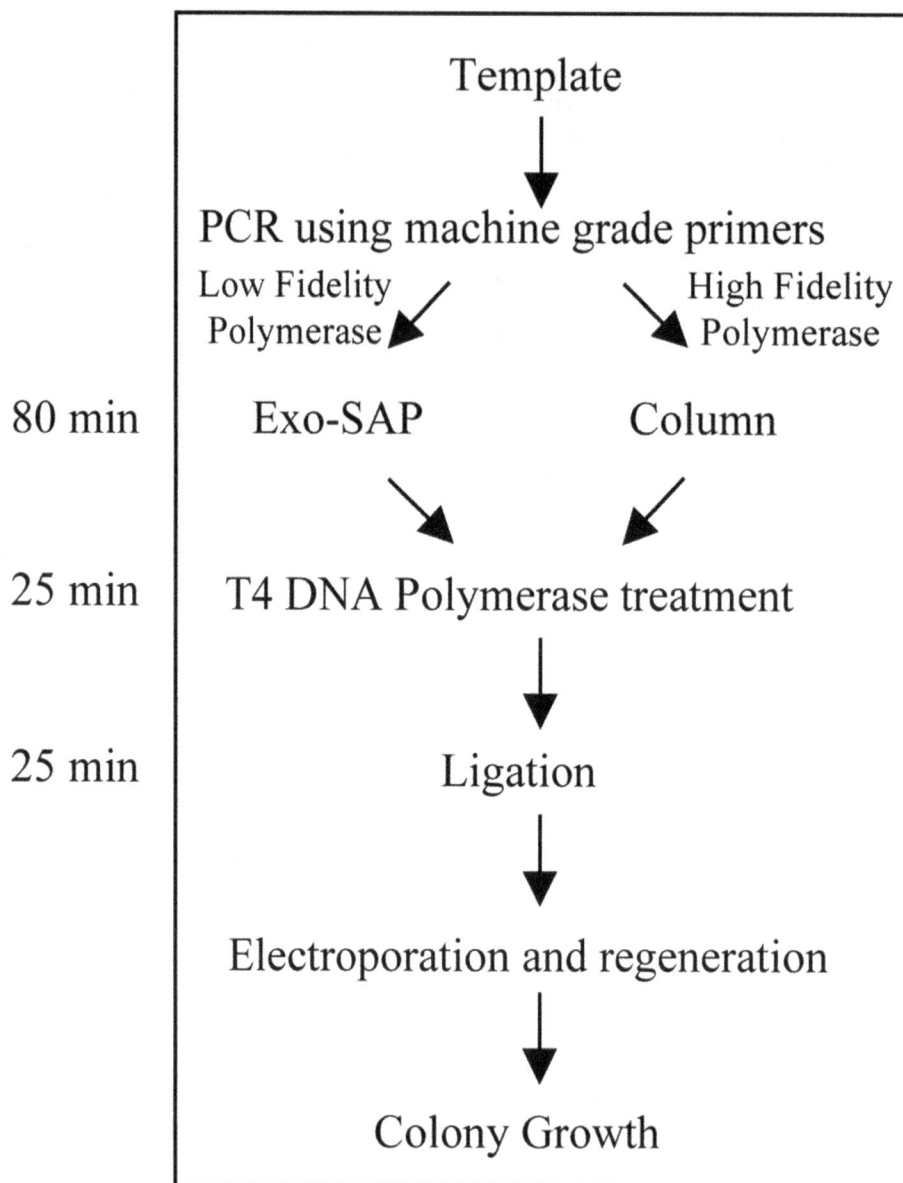

Figure 3. Flowchart and timing of the experimental steps for "Time Saver" cloning strategy.

DMSO. The PCR products were processed using Exo-SAP clean-up and the single tube strategy to obtain inserts with four base-long overhangs. For each target gene, colony PCR and sequencing analysis indicated that more than 95% of the colonies were recombinants (Fig. S6). High-throughput expression analysis of selected clones revealed that most clones had a good level of expression (Table 2). Following the same strategy, 12 batches of 24 genes each (ranging from 200 bp to 1.5 kbp) were cloned with similar efficiencies (Table 3; data shown for selected clones).

Application for genome-scale library construction

The cloning strategy described in this paper is not only suitable for routine and high-throughput cloning, but was also employed to efficiently construct highly complex whole genome fragment libraries of *Mtb* [17]. For high efficiency cloning, the quality of the insert, particularly its purity and quantity, is highly critical for efficient ligation. Therefore, for the library scale protocol, ~10 µg

of amplified product was purified by the column method followed by treatment of ~5 µg of the purified insert with T4 DNA polymerase in the presence of dTTP to create overhangs. The T4 DNA polymerase-treated insert was further column purified. Prior to ligation, the concentration and purity of the insert was estimated by gel electrophoresis, and then the insert was ligated to the BsaI-digested vector in 3-fold molar excess. The insert preparation has been described in Fig. S8 with necessary details about the procedure. Two large gene fragment libraries MTBLIB25C01 and MTBLIB27C01 containing 100–300 bp and 300–800 bp fragments, respectively, of the *Mtb* genome were constructed by cloning randomly generated fragments between the PelB signal sequence (*PelBss*) and full-length trypsin-resistant *gIIIP* in a phagemid based phage display vector (pVCEPI23764; Fig. S5B). The cloning efficiency ranged from 5×10^7 to 1×10^8 transformants per µg vector DNA using cells with a transformation efficiency of $\sim 5 \times 10^9$ transformants/µg supercoiled DNA. The

Table 2. Cloning efficiency and expression data of 24 genes (174 to 396 bp) using single tube strategy.

S.No	Sample	Size in (bp)	Mol.wt. (kDa)	Recombinants/Total amplified (%)[a]	ExpressionResults[b]
1.	Rv0666	174	9.8	8/8 (100)	N
2.	Rv3250c	183	10.7	6/6 (100)	Y
3.	Rv2803c	216	11.5	8/8 (100)	Y
4.	Rv1211	228	11.8	7/7 (100)	Y
5.	Rv1134	237	12.2	8/8 (100)	Y
6.	Rv1298	243	12.7	8/8 (100)	Y
7.	Rv1335	282	13.5	8/8 (100)	Y
8.	Rv1738	285	14.6	8/8 (100)	Y
9.	Rv0287	294	13.8	7/7 (100)	Y
10.	Rv2117	294	14.8	8/8 (100)	Y
11.	Rv3905c	312	14.4	8/8 (100)	Y
12.	Rv1579c	315	15.2	5/5 (100)	Y
13.	Rv3065	324	15.0	8/8 (100)	N
14.	Rv2348	327	15.3	8/8 (100)	N
15.	Rv2919c	339	16.2	7/7 (100)	Y
16.	Rv2007	345	16.0	7/7 (100)	Y
17.	Rv0253	357	16.4	5/5 (100)	Y
18.	Rv3748	360	16.6	7/8 (100)	N
19.	Rv2446c	372	17.3	8/8 (100)	N
20.	Rv3923c	378	17.9	7/8 (87.5)	Y
21.	Rv3289c	378	17.1	8/8 (100)	N
22.	Rv3675	378	17.7	7/7 (100)	Y
23.	Rv3346c	393	18.0	7/7 (100)	Y
24.	Rv1224	396	18.0	6/6 (100)	Y

[a]as determined by colony PCR.
[b]Y-Yes; N-No; all clones contained correct DNA sequence.

library of 100–300 bp fragments contained ~5×10⁷ independent clones, while the library of 300–800 bp fragments contained ~1×10⁸ independent clones. Sequence analysis of randomly-selected transformants from both libraries revealed that nearly 97% of the clones were recombinants with nucleotide sequences aligning to the *Mtb* genome [17].

Table 3. Cloning efficiency and expression data of 12 genes (720–1320 bp) cloned using single tube strategy.

S.No	Sample	Size in (bp)	Mol.wt. (kDa)	Recombinants/Total amplified (%)[a]	ExpressionResults[b]
1.	Rv2018	720	26.0	7/8 (87.5)	Y
2.	Rv2603c	756	26.8	6/6 (100)	Y
3.	Rv3814c	786	27.2	7/8 (87.5)	N
4.	Rv0877	789	27.4	7/7 (100)	Y
5.	Rv0765c	828	29.0	8/8 (100)	Y
6.	Rv2905	945	33.4	6/7 (85.7)	Y
7.	Rv1207	957	33.0	7/8 (87.5)	Y
8.	Rv2560	978	33.1	7/8 (87.5)	N
9.	Rv2837c	1011	35.4	8/7 (100)	Y
10.	Rv1415	1278	46.1	7/8 (87.5)	Y
11.	Rv3703c	1281	47.1	8/8 (100)	Y
12.	Rv2836c	1320	48.4	6/8 (75)	Y

[a]as determined by colony PCR.
[b]Y-Yes; N-No; all clones contained correct DNA sequence.

Discussion

The proposed cloning strategy for PCR-amplified DNA employs unique combination of the type IIs restriction endonuclease BsaI to create a linearized vector with four base-long 5'-overhangs, and T4 DNA polymerase treatment of the insert in presence of a single dNTP to create vector-compatible four base-long overhangs. Thus, the preparation of inserts with precise overhangs is restriction endonuclease-free.

The vector design, insert preparation and overall cloning strategy described here meets most of the essential criteria for an efficient and high-throughput cloning strategy to allow rapid cloning of PCR-amplified DNA and expression of the encoded proteins. The system includes the following advantages: (i) the reagents for vector and insert preparation are commonly used enzymes available from multiple sources, (ii) the four base-long overhangs are designed to be non-palindromic, preventing self-ligation of the empty vectors, (iii) crude primers without any modifications and carrying an only seven base-long extra sequence are used for amplification, avoiding the synthesis of long primers and introduction of numerous extra residues with large hydrophobic side chains into the final product, (iv) the insert preparation requires simple enzymatic treatment that allows directional cloning in the vector without concatemerization of the inserts, and (v) the vector can be easily customized. Furthermore, using the proposed cloning strategy, one can process an unlimited number of samples that can be cloned in a single day (Fig. S4) with a high cloning efficiency of nearly 100% error-free recombinants. The main challenge for the proposed strategy was its compatibility with PCR products amplified using high fidelity enzymes with high 3'-exonuclease activity. However, this issue was resolved by the use of column purification to remove the DNA polymerase, without significantly extending the processing time. We have successfully employed this single tube strategy and the column method to clone nearly 400 genes of various sizes (200 bp to 1.5 kbp) encoding proteins of diagnostic importance.

In recent years, several rapid cloning strategies based on the use of type IIs restriction enzymes have been described [11,12]. However, these methods require digestion of the PCR product with restriction endonucleases, thus limiting their use to the inserts which are devoid of these sites [11]. Most methods require either purification of the amplified product or addition of chemicals to inhibit polymerase activity [12]. Additionally, the T4 DNA polymerase treatment has been used to generate compatible 3' overhangs in the vector and insert in different ways; however, an overlapping sequence of >15 bp is needed, thus requiring long primers for amplification of the insert [4]. In LIC method, a specially designed vector is used where, in recombinants, the insert is flanked by extra nucleotides leading to extra amino acid residues. In SLIC strategy, although the vector compatible insert is prepared by PCR using long overhangs, the duration of treatment with T4 DNA polymerase might have to be carefully controlled [4]. This deficiency seems to have been resolved in SLiCE, but the requirement for long extra sequences to be added to the primers is a matter of concern [5]. In the proposed method for insert preparation, treatment with T4 DNA polymerase is independent of time, as 5–60 min treatment produces similar efficiencies. This has been possible due to the presence of one of the predefined dNTP in the insert preparation reaction. In the example shown in this paper, dTTP was used as single nucleotide because the overhangs in the vector backbone contained only A, G and C bases and therefore the compatible ends in the insert had T, G and C bases. Encounter of T base by exonuclease activity of the T4 DNA polymerase would terminate the exonuclease activity in

the presence of dTTP. The presence of one or two of the four dNTPs to terminate the exonuclease would depend upon the overhangs in the vector and thus the sequence of overhangs in the insert. Thus, very clean and robust protocols described here to prepare vector and inserts each with precise four-base overhangs have resulted in the novel, simple and efficient cloning strategy, which is suitable for both high-throughput cloning of PCR-amplified DNA and also for high efficiency library scale applications.

The strategy of insert preparation by T4 DNA polymerase in the presence of single dNTPs can also be used to produce modules with compatible ends to ligate multiple fragments thus expanding the scope of previously described strategy which is based on the use of type IIs restriction endonucleases, which limits the use only to those modules that are devoid of those restriction sites [18,19]. However, this process will require, either use of primers carrying 5' phosphate group for amplification of DNA or phosphorylation of the amplified products before T4 DNA polymerase treatment.

Using our proposed strategy, it is theoretically possible to produce 256 different four base-long overhangs comprising three of the four bases, both in the vector and the insert. However, the sequences of the overhangs can be selected to avoid palindromic sequences that will not produce compatible ends, even if a few bases of the overhang in the vector become excised. In our proposed strategy, the sequences of the overhangs were selected to ensure this aspect in addition to incorporating sequences that encode for amino acids lacking side chains that provide minimum steric hindrance.

Thus, in summary, the proposed strategy for cloning PCR-amplified products is superior, less complex and highly versatile compared to previously described cloning strategies [11,12], as existing strategies require either the use of restriction enzymes to prepare the insert or long primers for amplification and mandatory purification of the amplified inserts, and have not been demonstrated for constructing large libraries. The versatility of the newly described strategy is also evident from the following brief description of its possible applications and advantages. The basic vector pVLExp carrying the ColE1 origin of replication has been systematically modified to include additional features, as follows:

(i)   The pVMExp series of vectors has been created by deleting the *rop* gene, which has increased the copy number to facilitate preparation of the plasmid. This change did not affect the basal expression level of proteins or the level of expression after induction. Fig. S7 shows the systematic changes that have been carried out to convert the pET11 backbone into smaller backbone of pVMExp vectors without loosing any function.

(ii)   The tags have been modified to contain a TEV protease site between the H10 tag and encoded protein. The H10 tag was also replaced with H10-MBP (maltose binding protein tag) and TEV protease site to improve the solubility of the expressed proteins. Protocols have been developed to remove the H10 and H10-MPB tags from the purified proteins to obtain tag-free proteins (pVL/pVMExp4337/pVMMBPExp4437).

(iii)   The pVMExp vector backbone has been further altered to replace the pBR322-derived ColE1 replicon with another origin of replication, such as CloDF13, RSF1030 or P15A, along with different antibiotic markers such as a chloramphenicol resistance gene or kanamycin resistance gene. The T7 promoter has been replaced by other promoters, such as the inducible *ara*BAD promoter (PBAD), tetracycline promoter (Ptet) or Lac promoter (Plac). Many of these new vectors have been used extensively for cloning *Mtb* toxin and anti-toxin genes and to carry out functional studies with the co-transformants by differential expression using arabinose and tetracycline [20]. The vectors mentioned above have also been used to clone and express several *Chandipura* virus proteins [21].

(iv)   The above-mentioned vectors are being modified to contain a stuffer encoding the *SacB* gene as a positive selection marker, with the aim of obtaining 100% recombinants.

Thus, the concept of creating 4-base overhangs using T4 DNA polymerase treatment in the presence of one or two specific dNTPs to create non-compatible, non-palindromic ends and enable efficient ligation to a BsaI-digested vector that carries compatible overhangs for directional cloning is already being practiced as a robust, simple, cost effective cloning strategy which has a high efficiency and is suitable for high-throughput applications.

Supporting Information

Figure S1 PCR amplification of four mycobacterial genes. Amplification of genes (1) Rv1827 (2) Rv3029c (3) Rv1077 (4) Rv1908c using (a) HPLC grade primers and (b) Molecular Biology grade primers.

Figure S2 Total cell expression of four mycobacterial genes by auto-induction method. Total cell culture of different clones of each gene Rv1827 (lane 1–8), Rv3029c (lane 9–16), Rv1077 (lane 17–20) and Rv1908c (lane 21–28) were analyzed on 8–20% gradient gel by SDS-PAGE. The protein bands were visualized with Coomassie blue R-250 staining. Arrowhead indicates the band for different expressed proteins.

Figure S3 PCR amplification of three mycobacterial genes. (A) Amplification of three genes. (B) Colony PCR results of three genes amplified using T7P (5′ TAATACGACTCACTA-TAGGGGA 3′) and T7Tn (5′ CAGCCAACTCAGCTTCCT-TTC 3′) primers. Eight clones were screened from each of the 3 mycobacterial genes cloned by colony PCR and analyzed on 1.2% agarose gel.

Figure S4 Colony PCR results of five scFv genes amplified using AraP51 (5′ GCATTTTTATCCATAAGAT-TAGCG 3′) and T7Tn primers. Eight clones were screened from each of the five scFv genes cloned by colony PCR and analyzed on 1.2% agarose gel.

Figure S5 The figure above summarizes features of the various vectors constructed for employing restriction enzyme-free cloning strategy. Figure A1–A5, pVMExp T7-promoter based vectors for protein expression; B, pVC vector for constructing gene fragment libraries for display on filamentous phages; C, pRAK/pCAK Arabinose promoter (*ara*BAD) based expression vectors.

Figure S6 Colony PCR results of 24 genes amplified using T7P and T7TN primers. Eight clones were screened from each of the 24 mycobacterial genes cloned (1/24 to 24/24) by colony PCR and then analyzed on 1.2% agarose gel.

Figure S7 Schematic representation of the modifications carried out in pET11a for construction of pVMExp vectors. The numbers in the brackets represent the size of the vector backbone (in bp) excluding the sequence located between T7 promoter-lac operator (T7lac) and T7 terminator (T7TN).

Figure S8 Insert preparation for library- scale cloning. Genomic DNA of *M. tuberculosis* H37Rv was fragmented by sonication to obtain fragments in the range of 100–1200 bp (A). The fragments of sizes differing by increments of 100 bp (eg. 100–200 bp, 200–300 bp) were purified from 1.2% Sea Plaque GTG agarose (Lonza, Rockland, ME, USA). Two mixtures of fragments were prepared namely, '100–300 bp' mix, which was prepared by mixing 100–200 and 200–300 bp fragments in molar ratio- 1:1.5, and '300–800 bp' mix, which was prepared by mixing 300–400, 400–500, 500–600, and 700–800 bp fragments in molar ratio-1:1.5:2:2.5. The larger fragments were added in higher molar ratios to compensate for their ligation efficiencies. 10 µg of each fragment mix was end-repaired and phosphorylated (B) using Quick Blunting kit (NEB) followed by ligation of two adaptors (C) with blunt end on one side but 4-base 5′ overhang on the other side to achieve directionality during adapter ligation. 5′ (D1) and 3′ (E1) adapter duplex were 34 and 33 bp long, respectively, encoding for sequences that served as spacer. The sense strand of D1 carried 5′ biotin. Adapter ligation was set up using 10 µg mixture of fragment with 30 mole excess each of adapter duplex D1 and E1. Following ligation, the unligated adapters were removed by Qiaquick PCR purification kit (Qiagen) followed agarose gel electrophoresis. The adapter ligated DNA was treated with BstI DNA polymerase (D), followed by isolation of single stranded DNA (ssDNA) carrying one strand of D1 and E1 adapters on either end of a fragment using streptavidin coated M280 beads magnetic beads (E). These steps were similar to the process described for pyrosequencing template preparation [22]. In the process here, the ssDNA was then subjected to PCR amplification using primers, which anneal to D1 and E1 sequences (F), and amplified double stranded DNA (dsDNA) was obtained using 1:40 dilution of single stranded DNA for 20 cycles in 50 ul reaction. After clean up ~5 µg dsDNA was treated with T4 DNA polymerase in the presence of dTTP, which resulted in 4- base long 5′ overhangs, CGGC and CCTC at the 5′ and 3′ ends of the fragments (H). the part sequence (7 bases) of the adapter is shown in H and I as the remaining portion of the adapter can vary depending upon the requirement. These two ends are compatible with BsaI-digested vector pVCEPI23764. The library was constructed by ligating 2 µg BsaI-digested vector with 3 molar excess of two types of fragment mixtures in total volume of 100 µl, which were incubated for 16 hrs at 16°C (J). The entire ligation reaction was electroporated in *E.coli* TOP10F' cells (4 µl ligation mix/100 µl cells). This produced CO1 cells (K) as described earlier [17]. The library sizes have been described in the methods.

Table S1 Primers used for the amplification of myco-bacterial genes.

Acknowledgments

Corresponding authors are thankful to Tara Nath and Sunita Sharma for carrying out initial experiments.

Author Contributions

Conceived and designed the experiments: VKC AG. Performed the experiments: VKC AG NS VV SD CK PG. Analyzed the data: VKC AG. Contributed to the writing of the manuscript: VKC AG NS VV.

References

1. Cohen SN, Chang AC, Boyer HW, Helling RB (1973) Construction of biologically functional bacterial plasmids in vitro. Proc Natl Acad Sci U S A 70: 3240–3244.

2. Zhou MY, Gomez-Sanchez CE (2000) Universal TA cloning. Curr Issues Mol Biol 2: 1–7.

3. Aslanidis C, de Jong PJ (1990) Ligation-independent cloning of PCR products (LIC-PCR). Nucleic acids research 18: 6069–6074.

4. Jeong JY, Yim HS, Ryu JY, Lee HS, Lee JH, et al. (2012) One-step sequence- and ligation-independent cloning as a rapid and versatile cloning method for functional genomics studies. Appl Environ Microbiol 78: 5440–5443.

5. Zhang Y, Werling U, Edelmann W (2012) SLiCE: a novel bacterial cell extract-based DNA cloning method. Nucleic acids research 40: e55.

6. Hartley JL, Temple GF, Brasch MA (2000) DNA cloning using in vitro site-specific recombination. Genome Res 10: 1788–1795.

7. Walhout AJ, Temple GF, Brasch MA, Hartley JL, Lorson MA, et al. (2000) GATEWAY recombinational cloning: application to the cloning of large numbers of open reading frames or ORFeomes. Methods Enzymol 328: 575–592.

8. Colwill K, Wells CD, Elder K, Goudreault M, Hersi K, et al. (2006) Modification of the Creator recombination system for proteomics applications–improved expression by addition of splice sites. BMC Biotechnol 6: 13.

9. Geng L, Xin W, Huang DW, Feng G (2006) A universal cloning vector using vaccinia topoisomerase I. Mol Biotechnol 33: 23–28.

10. Berrow NS, Alderton D, Owens RJ (2009) The precise engineering of expression vectors using high-throughput In-Fusion PCR cloning. Methods Mol Biol 498: 75–90.

11. Engler C, Kandzia R, Marillonnet S (2008) A one pot, one step, precision cloning method with high throughput capability. PloS one 3: e3647.

12. Kotera I, Nagai T (2008) A high-throughput and single-tube recombination of crude PCR products using a DNA polymerase inhibitor and type IIS restriction enzyme. Journal of biotechnology 137: 1–7.

13. Chowdhury PS, Kushwaha A, Abrol S, Chaudhary VK (1994) An expression system for secretion and purification of a genetically engineered thermostable chimera of protein A and alkaline phosphatase. Protein Expr Purif 5: 89–95.

14. Studier FW, Rosenberg AH, Dunn JJ, Dubendorff JW (1990) Use of T7 RNA polymerase to direct expression of cloned genes. Methods Enzymol 185: 60–89.

15. Studier FW (2005) Protein production by auto-induction in high density shaking cultures. Protein Expr Purif 41: 207–234.

16. Cline J, Braman JC, Hogrefe HH (1996) PCR fidelity of pfu DNA polymerase and other thermostable DNA polymerases. Nucleic acids research 24: 3546–3551.

17. Gupta A, Shrivastava N, Grover P, Singh A, Mathur K, et al. (2013) A novel helper phage enabling construction of genome-scale ORF-enriched phage display libraries. PloS one 8: e75212.

18. Sarrion-Perdigones A, Falconi EE, Zandalinas SI, Juarez P, Fernandez-del-Carmen A, et al. (2011) GoldenBraid: an iterative cloning system for standardized assembly of reusable genetic modules. PloS one 6: e21622.

19. Weber E, Engler C, Gruetzner R, Werner S, Marillonnet S (2011) A modular cloning system for standardized assembly of multigene constructs. PloS one 6: e16765.

20. Gupta A (2009) Killing activity and rescue function of genome-wide toxin-antitoxin loci of Mycobacterium tuberculosis. FEMS microbiology letters 290: 45–53.

21. Kumar K, Rana J, Sreejith R, Gabrani R, Sharma SK, et al. (2012) Intraviral protein interactions of Chandipura virus. Archives of virology 157: 1949–1957.

22. Margulies M, Egholm M, Altman WE, Attiya S, Bader JS, et al. (2005) Genome sequencing in microfabricated high-density picolitre reactors. Nature 437: 376–380.

6

Serotypes and Genotypes of Invasive *Streptococcus pneumoniae* Before and After PCV10 Implementation in Southern Brazil

Juliana Caierão[1], Paulina Hawkins[2], Fernando Hayashi Sant'anna[1], Gabriela Rosa da Cunha[1], Pedro Alves d'Azevedo[1], Lesley McGee[3], Cícero Dias[1]*

1 Federal University of Health Science of Porto Alegre, Rio Grande do Sul, Brazil, 2 Emory University, Atlanta, Georgia, United States of America, 3 Centers for Disease Control and Prevention, Atlanta, Georgia, United States of America

Abstract

To reduce the burden of pneumococcal diseases, different formulations of pneumococcal conjugate vaccines (PCV) have been introduced in many countries. In Brazil, PCV10 has been available since 2010. We aimed to analyze the serotype and genetic composition of invasive pneumococci from Brazil in pre- and post- vaccination periods (2007–2012). Antibiotic susceptibility was determined and genotypes of macrolide and fluoroquinolone resistance were characterized. The genotypes of isolates of the most frequent serotypes were determined by multilocus sequence typing. The study included 325 isolates, which were primarily recovered from blood. The most common serotypes recovered were 14, 3, 4, 23F, 7F, 9V, 12F, 20, 19F, 8, 19A, and 5. Thirty-eight pneumococci (11.7%) were from children ≤5 years old. Considering the overall population, PCV10 and PCV13 serotype coverage was 50.1% and 64.9%, respectively. During the pre-vaccine period, isolates with serotypes belonging to the PVC10 represented 51.5% (100/194), whereas in the post vaccine they represented 48.0% (63/131). PCV13 serotypes represented 67.5% (131/194) and 59.2% (77/131) of total for pre- and post-vaccination periods, respectively. Seventy different sequence types [STs] were found, accounting for 9 clonal complexes [CCs] and 45 singletons. Eight STs (156, 180, 218, 8889, 53, 191, 770, and 4967) represented the majority (51.5%) of isolates. Fifty STs were associated with the pre-vaccination period (27 exclusive) and 43 (20 exclusive) with the post-vaccination period; 23 STs were identified in both periods. Some serotypes were particularly clonal (7F, 8, 12F, 20). Non-susceptibility to penicillin was associated with serotype 19A, CC320. Erythromycin resistance was heterogeneous when considering serotype and ST. A single serotype 23F (ST4967) isolate was resistant to levofloxacin. Continued surveillance is required to determine vaccine impact and to monitor changes in pneumococcal population biology post-PCV10 introduction in Brazil.

Editor: Herminia de Lencastre, Rockefeller University, United States of America

Funding: Funding from Coordenação de Aperfeiçoamento de Pessoal de Nível Superior (CAPES) CNPq: Edital MCT/CNPq14/2008, 473607/2008-5. The funders had no role in study design, data collection and analysis, decision to publish, or preparation of the manuscript.

Competing Interests: The authors have declared that no competing interests exist.

* Email: cicero@ufcspa.edu.br

Introduction

Infections associated with *Streptococcus pneumoniae* are a global public health problem. Invasive pneumococcal diseases (IPD), such as meningitis and bacteremia, are especially associated with high morbidity and mortality [1]. About 3,600 cases of meningitis, 14,000 cases of bacteremia, and between 200,000 and 300,000 cases of pneumonia associated with pneumococci occur annually in Latin America [2]. In Brazil, from 2000 to 2008, pneumococcal meningitis represented 11% of all bacterial meningitis, with an incidence of 9.5 cases/100,000 inhabitants in patients less than one year of age. Official data from Brazil demonstrate that in the same period 7,129,291 hospitalizations were associated with complications of community-acquired pneumonia where *S. pneumoniae* was the major etiological agent (www.datasus.gov. br). In addition, Andrade and co-workers estimated the incidence of IPD in children less than three years of age from Brazil (2007–2009) at 57.5/100,000 inhabitants [3].

To reduce the burden of pneumococcal disease, different formulations of pneumococcal conjugate vaccines have been introduced in many countries since 2000. In Brazil, a 10-valent formulation (PCV10) has been available for children less than 2 years of age since 2010 as part of the National Childhood Immunization Program, which is funded by Brazilian Federal Government. Although the benefits of PCV vaccination have been well documented in the countries in which it has been introduced [4,5,6,7,8,9,10,11,12,13,14], the success of the vaccine is influenced by the pneumococcal population, particularly serotype distribution in both colonization and invasive disease [15]. Indeed, while IPD associated with vaccine serotypes decreased significantly, some investigators detected a worrisome increase of IPD caused by serotypes not included in vaccine formulas [10,15,16].

Monitoring the epidemiology and understanding the dynamics of the pneumococcal population are essential to evaluating the efficacy of vaccines in each region, as well as indicating any need to modify vaccine formulations. In Brazil, a well-established program monitors the distribution of serotypes in invasive infections [17]; however, published studies additionally providing genotypes of *S. pneumoniae* are limited, especially for IPD. In the present study, we determined serotypes and genotypes of invasive pneumococci isolated from South Brazilian hospitals in periods pre- and post- vaccination (2007–2012) to identify possible vaccination- induced changes in the pneumococcal population.

Materials and Methods

Study setting and design

This is a retrospective study with *S. pneumoniae* recovered from patients with IPD attending three different hospitals (including two of the major tertiary hospital complexes of the city) in Porto Alegre, South Brazil, from 2007 to 2012. Porto Alegre has around 4 million inhabitants in its metropolitan area, and it is a major regional reference center for healthcare. Vaccination was introduced in Porto Alegre in August 2010, in a three-dose scheme (2, 4 and 6 months) plus a booster at 12 months. Catch-up campaign at time of introduction was as follows: 4 doses for aged 3–7 months; 3 doses for aged 8–9 months; 2 doses for aged 10–11 months; 1 dose for aged 12–23 months. In this context, pre- and post-vaccination periods in this study were considered years 2007–2010 and 2011–2012, respectively. Few data are available about PCV10 uptake in our region. Results of a recent study [14] demonstrated that from 2010 to 2011 PCV10 uptake in Porto Alegre was around 80%.

Microbiological identification, serotyping and antimicrobial susceptibility

Isolates were maintained at −80°C and their identification was confirmed by standard methods: Gram stain, colony morphology, optochin susceptibility and bile solubility [18]. Serotypes were determined by Quellung reaction, using pool, type and factor-specific antisera at the Centers for Disease Control and Prevention (CDC), Atlanta, USA. Broth microdilution was used to determine Minimal Inhibitory Concentration (MIC). The following antimicrobials were evaluated: penicillin, ceftriaxone, vancomycin, meropenem, erythromycin, levofloxacin, tetracycline, clindamycin and trimethoprim-sulfamethoxazole. Results were interpreted according to the Clinical and Laboratory Standards Institute [19]. The reference strain *S. pneumoniae* ATCC 49619 was used for quality control.

Genetic analysis of macrolide and quinolone resistance

Isolates presenting MIC≥0.5 µg/mL for erythromycin were submitted to a duplex PCR method for the detection of *erm*B and mef(A), according to Widdowson & Klugman (1998) [20]. To determine mutations present in pneumococci with levofloxacin MIC≥4 µg/mL, quinolone resistant determinant regions (QRDRs) of *gyr*A and *par*C were amplified and sequenced as described elsewhere [21].

Molecular typing

Multilocus sequence typing (MLST) was performed according to adapted method described by Enright & Spratt (1998) [22] and detailed at http://www.cdc.gov/ncidod/biotech/strep/alt-MLST-primers.htm for the most frequent and/or relevant serotypes. Allele sequences were edited and complementary sense and antisense fragments were aligned using CodonCode Aligner. Allele profile and Sequence Type (ST) were obtained from the database of the MLST web site (http://pubmlst.org/spneumoniae/). In cases of new allele or ST, data were submitted for the approval of the curator before being published. eBURST groups were defined using the most stringent (conservative) definition of eBURSTv3: all ST assigned to the same group must share identical alleles at 7/7 or 6 of the 7 loci with at least one other member of the group. For clonal complexes (CCs), a cut-off point of five identical loci to the predicted founder of its eBURST group was used.

Statistical and diversity analysis

Statistical and diversity analysis were performed using PAST software (version 2.17c). Simpson's index of diversity and Shannon index were calculated in order to estimate diversity. The significance of the difference of the diversity indices between two groups was assessed by bootstrapping, as described in the PAST manual. The odds ratio (OR) for each specific serotype, compared to the total number of isolates, was determined; 95% confidence intervals (CIs) were calculated. P values<0.05 were considered to be statistically significant.

Ethical considerations

This retrospective study was approved by the Research Ethics Committee of the Grupo Hospitalar Conceição (Project number 11–205), recognized by Comissão Nacional de Ética em Pesquisa (CONEPE) and by the Office for Human Research Protection

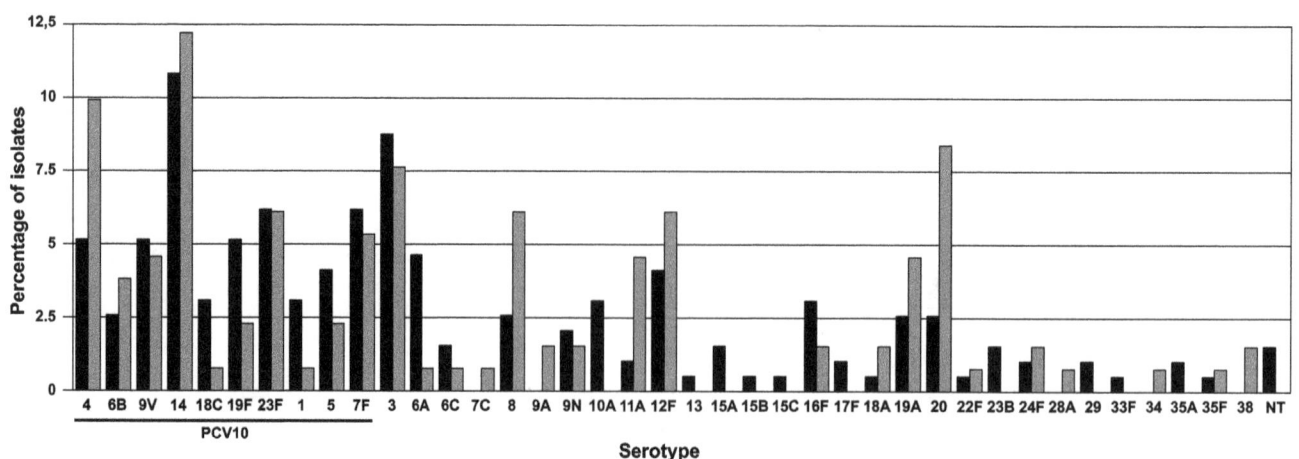

Figure 1. Distribution of serotypes in pre- and post-vaccination period among pneumococci recovered from IPD, during 2007–2012. Black bars represent pre-PCV10 and gray bars post-PCV10.

Table 1. Distribution of serotypes belonging to PCV10 among invasive pneumococci recovered from 2007 to 2012 (the distribution of all serotypes is demonstrated on the Table S1).

Serotype	N (%)	Pre*	Post**	OR#	95% CI	P-value
14	37 (11.4)	21	16	0.7992	0.3998–1.5976	0.5260
4	23 (7.1)	10	13	0.4534	0.1925–1.0682	0.0704
23F	20 (6.2)	12	8	0.9335	0.3706–2.3516	0.8840
7F	19 (5.8)	12	7	1.0760	0.4119–2.8106	0.8812
9V	16 (4.9)	10	6	1.0439	0.3698–2.9467	0.9354
19F	13 (4.0)	10	3	2.1404	0.5774–7.9334	0.2549
5	11 (3.4)	8	3	1.6944	0.4409–6.5113	0.4426
6B	10 (3.1)	5	5	0.6154	0.1745–2.1701	0.4502
18C	7 (2.0)	6	1	3.8351	0.4562–32.2393	0.2159
1	7 (2.0)	6	1	3.8351	0.2265–6.9555	0.2159
Total	**163**	**100**	**63**			

*Pre-vaccination period;
**post-vaccination period;
#OR: odds ratio.

Table 2. PCV10 coverage over the years of the study (2007–2012), considering epidemiologically relevant age groups.

year	Age (n\%)	PCV10	Non-PCV10	p-value
2007	All (40; 100%)	27	13	
	≤5 years old (3; 7.5%)	2	1	0.974
	≥6 years old (37; 92.5%)	25	12	
2008	All (10; 100%)	3	7	
	≤5 years old (1; 10%)	0	1	0.570
	≥6 years old (8; 80%)	2	6	
	Unknown age (1; 10%)	1	0	
2009	All (72; 100%)	32	40	
	≤5 years old (6; 9.72%)	3	3	0.645
	≥6 years old (62; 86.1%)	25	37	
	Unknown age* (4; 4.16%)	4	0	
2010	All (23; 100%)	10	13	
	≤5 years old (5; 26.1%)	3	2	0.374
	≥6 years old (16; 69.6%)	6	10	
	Unkown age* (2; 4.34%)	1	1	
2011	All (119; 100%)	54	65	
	≤5 years old (18; 15.0%)	12	6	0.043
	≥6 years old (100; 84.8%)	41	59	
	Unknown age (1; 0.8%)	1	0	
2012	All (61; 100%)	33	28	
	≤5 years old (5; 8.19%)	1	4	0.110
	≥6 years old (56; 91,8%)	32	24	
Total	All (325; 100%)	159	166	
	≤5 years old (38; 11.7%)	21	17	0.336
	≥6 years old (279; 85.8%)	131	148	
	Unknown age* (8; 2.5%)	7	1	

*included one isolate each year (2009 and 2010) recovered from a patient identified as "pediatric": serotype 14 for both year.

Table 3. Serotypes and STs observed among 231 pneumococci recovered from IPD (2007–2012).

Serotype (n)	ST dominant (n)	Other STs (n)
14 (31)	156 (22)	13 (5), 166 (1), 370 (1), 8881 (1), 8894 (1)
3 (24)	180 (17)	505 (3), 1176 (3), 72 (1)
12F (15)	218 (15)	
20 (15)	8889 (15)	
4 (14)	770 (9)	7026 (4), 8893 (1),
9V (13)	156 (9)	162 (4)
8 (13)	53 (11)	404 (1), 8885 (1)
19F (12)	271 (4), 177 (4)	733 (1), 763 (1), 878 (1), 8890 (1)
23F (12)	4967 (7)	338 (2), 37 (1), 5406 (1), 8895 (1)
7F (12)	191 (10)	218 (2)
19A (11)	2878 (4)	320 (2), 199 (1), 733 (1), 4967 (1), 8800 (1), 8884 (1)
6B (6)	748 (2)	553 (1), 724 (1), 8897 (1), 8896 (1)
5 (6)	289 (6)	
11A (5)	62 (4)	8882 (1)
9N (5)	66 (4)	8883 (1)
6A (4)		1939 (1), 8888 (1), 8879 (1), 5576 (1)
16F (4)	7438 (3)	733 (1)
1 (3)	304 (3)	
18C (3)	193 (2)	280 (1)
24F (3)	72 (3)	
10A (3)	742 (2)	741 (1)
38 (2)		393 (1), 310 (1)
18A (2)		8891 (1), 5063 (1)
35A (2)	2104 (2)	
6C (2)		172 (1), 3930 (1)
9A (2)		156 (1), 5406 (1)
15B (1)	766 (1)	
22F (1)	6403 (1)	
23B (1)	8886 (1)	
17F (1)	739 (1)	
28A (1)	494 (1)	
13 (1)	71 (1)	
7C (1)	737 (1)	

(OHRP/USDHHS). No clinical data of patients were used in this study (informed consent not applied). Patient records/information was anonymized and de-identified prior to analysis.

Results

A total of 325 non-duplicate pneumococci recovered from patients with IPD were included in the study. The most common isolation site was blood (n = 255, 78.5%), followed by cerebrospinal fluid (n = 54, 16.6%), pleural fluid (n = 12, 3.7%), ascites (n = 2, 0.6%), peritoneal fluid (n = 1, 0.3%) and joint fluid (n = 1, 0.3%). Patient age ranged from 0 to 94 years old, with an average of 45.2 years: 29 (8.9%), 38 (11.7%) and 72 (22.2%) were ≤2, ≤5 (0 to 60 months) and ≥65 years old, respectively. No age records were found for 6 patients and for two other no specific age was reported, although they were identified as "pediatric". Among all isolates, 194 (59.7%) were recovered in the pre-vaccination period (2007–

2010) and 131 (40.3%) were recovered post-vaccine introduction (2011–2012).

Forty serotypes were identified in total. *S. pneumoniae* infections were mostly associated with the following serotypes: 14 (11.4%), 3 (8.3%), 4 (7.1%), 23F (6.1%), 7F (5.8%), 9V (4.9%), 12F (4.9%), 20 (4.9%), 19F (4%), 8 (4%), 19A (3.4%) and 5 (3.4%). Three pneumococci were non-typable by Quellung reaction. Serotypes 10A, 13, 15A, 15B/C, 17F, 23B, 29, 33F and 35A were only found in the pre-vaccine period, and 7C, 9A, 28A and 38 were only found post-vaccine.

Figure 1 shows the distribution of serotypes before and after PCV10 introduction. It was observed some increase or decrease in the frequency of specific serotypes (Table S1). However, none was statistically significant, not even in the vaccinated population (children less than 2 years old).

Overall, PCV10 coverage was 50.1% and there was no statistical significance comparing PCV10 coverage before and

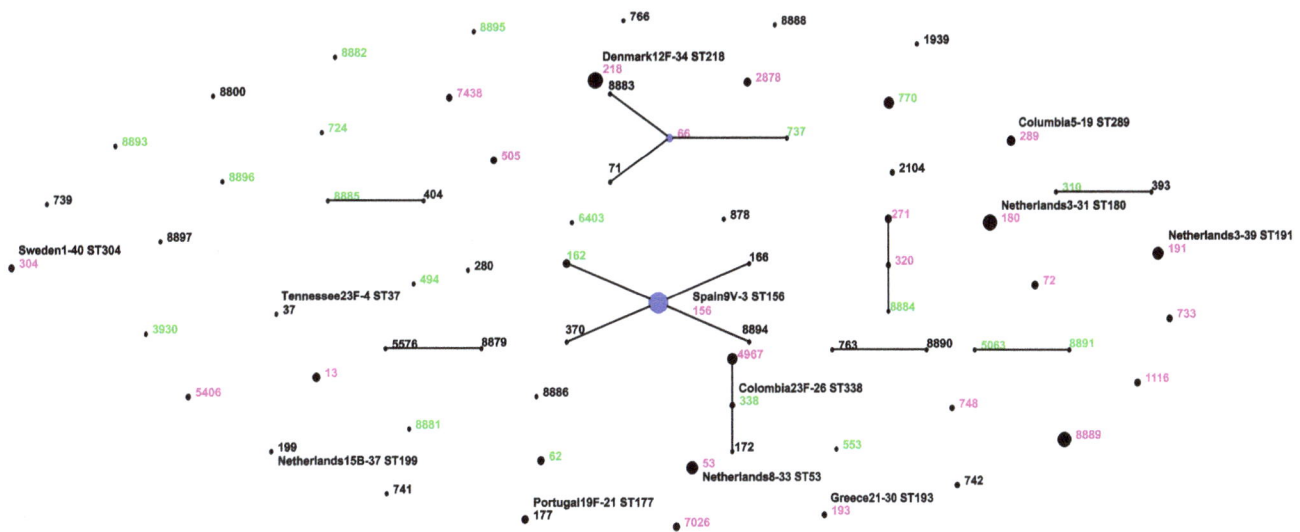

Figure 2. Population structure of invasive *S. pneumoniae* collected from 2007 to 2012. Black circles: pre-vaccine isolates; Green circles: post-vaccine isolates; pink circles: isolates present in both pre and post vaccine years; Blue circles: the ancestor of the CC.

after vaccination: 51.5% and 48.1%, respectively (p = 0.5411). Table 1 presents the occurrence of each PCV10 serotype before and after vaccination and Table 2 shows the PCV10 coverage over the years of the study, considering epidemiologically relevant age groups. PCV13 coverage among the general population was 64.9%.

To better determine if serotype distribution changed after PCV10 introduction, we compared pre (2009) and post-PCV10 (2012) periods. In 2009, among 72 isolates, 33 belonged to PCV10 serotypes (45.8%), while among 61 pneumococci recovered in 2012, 30 were from serotypes included in PCV10 (50.8%), which was not a statistically significant difference (p = 0.566).

Stratifying our population by age, PCV10 and PCV13 coverage was 55.3% and 73.7% among children ≤5 years old, respectively; and 49.4% and 63.8% among those >5 years of age, respectively. The occurrence of PCV10 serotypes did not vary statistically in pre- and post-vaccination era, neither among children under 5 years old (52.4% vs 58%), nor under 2 years old (58.8% vs 58.3%).

Two-hundred thirty-one pneumococci, belonging to the 33 most relevant and/or frequent serotypes, were characterized by MLST. Our population was composed of 70 different STs, with 9 CCs (composed of 25 STs) and 45 singletons (Table 3; Figure 2). Eight STs (156, 180, 218, 8889, 53, 191, 770 and 4967) represented 51.5% (119/231) of all isolates. Most serotypes were associated with more than one ST and a few STs were associated with more than one serotype (Table 3): ST72 (3 and 24F), ST156 (14, 9V and 9A), ST218 (12F and 7F), ST733 (16F, 19A and 19F), ST4967 (19A and 23F) and ST5406 (23F and 9A). Fifty STs were found in the pre-vaccination period (with 27 exclusively related to this period), while 43 STs (20 exclusive) were recovered from IPD after PCV10 implementation; 23 STs were identified in both periods (Table 4). Considering Simpson's index of diversity (94.8% and 94.9%, for pre and post, respectively) and Shannon index (3.47 and 3.33 before and after, respectively), genetic diversity was similar in both periods (p = 0.996 and p = 0.348 for Simpson's and Shannon indexes, respectively), which is also true for PCV10 (32 STs identified) versus non-PCV10 (43 STs characterized) serotypes (Simpson's index: 89.44% vs 91.4%; p = 0.189).

Among the 70 STs identified, 17 (24.3%) were new in the MLST website database, 10 of which were presented in our population as singletons and 7 grouped in CCs (Figure 2 and Table 4). Eight of the new STs were found prior to vaccination; 8 were isolated after PCV10 use and 1 was associated with both periods (ST8889). New alleles were found in 4 of the 17 new STs. Comparing our population to the entire MLST database, we observed 7 previously unreported combinations of STs/serotypes, suggesting possible capsular switching; 5 (83.3%) were associated with non-PCV10 serotypes (Table 5).

Three hundred and twenty two *S. pneumoniae* were viable for susceptibility testing. Overall, 42.8% (n = 138) had MICs greater than 0.06 μg/mL for penicillin (meningitis breakpoints) with a great ST variability (42 STs). Isolates with ceftriaxone MICs greater than 0.5 μg/mL (18.6%; n = 60) had a similar variability, with 15 STs identified. Considering non-meningitis breakpoints, 2 and 15 pneumococci were non-susceptible to penicillin (MIC = 4 μg/mL) and ceftriaxone (MIC = 2 μg/mL), respectively. Only one isolate from an adult (34 years old) was resistant to levofloxacin (MIC = 16 μ/mL), which was serotyped as 23F and sequence typed as ST4967. Non-susceptibility to erythromycin was observed in 7.9% (18/228) (among these, 11.1% (2/18) were intermediate and 88.9% (16/18) presented full resistance). The characteristics of these isolates, including STs and serotypes, are presented in Table 6. Despite the association of various distinct STs with the resistant isolates, three major CCs were present among these pneumococci (CC320, CC156 and CC338), grouping 74.3% of them. Serotypes 14 (n = 14) and 19A (n = 7) represented 58.3% of this specific population. Despite its association with resistant pneumococci, serotype 19A (n = 11) was also associated with susceptible isolates and represented a considerable genetic diversity (including two new STs): ST 2878 (n = 4), ST 320 (n = 2), ST 4967 (n = 1), ST 199 (n = 1), ST 733 (n = 1), ST 8800 (n = 1; new) and ST 8884 (n = 1; new).

One isolate was resistant to levofloxacin (1/322, 0.3%), and sequencing of *gyrA* and *parC* identified point mutations, Ser^{81}-Phe in *gyrA* and Ser^{79}-Phe and Lys^{137}-Asn in *parC*. Resistance to macrolides was determined for 311 isolates and 8.0% (25/311) were non-susceptible (2 intermediate and 23 fully resistant), which was associated with *mef*(A) (11/25; 44.0%), *ermB* (9/25; 36.0%)

Table 4. New STs and their relationship with serotypes and STs in the database at MLST website (MLST database – accessed 19th October, 2013).

Period*	ST	Serotype	SLV ST**	Serotype	Country
pre	8800	19A	43	14, 19F	Many
			424	19F	Many
			773	14	Colombia
			1113	7F	Uruguay
			4860	14	Brazil
	8879	6A	5576	6A	Brazil
	8883	9N	66	9N and others	Many, including Brazil
	8886	23B	387	23F	Brazil
			2777	6C	Brazil
			5962	23B	Brazil
			4431	23B	Brazil
	8888	6A	759	6A	Brazil
			757	6A	Brazil
	8890	19F	763	19F	Brazil, South Africa
	8894	14	156	14 and others	Many including Brazil
			162	14 and others	Many including Brazil
	8897#	6B			
pre/post	8889	20	235	20 and 7F	Spain, Austria, Poland
			8816	N/A	UK
			1345	20	UK
post	8881##	14			
	8882	11A	3184		
	8884	19A	320	19A and others	Many including Latin America
	8885	8	404	8	Europe and Brazil
			1629	8	UK
			7847	8	Scotland
	8891	18A	5063	18C	Brazil
	8893	4	5670	4	Brazil
	8895	23F	242	23F	Many including Brazil
	8896	6B	4977	6B	Brazil

*Pre: pre-vaccination; Post: post-vaccination;
**More frequent ST and/or the ones isolated in Brazil;
#8897 is DLV of STs 2781 (serotype 6B) and 8248 (serotype 6A/B), isolated previously in Brazil;
##8881 is DLV of STs 9 (serotype 14) and 15 (serotype 14). The first ST was previously isolated in many countries, as the second one, which was also related to Brazilian isolates.

and both *ermB* and *mefA* (5/25; 20.0%). All *ermB* positive isolates were also resistant to clindamycin (MIC≥1 µg/mL), regardless of *mefA* occurrence.

Discussion

The present study generates important data about the molecular epidemiology of circulating pneumococci in Brazil. Most studies previously published from Brazil have not focused on evaluating the molecular epidemiology of the pneumococcal population or, if they have, it has related only to pneumococci from carriage [23]. To our knowledge, only two studies present characteristics of isolates recovered from patients with IPD in Brazil. Zemlicková and co-workers (2005) [24] evaluated invasive pneumococci from five different Latin American countries, including only 41 isolates from Brazil. The second study characterized only meningitis isolates and molecular analysis was focused on a specific population of 107 *S. pneumoniae* resistant to penicillin [25]. In our study, we included pneumococcal isolates from all invasive specimens over a six-year period regardless of age or antibiotic susceptibility to enable us to generate a more comprehensive scenario of molecular characteristics of pneumococci associated with IPD in our region and country.

This study was not designed to evaluate the efficacy of PCV10. However, we understand our data are relevant to provide an initial assessment of the impact of PCV10 use in Brazil. After the introduction of PCV10, a significant decline in hospitalizations for pneumonia was noted in many Brazilian cities, but not in Porto

Table 5. Isolates representing possible capsular switching associated with pre-existing STs, considering MLST database (accessed 19th October, 2013).

Our Study						Previous studies		
ST	No. isolates	Period	Age (yrs)	Serotype	Site	Country	Serotype	Site
4967	1	Pre	37	19A	Blood	Brazil	23F	ND*
5406	1	Pre	<1	23F	Blood	Brazil	29	blood
	1	Post	80	9A	Blood			
166	1	Pre	<1	14	Blood	Many**	6B, 9, 9A, 9V, 11, 11A, 19A	Invasive and non-invasive
2104	2	Pre	49	35A	pleural	Spain	19F	blood
71	1	Pre	66	13	Blood	United Kingdom	22	CSF

*ND: Not defined;
**Taiwan, Korea, South Korea, Malaysia, France, Germany.

Alegre. This difference may in part be explained by the lower vaccination uptake (~80%) compared to the other cities (>90%), as well as the delay in start date for vaccine use in this city [14]. Two years after the introduction of PCV10 our data suggest that this region seems to still have a transitional pneumococcal population and accumulating more recent data post 2012 would permit a better understanding on vaccine impact. The absence of PCV10 influence on the serotype distribution during the study period may be justified by the characteristics of our population, which was mainly composed of adults. Indeed, as we only evaluated two years post-vaccination (and considering the lower vaccine uptake observed in Porto Alegre), there may not have been sufficient time to observe the herd protection effect in this study population. Also, the very low number of isolates from children ≤ 5 years of age (38 patients) and the generally small number of isolates per year may have compromised statistical analysis. A well-designed case-control study [26], performed only with children and focusing in evaluating effectiveness of vaccine in Brazil recently demonstrated that PCV10 prevents invasive disease caused by vaccine serotypes. However, evaluation of the serotype distribution showed that serotypes 14, 6B, 23F, 18C and 19F remain among the most frequent serotypes causing invasive disease two years after vaccine introduction.

Although a few STs (ST53, ST156, ST180, ST191, ST218, ST770, ST4967 and ST8889) represented almost half of the isolates, our population presented a high degree of genetic diversity in both pre- and post-vaccination periods (Simpson's index around 95%), which is similar to data from some reports [27,28] but higher than others [29]. Several serotypes were especially clonal and primarily associated with one ST, such as 7F (ST191) and the non-PCV10 serotypes: 3 (ST180), 12F (ST218), 8 (ST53) and 20 (ST8889).

Serotype 3 has been associated with increased mortality in different regions [30,31,32]. Inverarity et al. (2011b) [33] demonstrated that mortality was strongly associated with ST180. The high frequency of isolation of this serotype (and ST) and its relation with mortality necessitates continued surveillance to monitor for increases in this serotype post-PCV10 as protection against serotype 3 may be an important reason to consider the use of PCV13 in our region.

Serotype 12F has been shown to cause outbreaks in human populations with identifiable risk factors [34,35,36,37]. This serotype has a high case:carriage ratio (CCR), i.e., it is a hyper invasive serotype, and is rarely found in the nasopharynx [23,35,36,37]. In our population, this serotype was among the most frequent serotypes, presented a highly clonal occurrence and was exclusively distributed among adults.

All but two pneumococci from serotype 8, which also presents high CCR, were characterized as ST53 (PMEN clone Netherlands[8]-33), a well-recognized virulent clone [38,39,40] The remaining isolates were identified as ST404 and as a new one, ST8885, a SLV of ST404, which has already been isolated in different European countries and previously in Brazil [41] (Table 4).

Although somewhat controversial, serotype 7F also appears to be associated with high case-fatality [7,9]. Some authors have observed serotype 7F as one of the main serotypes associated with replacement following PCV7 introduction, through clonal expansion [28,29]. Pichon and co-workers (2013) [28] demonstrated ST191 (serotype 7F) as the most prevalent clone causing meningitis 3 years after the introduction of PCV7 in England and Wales. From reported studies, serotype 7F seems to be very rare in the nasopharynx of Brazilian children [23,42,43]. All but 2 (ST218) of our serotype 7F invasive pneumococci belonged to

Table 6. Characteristics of pneumococci recovered from IPD showing non-susceptibility to erythromycin, levofloxacin, ceftriaxone and penicillin.

Serotype (n)	ST (n)	CC	Susceptibility profile (µg/mL) (n/genotype)*
14 (11)	156 (5)	156	CRO = 2 (4)
			CRO = 2; ERY = 8 (1/ermB−; mef(A)+)
	13 (5)	singleton	ERY = 8 (5/ermB−; mef(A)+)
	370 (1)	156	CRO = 2 (1)
	8881 (1)	singleton	ERY = 8 (1/ermB−; mef(A)+)
19F (5)	271 (4)	320	ERY>32 (2/ermB+; mef(A)−)
			ERY>32; CRO = 2 (2/ermB+; mef(A)+)
	8890 (1)	320	ERY>32 (1/ermB+; mef(A)+)
19A (3)	320 (2)	320	ERY>32; CRO = 2 (1)
			ERY>32; PEN = 4; CRO = 2 (1)
	8884 (1)	320	ERY>32; PEN = 4; CRO = 2 (1/ermB+; mef(A)+)
23F (2)	4967 (2)	338	ERY = 0.5 (1/ermB−; mefE+)
			LEV = 16 (1/Ser^{81}-Phe in gyrA; Ser^{79}-Phe and Lys^{137}-Asn in parC.)
3 (1)	180 (1)	singleton	ERY = 0.5 (1/ermB+; mef(A)−)
4 (1)	770 (1)	singleton	ERY = 0.5 (1/ermB−; mef(A)+)
6C (1)	172 (1)	338	CRO = 2 (1)
9A(1)	156 (1)	156	CRO = 2 (1)

*Genotype for erythromycin and levofloxacin resistance only; Penicillin (PEN), ceftriaxone (CRO), erythromycin (ERY) and levofloxacin (LEV) MICs.

ST191, and the ST218 isolates are possibly capsular switch variants of serotype 12F.

ST8889, a new ST reported in this study, was exclusively related to serotype 20, and appears to be well established in Brazil, being detected over the 6 year period of this study. Seotype 20 is not part of any conjugated vaccine, although it is among the 23 selected serotypes of the polysaccharide vaccine. and little is known about its virulence and invasive characteristics. In the present study, this serotype was exclusively associated with IPD in adults, both before and after vaccination. ST8889 is a single locus variant (SLV) of three other STs recovered in Europe (Table 4): ST235, ST1345 and ST8816 and is a double locus variant (DLV) of an ST described in Brazil in 2001, from a meningitis case (ST762). Of note, serotype 20 seems to rarely colonize Brazilian children [23,43,44].

ST770 has already been described in Brazil as serotypes 4 and 23F. Interestingly, ST770 is a DLV of ST758, observed in two Brazilian meningitis cases in 2001, both described as serotype 20. Similarly, ST770 is a SLV of other STs from meningitis described in Brazil: ST8169 (serotype 4) in 2001 and ST772 (non-typable) in 2006, as demonstrated by data recovered from the MLST database.

CC156 has frequent occurrence around the world [45], including Latin America and Brazil [23,24,25]. ST156 (PMEN clone Spain^{9V}-3), a member of this CC and the most frequent ST among our population with both serotypes 9V and 14, is globally associated with important resistance profiles, including in Brazil [25]. We observed one isolate (recovered after vaccine introduction) with ST156 classified as serotype 9A, which has been previously described in the MLST database (Qatar and Lebanon, in 2006). We did not observe any PCV10 effect on this CC, as the occurrence before and after vaccination was quite similar in our population.

Our study shows a high genetic diversity including a considerable number of STs not previously reported. Moreover, the results show new capsular and clone combinations (suggesting capsular switching). Pneumococci with new combinations were mostly recovered during the pre-vaccine period, highlighting the natural diversity in the pneumococcal population and that the occurrence of factors other than selective pressure from vaccination encourages diversity.

The increase of serotype 19A in both carriage and invasive disease after PCV7 introduction is well documented [46]. However, this serotype also increased in regions without vaccine selective pressure, suggesting the participation of other factors, such as dissemination of some specific clones and antimicrobial pressure [47] in this process. Serotype 19A is frequently associated with ST199 and ST320, a DLV of ST236 (PMEN clone Taiwan^{19F}-14) [46,48]. ST320 is globally disseminated and strongly associated with penicillin resistance [47,49,50,51]. Recently, it was observed that the genetic background of ST320 provides advantages associated with improved colonization in the nasopharynx when compared to ST199 [48]. This advantage may be responsible in part for the rapid shift of ST199 to ST320 in USA soon after PCV7 [47]. However, in Germany, a country with low antibiotic pressure, the increase in 19A was attributed mainly to CC199 which increased and not CC320, suggesting multiple factors in play [52].

Both, ST320 and ST199 were recognized in the present work, although in low numbers. Despite the genetic diversity observed among serotype 19A in our population, ST2878 was the most frequent (36.4%). This ST had been documented in Brazil several other times and in different regions, with the first isolate recovered in 1998 (MLST database). Our study demonstrates the presence of ST2878 up to 2012, suggesting the widespread dissemination of this clone in Brazil.

CC320/271and CC156 were common among our penicillin non-susceptible pneumococci Although the proportion of isolates that were CC320 was considerably lower than for CC156, MIC values for penicillin among the CC320 isolates were higher (range 1 to 4 µg/mL). ST320 (serotype 19A) and ST271 (serotype 19F) are two well-recognized clones in CC320/271, while ST8884 (serotype 19A), a SLV of ST320, is newly described in this study. Our isolates belonging to ST8884 were multidrug-resistant, with resistance to tetracycline, erythromycin (with *ermB* and *mefA* genes) and non-susceptibility to β-lactams. Systematic molecular surveillance may be important to generate more data about this ST and its dissemination profile.

The prevalence of penicillin resistance seems to be variable, with reported rates as low as 0% [53,54,55] and as high as 27.9% [25,54] in Brazil. It is important to point out that previously published rates used the 2008 CLSI breakpoints to determine β-lactams susceptibility, and in this study, penicillin MICs ≥4 µg/mL were not observed. Non-susceptibility is mostly associated with serotypes 14, 19F, 23F, 6B and 19A [24,25,54] in Brazil. Although Zemlicková et al. (2005) [24] demonstrated a heterogeneous pre-vaccination population of penicillin-resistant pneumococci, Barroso et al. (2012) [25] suggested a major role of clone Spain[9V]-3 among pneumococci with this resistance. Even though this clone was commonly observed in our population, we did not observe an association with non-susceptible (MIC = 4 µg/mL) isolates, which were exclusively characterized as serotype 19A.

This study described pneumococcal population dynamics of isolates causing IPD in South Brazil, with a particular focus on the period after PCV10 implementation. Limitations of the study included few patients under five years of age (the target group for PCV10). In addition, given the relatively recent introduction of PCV10, our data reflect early effects of the immunization program. Continued surveillance and further studies need to be conducted to determine the full impact of vaccine in all age groups and to follow changes in pneumococcal population biology post-PCV10 introduction in Brazil.

Author Contributions

Conceived and designed the experiments: JC CD LMG. Performed the experiments: JC PH GRDC. Analyzed the data: JC PH LMG FHS CD. Contributed reagents/materials/analysis tools: PADA LMG. Contributed to the writing of the manuscript: JC CD LMG.

References

1. Welte T, Torres A, Nathwani D (2012) Clinical and economic burden of community-acquired pneumonia among adults in Europe. Thorax 67: 71–79.
2. Quadros CA (2009) From global to regional: The importance of pneumococcal disease in Latin America. Vaccine 27S: C29–C32.
3. Andrade AL, Oliveira R, Vieira MA, Minamisava R, Pessoa Jr V, et al. (2012) Population-based surveillance for invasive pneumococcal disease and pneumonia in infants and young children in Goiânia, Brazil. Vaccine 30: 1901–1909.
4. Casado-Flores J, Rodrigo C, Aristegui J, Martinon JM, Fenoll A, et al. (2008) Decline in pneumococcal meningitis in Spain after introduction of the heptavalent pneumococcal conjugate vaccine. Pediatr Infect Dis J 27: 1020–1022.
5. Ardanuy C, Tabau F, Pallares R, Calatayud L, Domínguez MA, et al. (2009) Epidemiology of invasive pneumococcal disease among adult patients in Barcelona before and after pediatric 7-valent pneumococcal conjugate vaccine introduction, 1997–2007. Clin Infect Dis 48: 57: 64.
6. Centers for Disease Control and Prevention (2009) Pneumonia hospitalizations among young children before and after introduction of pneumococcal conjugate vaccine–United States, 1997–2006. MMWR Morb Mortal Wkly Rep 58: 1–4.
7. Rückinger S, van der Linden M, Reinert RR, von Kries R, Burckhardt F, et al. (2009) Reduction in the incidence of invasive pneumococcal disease after general vaccination with 7-valent pneumococcal conjugate vaccine in Germany. Vaccine 27: 4136–4141.
8. Rodenburg G, de Greef S, Jansen AG, de Melker HE, Schouls LM, et al. (2010) Effects of pneumococcal conjugate vaccine 2 years after its introduction, the Netherlands. Emerg Infect Dis 16: 816–823.
9. Harboe ZB, Valentiner-Branth P, Benfield TL, Christensen JJ, Andersen PH, et al. (2010) Early effectiveness of heptavalent conjugate pneumococcal vaccination on invasive pneumococcal disease after the introduction in the Danish Childhood Immunization Programme. Vaccine 28: 2642–2647.
10. Pilishvili T, Lexau C, Farley MM, Hadler J, Harrison LH, et al. (2010) Sustained reductions in invasive pneumococcal disease in the era of conjugate vaccine. J Infect Dis 201: 32–41.
11. Vestrheim DF, Hoiby EA, Bergsaker MR, Ronning K, Aaberge IS, et al. (2010) Indirect effect of conjugate pneumococcal vaccination in a 2+1 dose schedule. Vaccine 28: 2214–2221.
12. Bettinger JA, Scheifele DW, Kellner JD, Halperin SA, Vaudry W (2010) The effect of routine vaccination on invasive pneumococcal infections in Canadian children, Immunization Monitoring Program, Active 2000–2007. Vaccine 28: 2130–2136.
13. Van der Linden M, Weiss S, Falkenhorst G, Siedler A, Imöhl M, von Kries R (2012) Four years of universal pneumococcal conjugate infant vaccination in Germany: impact on incidence of invasive pneumococcal disease and serotype distribution in children. Vaccine 30: 5880–5885.
14. Afonso ET, Minamisava R, Bierrenbach AL, Escalante JJC, Alencar AP, et al. (2013) Effect of 10-Valent pneumococcal vaccine on pneumonia among children, Brazil. Emerg Infect Dis 19: 589–597.
15. Weinberger DM, Malley R, Lipstich M (2011) Serotype replacement in disease after pneumococcal vaccination. Lancet 378: 1962–1973.
16. Miller E, Andrews NJ, Waight PA, Slack MP, George RC (2011) Herd immunity and serotype replacement 4 years after seven-valent pneumococcal conjugate vaccination in England and Wales: an observational cohort study. Lancet Infect Dis 11: 760–68.
17. PAHO (2013) Informe Regional de SIREVA II, 2012: Datos por país y por grupos de edad sobre las características de los aislamientos de *Streptococcus pneumoniae, Haemophilus influenzae y Neisseria meningitidis,* en procesos invasores.
18. Spellerberg B & Brandt C (2011) Streptococcus. In: Versalovic J, Carroll KC, Funke G, Jorgensen JH, Landry ML, Warnock DW. Manual of Clinical Microbiology, ed., American Society for Microbiology, Washington, DC 331–349.
19. CLSI (2013) Performance Standards for Antimicrobial Susceptibility testing. Twenty-third informational Supplement. M100-S23. Wayne, PA: Clinical and Laboratory Standards Institute.
20. Widdowson CA, Klugman KP (1998) Emergence of the M phenotype of the erythromycin-resistant pneumococci in South Africa. Emerg Infect Dis 4: 277–281.
21. Brueggemann AB, Coffman SL, Rhomberg P, Huyn H (2002) Fluoroquinolone resistance in *Streptococcus pneumoniae* in United States since 1994–1995. Antimicrob Agents Chemother 46: 680–8.
22. Enright MC, Spratt G (1998) A multilocus sequence typing scheme for *Streptococcus pneumoniae*: identification of clones associated with serious invasive disease. Microbiology 144: 3049–3060.
23. Pimenta FC, Carvalho M da G, Gertz RE, Bastos-Rocha CGB, Oliveira LSC, et al. (2011) Serotype and genotype distriution of pneumococcal carriage isolates recovered from Brazilian children attending day-care centers. J Medical Microbiol 60: 1455–1459.
24. Zemlicková H, Crisóstomo I, Brandileone MC, Camou T, Castañeda E, et al. (2005) Serotype and clonal types of penicillin-susceptible *Streptococcus pneumoniae* causing invasive disease in children in five Latin American countries. Microb Drug Resistance 11: 195–204.
25. Barroso DE, Godoy D, Castañeiras TM, Tulenko MM, Rebelo MC, et al. (2012) β-lactam resistance, serotype distribution, and genotypes of meningitis-causing *Streptococcus pneumoniae*, Rio de Janeiro, Brazil. Pediatr Infect Dis J 31: 30–36.
26. Domingues CM, Verani JR, Montenegro Renoiner El, de Cunto Brandileone MC, Flannery, et al. (2014) Effectiveness of tem-valent pneumococcal conjugate vaccine against invasive pneumococcal disease in Brazil: a matched case-control study. Lancet Respir Med 2: 464–471.
27. Muñoz-Almagro C, Ciruela P, Esteva C, Marco Francesc, Navarro M, et al. (2011) Serotypes and clones causing invasive pneumococcal disease before the use of new conjugate vaccines in Catalonia, Spain. J Infect 63: 151–162.
28. Pichon B, La Dhani SN, Slack MPE, Segonds-Pichon A, Andrews NJ, et al. (2013) Changes in the molecular epidemiology of *Streptococcus pneumoniae* causing meningitis following the introduction of pneumococcal conjugate vaccination in England and Wales. J Clin Microbiol 51: 820–827.
29. Aguiar SI, Brito MJ, Gonçalo-Marques J, Melo-Cristino J, Ramirez M (2010) Serotypes 1, 7F and 19A became the leading causes of pediatric invasive

pneumococcal infection in Portugal after 7 years of heptavalent conjugate vaccine use. Vaccine 28: 5167–5173.

30. Brueggemann AB, Peto TEA, Crook DW, Butler JC, Kristinsson KG, et al. (2004) Temporal and geographic stability of the serogroup specific invasive disease potential of *Streptococcus pneumoniae* in children. J Infect Dis 190: 1203–1211.

31. Harboe ZB, Thomsen RW, Riis A, Valentiner-Branth P, Christensen JJ, et al. (2009) Pneumococcal serotypes and mortality following invasive pneumococcal disease: a population-based cohort study. PloS Medicine 6: 1–13.

32. Jansen AGSC, Rodenurg GR, van der Ende A, van Alphen L, Veenhoven RH, et al. (2009) Invasive pneumococcal disease among adults: association among serotypes, disease characteristics and outcome. Clin Infect Dis 49: e23–29.

33. Inverarity D, Lamb K, Diggle M, Roertson C, Greenhalgh D, et al. (2011) Death or survival from invasive pneumococcal disease in Scotland: associations with serogroup and multilocus sequence types. J Med Microbiol 60: 793–802.

34. Rakov AV, Ubukata K, Robinson DA (2011) Population structure hyperinvasive serotype 12F, clonal complex 218 *Streptococcus pneumoniae* revealed by multilocus *boxB* sequence typing. Infect Genet Evol 11: 1929–1939.

35. Zulz T, Wenger JD, Rudolph K, Robinson A, Rakov AV, et al. (2013) Molecular characterization of *Streptococcus pneumoniae* serotype 12F isolates associated with rural community outbreaks in Alaska. J Clin Microbiol 51: 1402–1407.

36. Sleeman KL, Griffiths D, Shackley F, Diggle L, Gupta S, et al. (2006) Capsular serotype-specific attack rates and duration of carriage of *Streptococcus pneumoniae* in a population of children. J Infect Dis 194: 682–688.

37. Flasche S, van Hoek AJ, Sheasby E, Waight P, Andrews N, et al. (2011) Effect of pneumococcal conjugate vaccination on serotype-specific carriage and invasive disease in England: a cross-sectional study. PloS Medicine 8: e1001017.

38. Birtles A, McCarthy N, Sheppard CL, Rutter H, Guiver M, et al. (2005) Multilocus sequence typing directly on DNA from clinical samples and a cultured isolate to investigate linked fatal pneumococcal disease in residents of a shelter for homeless men. J Clin Microbiol 43: 2004–2008.

39. Lin CJ, Chen PY, Huang FL, Lee T, Chi CS, Lin CY (2006) Radiographic, clinical, and prognostic features of complicated and uncomplicated community-acquired lobar pneumonia in children. J Microiol Immunol Infect; 39: 489–95.

40. Jefferies JM, Johnston CH, Kirkham LA, Cowan GJ, Ross KS, et al. (2007) Presence of nonhemolytic pneumolysin in serotypes of *Streptococcus pneumoniae* associated with disease outbreaks. J Infect Dis 196: 936–944.

41. Inverarity D, Diggleb M, Ure R, Johnsonc P, Altstadtd P, et al. (2011) Molecular epidemiology and genetic diversity of pneumococcal carriage among children in Beni State, Bolivia. Trans R Soc Trop Med and Hyg 105: 445–451.

42. Laval CB, de Andrade AL, Pimenta FC, de Andrade JG, de Oliveira RM, et al. (2006) Serotype of carriage and invasive isolates of *Streptococcus pneumoniae* in Brazilian children in the era of pneumococcal vaccines. Clin Microbiol Infect 12: 50–55.

43. Neves FPG, Pinto TCA, Corrêa MA, Barreto RA, Moreira LSG, et al. (2013) Nasopharyngeal carriage, serotype distribution and antimicrobial resistance of *Streptococcus pneumoniae* among children from Brazil before the introduction of the 10-valent conjugate vaccine. BMC Infect Dis 13: 318.

44. Lucarevschi BR, Baldacci ER, Bricks LF, Bertoli CJ, Teixeira LM (2003) Oropharyngeal carriage of *Streptococcus pneumoniae* by children attending day care centers in Taubate, SP: correlation betwwn serotypes and the conjugated heptavalent pneumococcal vaccine. J Pediatr 79: 215–220.

45. McGee L, McDougal L, Zhou J, Spratt BG, Tenover FC, et al. (2001) Nomenclature of major antimicrobial-resistant clones of *Streptococcus pneumoniae* defined by the pneumococcal molecular epidemiology network. J Clin Microiol 39: 2565–2571.

46. Moore MR, Gertz RE, Woodury RL, Barkocy-Gallagher GA, Schaffner W, et al. (2008) Population snapshot of emergent *Strpetococcus pneumoniae* serotype 19A in the United States, 2005. J Infect Dis 197: 1016–1027.

47. Willems RJL, Hanage WP, Bessen DE, Feil EJ (2011) Population biology of Gram-positive pathogens: high-risk clones for dissemination of antibiotic resistance. FEMS Microbiol Rev 35: 872–900.

48. Hsieh YC, Lin TL, Chang KY, Huang YC, Chen CJ, et al. (2013) Expansion and evolution of *Streptococcus pneumoniae* serotype 19A ST320 clone as compared to its ancestral clone, Taiwan 19F-14 (ST236). J Infect Dis 15: 203–210.

49. Ardanuy C, Rolo D, Fenoll A, Tarrago D, Calatayud L, et al. (2009) Emergence of a multidrug-resistant clone (ST320) among invasive serotype 19A pneumococci in Spain. J Antimicrob Chemother 64: 507–510.

50. Shin J, Baek JY, Kim SH, Song JH, Ko KS (2011) Predominance of ST320 among *Streptococcus pneumoniae* serotype 19A isolates from 10 Asian countries. J Antimicrob Chemother 66: 1001–1004.

51. Gene A, del Amo E, Iñigo M, Monsonis M, Pallares R, et al. (2013) Pneumococcal serotypes causing acute otitis media among children in Barcelona (1999–2011): emergence of the multirresistant clone ST320 of serotype 19A. Pediatr Infect Dis J 32: e128–133.

52. van der Linden M, Reinert RR, Kern WV, Imöhl M (2013) Epidemiology of serotype 19A isolates from invasive pneumococcal disease in German children. BMC Infect Dis 13: 70. doi: 10.1186/1471-2334-13-70.

53. Mendes C, Marin ME, Quiñones F, Sifuentes-Osornio J, Siller CC, et al. (2003) Antibacterial resistance of community-acquired respiratory tract pathogens recovered from patients in Latin America: results from the PROTEKT surveillance study (1999–2000). Braz J Infect Dis 7: 44–61.

54. Alvares JR, Mantese OC, Paula AD, Wolkers PC, Almeida VV, et al. (2011) Prevalence of pneumococcal serotypes and resistance to antimicrobial agents in patients with meningitis: tem-year analysis. Braz J Infect Dis 15: 22–27.

55. Rossi F, Franco MR, Rodrigues HM, Andreazzi D (2012) *Streptococcus pneumoniae*: susceptibility to penicillin and moxifloxacin. J Brasi Pneumol 38: 66–71.

7

Differentiation of Human Umbilical Cord Matrix Mesenchymal Stem Cells into Neural-Like Progenitor Cells and Maturation into an Oligodendroglial-Like Lineage

Cristiana Leite[1], N. Tatiana Silva[1], Sandrine Mendes[2], Andreia Ribeiro[2], Joana Paes de Faria[3], Tânia Lourenço[1], Francisco dos Santos[4¤], Pedro Z. Andrade[4¤], Carla M. P. Cardoso[5], Margarida Vieira[5], Artur Paiva[2], Cláudia L. da Silva[4], Joaquim M. S. Cabral[4], João B. Relvas[3], Mário Grãos[1]*

1 Biocant - Technology Transfer Association, Biocant Park, Cantanhede, Portugal, 2 Blood and Transplantation Center of Coimbra, Portuguese Institute of the Blood and Transplantation, Coimbra, Portugal, 3 Instituto de Biologia Molecular e Celular, Porto, Portugal, 4 Institute for Biotechnology and Bioengineering and Department of Bioengineering, Instituto Superior Técnico, Universidade de Lisboa, Lisboa, Portugal, 5 Crioestaminal Saúde e Tecnologia, S.A., Biocant Park, Cantanhede, Portugal

Abstract

Mesenchymal stem cells (MSCs) are viewed as safe, readily available and promising adult stem cells, which are currently used in several clinical trials. Additionally, their soluble-factor secretion and multi-lineage differentiation capacities place MSCs in the forefront of stem cell types with expected near-future clinical applications. In the present work MSCs were isolated from the umbilical cord matrix (Wharton's jelly) of human umbilical cord samples. The cells were thoroughly characterized and confirmed as *bona-fide* MSCs, presenting *in vitro* low generation time, high proliferative and colony-forming unit-fibroblast (CFU-F) capacity, typical MSC immunophenotype and osteogenic, chondrogenic and adipogenic differentiation capacity. The cells were additionally subjected to an oligodendroglial-oriented step-wise differentiation protocol in order to test their neural- and oligodendroglial-like differentiation capacity. The results confirmed the neural-like plasticity of MSCs, and suggested that the cells presented an oligodendroglial-like phenotype throughout the differentiation protocol, in several aspects sharing characteristics common to those of *bona-fide* oligodendrocyte precursor cells and differentiated oligodendrocytes.

Editor: Rafael Linden, Universidade Federal do Rio de Janeiro, Brazil

Funding: Authors acknowledge "ISOCORD" project, led by Crioestaminal - Saúde e Tecnologia S.A., supported by the Portuguese program QREN/Mais Centro-Programa Operacional Regional do Centro (project nr. CENTRO – 07-0202 –FEDER –005547) and co-funded by the European Regional Development Fund (ERDF). This work was partially funded by the ERDF through Programa Operacional Factores de Competitividade - COMPETE and by national funds by FCT - Fundação para a Ciência e a Tecnologia (Portuguese Foundation for Science and Technology) through grant FCOMP-01-0124-FEDER-021150 - PTDC/SAU-ENB/119292/2010, which included a research fellowship awarded to TL, and COMPETE funding (Project "Stem cell based platforms for Regenerative and Therapeutic Medicine", Centro-07-ST24-FEDER-002008). FCT financially supported FS, PZA, and JPF through fellowships SFRH/BPD/81861/2011, SFRH/BPD/81909/2011 and SFRH/BPD/34834/2007, respectively. The funders had no role in study design, data collection and analysis, decision to publish, or preparation of the manuscript.

Competing Interests: CMPC and MV are employees of Crioestaminal- Saúde e Tecnologia S.A., a company that provides the service of cryopreservation of stem cells from human umbilical cord blood and matrix. These authors contributed to obtaining the umbilical cord samples and participated in study design related to the isolation and phenotypic characterization of the undifferentiated cells.

* Email: mgraos@biocant.pt

¤ Current address: Cell2B Advanced Therapeutics SA, Biocant Park, Cantanhede, Portugal

Introduction

Mesenchymal stem cells (MSCs), also known as mesenchymal stromal cells, are defined as multipotent adult stem cells, possessing self-renewal capacity and multilineage differentiation potential [1,2]. MSCs were originally identified in the bone marrow [3], but more recently, cells with characteristics similar to MSCs have been identified in many other locations, such as perivascular regions of multiple organs and tissues (like the fat tissue) [4] and several regions of the umbilical cord, namely the umbilical cord matrix (also known as the Wharton's jelly) [5].

MSCs have been characterized as a safe, available, low-immonogenic and clinically promising adult stem cell type [1,5,6]. Several reports in the literature have shown the potential of MSCs to differentiate into neural stem-like cells [7–9]. Despite controversy about MSCs (a mesenchymal cell type) differentiating into neural-like cellular fates, compelling evidence has shown that indeed MSCs express neuroectodermal markers, like nestin [8,10–13] and have at least a partial neural crest, neuroepithelial origin [14,15], suggesting plasticity towards neural-like lineages, opening research avenues for the treatment of distinct neurodegenerative diseases [16,17]. MSCs have been rather explored in terms of neuronal-like differentiation [8,13,18–20], but the first reports addressing oligodendrocyte-like specification were only published recently [21,22]. Nevertheless, further studies are required to fully address this potential.

Demyelination of the central nervous system (CNS) is caused by loss of oligodendrocytes (OLs) and may occur as a result of traumatic injury or non-traumatic neurodegenerative diseases, like multiple sclerosis (MS). Remyelination of the affected areas is typically low and demyelinated areas become inflamed and populated by astrocytes, causing the formation of scar tissue [23]. Stem cell-based approaches that allow for a quicker and more robust remyelination of the affected areas are considered promising for the treatment of demyelinating diseases. However, despite recent advances regarding oligodendroglial differentiation of pluripotent stem cells (namely human embryonic stem cells - hESCs [24,25] and induced pluripotent stem cells - iPSCs [26]), these are not yet considered safe for application in a clinical setting. Hence, the current lack of appropriate and safe cell sources hamper the use of stem cell-based approaches for the treatment of demyelinating diseases in the clinic.

The objectives of the present work were to thoroughly characterize human MSCs isolated from the umbilical cord matrix (UCM) and assess whether these cells possessed neural- and more specifically, oligodendroglial-like differentiation capacity. The results presented here suggest that umbilical cord matrix mesenchymal stem cells (UCM-MSCs) possess a certain degree of plasticity to differentiate into neural-like cells, and subsequently into cells with phenotypic characteristics of oligodendrocyte precursors and immature oligodendrocytes. Despite the need for testing further differentiation protocols and to perform *in vivo* functional studies to assess the full potential of these cells, the results presented here are promising in the context of cell-based therapeutic strategies for demyelinating diseases.

Materials and Methods

Isolation and culture of human mesenchymal stem cells (MSCs) from the umbilical cord matrix (UCM)

Human umbilical cords were obtained after birth from healthy donors, with written informed consent of the parent(s) and the study was approved by the Ethics Committee of Maternidade de Bissaya Barreto – Centro Hospitalar de Coimbra (ref. 356/Sec). Samples were stored at room temperature (RT) in sterile 50 ml conical tubes (VWR International) for 12 to 48 h before tissue processing. The isolation procedure of MSCs was adapted from a protocol described by Reinisch *et al.* [27]. Each umbilical cord unit was manipulated under sterile conditions using a class-II biosafety cabinet (Heraeus HS-18) and cut into sections of about 5 cm. The pieces were washed 2 or 3 times using sterile PBS (Sigma-Aldrich), to remove the blood. The umbilical vein was washed with sterile PBS using a 10 ml sterile syringe to remove blood and blood clots. Subsequently, the vein and arteries were removed to avoid endothelial cell contamination. The Wharton's jelly (WJ) was then cut into fragments of 2–5 mm with the help of sterile scalpel and forceps (Fine Science Tools). Groups of approximately 24 fragments were transferred to each 55 cm^2 tissue culture (TC) plate (Corning) and left to dry for 5 min inside the biosafety cabinet to promote adhesion of the fragments to the polystyrene surface. Once the cord fragments were properly adherent to the plastic, MSC proliferation medium [Alpha-MEM without nucleosides (Life Technologies, cat. no. 22561-021) supplemented with 10% volume/volume (v/v) MSC-qualified Fetal Bovine Serum (FBS) (Hyclone), 10 U/ml of Penicillin, 10 µg/ml Streptomycin and 2.5 µg/ml Amphotericin B (all from

Life Technologies)] was added to the culture plate, until all the fragments were completely covered. The fragments were then cultured for 10 days in an incubator (Shel Lab) at 37°C with 5% CO$_2$/95% air and 95% humidity, until MSCs started migrating out of the UCM pieces and forming well defined colonies. Then, UCM fragments were removed from the tissue culture plate and cells were passaged.

For the passaging, cells were detached and dissociated using Trypsin (500 µg/ml)-EDTA (200 µg/ml) solution (Life Technologies). Trypsin was inactivated using 10x the volume of MSC proliferation medium, centrifuged (300×g, for 5 min at RT), counted and seeded at a density of 3,000 cells/cm^2 in MSC proliferation medium and maintained in a CO$_2$ incubator (as above) until sub-confluence (80–90% confluence).

For cryopreservation, cells were treated in a similar way as described for the passaging procedure, except that cells were resuspended in cryopreservation solution (FBS supplemented with 10% DMSO – Sigma Aldrich) into cryotubes and transferred to a cell freezing container - Mr. Frosty - (both from Thermo Scientific), that was relocated inside a −80°C freezer to achieve a rate of cooling close to 1°C/minute. After 16 h, the cryotubes were transferred to a liquid nitrogen cryotank for long-term storage.

Proliferation kinetics of UCM-MSCs

MSCs isolated from 3 independent cord samples were continuously cultured from P2 to P8 and counted once they reached 80% confluence at each passage. The population doubling (PD) rate was determined at each passage using the equation $N_H/N_I = 2^X$, or $[\log_{10}(N_H) - \log_{10}(N_I)]/\log_{10}(2) = X$, where N_I represents the number of cells plated at each passage, N_H the number of cells harvested at the end of each respective passage and X the population doubling (PD), as described [28].

The PD for each passage was calculated and added to the PD of the previous passages to generate data for cumulative population doublings (CPD). In addition, the generation time (GT) - average time between two cell doublings - was calculated from P2 to P8 using the following formula, as described [29]: $X = [\log_{10}(2) \times \Delta t]/[\log_{10}(N_H) - \log_{10}(N_I)]$.

The total number of cells (TNC) was determined at each passage (P1-P8) by cumulative counting of the cells once they reached a confluence of 80%, using the formula: $X = N_H \times B/N_I$, in which B represents the total number of cells in the previous passage. TNC accounts for the theoretical number of cells that could be obtained if no cells were discarded between each passage.

Colony-forming unit-fibroblast (CFU-F) assay

The colony-forming unit-fibroblast (CFU-F) assay of MSCs was determined for 3 independent cord samples at P2 and P8. Cells were seeded at 3 cells/cm^2 on 55 cm^2 TC dishes in MSC proliferation medium and cultured for 15 days. One third of the medium was replaced twice a week. Cells were fixed with 4% paraformaldehyde (PFA) (Sigma-Aldrich) for 20 min at RT and stained with Giemsa solution (Sigma-Aldrich). Individual colonies were counted manually.

Multilineage differentiation of UCM-MSCs

Chondrogenic, adipogenic and osteogenic differentiation capacity of isolated MSCs was assessed using early passage (P2-P3) MSCs. For osteogenic and adipogenic differentiation, cells were plated at 1,000 cells/cm^2 in the presence of Dulbecco's modified Eagle's medium (DMEM) (Life Technologies) supplemented with 10% MSC-qualified FBS (Hyclone) until reaching 80–100% confluence. Medium was then removed and StemPRO (Life

Technologies) osteogenic or adipogenic differentiation medium was added and cells were cultured for 14 days. For chondrogenic differentiation, the cells were plated on low-attachment plates (Corning). A pellet of 5×10^5 to 1×10^6 cells was resuspended and droplets of this suspension were plated on the surface of each well. After plating, StemPRO chondrogenic differentiation medium was added and cells were cultured for 14 days.

For osteogenic differentiation assessment, alkaline phosphatase (ALP) and Von Kossa staining were performed. For ALP staining, the cells were washed with PBS and fixed in 10% cold neutral-buffered formalin (Sigma-Aldrich) for 15 min.

Cells were then washed and kept in distilled water for another 15 min and then stained with a 0.1 M Tris-HCl solution (Sigma-Aldrich) containing the substrate Naphtol AS MX-PO4 (0.1 mg/ml) (Sigma-Aldrich) for ALP in dimethylformamide (Fischer Scientific) and 0.6 mg/ml Red Violet LB salt (Sigma-Aldrich) for 45 min. The excess staining was removed by washing 3 times with distilled water. For the Von Kossa staining, the cells were washed with PBS and stained with 2.5% (w/w) silver nitrate (Sigma-Aldrich) for 30 min at RT and then washed 3 times with distilled water.

For adipogenic differentiation assessment, cells were washed with PBS and fixed with PFA using a 2.5% solution for 30 min at RT. Cells were washed once in distilled water and incubated with 0.3% Oil Red-O solution (Sigma-Aldrich) for 1 h at RT, and then washed twice with distilled water.

For chondrogenic differentiation assessment, cells were washed once with PBS and fixed with 2% PFA solution for 30 min at RT. The cells were then washed with PBS and stained with 1% Alcian Blue solution (Sigma-Aldrich) prepared in 0.1 M HCl for 30 min. The excess was removed by washing 3 times with PBS and the cells were then maintained in distilled water.

Immunophenotypic characterization of UCM-MSCs

The immunophenotypic characterization of UCM-MSCs [30] was performed at passages 2 and 8. Cells were dissociated using StemPro Accutase (Life Technologies), labeled with antibodies against the indicated antigens and analyzed by flow cytometry (FACS Canto II, Becton-Dickinson). The following antibodies were used for the labeling: mouse PO anti-human CD45 IgG1, clone HI30 (5 µL/test) from Life Technologies; mouse APC anti-human CD90 IgG1, clone 5E10 (0.2 mg/ml) from BD Pharmingen; mouse PE anti-human CD105 IgM, clone ALB9 (10 µL/test) from Beckman Coulter; mouse PB anti-human CD11b IgG1, clone ICRF44 (0.2 mg/ml) from BD Pharmingen; mouse Pe-Cy7 anti-human CD13 IgG1, clone WM15 (2.5 µL/test) from BD Pharmingen; mouse PerCP-Cy5.5 anti-human CD34 IgG1, clone 8G12 (5 µL/test) from BD Pharmingen; mouse PE anti-human CD73 IgG1, clone AD2 (10 µL/test) from BD Pharmingen; mouse PE anti-human NFGR IgG1, clone C40-1457 (10 µL/test) from BD Pharmingen.

Neural-like induction of UCM-MSCs

Neuroectodermal-like induction of UCM-MSCs. UCM-MSCs in P3-P7 were dissociated using Trypsin (500 µg/ml)-EDTA (200 µg/ml) solution, counted and seeded on TC dishes at a density of 12,500 cells/cm^2 in MSC proliferation medium. On the following day, cells were washed with PBS and medium was replaced by neuroectodermal-induction medium [22] composed by DMEM/F12 (Life Technologies) supplemented with N2 (Life Technologies), 10 ng/ml human recombinant epidermal growth factor (EGF) (Peprotech), 10 U/ml of Penicillin, 10 µg/ml Streptomycin and 2.5 µg/ml Amphotericin B. Cells were incubated for 3 days at 37°C, 5% CO$_2$/95% air and 95% humidity.

Neural stem cell (NSC)-like induction of Neuroectodermal-like induced MSCs (niMSCs). Neuroectodermal-like induced MSCs (niMSCs) were dissociated using Trypsin (500 µg/ml)-EDTA (200 µg/ml) solution and seeded at a density of 60,000 cells/cm^2 on non-treated polystyrene (non-tissue culture) aseptic plates (Gosselin) previously treated for 30 min under the UV light of the laminar flow cabinet. Cells were maintained in NSC induction medium [22] [Neurobasal medium (Life Technologies) supplemented with B27 (Life Technologies) and 20 ng/ml of human recombinant EGF and basic fibroblast growth factor (bFGF) (Peprotech), 10 U/ml of Penicillin, 10 µg/ml Streptomycin and 2.5 µg/ml Amphotericin B] in an incubator at 37°C, 5% CO$_2$/95% air and 95% humidity for 18 days with medium replacement every 3–4 days.

Oligodendrocyte precursor cell (OPC)-like induction of NSC-like cells. NSC-like cells were dissociated into single cells using StemPro Accutase, counted and seeded at 20,000 cells/cm^2 onto uncoated 6-well TC plates (Corning) or similar plates coated with human purified fibronectin (Roche) or laminin-2/merosin (Millipore). The coatings were performed by covering the well with 10 µg/ml of fibronectin or laminin-2 in PBS at 37°C for 4 h. Plates were then blocked with 0.3% Bovine Serum Albumin (BSA) in PBS for 30 min at 37°C and washed 3 times with PBS before use. The cells were cultured in NSC induction medium [22] with 10 ng/ml of EGF and bFGF. At the end of 3 days, half the medium was replaced by OPC induction medium composed by Neurobasal medium supplemented with B27, 10 ng/ml of human recombinant bFGF, platelet-derived growth factor-AA (PDGF-AA) and 100 ng/ml of Sonic Hedgehog (SHH) (both from Peprotech). At the end of 3 days, the entire medium was replaced by OPC induction medium. Cells were in culture for an additional 12 days period with complete medium changes every 3–4 days.

Maturation of OPC-like cells into Oligodendrocyte (OL)-like cells. OPC-like cells were dissociated into single cells using StemPro Accutase, counted and seeded at 20,000 cells/cm^2 on 96 well TC plates (Corning), coated with 0.1 mg/ml poly-D-lysine (PDL) (Sigma-Aldrich) and 10 µg/ml laminin-2. PDL coating was performed overnight at 37°C by covering the well with 0.1 mg/ml of PDL in PBS and washed 3 times with PBS. Laminin coating was performed by covering the PDL coated well with 10 µg/ml of laminin in PBS and incubated at 37°C for 2 h. The wells were washed 3 times with PBS before use. Cells were cultured in OL differentiation medium, composed by DMEM/F12 supplemented with N2 and 0.5% FBS. T3 (Sigma-Aldrich) and F3/contactin (R&D Systems), were added to the OL differentiation medium in the following conditions: 30 ng/ml of T3 or 10 nM of F3 or both, for a 10 day period. Alternatively, 30 ng/ml of T3 were added in the first 7 days and both factors for the last 3 days of differentiation. The medium was replaced every 3–4 days.

Rat primary astrocytes obtained as described [31,32] were used as a positive control for astroglial markers assessed by immunocytochemistry analysis. Procedures were performed according to the European Union Directive 86/609/EEC and the legislation Portaria n. 1005/92, issued by the Portuguese Government for the protection of animals used for experimental and other scientific purposes.

Immunocytochemistry analysis and F-actin staining

Cells were fixed with 4% PFA for 20 min at RT, rinsed with PBS and permeabilized with PBS with 0.1% Triton X-100 for 20 min, except when using anti-human Golgi antibody, where permeabilization was performed using cold acetone for 15 seconds. Cells were incubated in blocking solution (PBS with 0.1% BSA) for 30 min at RT. Cells were incubated with primary antibodies diluted in blocking solution (see below), overnight at 4°C in humidified conditions. The cells were washed with PBS and then incubated with the appropriate fluorescently-labeled secondary antibodies (see below) in PBS with 0.1% BSA for 1 h at RT. Cells were washed and fixed with 4% PFA for 5 min, to crosslink antigen/antibody complexes, washed 3 times and then incubated for 4 min with 200 ng/ml of DAPI (Sigma-Aldrich). To stain cells for polymerized actin (F-actin), a solution of 4 μM FITC-phalloidin (Sigma-Aldrich) in PBS was incubated for 20 min at RT at the end of the immunocytochemistry protocol (after the incubation with the secondary antibodies), and then washed 3 times with PBS to remove unbound reagent.

The following primary antibodies were used: mouse anti-nestin, clone 10C2 (1:400) from Millipore; mouse anti-O4, clone O4 (1:400) from R&D Systems; mouse anti-A2B5, clone 105 (1:500) from R&D Systems; rabbit anti-galactocerebroside (GalC) (1:50) from Millipore; rabbit anti-MBP (1:200) from Sigma-Aldrich; rabbit anti-GFAP (1:500) from Dako Cytomation; rabbit anti-β-III-tubulin IgG1, clone TUJ1 1-15-79 (1:2000) from Covance and mouse anti-human Golgi, clone 371-4 (1:30) from Chemicon. The following secondary fluorescently-labeled antibodies were used according to the host species of each primary antibody: Alexa Fluor 488 donkey anti-mouse IgG (H+L) (1:200) and Alexa Fluor 568 goat anti-rabbit IgG (H+L) (1:200), both from Life Technologies.

Fluorescence microscopy was performed using a Zeiss Axiovert 200 M microscope using AxioVision Release 4.8 software (Zeiss) for image acquisition. Exposure time was the same for each marker analyzed and for each independent experiment.

Image analysis and fluorescence quantification

Mean fluorescence intensity (MFI) was quantified using ImageJ software (NIH) for each marker analyzed. In detail, for each marker and for each independent experiment (consisting of cells in all differentiation steps), the intensity of the background and the signal were obtained by determining the background and signal threshold levels (Threshold tool) of at least 3 different fields belonging to the differentiation step with the highest signal intensity. The average threshold levels were then calculated and applied (using the 'Set threshold' tool) to all images under analysis (for each marker in each independent experiment). After setting the thresholds, the MFI of the background and the signal were obtained using the Measure tool and the first was subtracted to the latter, analyzing at least 3 fields for each differentiation step. The resulting values were then averaged and subjected to statistical analysis.

RNA isolation and Real-Time RT-PCR

Total RNA was extracted from cells in culture in different stages of differentiation using the RNeasy Mini Kit (Qiagen) and treated with DNase-I (Qiagen). 250 ng of RNA were used in each duplicate per sample to synthetize cDNA using the SuperScript II Reverse Transcriptase (Life Technologies). Quantitative PCR analysis was done in each duplicate using the total amount of cDNA obtained in each reaction using Power Sybr Green PCR Master Mix (Life Technologies) using a real-time PCR system (7500 Fast real-time PCR System and 7500 Software V2.0.4, Applied Biosystems) and values were normalized to the levels of *actin* and undifferentiated MSCs were used as the control sample. The PCR cycling parameters were 94°C for 5 min; 30 cycles of 30 seconds at 94°C, 1 minute at 60°C, and 1 minute at 72°C; and final extension at 72°C for 10 min. The primers were as follows: *actin* forward, 5′-CAGAAGGATTCCTATGTGGGC-3′, reverse, 5′- GAGGGCATACCCCTCGTAGAT-3′; *fibronectin*,

forward, 5′-GAGATCAGTGGGATAAGCAGCA-3′, reverse, 5′- CCTCTTCA-TGACGCTTGTGGA-3′; *sox2*, forward, 5′-CAGGAGAACCCCAAGATGC-3′, reverse, 5′-GCAGCCGCT-TAGCCTCG-3′; *nestin*, forward, 5′-CAGCTGGCGCACCTC-AAGATG-3′, reverse, 5′- AGGGAAGTTGGGCTCAGGACT-GG-3′; *mbp*, forward, 5′-CTGGGCAGC-TGTTAGAGTCC-3′, reverse, 5′-TGGAGCAAAGGTTTGGTGTC-3′; *gfap*, forward, 5′-CTGTTGCCAGAGATGGAGGTT-3′, reverse, 5′-TCATC-GCTCAGGAGGTCCTT-3′; *neurofilament*, forward, 5′-GAGC-GCAAAGACTACCTGAAGA-3′, reverse, 5′-CAGC-GATTTC-TATATCCAGAGCC-3′.

Co-culture of OPC/OL-like cells with mouse DRG neurons

Purified dorsal root ganglion (DRG) neurons were obtained by a culture system previously described [33] with minor modifications. Briefly, DRGs were dissected from E14-E16 C57BL/6 mice and explants were transferred to 22-mm glass coverslips previously coated with 0.1 mg/ml poly-D-lysine for 1 h at RT followed by 10 μg/ml laminin-1 (Sigma-Aldrich) for 1 h at 37°C. DRG maturation medium was composed by DMEM high glucose supplemented with 10% FBS, 50 ng/ml Nerve Growth Factor (NGF) and 10 U/ml of Penicillin/10 μg/ml Streptomycin (all from Life Technologies). After 2 to 3 days of culture, 0.5 μM Cytosine Arabinose (AraC; Sigma-Aldrich) was added to inhibit proliferating cells. After a 10 day period of DRG maturation, OPC-like cells derived from UCM-MSCs were plated on top of the neurons at a density of 20,000 cells/cm² and the culture remained in OL-like maturation medium supplemented with T3 [as described in the section *Maturation of OPC-like cells into Oligodendrocyte (OL)-like cells*]. Medium was changed every 3 or 4 days. By the end of 14 days, cells were fixed and immunocyto-chemistry analysis was performed.

Statistical analysis

Statistical analysis was performed by Kruskal-Wallis one-way ANOVA followed by Dunn's multiple comparison test or two-tailed Mann-Whitney test (when data was not normal or homoscedastic), or by repeated measures one-way ANOVA followed by Tukey's test (when a parametric analysis of paired data was appropriate), as indicated. Analysis was done using the software GraphPad Prism 5. Values represent mean ± SEM of at least 3 independent experiments (*$P<0.05$; **$P<0.01$ and ***$P<0.001$ for statistically significant differences).

Protein extraction and quantification, SDS-PAGE, and immunobloting

To obtain protein extracts from MSCs, cells from one TC plate with 21 cm² were detached using StemPro Accutase. The enzyme was inactivated with PBS and then cells were centrifuged (300×*g*, for 5 min at RT). The pellet was washed with PBS and centrifuged again. The supernatant was discarded and the pellet was frozen in liquid nitrogen for 20 seconds and stored at −80°C until further processing. To prepare protein extracts, frozen cell pellets were incubated with 200 μL of cold (4°C) extraction buffer [50 mM Tris-HCl pH 7.4 supplemented with cOmplete, EDTA-free protease inhibitor cocktail (Roche)]. Cells were disrupted using a 200 μl micropipette by repeatedly pipetting up and down on ice and then sonicated on ice, using an ultrasonic cell disrupter (VibraCell - model VCX 750, Sonics & Materials, Inc.) for 1 min (cycles of 1 s pulse interspaced by 1 s) with an amplitude of 40%, to fully disrupt the membrane structure. After centrifugation (at 20,000×*g* at 4°C for 30 min), the supernatants were retrieved and the protein was precipitated by adding 6 volumes of cold acetone

($-20°C$). Samples were then vortexed and centrifuged (at $20,000 \times g$ at $4°C$ for 30 min). The pellets were washed with 90% (v/v) cold acetone and then allowed to dry at RT for 15 min. Pellets were resuspended in 100 µl of 1x sample buffer [58.3 mM Tris/HCl, pH 6.8, 17 µg/ml SDS (BioRad), 50 µl/ml glycerol (GE Healthcare), 15.5 µg/ml DTT (Bioron), 20 µg/ml Bromophenol Blue (GE Healthcare)] and heated at $95°C$ for 5 min. Protein was quantified using the 2D quant kit (GE Healthcare), according to manufacturer's instructions.

To obtain protein extracts from brain cortex to be used as a positive control for MBP, the brain of one adult female Wistar rat was extracted and stored at $-80°C$ until further processing (procedures were performed according to the European Union Directive 86/609/EEC and the legislation Portaria n. 1005/92, issued by the Portuguese Government for the protection of animals used for experimental and other scientific purposes). To obtain protein extracts, the brain was thawed with cold extraction buffer and then transferred to one centrifuge tube with 750 µL of extraction buffer. The tissue was homogenized using an ultrasonic cell disrupter microtip (VibraCell - model VCX 130, Sonics & Materials, Inc.) with an amplitude of 40–60% (cycles of 1 s pulse interspaced by 1 s) until homogenized. After centrifugation (at $5,000 \times g$ at $4°C$ for 5 min), the supernatant was collected. Next, protein was precipitated (200 µL of the total protein extract) and further processes as described to obtain protein extracts from MSCs.

Protein samples (15 µg/lane for rat brain cortex and 35 µg/lane for MSCs extracts) were separated by SDS-PAGE on SDS/discontinuous 4–12.5% (w/v) acrylamide–bisacrylamide (Bio-Rad) gels at constant voltage (80V while samples were in the stacking gel and 120V after entering the resolving gel, for about 1 h 30 min) using a Mini-Protean III (BioRad) apparatus, with running buffer (BioRad).

Blotting was done using polyvinylidene fluoride (PVDF) membranes (BioRad). Membranes were activated by rinsing in methanol, washed with deionized water and then kept in transfer buffer (BioRad). Protein transfer (western-blot) was performed using a Trans-Blot Turbo transfer system (BioRad) according to the manufacturer's instructions, using Trans-Blot Turbo transfer buffer (BioRad).

For immunoblotting, membranes were blocked with blocking solution [PBS-0.1% (v/v) Tween 20 (GE Healthcare) and 5 g/100 ml non-fat dried milk] for 1 h at RT. Incubation with anti-MBP antibody was performed with gentle agitation, overnight at $4°C$, followed by 1 h at RT, while staining with anti-GAPDH antibody occurred for 2 h at RT. Membranes were washed with PBS-0.1% (v/v) Tween 20 and then incubated with the respective secondary antibody diluted in blocking solution (1 h at room temperature) and washed with PBS-0.1% (v/v) Tween 20.

Primary antibodies used were diluted in blocking solution and were rabbit monoclonal (clone EP1448Y) anti-MBP (Abcam, Ab53294) diluted 1:1,000 and mouse monoclonal anti-GAPDH (Santa Cruz Biotechnology, SC-365062) diluted 1:500. The respective secondary antibodies were also diluted in blocking solution and were goat anti-rabbit antibody conjugated with alkaline phosphatase (Jackson ImmunoResearch Laboratories, Inc.) diluted 1:6,000 and donkey anti-mouse antibody conjugated with Alexa 488 (Life Technologies) diluted 1:500.

An ECF (enhanced chemiflorescence) kit (GE Healthcare) was used for detection of the alkaline phosphatase-conjugated antibody, according to the manufacturer's instructions. Alkaline phosphatase activity (using ECF) and Alexa 488 fluorescence was visualized on a Molecular Imager FX system (BioRad) using the software Quantity One (BioRad). The integrated density of the western-blot bands was obtained using Image J software, using the 'Threshold' tool, followed by the command 'Analyze Particles'.

Results

Isolation, expansion and characterization of human umbilical cord matrix mesenchymal stem cells (UCM-MSCs)

Human MSCs were isolated from 12 umbilical cords, as described in the 'Materials and Methods' section, with a success rate of 100% (in agreement to what has been previously reported [34]). At the end of 5 to 10 days in culture, several Wharton's jelly fragments were still attached to the tissue culture (TC) dishes and showed cells migrating from the tissue. Colonies of cells displaying an MSC-like phenotype, with spindle-shaped morphology could be readily identified by phase-contrast microscopy (Figure S1). Cells proliferated rapidly and formed compact colonies by the end of 10-14 days in vitro. Then, the cells were detached and dissociated and by the end of Passage 1 (P1) constituted a homogeneous monolayer with MSC-like morphology (adherent, spindle-shaped fibroblastoid-like cells). The cells were then either cryopreserved or expanded until passage 8 (P8) and further characterized.

UCM-MSCs have been described to proliferate readily in vitro [34,35], hence we sought to thoroughly characterize their proliferation kinetics, namely the total number of cells - TNC (Figures 1A, B), generation time - GT (Figure 1C), population doubling - PD (Figure 1D) and cumulative population doubling - CPD (Figure 1E) of cells isolated from 3 randomly selected independent samples (CM#2, CM#3 and CM#7) between passages 2 and 8 (P2-P8). Notably, by the end of P4 (17 to 21 days after the initial isolation of the UCM explants), the total number of cells obtained from each sample (i.e., if no cells had been discarded until that point) had surpassed 1×10^9 cells (Figures 1A, B), well above what is considered relevant for clinical applications [29] (about 1–2 million cells per kg of body weight) and at passage 8 the TNC was 5.98×10^{13} ($\pm 4.76 \times 10^{13}$). On average, between P2 and P8, the generation time ranged from 1.02 ± 0.071 (SEM) to 1.55 ± 0.316 (SEM) days (Figure 1C). Moreover, the colony-forming unit-fibroblast (CFU-F) capacity of the cells (Figure 1F) was maintained throughout P2 to P8 [48.5 ± 7.52 (SEM) and 43.8 ± 16.20 (SEM)]. For both types of assays (GT and CFU-F), no statistically significant differences were found between the time points, indicating that neither the generation time nor the CFU-F capacity were significantly affected by passaging the cells until P8. Hence, the cells showed high proliferative capacity typical of bona-fide MSCs, maintained a short generation time from passages 2 to 8 and reached a clinically relevant number (superior to 1×10^9 cells) within a time-frame of 17 to 21 days.

Multilineage differentiation capacity and immunophenotypic characterization of UCM-MSCs

MSCs are multipotent stem cells that can differentiate in vitro into chondrocytes, osteocytes and adipocytes [2]. The multilineage differentiation potential of UCM-MSCs was demonstrated in culture (Figure 2A-C), under conditions that favor osteogenic, adipogenic or chondrogenic differentiation (Materials and Methods section). Chondrogenic induction was observed by Alcian Blue staining (Figure 2A), while osteogenic differentiation was evident by an increase in alkaline phosphatase (ALP) activity (reddish areas) and enhanced mineralization showed by von Kossa staining (dark areas), as shown in Figure 2B. Adipogenic induction, although less efficient, was visible by the cellular accumulation

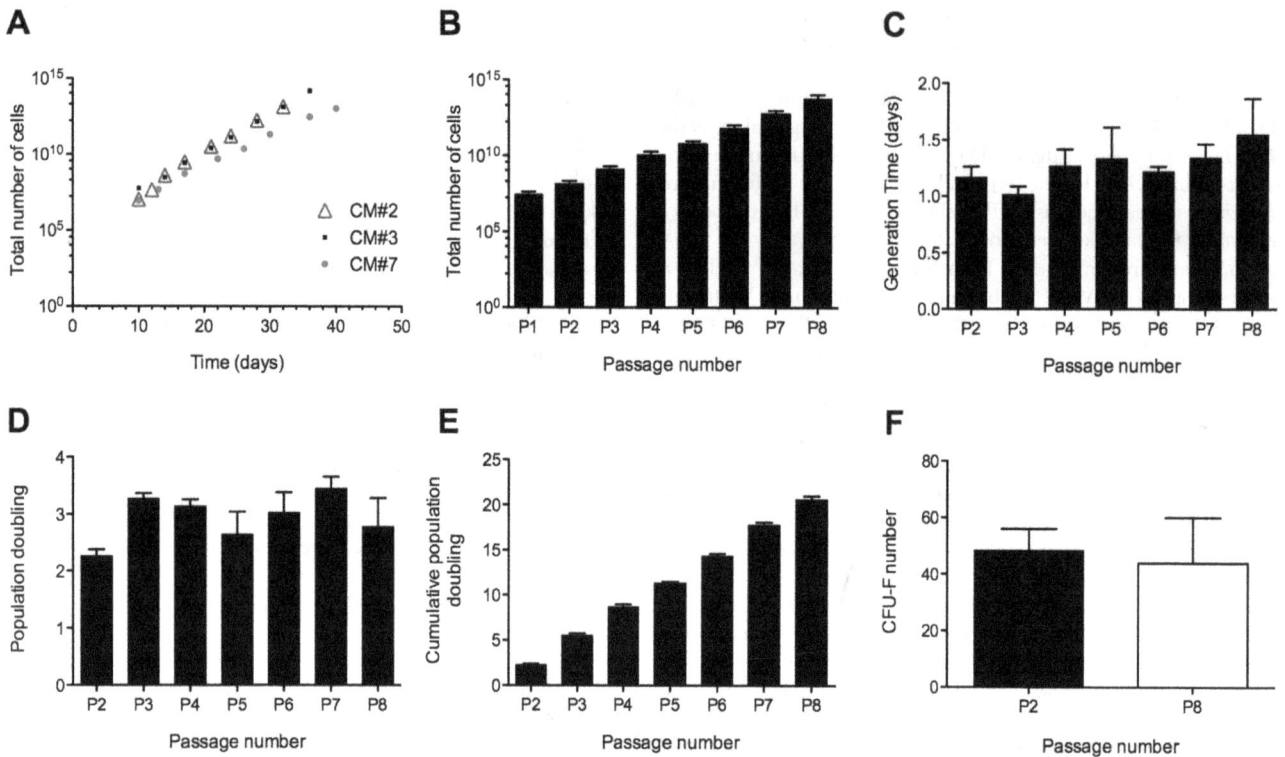

Figure 1. Proliferation and colony-forming capacity of human umbilical cord mesenchymal stem cells (hUCM-MSCs) *in vitro*. Total number of cells (hUCM-MSCs) along time (days) and passage number (*A, B*), generation time (*C*), population doubling (*D*), cumulative population doubling (*E*) from passage 1 (P1) or P2 until P8, as indicated, and colony forming unit-fibroblast (CFU-F) number at P2 and P8 (*F*). Cells were isolated from 3 distinct human umbilical cord matrix samples (CM#2, #3 and #7). From P2 onwards (inclusively), cells were plated at a fixed density of 3,000 cells/cm^2, allowed to proliferate until sub-confluence and re-plated in the same way (*A-E*), or plated at 3 cells/cm^2 at P2 and P8 and cultured for 15 days for the CFU-F study (*F*). The total number of cells (TNC) was determined at each passage (P1-P8) by cumulative counting of the cells once they reached a confluence of 80% (*A, B*). The TNC designates the theoretical number of cells that could be obtained if no cells were discarded between each passage. The observed mean generation time (GT) was between 1.02 (±0.071) and 1.55 (±0.316) days (*C*), and no statistically significant differences were found in GT from passages 2 to 8 (Kruskal-Wallis one-way ANOVA followed by Dunn's Multiple Comparison test). The CFU-F capacity was maintained from P2 to P8 (*F*) and no statistically significant differences were found (two-tailed Mann-Whitney test). Bars represent mean ± SEM (*B-F*).

of lipid-rich vacuoles that stained with Oil Red-O (Figure 2C). These results indicate that UCM-MSCs possess the multilineage differentiation capacity characteristic of MSCs [2].

Flow cytometry analysis showed that UCM-MSCs were positive for CD13, CD73, CD90 and CD105, while cells did not express CD34 and CD45 (hematopoietic lineage markers), CD11b and NGFR (Figure 2D-K and Table 1 represent the analysis of one sample in passage 2, while data from 3 independent samples at passage 2 and 8 is presented in Figure S2), showing that this phenotypic profile was consistent between different donors and with the MSC phenotype previously described by us [36] and others [30,37,38].

Differentiation of UCM-MSCs into an oligodendrocyte-like lineage

After assuring the fidelity of the UCM-MSCs obtained, we focused on the differentiation of these cells into oligodendrocyte-like cells using a stepwise protocol, based on the literature for neural- and oligodendroglial-like induction of distinct types of stem cells, including embryonic stem cells [24,25,39] and MSCs [7,9,22]. MSCs were subjected to distinct soluble factors, culture surfaces and coating conditions during the differentiation process (Figure S3). Briefly, undifferentiated MSCs (Figures 3A and 4A-C) were treated with epidermal growth factor (EGF) for 3 days for

generation of neuroectodermal-induced MSCs (niMSCs) (Figure 3B). Next, niMSCs were cultured on non-treated polystyrene in the presence of EGF and basic fibroblast growth factor (bFGF) for induction into neural stem-like (NSC-like) cells (Figures 3C, D, F and 4D-F). The latter were then cultured on tissue culture polystyrene using different coating conditions (Figure S3) for the generation of oligodendrocyte precursor-like (OPC-like) cells (Figure 4G-O). Finally, these progenitor cells were seeded on poly-D-lysine (PDL) and laminin-2/merosin (MN) double-coated surfaces for the generation of oligodendrocyte-like (OL-like) cells (Figures 5 and 6).

Induction of MSCs into neural-like precursor cells

MSCs used for differentiation experiments were taken from cryopreserved cell batches between passages 3 and 7 obtained from 3 independent donors, characterized as described above (Figures 1 and 2). After 3 days under neuroectodermal-inducing conditions (in the presence of N2 and EGF) the niMSCs kept the spindle-shape morphology (Figure 3B) characteristic of MSCs (Figure 3A) and seemed to maintain proliferation.

In order to induce the differentiation of niMSCs into a NSC-like phenotype, cells were seeded on non-treated polystyrene plates (sterile, non-tissue culture 'bacterial dishes') and cultured up to 18 days (Figure 3C,D). The induction medium contained the soluble

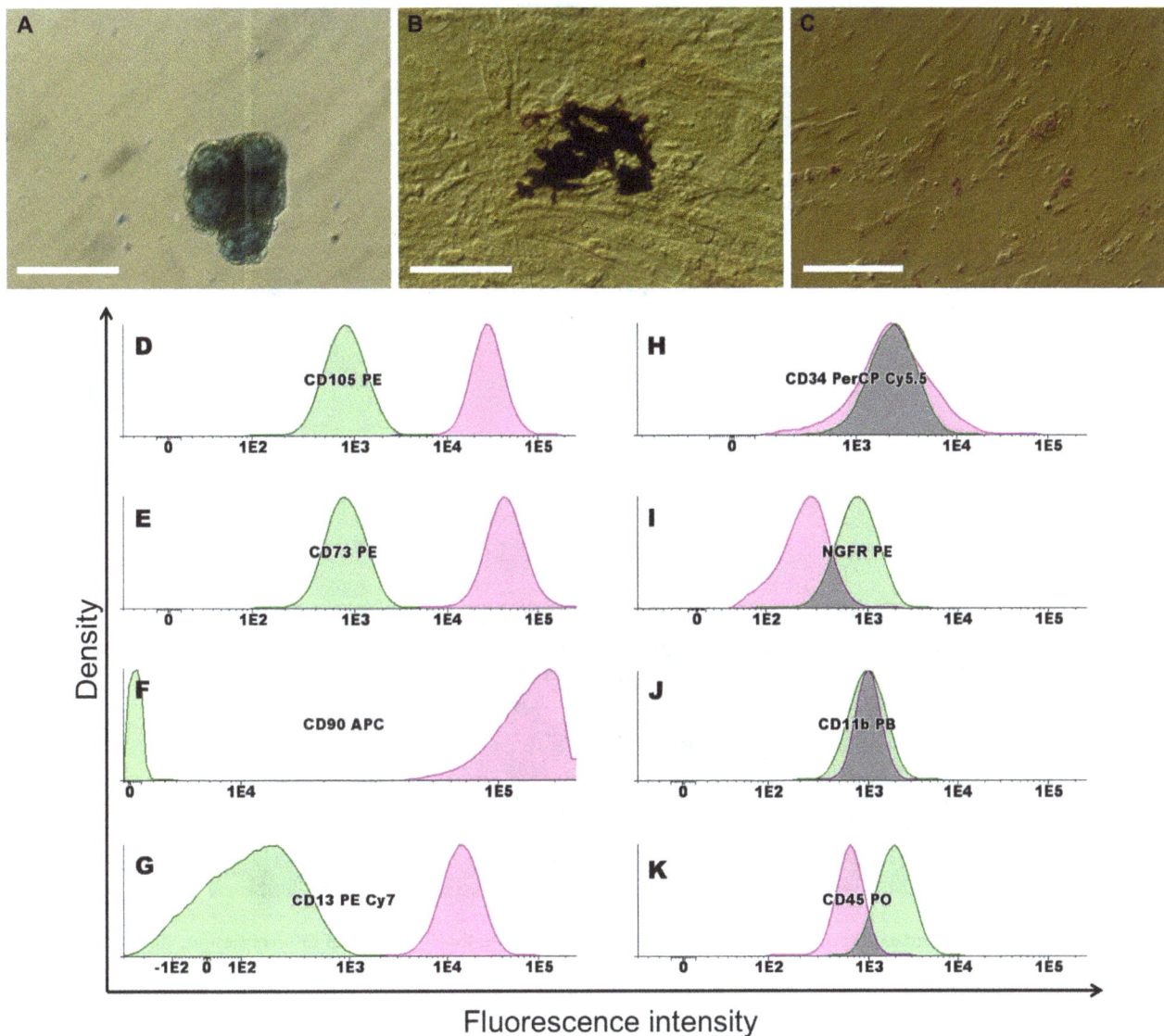

Figure 2. Multilineage differentiation potential and immunophenotypic characterization of hUCM-MSCs. The multilineage differentiation potential of hUCM-MSCs was demonstrated after 14 days in culture, under conditions that favour chondrogenic, osteogenic or adipogenic differentiation. Phase contrast images of cells stained with Alcian Blue for chondrogenesis (A), ALP and von Kossa for osteogenesis (B) or Oil Red-O for adipogenesis (C). Scale bar represents 100 µm. For immunophenotypic characterization, hUCM-MSCs at passage 2 were dissociated using Accutase (Life Technologies), labelled with antibodies against the indicated antigens and analysed by flow cytometry. Cells were positive for CD105 (D), CD73 (E), CD90 (F) and CD13 (G) and negative for CD34 (H), NGFR (I), CD11b (J) and CD45 (K) (pink lines) when compared with unlabelled MSCs (green lines), as depicted in the histograms. Histograms were obtained from one sample (UCM#2) at passage 2 and are representative of 3 independent samples at P2 and P8.

Table 1. Summary of the flow cytometry analysis of UCM-MSCs.

Positive markers	Negative markers
CD13	CD11b
CD73	CD34
CD90	CD45
CD105	NGFR

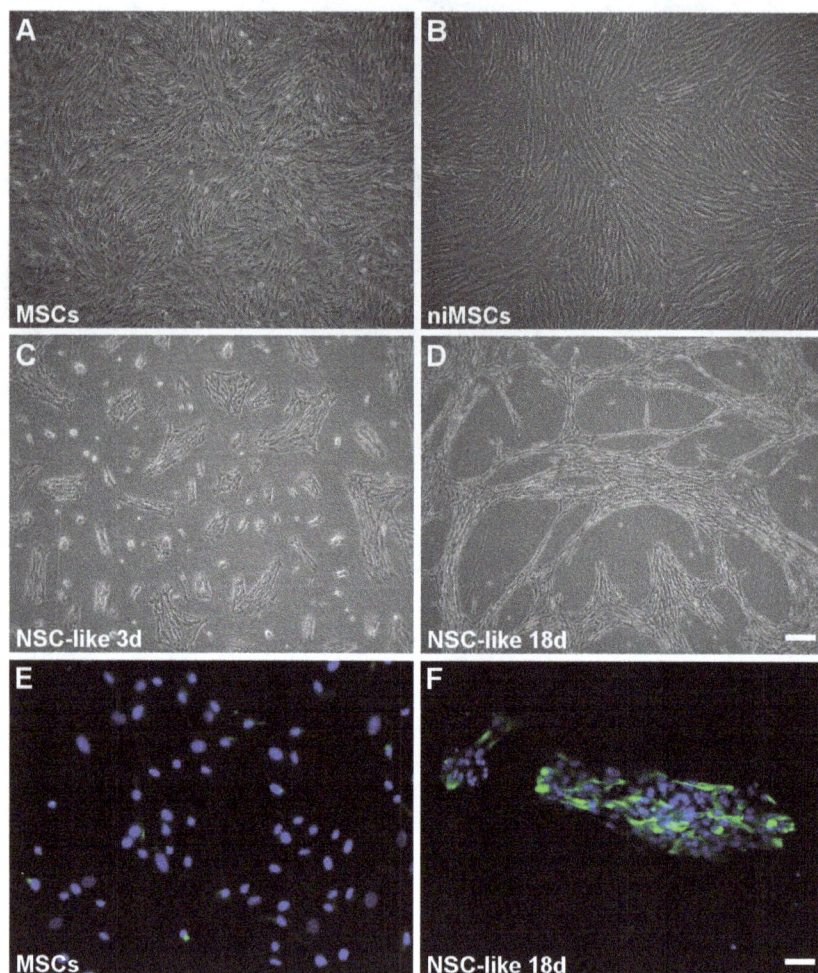

Figure 3. Cell morphology and expression of nestin in hUCM-MSC and during the first steps of differentiation. Phase contrast images of undifferentiated hUCM-MSCs (*A*), neuroectodermal-like induced MSCs – niMSCs (*B*) and NSC-like cells after 3 (*C*) and 18 days (*D*) in culture in NSC induction medium. Immunofluorescence microscopy images for the neural precursor marker nestin (in green) in undifferentiated MSCs (*E*) and NSC-like cells (*F*). Counterstaining of nuclei was performed with DAPI (in blue). In (*A-D*) the scale bar represents 200 μm and in (*E-F*) represents 50 μm. Images are representative of at least 3 independent experiments.

factors bFGF and EGF, described to favor neuroectodermal-like fate of MSCs [7,9]. After 3 days in culture (Figure 3C), cells had formed patches or small colonies that were maintained until 18 days in NSC-induction medium (Figure 3D). Although undifferentiated MSCs already expressed nestin [11] to a certain extent (Figures 3E and 4A), there was an increase in the expression of this NSC marker after the NSC-like induction phase, as addressed by immunocytochemistry (Figures 3F and 4D). The quantification of the mean fluorescence intensity (MFI) of immunofluorescence images (see Materials and Methods) was addressed for distinct neural markers at the different stages of differentiation under study (Figure 7A-D). The MFI of nestin in cells after NSC-like induction (NSC-like) showed a significant increase compared to undifferentiated MSCs (Figure 7A).

No substantial expression of the oligodendrocyte precursor markers A2B5 or O4 (the latter being also present in mature oligodendrocytes) could be detected at the NSC-like stage or in undifferentiated MSCs (Figures 4 E/F and B/C, respectively), and the faint staining visible in the immunocytochemistry images seem to represent background staining, as will become evident when comparing MSCs with cells at more mature differentiation stages (Figures 4, 6 and 7B). Interestingly, galactocerebroside (GalC, a

marker of mature oligodendrocytes) seems to be already expressed by undifferentiated MSCs (Figure 6S), confirming observations made by others [7]. However, its expression was significantly increased after NSC-like induction (Figure 7C). The expression of myelin basic protein (MBP), a protein expressed by mature oligodendrocytes, also showed a tendency to increase during the NSC-like differentiation stage (Figure 7D). Nevertheless, MBP could already be detected in low amounts in undifferentiated MSCs, as determined by immunocytochemistry (Figures 6T and 7D) and semi-quantitative RT-PCR (not shown), in agreement with results by others [7].

Oligodendrocyte precursor cell (OPC)-like induction

At the end of the NSC-like induction protocol, cells were detached and seeded on 6-well plates coated with fibronectin (FN), described to play an essential role in survival and proliferation signaling during the OPC differentiation stage [40]. In parallel, cells were also seeded on laminin-2/merosin (MN) coated wells, as previously described by Zhang and colleagues [22] for the differentiation of MSC into OPC-like cells. As control, cells were also plated on non-coated (NC) tissue culture wells.

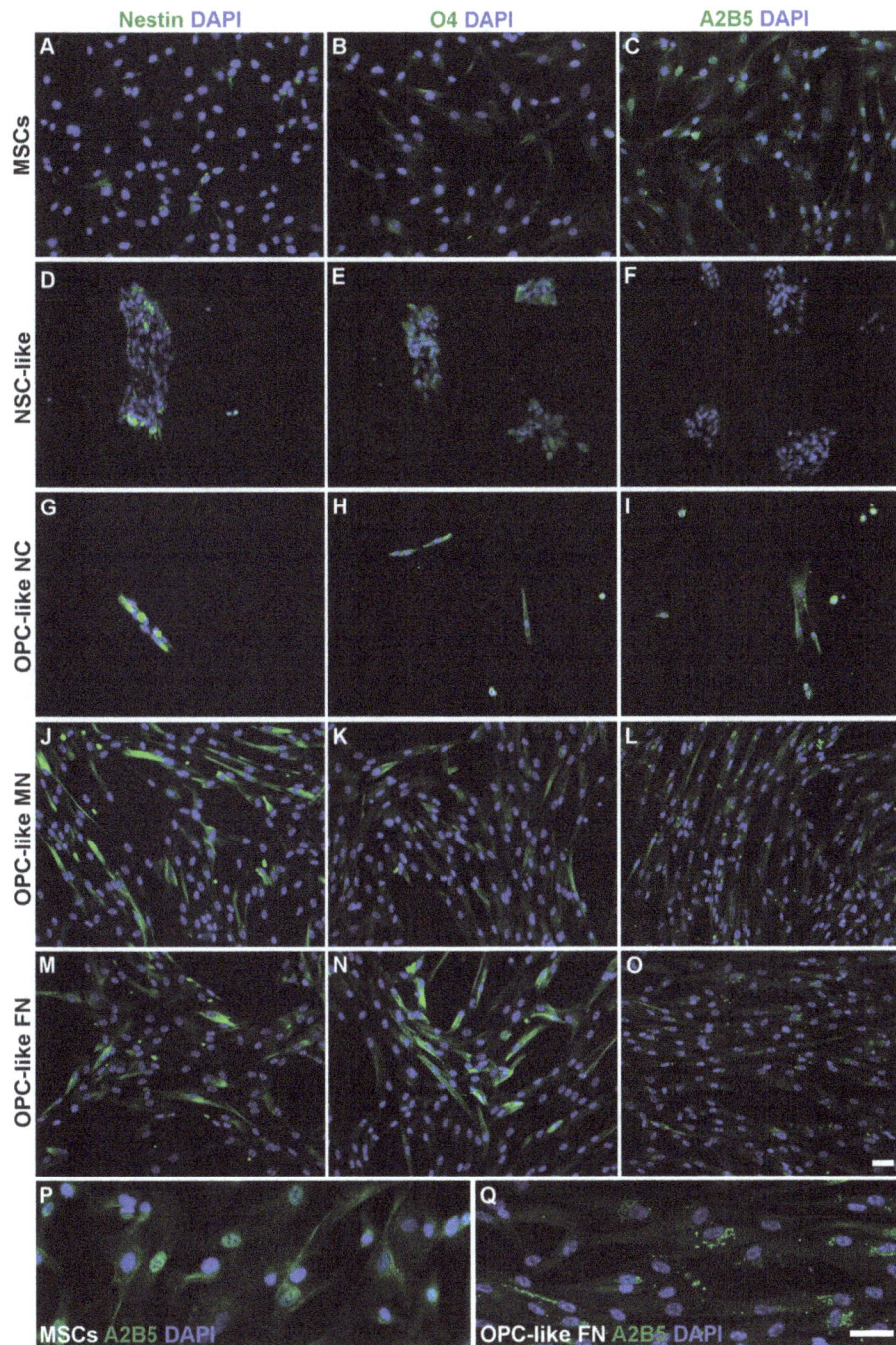

Figure 4. Expression of neural- and oligodendroglial-precursor markers in hUCM-MSCs, NSC- and OPC-like cells. The expression of nestin (neural precursor marker), A2B5 (OPC marker) and O4 (OPC/OL marker) in undifferentiated hUCM-MSCs (*A-C*), after neural-like induction (*D-F*) and OPC-like induction using non-coated (*G-I*), laminin-2 (merosin-MN)-coated (*J-L*) and fibronectin (FN)-coated (*M-O*) tissue culture polystyrene was determined by immunofluorescence microscopy (in green). (*P*) and (*Q*) are inserts of (*C*) and (*O*) images, respectively, to highlight the presence of the typical OPC punctate and perinuclear distribution of A2B5 in OPC-like cells (*Q*), in contrast with control hUCM-MSCs (*P*). Counterstaining of nuclei was performed with DAPI (in blue) and the scale bars represent 50 μm. Images are representative of at least 3 independent experiments.

By the end of the OPC-like induction stage, the cells cultured on FN- or MN-coated plates survived well and maintained the spindle shape morphology, when compared with cells cultured on uncoated wells, which did not attach well to the surface and only some small adherent clusters and single cells survived until the end of this stage (Figure 4G-O).

Figure 4 shows that there seemed to be a tendency for increased expression of Nestin, O4 and A2B5 in OPC-like cells (regardless of the coating conditions used) when compared with undifferentiated MSCs. The quantification of the MFI of Nestin and O4 (Figure 7A,B) showed that despite not being statistically significant, there was a slight increase of these markers in OPC-like cells cultured in FN when compared with MSCs. RT-PCR analysis

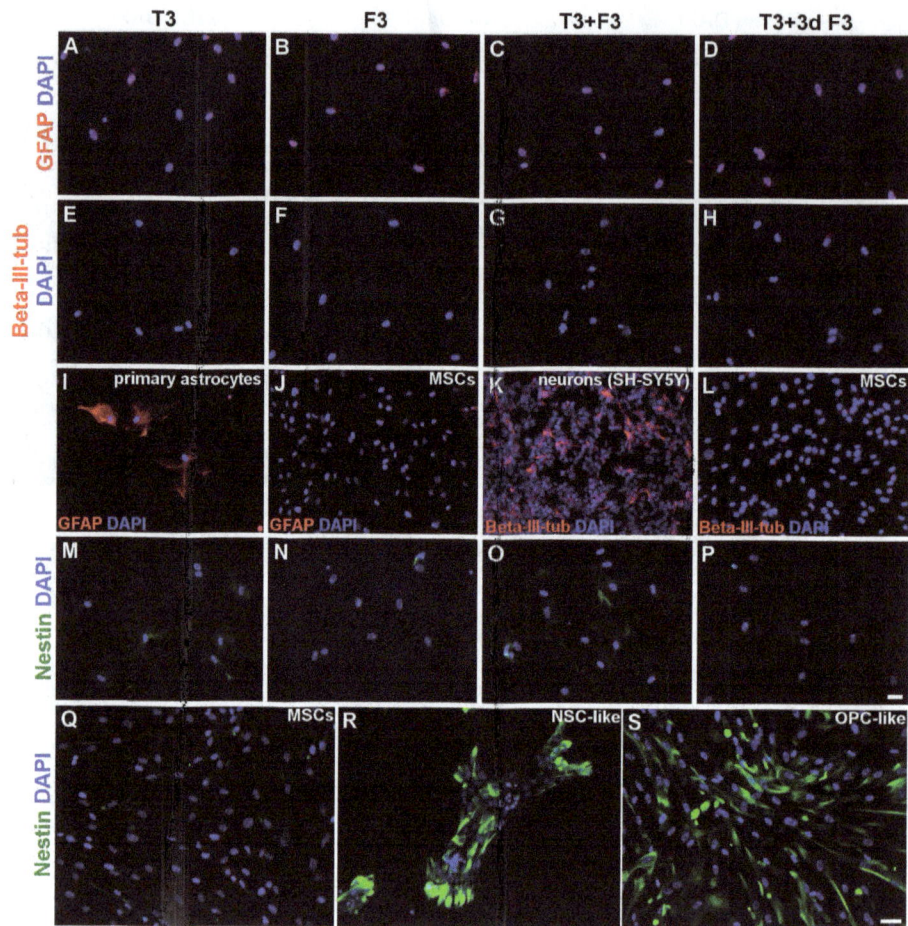

Figure 5. Expression of neural precursor, astrocytic and neuronal markers by OL-like cells. The expression of the astrocytic marker GFAP (*A-D and I-J*), the neuronal marker beta-III-tubulin (*E-H and K-L*) and the neural precursor marker nestin (*M-P and Q-S*) in OL-like cells derived from OPC-like cells cultured on fibronectin (*A-H and M-P*) was determined by immunofluorescence microscopy. OL-like cells were differentiated for 10 days on laminin-2 (merosin)-coated wells, in the presence of T3 (thyroid hormone), F3 (contactin), T3+F3 or 7 days T3 followed by 3 days T3+F3 (T3+3d F3), as indicated. It was evident that the neural precursor marker nestin decreased in the OL-like stage of differentiation (*M-P*), as compared to NSC- or OPC-like cells (*R and S, respectively*), indicating maturation of the OL-like cells. It was also clear that the astrocytic marker GFAP was present in rat astrocytes as expected (*I*), but essentially absent in OL-like cells (*A-D*) or hUCM-MSCs (*J*). Similarly, the neuronal marker beta-III-tubulin was detected in the neuronal cell line SH-SY5Y (*K*) but undetected in OL-like cells (*E-H*) or hUCM-MSCs (*L*). Counterstaining of nuclei was performed with DAPI (in blue) and the scale bars represent 50 μm - scale bar in image (*P*) is representative for images (*A-P*) and scale bar in image (*S*) is representative for images (*Q-S*). Images are representative of at least 3 independent experiments.

showed that nestin was in fact upregulated in terms of gene expression in OPC-like cells when compared to undifferentiated MSC (Figure 7E). The presence of Nestin in OPC-like cells is in agreement with the literature, since it has been reported that OPCs may express nestin [41,42]. Nonetheless, there seems to be a peak of Nestin expression during the NSC-like stage, and then a significant decrease at the OPC-like stage (Figure 7A), which is consistent with a NSC-like stage followed by an OPC-like stage of differentiation. Although MFI analysis of the levels of O4 between OPC-like cells and undifferentiated MSCs showed that these differences were not statistically significant (Figure 7B), there was a trend of increased expression of O4 on OPC-like cells when comparing with MSCs (Figure 4B and N), indicating at least some degree of oligodendroglial-like commitment of the cells. Moreover, the presence of A2B5 was visible in OPC-like cells (especially when cultured on FN or MN-coated surfaces) when compared to MSCs (Figure 4C, I, L, O, P and Q). Although it was not possible to quantify the MFI of this marker (due to its punctate distribution), it can be readily perceived in Figures 4P and Q (which show a

higher magnification of the images depicted in Figures 4C and O, respectively) that the punctate and perinuclear distribution of A2B5 similar to that reported in *bona-fide* OPCs [42] was only visible in OPC-like cells and not in MSCs.

Overall, the protocol tested using FN or MN-coated tissue culture polystyrene (TCPS) seemed to favor a neural- and oligodendroglial precursor-like fate of MSCs.

Oligodendrocyte (OL)-like induction

OPC-like cells were detached and dissociated from the non-coated, fibronectin- or laminin-coated TCPS and re-seeded on PDL and MN double-coated 96 well plates, in distinct OL differentiation media (see Materials and Methods). The presence of merosin is known to favor oligodendrocyte differentiation [43], hence this extracellular matrix (ECM) protein was chosen as a coating element for the final stage of the protocol.

In order to induce the differentiation of OPC-like cells into OL-like cells, the thyroid hormone T3 (known to induce the differentiation of oligodendrocyte progenitor into mature OLs

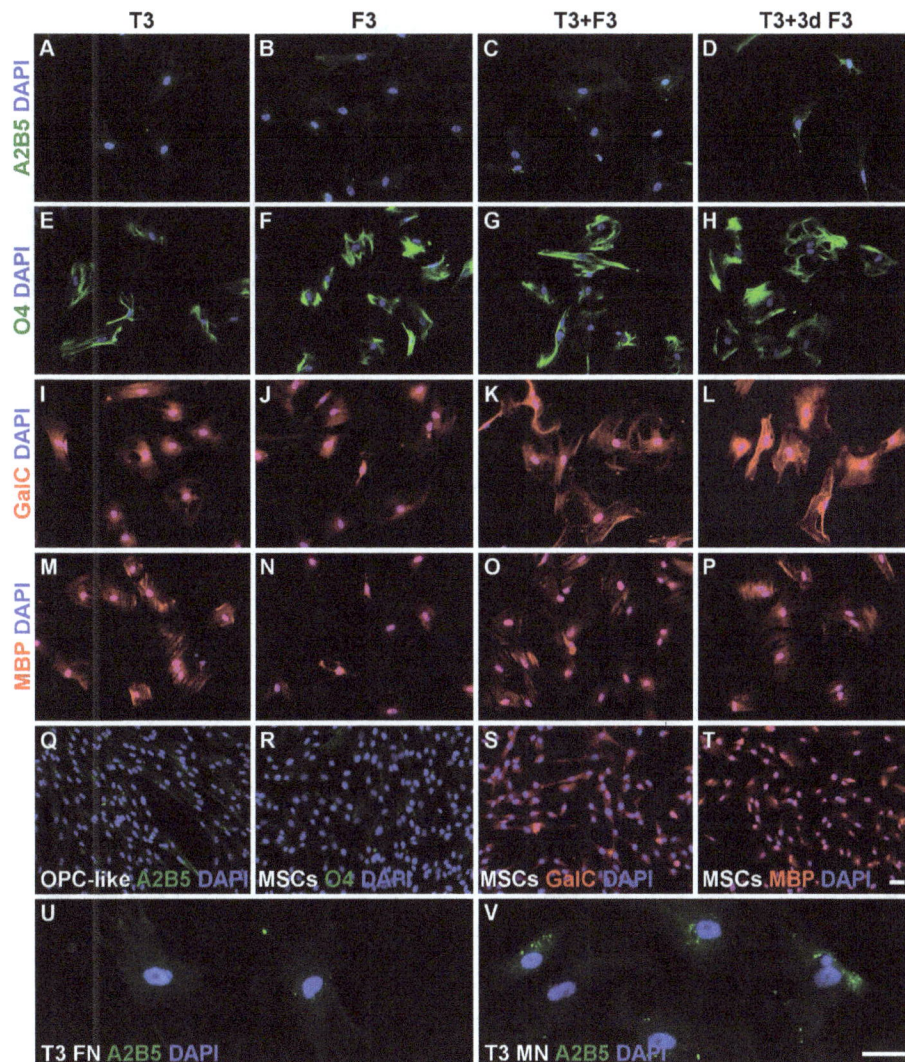

Figure 6. Expression of oligodendroglial markers by OL-like cells. The expression of the OPC marker A2B5 (*A-D*), the OPC/OL marker O4 (*E-H*) and the OL markers GalC (*I-L*) and MBP (*M-P*) in OL-like cells derived from OPC-like cells cultured on fibronectin was determined by immunofluorescence microscopy. OL-like cells were differentiated for 10 days on laminin 2 (merosin)-coated wells, in the presence of T3 (thyroid hormone), F3 (contactin), T3+F3 or 7 days T3 followed by 3 days T3+F3 (T3+3d F3), as indicated. It could be observed that the OPC marker A2B5 greatly decreased in the last stage of differentiation (*A-D*), as compared to OPC-like cells (*Q*), indicating maturation of the OL-like cells. The decrease was also more evident when OL-like cells were derived from OPC-like cells cultured on fibronectin (FN)-coated wells (*U*) than in laminin-2/merosin (LM/MN)-coated wells (*V*). The OPC/OL marker O4 was expressed at high levels in OL-like cells (*E-H*), in contrast to the low levels in hUCM-MSCs (*R*). The OL marker GalC was more highly expressed in OL-like cells (*I-L*) as compared to MSCs (*S*), while MBP seemed to be expressed (*M-P*) at similar levels (*T*). Counterstaining of nuclei was performed with DAPI (in blue) and the scale bars represent 50 μm - scale bar in image (*T*) is representative for images (*A-T*) and scale bar in image (*V*) is representative for images (*U-V*). Images are representative of at least 3 independent experiments.

[44]), F3/Contactin (a ligand known to induce the maturation of oligodendrocytes [21,22,45]), or T3 combined with F3 (T3F3) were added to the basal differentiation media and incubated for 10 days. It was also tested the effect of T3 in the first 7 days of culture and T3 together with F3 during the last 3 days of differentiation (7dT3+3dT3F3).

Immunocytochemistry analysis of nestin in OL-like cells previously cultured on FN-coated TCPS during the OPC-like stage (the preceding differentiation step) showed that the expression of this neural stem/progenitor marker decreased significantly during the final stage of differentiation in all the conditions tested in the final differentiation stage when compared with NSC-like cells, or when compared with OPC-like cells when differentiated terminally in the presence of 7dT3+3dT3F3 or T3

(Figures 5M-S and 7A). Nestin levels peaked at the NSC-like stage and then decrease during the final differentiation step (OL-like stage), to levels similar to those of undifferentiated MSCs (Figures 5M-S and 7A). These results support the idea that the cells underwent a neural progenitor-like state during the NSC- and OPC-like stages of the differentiation protocol and then differentiated into a more mature phenotype at the OL-like stage (similar to what has been described for *bona-fide* OLs [46]).

Reinforcing this idea was the behavior of the OPC marker A2B5 [46], which decreased during the final stage of differentiation (OL-like cells) when compared with OPC-like cells (Figure 6 A-D, Q, U, V). This effect was more evident in oligodendrocyte-like cells that had been previously cultured on FN-coated dishes during the OPC-like differentiation stage than in cells that had

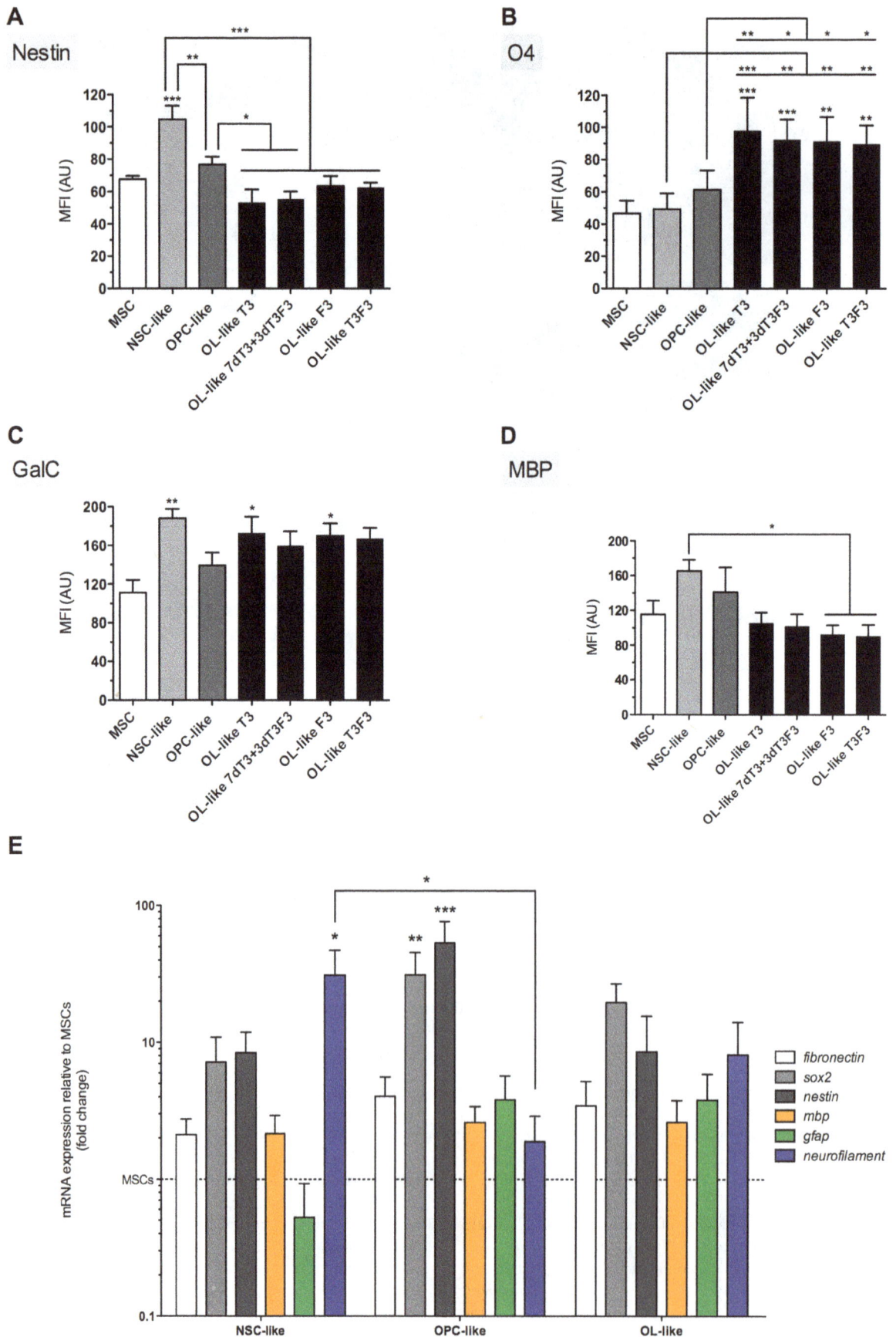

Figure 7. Quantification of mean fluorescence intensity (MFI) of immunofluorescence images and mRNA expression of differentiation stage markers. MFI values for nestin (*A*), O4 (*B*), GalC (*C*) and MBP (*D*) of immnunofluorescence images (representative images on Figures 4 to 6) were quantified using Image J software. The graphics represent results regarding cells cultured during the OPC-like differentiation stage on fibronectin-coated TC polystyrene. OL-like cells were differentiated for 10 days on laminin 2 (merosin)-coated wells, cultured in the presence of T3, F3, T3F3 or 7dT3+3dT3F3, as indicated. Bars represent mean ± SEM of at least 3 independent experiments. Statistical analysis was performed by repeated measures one-way ANOVA followed by Tukey's Multiple Comparison Test. Statistically significant differences for each marker between MSCs and cells at other differentiation stages were indicated on top of the corresponding graphic bars, while differences between other conditions were represented using connectors (*$P<0,05$; **$P<0,01$ and ***$P<0,001$). Real time RT-PCR analysis (*E*) was performed for genes representative of neural progenitors (*sox2* and *nestin*), neurons (*neurofilament*), astrocytes (*gfap*) and oligodendrocytes (*mbp*), while *fibronectin* was used as a mesenchymal marker. Bars represent mean ± SEM of at least 3 independent experiments and are expressed as fold change of $2^{-\Delta\Delta Ct}$ using *actin* as a reference gene and undifferentiated MSCs as the control condition. Statistical analysis was performed by Kruskal-Wallis one-way ANOVA followed by Dunn's multiple comparison test (*$P<0,05$, **$P<0,01$ and ***$P<0,001$).

previously been cultured on MN-coated TCPS (Figure 6U, V). The loss of this OPC marker indicates that OL-like cells derived from OPC-like cells cultured on FN-coated wells might be more mature than those obtained from OPC-like cells cultured on MN-coated or non-coated wells, in agreement with reports highlighting the important role of fibronectin during the OPC stage of differentiation [40].

To further characterize the OL-like cells obtained, we screened for the presence of neural maturation markers for astrocytic (GFAP), neuronal (beta-III-tubulin) and oligodendroglial (O4, GalC and MBP) lineages. The images of OL-like cells presented in Figures 5 and 6 were obtained from cells that had previously been cultured on FN-coated TCPS during the OPC-like state, and from this point onwards will be considered as the main condition under analysis during the final stage of differentiation, unless stated otherwise.

The expression of GFAP was negative in OL-like cells in all differentiation conditions (Figure 5 A-D), similar to what had been observed in undifferentiated MSCs (Figure 5J) and in contrast to what was shown in Figure 5I, where primary rat astrocytes were used as positive control for this marker. Cells that were cultured on laminin-2 or non-coated wells during the OPC-like stage showed some expression of this astrocytic marker in the final differentiation stage (OL-like stage) - data not shown. These results further supported the idea that the oligodendroglial-like fate of MSCs seems to be favored by the presence of FN during the OPC-like stage.

In Figure 5E-H, it can be observed that the expression of beta-III-tubulin was essentially absent in OL-like cells obtained in the presence of any of the distinct differentiation media, although there was some expression in cells that were cultured on non-coated wells during the OPC-like stage (data not shown). In Figure 5K, the SH-SY5Y cell line was used as a positive control for this neuronal marker, which was absent in undifferentiated MSCs (Figure 5L).

The expression of O4, an OPC marker which is maintained in mature oligodendrocytes [46], increased during the differentiation of MSCs into OL-like cells (Figure 6E-H). For cells cultured on fibronectin during the OPC-like stage, the quantification of the MFI of O4 showed a significant increase between MSCs, NSC-like or OPC-like cells and the cells obtained during the final stage of differentiation (OL-like cells) cultured in the presence of any combination of differentiation factors (Figure 7B). Similar results were obtained for cells previously cultured on non-coated TCPS, or for cells maintained on MN-coated TCPS during the OPC-like stage when finally differentiated in presence of F3 only (data not shown).

GalC seemed to be already expressed at some extent by undifferentiated MSCs (Figure 6S). However, the expression of this oligodendrocyte marker increased in oligodendrocyte-like cells differentiated under all tested conditions (Figure 6I-L). The MFI

of GalC showed a tendency to increase between MSCs and OL-like cells in all conditions. Compared to MSCs, the level of GalC expressed by OL-like cells was significantly increased only in OL-like cells cultured in the presence of T3 or F3 (Figure 7C). Unexpectedly, the presence of GalC was also higher in NSC-like cells, when compared with undifferentiated MSCs. Nevertheless, only during the final stage of differentiation (in OL-like cells) it was possible to observe cells with a branched morphology similar to oligodendrocytes and strongly positive for GalC, like those terminally differentiated in the presence of T3 or F3 (asterisk in Figures 6I and J, respectively). For cells cultured on non-coated wells during the OPC-like stage, there was a significant increase in the expression of GalC between OL-like cells (cultured in the presence of any of the final differentiation factors) and MSCs, but no significant differences were found between OL-like cells derived from OPC-like cells cultured on laminin-coated wells and MSCs (data not shown).

Immunocytochemistry analysis of MBP showed that this mature oligodendrocyte marker was present in OL-like cells (Figure 6M-P) after the final differentiation step, however, the levels of MBP were similar to those observed in undifferentiated MSCs, as illustrated in Figure 6T (the presence of MBP in MSCs was confirmed by western-blot analysis – Figure S4). The quantification of the MFI throughout the distinct stages of the differentiation protocol showed that the expression of MBP had a tendency to increase (although the differences were not statistically significant) during the NSC- and OPC-like stages of differentiation (Figure 7D). In fact, there was a significant decrease in the expression of MBP in the final step of differentiation, between NSC-like and OL-like cells cultured in the presence of F3 or T3F3, but not T3 or 7dT3+3dT3F3 (Figure 7D). Also, for cells cultured on uncoated surfaces during the OPC-like stage, there was a significant decrease in the expression of MBP between NSC-like cells and oligodendrocyte-like cells cultured in the presence of F3 and between OPC-like cells and OL-like cells cultured in the presence of T3 or F3 (not shown).

The expression of the oligodendrocyte markers MBP, GalC and O4, together with the downregulation of nestin and A2B5 in the OL-like differentiated cells, supports the idea that MSCs could be induced to differentiate into OL-like cells, sharing the behavior of several markers expressed by *bona-fide* oligodendrocytes and its precursors [46].

The more evident downregulation of A2B5 and the absence of neuronal (beta-III-tubulin) and astrocytic (GFAP) markers observed in OL-like cells derived from OPC-like cells cultured on fibronectin, suggests that these coating conditions (during the OPC-like stage) are suitable for the differentiation of MSCs into oligodendrocyte-like cells. Hence, UCM-MSCs could be differentiated into NSC-like cells, OPC-like cells and finally into oligodendrocyte-like cells, with expression of specific markers for each stage of differentiation. Figure 6 I and J (staining for GalC) illustrates terminally differentiated OL-like cells displaying a branched morphology (asterisks within images), resembling oligodendrocytes. Despite several cells with this morphology could be observed after the final differentiation step, they represented a minority of the cells.

Pattern of expression of MBP

Although we could not observe an increase of MBP expression in OL-like cells when compared with undifferentiated MSCs, it could be observed that there were distinct patterns of expression of MBP in the differentiated cells. One of the patterns was common to all stages of differentiation, including undifferentiated MSCs (Figure 6T), which was a strong staining in the nuclear area and more diffuse in cytosolic areas surrounding the nucleus (e.g.: Figure 8A, MBP panel, arrows). The other patterns suggested a structured distribution of MBP, resembling cytosolic filaments organized in a parallel manner, or at the periphery of the cells (e.g.: Figure 8A, MBP panel, arrow heads) and were present only in OL-like cells.

The quantification of the percentage of OL-like cells expressing structured MBP showed that the parallel distribution pattern of MBP could be observed in between approximately 25% and 45% of the cells (Figure 8B, left) and the occurrence of structured MBP at the periphery of the cells (Figure 8B, right) was present in between about 18% and 34% of the cells, depending on the differentiation medium used. Although the combined presence of T3 and F3 seemed to favor the appearance of patterned MBP, we could not find statistically significant differences between the distinct differentiation conditions.

The structured distribution of MBP resembled the pattern of polymerized actin filaments (F-actin). It is known that, in mature oligodendrocytes, MBP associates with F-actin and that this interaction seems to be important for the myelinating activity of OLs [47,48].

In order to test whether the structured MBP was coincident with polymerized actin filaments, the cells were analyzed for the presence of F-actin using fluorescently-labeled phalloidin. The immunocytochemistry results showed that whenever MBP could be detected in a structured manner (either with the parallel or the peripheral distribution), it always co-localized with F-actin (Figure 8A, merge panel, arrow heads), in contrast to the diffuse MBP, which did not (e.g., Figure 8A, merge panel, arrows). Cells with oligodendrocyte-like morphology could also be observed, displaying co-localization of MBP with F-actin (e.g.: Figure 8A, asterisk).

Real-time RT-PCR analysis

In order to further characterize the cells obtained at each distinct phase of the differentiation process, cells were analyzed at the transcript level by real-time RT-PCR. *Actin* was used as a house-keeping gene, *fibronectin* as a mesenchymal marker and several neural genes, representative of neural progenitors (*sox2* and *nestin*), neurons (*neurofilament*), astrocytes (*gfap*) and oligodendrocytes (*mbp*), were analyzed.

Data presented in Figure 7E shows that comparing to undifferentiated MSCs, the mesenchymal marker *fibronectin* remained essentially unchanged throughout the differentiation process, while the neural progenitor genes *sox2* and *nestin* were upregulated during the OPC-like stage of differentiation, then returning to lower levels after the OL-like stage. The presence of *nestin* during the OPC differentiation stage of oligodendroglial lineages has been known for some time [41,42], whereas the expression of *sox2* has only been reported more recently to occur in *bona-fide* OPCs, but not by mature oligodendrocytes [49–51]. All other markers did not show any statistically significant

difference comparing with MSCs, except for neurofilament, that exhibited a peak at the NSC-like stage and then, in subsequent differentiation steps, decreased to levels similar to those already expressed by MSCs, suggesting that an eventual neuronal-like phenotypic tendency was lost during the later differentiation steps.

Overall, the RT-PCR data seems to be in agreement with the immunocytochemistry results, suggesting that the NSC/OPC-like differentiation stages are associated with the increased expression of neural progenitor markers, while the oligodendroglial maturation marker MBP seemed to be expressed by OL-like cells at levels similar to those already present in undifferentiated MSCs.

Co-culture experiments

In order to address the interaction between the obtained OL-like cells and neuronal cells, a co-culture experiment was set up [33]. Mouse dorsal root ganglion (DRG) explants were cultured for 10 days until the axons were well developed, and then MSC-derived cells at the end of the OPC-like stage were added to the DRG culture and allowed to be in co-culture for further 14 days in OL-like maturation medium (Materials and Methods).

By the end of the co-culture period, it could be observed that OL-like cells typically tended to be present in the vicinity of neurons (Figure 9A - neurons stained in red for beta-III-tubulin and OL-like cells stained in green for human Golgi). This observation suggests that there might occur a positive crosstalk between both types of cells (neurons and OL-like cells).

Occasionally, OL-like cells with branched morphology could be observed (Figure 9B-D, asterisk). Although there was no evidence of robust structures resembling myelin sheaths wrapping neurons, in contrast to those described for *bona-fide* oligodendrocytes [33], in some cases there were apparent contact points between branches of OL-like cells and axons of DRGs (Figure 9C, D – arrows). These results suggest that although there was no evidence for robust axonal ensheathment, there might be physical interactions between OL-like cells and neuronal axons.

Overall, the results indicate that MSCs are prone to differentiate into neural-like cells, namely oligodendrocyte-like cells, expressing some of the typical oligodendroglial lineage markers along the distinct differentiation steps. The OL-like cells obtained expressed some of the mature oligodendrocyte markers (e.g.: MBP, GalC), but seemed to be immature comparing to actual *bona-fide* oligodendrocytes in terms of morphology and function. Although some cells did present an OL-like morphology and seemed to establish contacts with axons of neuronal cells, there was no evidence for robust myelination in OL-like/DRG co-culture experiments. Nevertheless, MSCs seem to have the potential to differentiate into oligodendroglial-like lineages and with improvements to the protocol, a more mature phenotype might be attained, with a more conclusive functional phenotype.

Discussion

Human umbilical cord matrix mesenchymal stem cells (UCM-MSCs) were isolated from 12 umbilical cords with 100% efficiency using a protocol based on that previously described by Reinisch and colleagues [27]. The cells exhibited low generation time and proliferated readily up to at least 8 passages - P8 (Figures 1C-E), reaching a total number of cells over 1×10^9 after 4 passages, within 17 to 21 days after explant isolation (Figures 1A, B). Nevertheless, it might have been possible to further lower the number of passages and time required to reach this number of cells, which is well above what is considered to be a therapeutic dose of at least 2×10^6 MSCs/kg of body weight for infusion [29], by increasing the amount of umbilical cord tissue processed, since

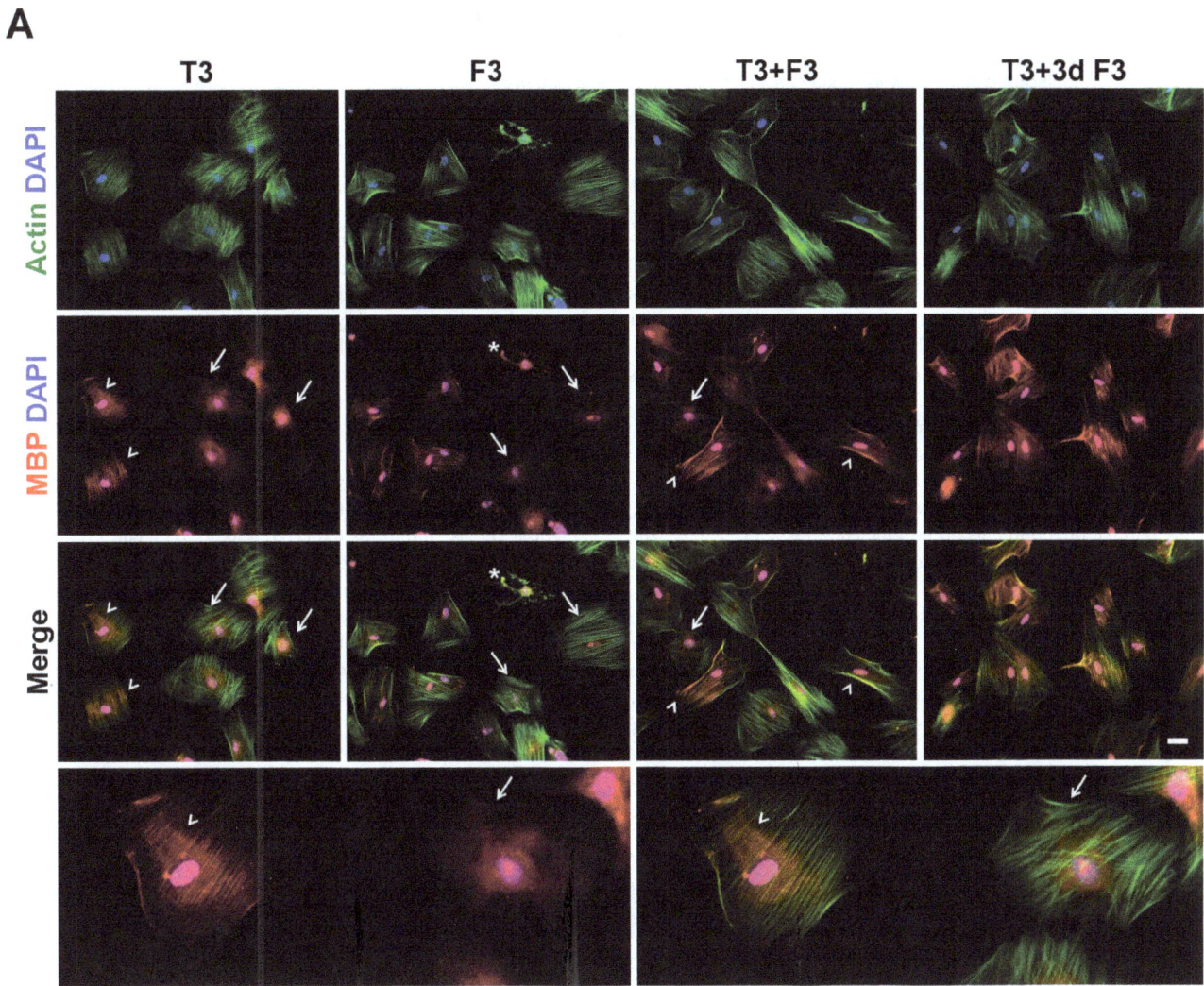

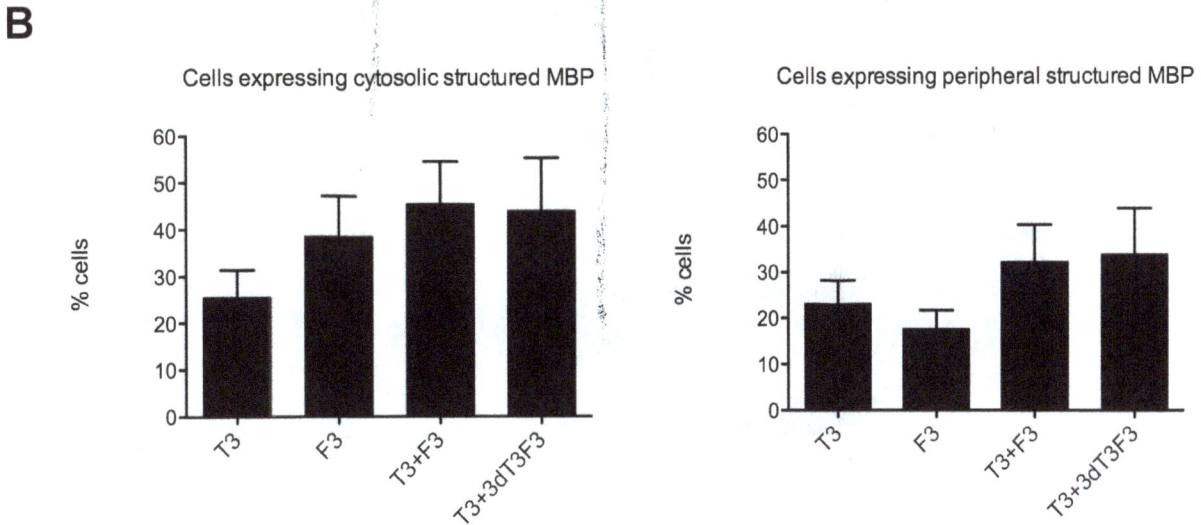

Figure 8. Co-localization of MBP and F-actin in oligodendrocyte-like cells. Fluorescence microscopy images of OL-like cells (*A*) stained with fluorescent phalloidin (in green) to assess the presence of polymerized actin (F-actin) and an antibody against MBP (in red) and both channels merged, as indicated. Cells were derived from OPC-like cells cultured on FN-coated wells. Cells showing structured MBP and with partial co-localization with F-actin (arrow heads) or with diffuse MBP and no co-localization with F-actin (arrows) were visible after OL-like differentiation in the presence of T3, F3, T3+F3 or 7 days T3 followed by 3 days T3+F3 (T3+3d F3), as indicated. The bottom panel is a magnification of the T3 condition.

Images are representative of at least 3 independent experiments and scale bar corresponds to 50 μm. The percentage of cells expressing structured MBP was calculated (*B*) under two categories: cells expressing structured cytoplasmic (left) or peripheral (right) MBP. Values were expressed as percentage of total cells and bars represent mean ± SEM percentage of cells present in at least 14 fields, belonging to 3 independent experiments. No statistically significant differences were found between the distinct differentiation conditions (Kruskal-Wallis one-way ANOVA followed by Dunn's multiple comparison test).

typically only a small part of each umbilical cord sample was processed. Importantly, the generation time did not increase significantly when cells where passaged until P8, which corresponded to about 20 population doublings (20.6±0.4) - Figure 1C and D -, indicating that the cells did not reach senescence until this number of duplications, in accordance with recent literature that showed that UCM-MSCs could be kept in proliferative conditions *in vitro* until approximately 33 cumulative population doublings (33.7±2.1) before entering replicative senescence [35].

The cells were also able to form colonies with similar frequency (no statistically significant differences were found) at passages 2 and 8 (Figure 1F), further indicating that the cells were proliferative, healthy and maintained stemness for at least 8 passages (or ~20 population doublings). The efficiency for CFU-F (i.e., the number of colonies formed divided by the total number of cells initially plated) at P2 and P8 was 29.3±4.7 and 26.7±9.9, respectively, comparable to the efficiency described in the literature, of 35.2±2.69 [52]. The slightly higher efficiency reported [52] compared to our study might be explained by the fact that the cell density used in the assay described by Hou *et al.* was substantially higher (50 cells/cm^2) than the one used in our study (3 cells/cm^2). By having a higher cell density, we may speculate that the autocrine signaling is favored and a certain

threshold to trigger proliferation and colony formation is attained faster, compared to our protocol.

The MSCs obtained from the umbilical cord matrix were able to undergo chondrogenic, osteogenic and adipogenic fates (Figures 2A-C), the typical multilineage differentiation lineages reported for this stem cell type [2]. Nevertheless, we observed low adipogenic induction, a characteristic which has already been reported in the literature for UCM-MSCs, when compared with MSCs from other sources, such as the bone-marrow [53].

The immunophenotypic characterization of the UCM-MSCs obtained further validated their genuine MSC identity, since the cells were positive for CD13, CD73, CD90 and CD105 (Figures 2D-G) and did not expressed CD34, CD45 and CD11b (Figures 2H, K and J), as expected [30,36,37]. The presence of nerve growth factor receptor (NGFR, also known as CD271) has been reported in small populations of MSCs from distinct sources, including the bone marrow, umbilical cord matrix and adipose tissue, mostly in fresh samples but not in cultured cells [38,54,55]. Regarding the umbilical cord matrix, NGFR has been recently detected to be weakly expressed *in situ* in fresh umbilical cord samples and almost undetected after culture [56]. Since our immunophenotypic analysis was performed only after culture (at P2 and P8), the absence of this marker is therefore not surprising (Figure 2I).

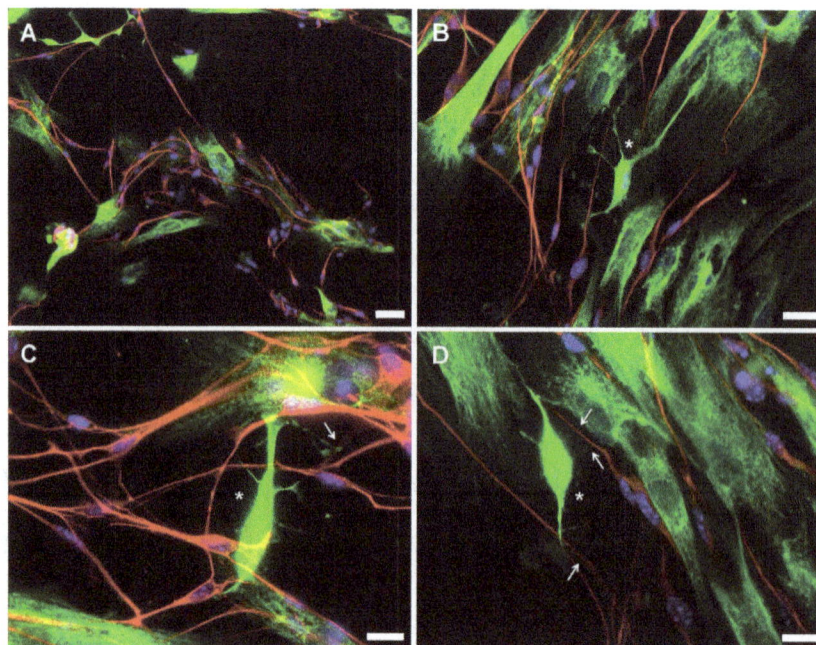

Figure 9. Immunofluorescence microscopy images of OL-like cells derived from hUCM-MSCs in co-culture with mouse dorsal root ganglion (DRG) neurons. The co-cultures were stained using an anti-human Golgi antibody (green) to specifically identify the human cells (OL-like cells derived form MSCs) and an anti-beta-III-tubulin antibody (red) to label the mouse DRG neurons. The images acquired in the green channel (anti-human Golgi antibody) were deliberately slightly overexposed to allow for a better understanding of the cellular morphology, which did not affect the identification of human versus mouse cells, as evidenced by the lack of green signal in the mouse DRG neurons (in red). Counterstaining of the nuclei was performed using DAPI (blue). It was apparent that the OL-like cells and DRG neurons tend to cluster together, as illustrated in a lower magnification image (*A*). Higher magnification images (*B-D*) suggest the existence of contact points (arrows) between branches of OL-like cells displaying an immature oligodendrocyte-like morphology (asterisks) and neurites. Scale bars correspond to 50 μm (*A*) or 20 μm (*B-D*).

Verifying that the UCM-MSCs isolated were adherent to plastic in culture conditions, presented a typical MSC morphology (Figure 3A and Figure S1) and immunophenotype (Figures 2D-K and Figure S2) and were able to differentiate *in vitro* into osteoblasts, adipocytes and chondroblasts (Figure 2A-C), all the criteria were met [30] to define these cells as *bona-fide* mesenchymal stem cells.

MSCs are plastic stem cells [5] and bone-marrow-derived MSCs were reported to possess, at least partially, a neuroectodermal origin [14,15]. Moreover, several reports in the literature have shown the potential of MSCs to differentiate into neural-like cells [7,9], mostly towards neuronal- and astrocyte-like cells [8,12,18,19], while the differentiation of MSCs towards oligodendroglial-like cell types has remained poorly explored (detailed below). Hence, we sought to explore the capacity of UCM-MSCs to differentiate into oligodendroglial-like lineages.

Several reports have shown the induction of pluripotent stem cells (ESCs [24,25,39,57] and iPSCs [26] into oligodendrocyte lineages. Nevertheless, there are concerns regarding the clinical application of cells derived from pluripotent stem cells, like the risk of teratoma formation [58]. Another drawback is low engraftment due to immunological rejection of the transplanted cells by the host immune system, although it was shown that short-term treatments with immunosuppressive drugs in mice enhanced the engraftment of unrelated pluripotent stem cells [59].

Alternatively, MSCs are generally regarded as a safe, non-tumorigenic and low-immunogenic adult stem cell type (with several clinical trials ongoing) [60], that seem to have the potential to differentiate into neural-like cells [7,9], hence could represent an interesting alternative to pluripotent stem cells for clinical use. The contribution for the establishment of a robust oligodendroglial-like differentiation protocol of MSCs has an important impact for the treatment of demyelinating diseases and regenerative medicine in general.

Few reports have explored the potential of human MSCs to differentiate into oligodendrocyte-like lineages. In one of the first reports, human CD90+ bone-marrow MSCs [21] were differentiated into cells expressing oligodendrocyte markers (O4, NG2, MBP), but also astroglial (GFAP) and, to a lesser extent, neuronal (beta-III-tubulin) markers, leaving unclear what was the true identity of those cells. Nevertheless, some cells seemed to adopt a myelinating oligodendroglial-like behavior in an *in vivo* myelination mouse retina model. Shortly after, Kennea and colleagues [61] obtained cells displaying some of the typical OL-like characteristics by differentiating human MSCs. However, the cells had fetal origin, which represent an ethical, technical and potentially safety drawback. Moreover, the protocol developed required either the use of conditioned medium obtained from the B104 rat neuroblastoma cell line or the overexpression of the pro-oligodendrocyte Olig-2 gene delivered by lentiviral transduction, further raising difficulties for possible future therapeutic applications. More recently, UCM-MSCs were differentiated into OL-like cells [22], however this study focused mostly on the secretome and neurotrophic effect of the oligodendrocyte progenitor-like cells obtained.

Taking into account several protocols available in the literature describing neural-like and oligodendroglial-like lineage induction of distinct stem cell types [7,9,22,24,39,57], we designed and optimized a step-wise protocol, testing distinct media, culture substrates and coating conditions for this endeavor (Figure S3).

Undifferentiated MSCs expressed nestin to a certain extent [11] (Figure 3E) and the presence of bFGF and EGF was reported to enhance the expression of this neural marker and confer cells with a neural precursor-like phenotype, as observed by the formation of neurosphere-like structures when cultured in low attachment conditions [9]. We used a similar approach (data not shown), but in parallel to the neurosphere-like approach also adopted by Zhang and colleagues for the differentiation of MSCs into OL-like cells [22], we decided to test a 2D approach (cells in monolayer), similar to what had been previously reported to obtain neural progenitors from ESCs [62]. Our approach was carried out using non-treated (non-tissue culture) polystyrene plates, that conferred some degree of attachment to the cells, but due to low adhesiveness, promoted the formation of compact 2D cell clusters or colonies that expressed high levels of nestin (Figures 3C and F). This allowed for a simpler culture system and avoided the difficulties of dissociating the 3D cell clusters, which resulted in considerable cell death. Moreover, during culture, substantial cell death occurred in the interior of the neurosphere-like structures due to the difficulty in controlling the sphere size (not shown), probably due to limited access of nutrients to those regions.

Compared to undifferentiated MSCs, NSC-like cells were strongly positive for the neural stem/progenitor marker nestin (Figures 3E and F, 4A and D and 7A). In this stage of differentiation, cells were essentially negative for OPC markers such as A2B5 and O4 (Figures 4E, F and 7B). Surprisingly, there was an increase in the expression of MBP and GalC during this differentiation step (Figures 7C and D), which may be a result of the fact that undifferentiated MSCs already expressed a basal level of MBP and GalC (Figures 6S and T and Figures 7C and D), in agreement with previous reports [7]. NSC-like cells also seemed to express other neural markers, such as β-III-tubulin and GFAP (data not shown) or neurofilament, as addressed by RT-PCR (Figure 7E). Again, in agreement with our observations (Figures 5J and L), previous reports stated that some undifferentiated MSCs expressed such neural markers [7], which may explain the observation of β-III-tubulin and GFAP during the neural-like commitment of the cells. Nevertheless, these neuronal and astrocytic markers were essentially absent or downregulated in the last stage of differentiation, especially in cells that were cultured on fibronectin during the OPC-like stage of differentiation (Figures 5A-H and Figure 7E).

During the OPC-like stage of differentiation, cells acquired a bipolar morphology and typical oligodendrocyte precursor lineage markers such as A2B5 and O4 (Figure 4). Moreover, the presence of the neural stem/progenitor marker nestin at both the protein (Figure 4) and mRNA level (Figure 7E) were consistent with the oligodendrocyte precursor-like stage of these cells [41,42]. However, the profile of mRNA and protein expression of nestin did not overlap completely. While the peak of nestin protein expression occurred at the NSC-like stage (Figure 7A), the peak of *nestin* mRNA happened at the OPC-like phase (Figure 7E). One hypothesis that can be considered to explain such result is the presence of PDGF in the culture media during the OPC-like induction step of the differentiation protocol, a growth factor that was shown to induce transcription of *nestin* [42]. Moreover, the levels of nestin protein were reported to be controlled at a post-translational level, namely through proteasome-mediated degradation upon differentiation of NSCs [63], which may account for the lower protein level of nestin in OPC-like cells compared with NSC-like cells despite the mRNA levels being higher in the first. Along with *nestin*, *sox2* was also upregulated during the OPC-like stage (Figure 7E) and although this marker was typically considered to be absent in OPCs [41], the expression of sox2 was very recently reported in oligodendroglial progenitors in the spinal cord of mice and rats by at least two independent research groups [49–51]. Interestingly, sox2 was reported as a possible regulator of *nestin* [64], which may explain the similar expression profile of both genes (Figure 7E).

The presence of ECM proteins like merosin (previously used by Zhang *et al.* [22]) or fibronectin seem to have favored equally well the survival of OPC-like cells, in contrast with uncoated dishes, in which very substantial cell loss was observed (Figure 4). Nevertheless, it became apparent during the OL-like stage of differentiation that the presence of FN during the OPC-like stage resulted in a subsequent more evident decrease of the OPC marker A2B5 in the OL-like cells, in comparison to OPC-like cells cultured in presence of merosin (Figure 6U and V), suggesting a higher maturation of the OL-like cells that were previously cultured in the presence of FN during the OPC-like stage.

Moreover, the OL-like cells obtained expressed typical oligodendrocyte differentiation markers. O4 and GalC (in case of differentiation in presence of T3 or F3), were upregulated in OL-like cells, compared to undifferentiated MSCs (Figures 7B and C). O4 is present throughout several stages of differentiation of *bona-fide* oligodendrocytes, namely from the OPC to the mature OL stage, whereas GalC is present only in differentiated OLs, although this marker can be detected in both immature and mature differentiated oligodendrocytes [46,65]. On the other hand, the OL-like cells obtained exhibited only low amounts of MBP (Figure 7D), a marker that is typically present only in mature OLs [46,65]. Taken together, these results suggest that the OL-like cells obtained have a differentiated, but not fully matured OL-like phenotype.

Despite the immature phenotype, several OL-like cells displayed a patterned MBP distribution that was consistent with F-actin co-localization, unlike that of MSCs (Figure 8). This co-localization pattern is similar to the co-localization that has been reported for *bona-fide* oligodendrocytes [47,48].

Moreover, co-culture experiments showed that OL-like cells were typically in close vicinity of neurons, suggesting that a positive crosstalk between both cell types might be present. The nature of such interactions may depend on soluble factors and/or direct cell-cell contact, although further experiments would be required to address that specific issue. It was also evident that some of the OL-like cells displayed a branched oligodendrocyte-like morphology (Figure 9), that suggested the existence of contact points between both cell types. Nevertheless, no clear signs of axonal ensheathment by OL-like cells were visible.

In summary, we have shown that cells isolated from the umbilical cord matrix are *bona-fide* mesenchymal stem cells. We have confirmed the neural-like plasticity of MSCs and explored their oligodendroglial-like commitment using a step-wise differentiation protocol. Cells displayed several neural and oligodendroglial markers at specific points of the differentiation protocol, suggesting an oligodendroglial-like specification of the cells along the process. Although a fully differentiated phenotype has not been reached, some typical and prominent features were similar to those of *bona-fide* oligodendrocytes.

Supporting Information

Figure S1 Isolation of MSCs from umbilical cord matrix explants. Proliferating MSCs with a fibroblastoid-like shape

could be readily identified migrating from umbilical cord matrix fragments after a 10 days culture period in proliferation medium (see Materials and methods). Scale bar corresponds to 200 μm.

Figure S2 Immunophenotype of UCM-MSCs. Immunophenotypic characterization by flow cytometry of three independent donor samples - UCM#2 at passage 2 (*A*) and P8 (*B*), UCM#3 at P2 (*C*) and P8 (*D*), and UCM#7 at P2 (*E*) and P8 (*F*).

Figure S3 Overview of the experimental conditions tested to differentiate hUCM-MSCs into oligodendrocyte (OL)-like cells. Schematics of the differentiation protocol through the different stages (S0 to S4), and respective nomenclature. Soluble factors (Sol. Factors), surfaces and alternative coatings used at the distinct steps of differentiation are indicated.

Figure S4 Assessment of the expression of MBP in hUCM-MSCs by western-blot analysis. Western-blot analysis was performed using antibodies against MBP and GAPDH (as loading control). MBP could be readily detected (band at ~33 kDa, as announced by the manufacturer of the antibody) in protein extracts of MSCs (35 μg of total protein per lane) and rat brain cortex extracts (15 μg of total protein per lane), the latter being used as a positive control (*A*). GAPDH was used as a reference protein to calculate the relative expression of MBP, based on the ratio of the integrated densities of the band of MBP divided by that of GAPDH, for each sample (*B*). It could be observed that despite expressing much less MBP than that found in rat brain cortex, MSCs expressed appreciable levels of MBP.

Table S1 Figure 1 data. Raw data used to produce Figure 1.

Table S2 Figure 7 data. Raw data of MFI values used to produce Figure 7.

Table S3 Figure 8 data. Raw data used to produce Figure 8B.

Acknowledgments

Sandro Pereira is acknowledged for discussion regarding results and statistical analysis. We thank Miranda Mele and Ivan Salazar at the laboratory of Prof. Carlos B. Duarte (CNC-University of Coimbra) for cultures of primary rat astrocytes and providing rat brain cortex tissue.

Author Contributions

Conceived and designed the experiments: CL JPF CMPC MV AP CLS JMSC JBR MG. Performed the experiments: CL NTS SM AR TL JPF FS PZA. Analyzed the data: CL NTS SM AR TL MG. Contributed reagents/materials/analysis tools: CMPC MV. Wrote the paper: CL NTS MG.

References

1. Caplan AI, Bruder SP (2001) Mesenchymal stem cells: building blocks for molecular medicine in the 21st century. Trends Mol Med 7: 259–264.
2. Pittenger MF, Mackay AM, Beck SC, Jaiswal RK, Douglas R, et al. (1999) Multilineage potential of adult human mesenchymal stem cells. Science 284: 143–147.
3. Friedenstein AJ, Chailakhjan RK, Lalykina KS (1970) The development of fibroblast colonies in monolayer cultures of guinea-pig bone marrow and spleen cells. Cell Tissue Kinet 3: 393–403.
4. Crisan M, Yap S, Casteilla L, Chen CW, Corselli M, et al. (2008) A perivascular origin for mesenchymal stem cells in multiple human organs. Cell Stem Cell 3: 301–313.

5. Troyer DL, Weiss ML (2008) Wharton's jelly-derived cells are a primitive stromal cell population. Stem Cells 26: 591–599.

6. Caplan AI (2009) Why are MSCs therapeutic? New data: new insight. J Pathol 217: 318–324.

7. Hermann A, Liebau S, Gastl R, Fickert S, Habisch HJ, et al. (2006) Comparative analysis of neuroectodermal differentiation capacity of human bone marrow stromal cells using various conversion protocols. JNeurosciRes 83: 1502–1514.

8. Wislet-Gendebien S, Hans G, Leprince P, Rigo JM, Moonen G, et al. (2005) Plasticity of cultured mesenchymal stem cells: switch from nestin-positive to excitable neuron-like phenotype. Stem Cells 23: 392–402.

9. Hermann A, Gastl R, Liebau S, Popa MO, Fiedler J, et al. (2004) Efficient generation of neural stem cell-like cells from adult human bone marrow stromal cells. J Cell Sci 117: 4411–4422.

10. Jang S, Cho HH, Cho YB, Park JS, Jeong HS (2010) Functional neural differentiation of human adipose tissue-derived stem cells using bFGF and forskolin. BMCCell Biol 11: 25.

11. Mendez-Ferrer S, Michurina TV, Ferraro F, Mazloom AR, Macarthur BD, et al. (2010) Mesenchymal and haematopoietic stem cells form a unique bone marrow niche. Nature 466: 829–834.

12. Wislet-Gendebien S, Leprince P, Moonen G, Rogister B (2003) Regulation of neural markers nestin and GFAP expression by cultivated bone marrow stromal cells. JCell Sci 116: 3295–3302.

13. Kondo T, Johnson SA, Yoder MC, Romand R, Hashino E (2005) Sonic hedgehog and retinoic acid synergistically promote sensory fate specification from bone marrow-derived pluripotent stem cells. Proc Natl Acad Sci U S A 102: 4789–4794.

14. Morikawa S, Mabuchi Y, Niibe K, Suzuki S, Nagoshi N, et al. (2009) Development of mesenchymal stem cells partially originate from the neural crest. BiochemBiophysResCommun 379: 1114–1119.

15. Takashima Y, Era T, Nakao K, Kondo S, Kasuga M, et al. (2007) Neuroepithelial cells supply an initial transient wave of MSC differentiation. Cell 129: 1377–1388.

16. Harris VK, Faroqui R, Vyshkina T, Sadiq SA (2012) Characterization of autologous mesenchymal stem cell-derived neural progenitors as a feasible source of stem cells for central nervous system applications in multiple sclerosis. Stem Cells Transl Med 1: 536–547.

17. Hayashi T, Wakao S, Kitada M, Ose T, Watabe H, et al. (2013) Autologous mesenchymal stem cell-derived dopaminergic neurons function in parkinsonian macaques. J Clin Invest 123: 272–284.

18. Arthur A, Rychkov G, Shi S, Koblar SA, Gronthos S (2008) Adult human dental pulp stem cells differentiate toward functionally active neurons under appropriate environmental cues. Stem Cells 26: 1787–1795.

19. Engler AJ, Sen S, Sweeney HL, Discher DE (2006) Matrix elasticity directs stem cell lineage specification. Cell 126: 677–689.

20. Cho KJ, Trzaska KA, Greco SJ, McArdle J, Wang FS, et al. (2005) Neurons derived from human mesenchymal stem cells show synaptic transmission and can be induced to produce the neurotransmitter substance P by interleukin-1 alpha. Stem Cells 23: 383–391.

21. Lu L, Chen X, Zhang CW, Yang WL, Wu YJ, et al. (2008) Morphological and functional characterization of predifferentiation of myelinating glia-like cells from human bone marrow stromal cells through activation of F3/Notch signaling in mouse retina. Stem Cells 26: 580–590.

22. Zhang HT, Fan J, Cai YQ, Zhao SJ, Xue S, et al. (2010) Human Wharton's jelly cells can be induced to differentiate into growth factor-secreting oligodendrocyte progenitor-like cells. Differentiation 79: 15–20.

23. Noseworthy JH, Lucchinetti C, Rodriguez M, Weinshenker BG (2000) Multiple sclerosis. NEnglJ Med 343: 938–952.

24. Kang SM, Cho MS, Seo H, Yoon CJ, Oh SK, et al. (2007) Efficient induction of oligodendrocytes from human embryonic stem cells. Stem Cells 25: 419–424.

25. Keirstead HS, Nistor G, Bernal G, Totoiu M, Cloutier F, et al. (2005) Human embryonic stem cell-derived oligodendrocyte progenitor cell transplants remyelinate and restore locomotion after spinal cord injury. JNeurosci 25: 4694–4705.

26. Wang S, Bates J, Li X, Schanz S, Chandler-Militello D, et al. (2013) Human iPSC-derived oligodendrocyte progenitor cells can myelinate and rescue a mouse model of congenital hypomyelination. Cell Stem Cell 12: 252–264.

27. Reinisch A, Strunk D (2009) Isolation and animal serum free expansion of human umbilical cord derived mesenchymal stromal cells (MSCs) and endothelial colony forming progenitor cells (ECFCs). J Vis Exp. (32), e1525, doi:10.3791/1525.

28. Cristofalo VJ, Allen RG, Pignolo RJ, Martin BG, Beck JC (1998) Relationship between donor age and the replicative lifespan of human cells in culture: a reevaluation. Proc Natl Acad Sci U S A 95: 10614–10619.

29. Bieback K, Hecker A, Kocaomer A, Lannert H, Schallmoser K, et al. (2009) Human alternatives to fetal bovine serum for the expansion of mesenchymal stromal cells from bone marrow. Stem Cells 27: 2331–2341.

30. Dominici M, Le BK, Mueller I, Slaper-Cortenbach I, Marini F, et al. (2006) Minimal criteria for defining multipotent mesenchymal stromal cells. The International Society for Cellular Therapy position statement. Cytotherapy 8: 315–317.

31. Kaech S, Banker G (2006) Culturing hippocampal neurons. Nat Protoc 1: 2406–2415.

32. Santos SD, Iuliano O, Ribeiro L, Veran J, Ferreira JS, et al. (2012) Contactin-associated protein 1 (Caspr1) regulates the traffic and synaptic content of alpha-amino-3-hydroxy-5-methyl-4-isoxazolepropionic acid (AMPA)-type glutamate receptors. J Biol Chem 287: 6868–6877.

33. Wang Z, Colognato H, Ffrench-Constant C (2007) Contrasting effects of mitogenic growth factors on myelination in neuron-oligodendrocyte co-cultures. Glia 55: 537–545.

34. Secco M, Zucconi E, Vieira NM, Fogaca LL, Cerqueira A, et al. (2008) Multipotent stem cells from umbilical cord: cord is richer than blood! Stem Cells 26: 146–150.

35. Scheers I, Lombard C, Paganelli M, Campard D, Najimi M, et al. (2013) Human umbilical cord matrix stem cells maintain multilineage differentiation abilities and do not transform during long-term culture. PLoS One 8: e71374.

36. Ribeiro A, Laranjeira P, Mendes S, Velada I, Leite C, et al. (2013) Mesenchymal stem cells from umbilical cord matrix, adipose tissue and bone marrow exhibit different capability to suppress peripheral blood B, natural killer and T cells. Stem Cell Research & Therapy 4: 125.

37. Ciavarella S, Dammacco F, De Matteo M, Loverro G, Silvestris F (2009) Umbilical cord mesenchymal stem cells: role of regulatory genes in their differentiation to osteoblasts. Stem Cells Dev 18: 1211–1220.

38. Turnovcova K, Ruzickova K, Vanecek V, Sykova E, Jendelova P (2009) Properties and growth of human bone marrow mesenchymal stromal cells cultivated in different media. Cytotherapy 11: 874–885.

39. Nistor GI, Totoiu MO, Haque N, Carpenter MK, Keirstead HS (2005) Human embryonic stem cells differentiate into oligodendrocytes in high purity and myelinate after spinal cord transplantation. Glia 49: 385–396.

40. Colognato H, Ramachandrappa S, Olsen IM, ffrench-Constant C (2004) Integrins direct Src family kinases to regulate distinct phases of oligodendrocyte development. J Cell Biol 167: 365–375.

41. Kondo T, Raff M (2004) Chromatin remodeling and histone modification in the conversion of oligodendrocyte precursors to neural stem cells. Genes Dev 18: 2963–2972.

42. Almazan G, Vela JM, Molina-Holgado E, Guaza C (2001) Re-evaluation of nestin as a marker of oligodendrocyte lineage cells. Microsc Res Tech 52: 753–765.

43. Buttery PC, Ffrench-Constant C (1999) Laminin-2/integrin interactions enhance myelin membrane formation by oligodendrocytes. MolCell Neurosci 14: 199–212.

44. Baas D, Bourbeau D, Sarlieve LL, Ittel ME, Dussault JH, et al. (1997) Oligodendrocyte maturation and progenitor cell proliferation are independently regulated by thyroid hormone. Glia 19: 324–332.

45. Hu QD, Ang BT, Karsak M, Hu WP, Cui XY, et al. (2003) F3/contactin acts as a functional ligand for Notch during oligodendrocyte maturation. Cell 115: 163–175.

46. Baumann N, Pham-Dinh D (2001) Biology of oligodendrocyte and myelin in the mammalian central nervous system. Physiol Rev 81: 871–927.

47. Boggs JM, Rangaraj G, Gao W, Heng YM (2006) Effect of phosphorylation of myelin basic protein by MAPK on its interactions with actin and actin binding to a lipid membrane in vitro. Biochemistry 45: 391–401.

48. Boggs JM, Rangaraj G, Heng YM, Liu Y, Harauz G (2011) Myelin basic protein binds microtubules to a membrane surface and to actin filaments in vitro: effect of phosphorylation and deimination. Biochim Biophys Acta 1808: 761–773.

49. Hoffmann SA, Hos D, Kuspert M, Lang RA, Lovell-Badge R, et al. (2014) Stem cell factor Sox2 and its close relative Sox3 have differentiation functions in oligodendrocytes. Development 141: 39–50.

50. Lee HJ, Wu J, Chung J, Wrathall JR (2013) SOX2 expression is upregulated in adult spinal cord after contusion injury in both oligodendrocyte lineage and ependymal cells. J Neurosci Res 91: 196–210.

51. Whittaker MT, Zai LJ, Lee HJ, Pajoohesh-Ganji A, Wu J, et al. (2012) GGF2 (Nrg1-beta3) treatment enhances NG2+ cell response and improves functional recovery after spinal cord injury. Glia 60: 281–294.

52. Hou J, Han ZP, Jing YY, Yang X, Zhang SS, et al. (2013) Autophagy prevents irradiation injury and maintains stemness through decreasing ROS generation in mesenchymal stem cells. Cell Death Dis 4: e844.

53. Corotchi MC, Popa MA, Remes A, Sima LE, Gussi I, et al. (2013) Isolation method and xeno-free culture conditions influence multipotent differentiation capacity of human Wharton's jelly-derived mesenchymal stem cells. Stem Cell Res Ther 4: 81.

54. Martins AA, Paiva A, Morgado JM, Gomes A, Pais ML (2009) Quantification and immunophenotypic characterization of bone marrow and umbilical cord blood mesenchymal stem cells by multicolor flow cytometry. Transplant Proc 41: 943–946.

55. La Rocca G, Corrao S, Lo Iacono M, Corsello T, Farina F, et al. (2012) Novel Immunomodulatory Markers Expressed by Human WJ-MSC: an Updated Review in Regenerative and Reparative Medicine. The Open Tissue Engineering and Regenerative Medicine Journal 5: 50–58.

56. Margossian T, Reppel L, Makdissy N, Stoltz JF, Bensoussan D, et al. (2012) Mesenchymal stem cells derived from Wharton's jelly: comparative phenotype analysis between tissue and in vitro expansion. Biomed Mater Eng 22: 243–254.

57. Brustle O, Jones KN, Learish RD, Karram K, Choudhary K, et al. (1999) Embryonic stem cell-derived glial precursors: a source of myelinating transplants. Science 285: 754–756.

58. Blum B, Bar-Nur O, Golan-Lev T, Benvenisty N (2009) The anti-apoptotic gene survivin contributes to teratoma formation by human embryonic stem cells. Nat Biotechnol 27: 281–287.

59. Pearl JI, Lee AS, Leveson-Gower DB, Sun N, Ghosh Z, et al. (2011) Short-term immunosuppression promotes engraftment of embryonic and induced pluripotent stem cells. Cell Stem Cell 8: 309–317.

60. Anzalone R, Lo Iacono M, Corrao S, Magno F, Loria T, et al. (2010) New emerging potentials for human Wharton's jelly mesenchymal stem cells: immunological features and hepatocyte-like differentiative capacity. Stem Cells Dev 19: 423–438.

61. Kennea NL, Waddington SN, Chan J, O'Donoghue K, Yeung D, et al. (2009) Differentiation of human fetal mesenchymal stem cells into cells with an oligodendrocyte phenotype. Cell Cycle 8: 1069–1079.

62. Conti L, Pollard SM, Gorba T, Reitano E, Toselli M, et al. (2005) Niche-independent symmetrical self-renewal of a mammalian tissue stem cell. PLoSBiol 3: e283.

63. Mellodew K, Suhr R, Uwanogho DA, Reuter I, Lendahl U, et al. (2004) Nestin expression is lost in a neural stem cell line through a mechanism involving the proteasome and Notch signalling. Brain Res Dev Brain Res 151: 13–23.

64. Jin Z, Liu L, Bian W, Chen Y, Xu G, et al. (2009) Different transcription factors regulate nestin gene expression during P19 cell neural differentiation and central nervous system development. J Biol Chem 284: 8160–8173.

65. Lee JC, Mayer-Proschel M, Rao MS (2000) Gliogenesis in the central nervous system. Glia 30: 105–121.

8

Generation of a Single Chain Antibody Variable Fragment (scFv) to Sense Selectively RhoB Activation

Patrick Chinestra[1], Aurélien Olichon[1,2], Claire Medale-Giamarchi[1,2], Isabelle Lajoie-Mazenc[1,2], Rémi Gence[1,2], Cyril Inard[1], Laetitia Ligat[4], Jean-Charles Faye[1], Gilles Favre[1,2,3]*

1 Inserm, UMR 1037-CRCT, GTPases Rho dans la progression tumorale, Toulouse, France, **2** Université Toulouse III-Paul Sabatier, Faculté des Sciences Pharmaceutiques, Toulouse, France, **3** Institut Claudius Regaud, Toulouse, France, **4** CRCT, plateau de protéomique, Toulouse, France

Abstract

Determining the cellular level of activated form of RhoGTPases is of key importance to understand their regulatory functions in cell physiopathology. We previously reported scFvC1, that selectively bind to the GTP-bound form of RhoA, RhoB and RhoC. In this present study we generate, by molecular evolution, a new phage library to isolate scFvs displaying high affinity and selectivity to RhoA and RhoB. Using phage display affinity maturation against the GTP-locked mutant RhoAL63, we isolated scFvs against RhoA active conformation that display K_d values at the nanomolar range, which corresponded to an increase of affinity of three orders of magnitude compared to scFvC1. Although a majority of these evolved scFvs remained selective towards the active conformation of RhoA, RhoB and RhoC, we identified some scFvs that bind to RhoA and RhoC but not to RhoB activated form. Alternatively, we performed a substractive panning towards RhoB, and isolated the scFvE3 exhibiting a 10 times higher affinity for RhoB than RhoA activated forms. We showed the peculiar ability of scFvE3 to detect RhoB but not RhoA GTP-bound form in cell extracts overexpressing Guanine nucleotide Exchange Factor XPLN as well as in EGF stimulated HeLa cells. Our results demonstrated the ability of scFvs to distinguish RhoB from RhoA GTP-bound form and provide new selective tools to analyze the cell biology of RhoB GTPase regulation.

Editor: Mitchell Ho, National Cancer Institute, NIH, United States of America

Funding: This work was supported by grants from Institut National de la Santé et de la Recherche Médicale (INSERM), Centre de Recherche de l'Institut Claudius Regaud. The funders had no role in study design, data collection and analysis, decision to publish, or preparation of the manuscript.

Competing Interests: The authors have declared that no competing interests exist.

* Email: favre.gilles@iuct-oncopole.fr

Introduction

The members of the large family of monomeric GTP-binding proteins, or small G proteins, function as molecular switches triggering signalling cascades involved in the regulation of a wide variety of cell processing. They serve as key regulators of extracellular-stimuli-transducers that mainly direct actin reorganisation, cell-cycle progression and gene expression [1] and have been implicated in cancer progression [2]. Monomeric GTPases cycle between an inactive GDP-bound to an active GTP-bound state that differ by the positioning of the switch I and switch II domains [3]. The active conformation interacts with effector proteins to induce downstream signalling events. Guanine nucleotide Exchange Factors (GEFs), promoting the release of bound GDP and its replacement by GTP, activate the Rho GTPases. GTPase-activating proteins (GAPs) accelerate the GTP hydrolysis and turn off the RhoGTPase to the inactivated GDP-bound form. RhoGTPases are anchored to membranes by prenylated carboxy terminal cysteine and are also regulated by Guanine nucleotide Dissociation Inhibitors (GDIs), which main known function is to maintain GTPases in soluble inactive complexes [4]. The Ras superfamily is structurally classified into seven families: Ras, Rho, Rab, Sar1/Arf, Ran, MIRO and RhoBTB3 [5]. Rho proteins comprise 20 members that differ from other GTPases by the presence of an insert loop. Among the

Rho proteins we focus on RhoA, RhoB and RhoC, which have long been confused in their biological activities because of their high amino acid sequence homology. Indeed, RhoB shares more than 80% homology [6] with RhoA and RhoC while RhoA and RhoC identity reaches 92%. However, it is now admitted that they differ in many biochemical characteristics and cellular functions. RhoA and RhoC are constitutively expressed while RhoB is an early inducible gene. RhoA and RhoC localize to the plasma membrane while RhoB has been found associated both to the plasma membrane and to the endosome [7,8] and more recently acting at the nuclear level [9]. Lastly, we [10] and others [11] have demonstrated that RhoB but not RhoA or RhoC displays gene suppressor activity in many cancer types and is critical to control cell survival upon genotoxic stress [12,13] or even in DNA damage response [14].

To date, the reference tool to evaluate the GTP-bound form of Rho in cell extracts is based on a pulldown assay relying on the Rho binding domain of rhotekin (RBD) as the bait [15]. One main caveat of this approach is that the RBD effector domain lacks of selectivity towards the three activated forms of RhoA, RhoB and RhoC homologues, and have low affinity to the Rho proteins. Another limitation resides in the poor stability of the RBD recombinant polypeptide which require to be purified only as a GST-fusion. There is a real need for reliable and selective tools, more versatile to investigate the cellular activation of RhoGT-Pases.

The detection of the level of single activated Rho is still challenging and would represent a significant progress in the study of their biological role. In this vein of research, we have previously reported the characterization of the scFvC1 conformational sensor selective of RhoA, RhoB and RhoC activated forms [16] but with a relative low affinity ($K_d = 3$ µM). We achieved a new scFvs library through molecular evolution of scFvC1 and performed affinity maturation selections with phage display technology. We obtained several scFvs exhibiting a strong improvement of affinity reaching the nanomolar range. Furthermore a substractive selection strategy led to the identification of scFvs discriminating RhoB from RhoA in their active conformation, despite a near 100% identity in the switch I and switch II domains [3]. Moreover, we demonstrated that these scFv selectively recognize cellular activated form of RhoB providing new tools to study RhoB functions.

Materials and Methods

Construction of a mutant library for scFv C1 antibody affinity maturation and cloning

The starting material for library construction was the pHEN C1 phagemid previously described [16]. Plasmid DNA was submitted to random mutagenesis by epPCR using GeneMorph II EZClone Domain Mutagenesis Kit (Stratagene) according to the high mutation rate protocol using the C1 specific upstream primer 5′TTATTACTCGCGGCCCAGCCGG3′ that hybridized just upstream of the NcoI restriction site of the pHEN vector and downstream primer 5′ GTGATGGTGATGATGATGTGC 3′. The mutated gene fragments were gel purified and digested with NcoI and NotI (New England Biolabs) followed by ligation into the corresponding sites of the pHEN phagemid containing an irrelevant scFv in order to avoid the presence of the native scFvC1 in the subsequent steps of affinity maturation. The library was subsequently transformed into electro competent XL1 blue *Escherichia coli* (*E.coli*) (Stratagene) and random clones sequenced with the primers LMB3 5′ACAGGAAACAGCTATGACC3′ and pHEN-SEQ 5′CTATGCGGCCCCATTCAG3′.

Affinity maturation, phage display selection and screening

Phage stocks of the library, phage display selection and screening techniques were performed as previously described in detail [17]. Briefly, we used mutant L63 of Rho (locked in GTP binding structure) expressed in BL21 *E.coli* strain as a GST-fusion protein captured in glutathione coated micro wells (Pierce). After blocking the well with PBS containing 3% non-fat dried milk powder (MPBS), phages (10^9 TU) were added and allowed to incubate for 30 min under stirring followed by 90 min without shaking at room temperature. After extensive PBS/0.1% Tween-20 and PBS washes used to minimise non-specific phage interaction, phages were eluted by adding triethylamine (1.4% in water) and subsequently neutralized by 1 volume of Tris-HCl (1 M, pH 7). Eluted phages were amplified by infecting exponentially growing XL1 blue *E.coli* and subsequently plated on ampicillin containing agar plates for CFU titer determination and further analysis by colony PCR and DNA sequencing.

Affinity maturation experiments were performed as described above by varying the washing stringency and by lowering the concentration of antigen. Briefly, for the first two rounds of selection, phages were incubated with saturating concentration of GST-RhoAL63-bound glutathione coated micro well (1 µg). In the first round, wells were washed 10 times with PBS/0.1% Tween-20 followed by 3 times with PBS while the second one consisted in 20 washes with PBS/0.1% Tween-20 followed by 5 washes with PBS. The subsequent two rounds were performed with decreasing concentrations of antigen determined by ELISA and corresponding to the concentration of antigen giving 70% and 30% of the maximal binding of phages from the previous round respectively. Consequently we coated wells with 50 ng/mL of antigen for the 3rd round and 5 ng/mL for the 4th round.

For the isolation of scFv against RhoB active conformation, the first round of selection consisted in incubating phages with 0.5 µg of the mutant GST-RhoBL63 bound to glutathione coated micro wells. The subsequent two rounds were conducted with a subtractive selection procedure consisting in a preincubation of phages with soluble GST-RhoAL63 (10 µg) during 1 h followed by another one hour in the presence of glutathione coated beads in order to eliminate RhoAL63 bound phages. After a brief centrifugation the remaining unbound phages were incubated with 0.5 µg of the mutant GST-RhoBL63 bound glutathione coated micro wells, washed and eluted as described above.

Affinity maturation and selection of conformation-specific scFv against the active form of RhoB were monitored by polyclonal phage ELISA on captured GST-Rho as previously described [17].

Antigen preparation for scFvs library selections

Recombinant GST fusion Rho proteins were expressed and purified in a protease deficient strain (*E. coli* BL21) as previously described [17].

The cDNAs encoding RhoAL63, RhoBL63 or RhoA wild type were inserted into the pHIS parallel2 vector from Dr. P. Sheffield [18], in-frame at the 3′ end of the 6xHIS tag. Recombinant proteins were expressed in a protease deficient strain (*E. coli* BL21) and were subsequently purified under native condition using Ni-NTA resins (QIAGEN) according to manufacturer's instructions. Purified recombinant proteins were resolved on a 12.5% SDS-PAGE gel and visualized by Coomassie staining. Measuring absorbance at 280 nm assessed protein concentrations.

Production and purification of single-chain Fv antibody

Soluble scFvs from selected clones were subcloned in fusion with the N-terminal domain of the phage P3 (NP3) protein necessary for their binding capabilities when expressed as soluble fragments as previously described for the scFvC1 [16]. Fusion scFvs were expressed in XL1 blue *E.coli* and were purified from periplasmic fraction using Ni-NTA resins (QIAGEN) as previously described [17]. Purified scFvs were then concentrated with the ProteoSpin CBED Micro Kit (Norgen) according to the protocol for acidic proteins. scFvs were resolved on a 10% SDS-PAGE gel and visualized by Coomassie staining. Measuring absorbance at 280 nm assessed protein concentrations.

For CBD pull down experiments, scFvs were expressed in fusion with the Chitin Binding Domain tag (CBD). The NP3 fragment from pHEN scFv-NP3 and the CBD fragment from the pTYB1 plasmid (New England Biolabs) were PCR amplified in order to introduce a BglII site at the 3′ and 5′ fragments extremities respectively with the following primers: P3-CBD-s 5′GTGCG-GCCGCACATCATCATCACC3′, P3-CBD-as 5′TAGATCA-GATCTGGATCCACGC- GGAACCAGAGAGCCGCCGCC-AGCATTGACAGG3′ and CBD-Bgl-s 5′TAGCTAAGATCT-GG- GATTACTTTATCTGATGATTCTGATC3′, CBD-Eco-as 5′TCAGTAGAATTCTTATCATTGAA-GCTGCCACAAG-GCAGG3′. The NP3 PCR fragment was digested NotI/BglII and the CBD PCR product by BglII/EcoRI. The NP3 fragment of the pHEN-scFv-NP3 was removed by NotI/EcoRI digestion and replaced with the two PCR amplified fragments in a trimolecular ligation in order to obtain the pHEN-scFv-NCBD plasmid.

scFv-NP3-CBD clones were expressed in XL1 blue E.coli as described above. Culture supernatants added to periplasmic extracts were incubated with chitin beads (New England Biolabs) by means of a peristaltic pump at 4°C according to the manufacturer's instructions. scFvs bound to chitin beads were stored at −80°C in TBS/50% glycerol until use.

ELISA experiments

For the screening of selected clones and the assessment of specificity, ELISA were performed as previously described [17]. Recombinant GST-Rho proteins were incubated on Reacti-Bind glutathione coated plates for 1 hour at room temperature. Phages or purified scFvs were revealed with antibodies anti-M13-HRP (GE Healthcare) or anti-c-myc-HRP (Novus Biologicals) respectively. Nucleotide loading of recombinant Rho from crude bacterial extracts was performed as previously described [17].

A method described by Friguet [19] was performed to determine K_d values. While coating GST-Rho on Reacti-Bind plates, purified scFvs (10^{-9}–10^{-8} M) were incubated until equilibrium (30 min, 23–25°C) with a range of concentrations of soluble 6xHis-RhoAL63 or 6xHis-RhoBL63. The antigen-scFvs complexes were then transferred to the GST-Rho-coated plates and incubated for 10 min (23–25°C). After washings, the free scFvs fraction was quantified using antibody anti-c-myc-HRP. Before performing these assays, parameters such as time of equilibrium, linearity of the curve and time of incubation for the quantification of the free scFvs fractions were determined as previously described by Martineau [20]. K_d values were determined by non-linear regression of the curves plotting the ratio A_0-A/A_0 against the range of concentrations of soluble antigens using Prism (Graphpad) software, where A_0 and A are the absorbance in absence and in presence of soluble antigens, respectively.

Surface Plasmon Resonance assays

All binding studies based on SPR technology were performed on BIAcore T200 optical biosensor instrument (GE Healthcare). Immobilization of anti-GST antibody (30 μg/ml) to capture GST fusion proteins (GST-RhoAL63, GST-RhoBL63 and GST-RhoC) was performed by amine coupling on Sensor Chip CM5 in HBS-P buffer (10 mM Hepes pH 7.4, 150 mM NaCl, 0.005% surfactant P20) (GE Healthcare). Capture of Rho-GST proteins was performed at a flow rate of 10 μl/min with a final Protein concentration of 10 μg/ml. Total amount of immobilized anti-GST antibody was 7500 RU and total amounts of captured GST fusion proteins were about 1000 RU. GST alone was captured on channel (Fc1) for non-specific binding measurements. Fc1 was used as a reference channel. Binding analyses were performed with antibodies at different concentrations over the immobilized GST fusion protein surface at 25°C for 2 minutes at a flow rate of 30 μl/min. A single-cycle kinetics (SCK) analysis to determine association, dissociation and affinity constants (ka, kd, and K_d respectively) was carried out by injecting different antibodies concentrations (100 nM–6.25 nM). Binding parameters were obtained by fitting the overlaid sensorgrams with the 1:1 Langmuir binding model of the BIAevaluation software version 2.0.

GST and CBD pull down experiments

HeLa S3 cells ((Cervical adenocarcinoma; ATCC, CCL-2.2) (3×10^6 cells) were seeded on to 145 cm^2 tissue-culture dishes and grown in Dulbecco's modified Eagle's medium (DMEM) (Cambrex Lonza) supplemented with 10% fetal calf serum. For nucleotide loading, cells extracts (see below) were incubated with

1 mM GDP or 100 μM GTPγS, 10 mM EDTA at 30°C for 45 min and supplemented with 60 mM MgCl2. For XPLN overexpression, cells were transfected with pRK5-XPLN (generously provided by Alan Hall) or pEGFP (Clontech) as a control of transfection using JetPRIME (Polyplus) method. For EGF stimulation experiments, HeLa cells were cultured in serum free media for 24 h before addition of EGF (Sigma) (2.5 ng/mL) for 10 min. Cells were scraped in 400 μL ice cold lysis buffer (50 mM Tris–HCl, pH 7.5, 500 mM NaCl, 10 mM MgCl2, 1% (v/v) Triton X-100, proteases inhibitors (Sigma-Aldrich)) mixed thoroughly and cleared by centrifugation at 16 000xg for 2 min at 4°C. An aliquot (5 or 10%) from each lysate was taken as input controls. Analysis of the level of activated Rho were performed by using the method initially described by Ren et al. [15] that is the GST fusion protein containing the Rho binding domain of the downstream effector rhotekin and adapted to RhoB [21]. Briefly, the Rho binding domain of rhotekin (RBD), an effector of Rho proteins that selectively binds to the GTP-loaded form, was expressed as a recombinant fusion with GST in E.coli and purified through binding to GST-Sepharose beads. Cells treated extracts were incubated with either 10–20 μg of GST-RBD or 0,5–1 μg of scFv-bound chitin beads and rotated for 45 min at 4°C. Beads were washed three times with ice-cold wash buffer (50 mM Tris-HCl, pH 7.5, 150 mM NaCl, 10 mM MgCl2, 1% (v/v) Triton X-100). Bound proteins were eluted from the beads with SDS-PAGE sample buffer at 95°C and separated on 12.5% SDS-PAGE for Western Blot analysis with anti-RhoA (Cell Signaling Technology) or anti-RhoB (Santa Cruz Biotechnology) antibodies followed by HRP-conjugated secondary antibodies (Bio-Rad). Visualization of proteins on Western blots was performed either with the ECL Western blotting substrate detection system (Pierce) or the ChemiDoc imaging system (BioRad).

Molecular modeling

All structures information were retrieved from the RCSB Protein Data Bank (www.rcsb.org): The PDB structure entry for active RhoAV14-GTPγS is 1A2B [3]. The PDB ID for Mg-Free form of RhoA-GDP is 1DPF [22] and 2FV8 for Mg-Free form of RhoB-GDP [23].

Molecular surface of RhoA and structural superposition of RhoA & RhoB were performed by the *UCSF-Chimera* Package (http://www.cgl.ucsf.edu/chimera) [47]: The molecular surface was created by the *MS/MS surface tool* with the parameter "Probe radius" set to 1.4 Å. The *Matchmaker tool* was used to generate the structural superposition with the following settings (Alignment algorithm: Needleman-Wunsh; Matrix: BLOSUM-62).

Results

Library construction and affinity maturation

The aim of our study was to generate scFv antibodies selective towards the active form of the small GTPases RhoA or RhoB. Our strategy is based on the improvement in terms of binding affinity and selectivity, through molecular evolution, of the previously described scFvC1 [16]. We demonstrated its ability to bind selectively to the active form of RhoA, RhoB and RhoC. To perform a molecular evolution of the scFvC1, diversity was introduced randomly by error-prone PCR (epPCR) into its coding DNA as previously described [24]. Mutation rate protocols of the GeneMorph II EZClone Domain Mutagenesis Kit (Stratagene), described in experimental procedures permitted us to obtain a library having a random nucleotide substitution frequency of 0.24%. This corresponded to a range of 1-5 amino acid substitutions taking into account that 20% of the clones sequenced

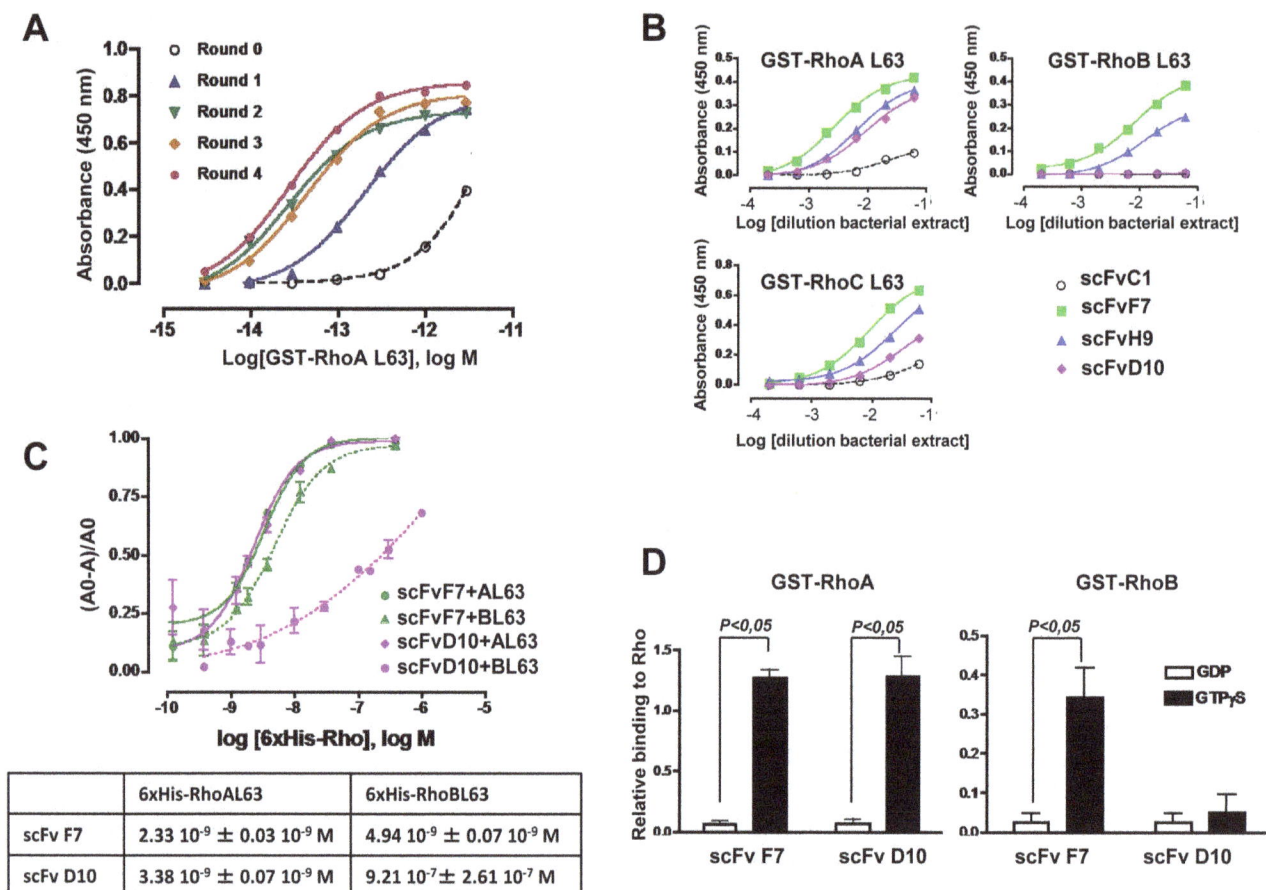

Figure 1. Affinity maturation revealed the possibility to obtain binders distinguishing RhoA and RhoC from RhoB active conformations. A, Improvement in apparent affinity throughout the rounds of selection was evaluated by a polyclonal phage ELISA on dilution of GST-RhoAL63 displayed in molar logarithmic scale (Log M). B, Three purified scFvs (F7, H9 and D10) were analyzed for their binding specificity towards L63 active mutants of recombinant GST-RhoA, RhoB and RhoC by ELISA. Purified scFvC1 was used as a control. C, Affinities of two scFvs (F7 and D10) for 6xHis-RhoAL63 (AL63), 6xHis-RhoBL63 (BL63) were measured by competitive ELISA as described in experimental procedures. K_d values were determined by nonlinear regression and listed in the insert table (mean ± SD, n = 3 each). D, The specificity of purified scFvF7 and scFvD10 for the active form of the recombinant wild type GST-RhoA and GST-RhoB loaded with either GDP or GTPγS were assessed by ELISA. Results are expressed as normalized absorbance of the scFvs to the total amount of coated GST-Rho quantified by the use of commercial antibodies (mean ± SD, Mann-Whitney test, n = 4 each).

did not exhibit any mutation. This frequency of mutation is located in the low range of libraries generally described for affinity maturation that is a range of mutation frequency of 0.4–0.8% [25,26], 0,5–3% [27], 0.1–0.8% [28]. This evolved scFvC1 library had an estimated size of 8.10^6 independent clones.

In a first selection strategy aiming of affinity maturation, the library was panned by increasing the washing steps number and by lowering the antigen concentration during the successive rounds of selection. We used GST-RhoAL63 as bait which presents very low kinetics of GTP hydrolysis in order to preserve the active conformation of the protein all along the selection procedure. The first round of panning was carried out with saturating antigen coating in order to remove nonfunctional clones formed by the mutagenesis process and also allowing the amplification of rare and poorly expressed scFvs. To select binders with improved affinity, the second round was performed in a similar way but with an increased washing stringency. We further assessed by ELISA the antigen concentrations giving 70% and 30% of the maximal binding of phage outputs from the previous rounds (data not shown), then applied this decrease corresponding to 50 ng/mL and 5 ng/mL respectively to rounds three and four. Such

decreased antigen concentrations, while keeping the input of phage constant, induced a competition between phages that display scFv very similar between each other [29]. We then compared the binding of the output phages obtained over the four selection rounds to GST-RhoL63 by polyclonal phage ELISA (Figure 1A). The results clearly indicated an increase of the signal, which can reflect an improvement of the affinity of the selected phage population. However, one can notice that the enrichment occurred as soon as the washing stringency was increased in the second round and that the effect of decreasing antigen concentrations appeared to be mild. At this stage, we cannot evaluate whether the antigen concentrations were not decreased enough to create sufficient competition or else the limit of the library was reached.

Characterization of the selected single chain antibodies

Screening of individual clones was carried out by phage ELISA and 21 out of 94 clones were chosen for further analysis given their high signal in ELISA (data not shown). Among the selected clones, DNA sequencing revealed that two of them did not exhibit any mutation, therefore corresponding to the original scFvC1. The

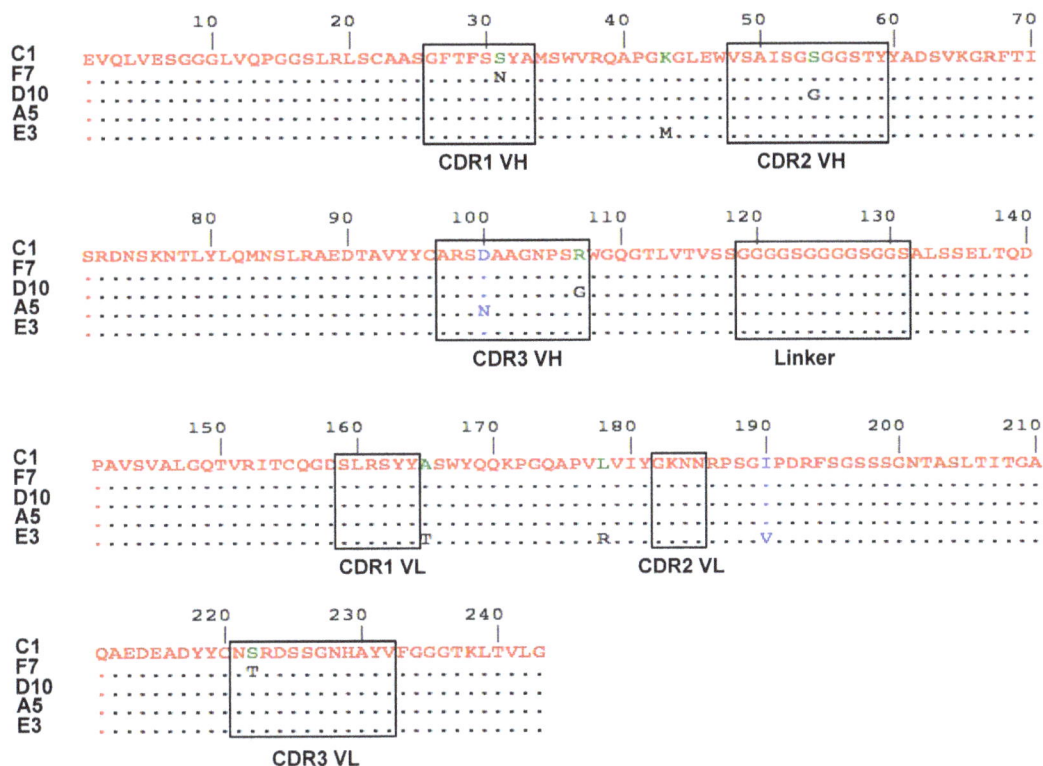

Figure 2. Protein sequence alignment of the characterized scFvs (F7, D10, A5, E3) showing the full sequence of scFvC1 and mutated aminoacid. Red to (-) correspond to aminoacid conservation. Blue amino acid correspond to mutation between strongly similar aminoacids. Green to black indicates aminoacid change between group of weakly conserved properties. Domains referred as complementary determining regions (CDR) of heavy chain (VH) and light chain (VL) are also indicated as well as the linker peptide between VH and VL.

other clones presented amino acid substitutions mainly located in the CDR of the VH and VL domains sometimes associated with mutations located in the framework or the linker (data not shown). These data strongly suggested that the majority of the clones selected could have interesting properties as the CDR of the antibodies mainly determines their affinity and specificity properties [30]. Sequence analysis revealed that one variant possessing the N104S substitution located within the CDR3 VH was represented seven times while three others exhibited the same mutation with one or two others additional mutations. Furthermore, four clones harbored the R107G substitution located within the CDR3 VH with or without additional mutations. These observations were in favor of a specific enrichment during biopanning.

In a first set of experiment, assessment of apparent affinity of the selected clones was performed by clonal phage ELISA (data not shown) as well as in their soluble forms against a range of antigen concentrations *i.e.* the locked GTP mutant of RhoA, RhoB and RhoC (Figure 1B). As expected, their signal intensities were dramatically higher than the one of the scFvC1, suggesting an affinity improvement particularly for the clone F7. Two clones including D10 were unable to bind to RhoB activated form but were binding to RhoA and RhoC activated forms with high apparent affinity.

We focused on F7 and D10 clones, which exhibited an apparent high affinity and different selectivities. ScFvF7 and scFvD10 were further analyzed in a competitive ELISA method that allow to assess the binding affinity [19,20]. As shown in Figure 1C, binding affinities to RhoAL63 and RhoBL63 were similar for the scFvF7 ($2.33.10^{-9}$ M and $4.94.10^{-9}$ M respectively), whereas scFvD10

recognized RhoBL63 with an IC$_{50}$ value of only $9.21.10^{-7}$ M, corresponding of more than 200-fold decrease in affinity compared to that to RhoAL63 ($3.38.10^{-9}$ M). Due to its very low affinity, scFvC1 could not be included in the above experiment. However, its affinity measured by Surface Plasmonic Resonance in a previous study (3.10^{-6} M) [16] permitted us to evaluate a gain of affinity of almost three orders of magnitude for scFvF7 and D10 towards RhoA. To ensure whether scFvF7 and D10 retain their respective selectivity towards the active form of the non mutated Rho, we next assayed their binding to RhoAwt and RhoBwt preloaded with GDP nucleotide or the slowly hydrolysable analogue of GTP, GTPγS. Results showed that the two scFvs selectively bound to recombinant RhoA-GTPγS, but that unlike scFvF7, scFvD10 failed to bind to the active form of wild type RhoB (Figure 1D). Aminoacid sequences of these two clones aligned to the scFvC1 are shown in Figure 2.

Selection of an scFv against RhoB active conformation

The results described above clearly demonstrated that the selection against active conformation of RhoA led to the isolation of evolved scFvC1 that can distinguish RhoA from RhoB active form. Therefore, we hypothesized that it could be possible to isolate from the scFvC1 library antibody fragments selective towards RhoB active conformation provided we used a suitable strategy of selection. To this aim, the scFvC1 library was first panned against GST-RhoBL63 with low washing stringency in order to amplify rare and poorly expressed scFvs. The two subsequent rounds of selection consisted in a counter-selection performed by pre-incubating the phages against GST-RhoAL63, thus removing scFvs able to bind to RhoA. The remaining

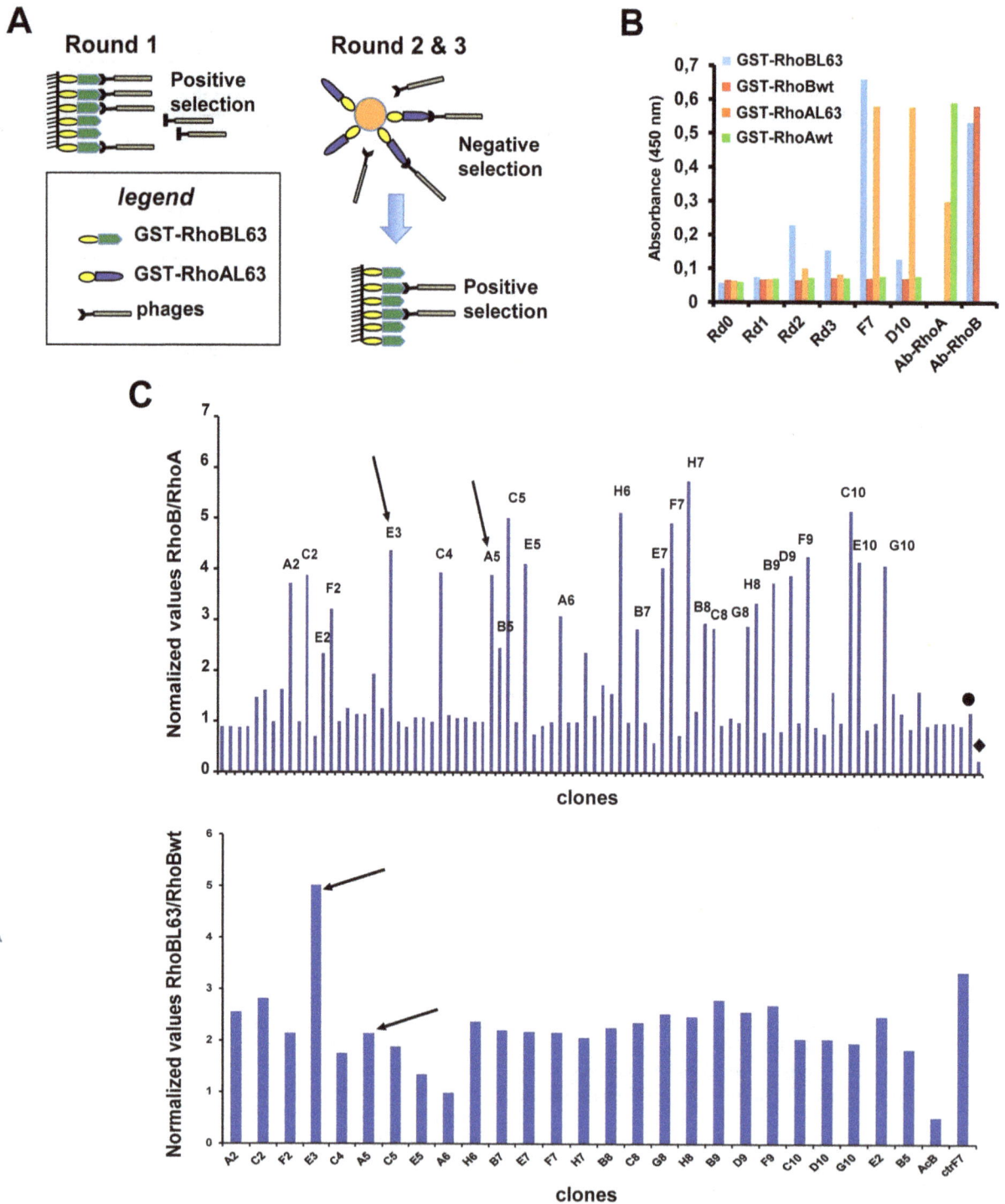

Figure 3. Selection of a RhoB active conformation specific scFv. A, Strategy of phage display selection. B, The enrichment of clones specific of the RhoB active form throughout the selection procedure was assessed by polyclonal phage ELISA on captured GST-Rho proteins from crude extract. Total amount of coated GST-Rho and active form of GST-RhoL63 were quantified with commercial anti-RhoA (Ab-RhoA) and anti-RhoB (Ab-RhoB) antibodies, and phageF7 (F7) and phageD10 (D10), respectively. GDP-bound GST-Rho (wt) was included as controls. C, (top panel) 88 individual clones were analyzed for their binding to GST-RhoBL63 and GST-RhoAL63 by phage ELISA. Results are expressed as the ratio of absorbance against GST-RhoBL63 *vs.* GST-RhoAL63. PhageF7 (black circle) and phageD10 (black diamond) were included as controls. Arrows indicate the clones E3 and A5 further selected. (Bottom panel) 26 clones were further analyzed for their binding to GST-RhoBL63 and GST-RhoBwt-GDP. Results are expressed as the ratio of absorbance against GST-RhoBL63 *vs.* GST-RhoBwt. Arrows indicate the clones E3 and A5. E3 was the best conformational sensor selective of active RhoB-GTP.

A

	6xHis-RhoBL63	6xHis-RhoAL63
scFv A5	$2.40\ 10^{-8} \pm 0.09\ 10^{-8}$ M	$2.45\ 10^{-8} \pm 0.14\ 10^{-8}$ M
scFv E3	$3.63\ 10^{-9} \pm 0.07\ 10^{-9}$ M	$2.29\ 10^{-8} \pm 0.09\ 10^{-8}$ M

B

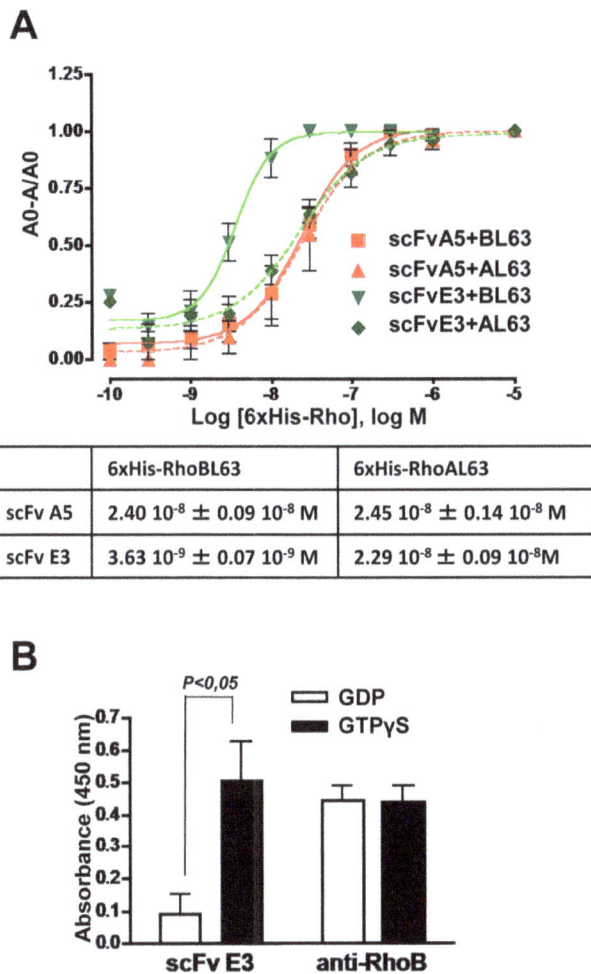

Figure 4. ScFvE3 preferentially binds to RhoB active conformation. A, Binding affinities of two selected scFvs (A5 and E3) for 6xHis-RhoAL63 (AL63), 6xHis-RhoBL63 (BL63) were measured by competitive ELISA on dilution of 6his-RhoL63 displayed in molar logarithmic scale (Log M), as described in experimental procedures. K_d values were determined by nonlinear regression and listed in the insert table (mean $\pm$ SD, n = 3 each). B, The specificity of the purified scFvE3 for the active form of the recombinant wild type GST-RhoB loaded with either GDP or GTPγS was assessed by ELISA. Results are expressed as absorbance at 450 nm (mean $\pm$ SD, Mann-Whitney test, n = 4).

unbound fraction was then panned against GST-RhoBL63 bound glutathione coated micro wells (Figure 3A) and amplified. Polyclonal ELISA using phages from each round of selection were showing an enrichment of phages able to selectively bind to RhoBL63 that was effective as soon as the second round (Figure 3B).

Afterwards, 88 individual clones were examined for their selective binding to RhoBL63 with respect to RhoAL63. 26 phages exhibiting a ratio greater than or equal to 3 were further analyzed for their capability to selective bind to RhoBL63 with respect to RhoBwt that is mainly in the GDP form (Figure 3C). DNA sequencing revealed that the D100N substitution located within the CDR3 VH with or without additional mutations was found in all but one selected clones. Furthermore, one clone having the unique D100N substitution was represented five times while two others clones were found identical. Only the scFvE3 differed from the others clones as it had 4 substitutions located

within the scaffold of the antibody fragment (Figure 2). We first measured the differential affinity to RhoBL63 *vs.* RhoAL63 of the scFvA5 displaying the unique D100N substitution and the scFvE3 devoid of this mutation by performing a competitive ELISA. Surprisingly, only the scFvE3 displayed a higher affinity towards RhoB ($3.63\ 10^{-9}$ M) than RhoA ($2.29\ 10^{-8}$ M), corresponding to a differential affinity factor of 6 in this assay (Figure 4A). At this stage, no explanation could be advanced concerning the fact that the D100N substitution shared by the majority of the selected clones did not result in an increase of affinity towards RhoB. We further focused on characterization of the scFvE3. Finally we confirmed its selectivity for the active form of RhoB by comparing its binding with non mutated RhoB loaded with nucleotides *i.e.* GDP or GTPγS (Figure 4B).

Kinetic Affinity measurements of selected scFvs

Molecular evolution led us to discover scFvs showing apparent high affinity towards the 3 Rho or more selectivity between RhoA/C and RhoB. To get insight into the interaction properties of scFvF7, D10 or E3 we performed real-time binding measurements by Surface Plasmonic Resonance (SPR). We confirmed affinities and selectivities measured by ELISA for the scFvs F7 and E3 towards RhoA and RhoB active mutants. Whereas scFvF7 harbored almost the same K_d values in the nanomolar range toward these 2 antigens, affinity of scFvE3 appeared to be even 10 times higher towards RhoB than RhoA mainly due to a faster dissociation rate on RhoA protein (Figure 5). However affinity measurements towards GST-RhoCL63 seemed to be unreliable as the number of resonance units was too low to give an accurate determination of the kinetic parameters. Finally, SPR affinity measurements of the scFvD10 were impeded by the poor stability of this peculiar scFv as resonance units signal were always low and K_d values could not even be determined for RhoC. The affinity towards RhoA was found ten times lower than the one obtained with ELISA. Nevertheless, a selectivity of the scFvD10 for RhoA vs RhoB was confirmed as no binding towards RhoB was observed.

Pulldown of endogenous RhoB activity

Rho proteins are subjected to multiple post-translational modifications such as prenylation [31,32], palmitoylation [33] as well as phophorylation [34,35]. Given that the selection of scFvs was carried out against bacterially expressed recombinant antigens devoid of any known post-translational modifications, there was a need to confirm that the selected scFvs were able to recognize active Rho derived from eukaryotic cells. To this end, scFvs were used in active Rho pull down experiment similarly as the conventional Rho-binding domain of Rhotekin (RBD) in GST pulldown. As GST fusion of scFvs were unstable in the secretion pathway, we first expressed them in fusion with the N-terminal of the Chitin Binding Domain (CBD) of the chitinase A1 which has been shown to bind with high affinity to chitin [36] and were purified onto chitin-coated beads. In order to artificially control the amount of antigen in the inactive or active conformation, HeLa cells extracts were first loaded either with GDP or GTPγS, then mixed with scFvs-bound beads referred as CBD-pulldown. Rho proteins specifically bound to the scFvs-beads were analysed by Western Blot with commercial anti-RhoA, anti-RhoB and anti-RhoC antibodies. The scFvF7, D10 and E3 interacted specifically with the GTPγS-loaded Rho proteins. Moreover, the selectivity that we observed on recombinant Rho proteins were fully retained on endogenous active form of each Rho, namely that the scFvF7 recognized RhoA, RhoB and RhoC, the scFvD10 RhoA and RhoC, the scFvE3 only RhoB (Figure 6A).

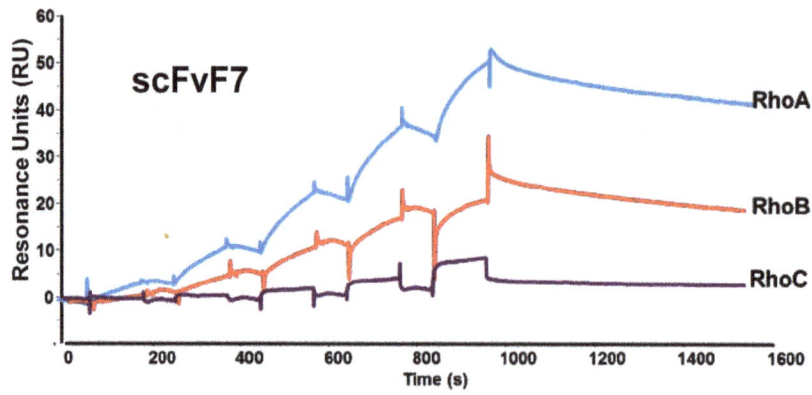

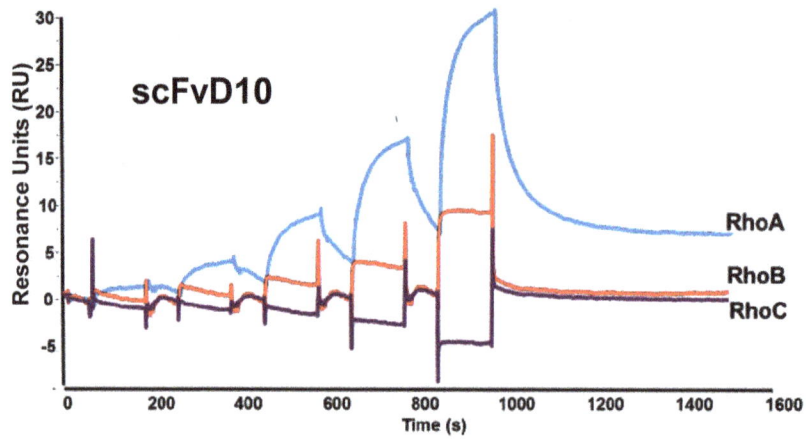

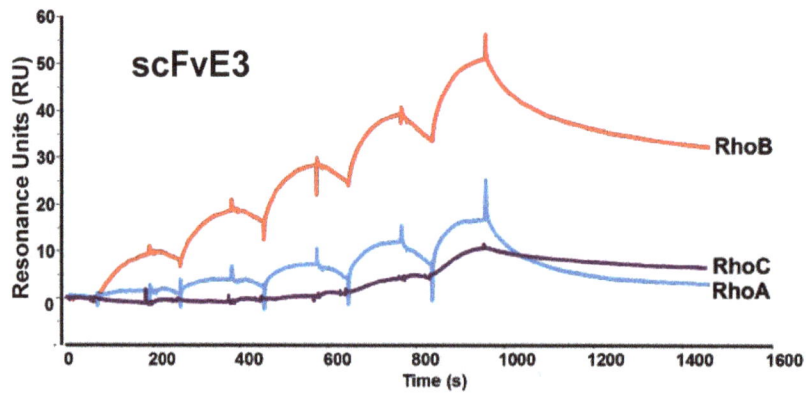

scFv	GST-Rho	Ka (1/MS)	Kd (1/s)	KD (M)
	RhoA	1.053^{+5}	2.479^{-4}	2.353^{-9}
F7	RhoB	1.222^{+5}	4.14^{-4}	3.39^{-9}
	RhoC	3.878^{+4}	9.948^{-5}	2.56^{-9} §
	RhoA	7.5^{+4}	2.974^{-3}	3.96^{-8}
D10	RhoB	ND	ND	ND
	RhoC	ND	ND	ND
	RhoA	4.765^{+5}	1.758^{-2}	3.69^{-8}
E3	RhoB	2.173^{+5}	6.828^{-4}	3.14^{-9}
	RhoC	1.758^{+3}	5.64^{-4}	3.18^{-7} §

Figure 5. Real-time binding of scFvs F7, D10 and E3 by Surface Plasmon Resonance on immobilized GST fusion Rho active mutant proteins. Single Cycle Kinetics analysis was performed on immobilized GST fusion proteins RhoAL63, RhoBL63 and RhoCL63 (1000 RU) with five injections of analyte at 6.25nM, 12.5nM, 25nM, 50nM, and 100nM. Analyte injections lasted for 120 s each and were separated by 180 s dissociation phases. An extended dissociation period of 10 min followed the last injection. The two sensograms recorded for a given analyte were fitted globally to a 1:1 interaction (data not show). Each sensogram represents a differential response where reference channel of immobilized GST protein has been substracted and is expressed in RU as a function of time in second. Bottom: table summarizing kinetic constant parameters. ND means not determined. § means that kinetics parameters were obtained by fitting curves with too low resonance units to give accurate values.

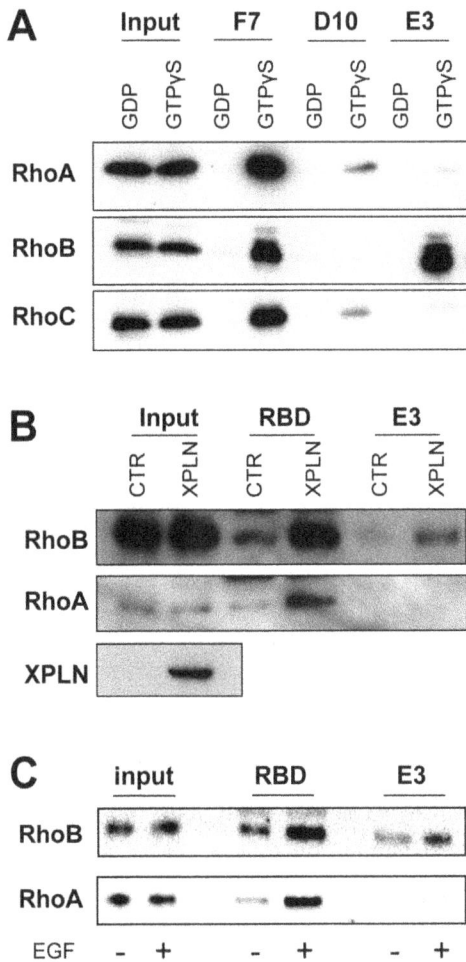

Figure 6. scFvE3 is a selective sensor of RhoB activation in HeLa cells. A, CBD-pulldown experiments on nucleotides loaded HeLa cell extracts showing the specificities of the selected scFvs. HeLa cell extracts were loaded with either GDP (1 mM) or GTPγS (100 μM) and incubated with scFvs F7, D10 and E3 fixed on chitin beads. CBD-pulldowns were analyzed by Western blotting using anti-RhoA, anti-RhoB and anti-RhoC antibodies. Total extract used for CBD-pulldown is indicated as *input* and examined by western blotting with the same antibodies. Western Blot is representative of 4 independent experiments. B, RhoB and RhoA activation were assessed by GST-pulldown (RBD) and CBD-pulldown (E3) experiments with cell lysate from HeLa cells transiently transfected with plasmids expressing Myc-tagged XPLN or GFP under the control of CMV promotor. XPLN was detected by using an anti-c-myc antibody. C, HeLa cells were serum-starved for 24 h and treated with EGF (2.5 ng/mL) for 10 min before lysis then RhoB and RhoA activation were assessed by GST-pulldown (RBD) and CBD-pulldown (E3) experiments. Beads-bound proteins were analyzed by Western blotting using anti-RhoA and anti-RhoB antibodies. Total cell extracts are indicated as *input* and examined by western blotting with the same antibodies. Western Blots are representatives of 2 independent experiments.

We next focused on the ability of scFvE3 to specifically bind to endogenous activated RhoB in the cellular complexity. This was first assessed by CBD-pulldown on HeLa cells transiently expressing the Myc-tagged XPLN that stimulates guanine nucleotide exchange on RhoA and RhoB, but not RhoC [37]. As a control, we performed an RBD-pulldown to specifically precipitate GTP-bound RhoB and RhoA from cell extracts. As shown in Figure 6B, the scFvE3 only interacted with GTP-bound RhoB whereas RhoA was also activated as revealed by the RBD interaction. These results were confirmed when HeLa cultured cells were stimulated with Epidermal growth factor (EGF) which has been shown to transitory activate RhoA and RhoB GTPases [21] (Figure 6C).

Discussion

Phage display technology has been proved to be effective in raising antibodies with markedly enhanced specificities as well as in improving antibodies binding affinities. We performed an affinity maturation by introducing random mutations within the scFvC1 coding DNA that recognizes GTP-bound form of RhoA, RhoB and RhoC. Affinity maturation can be conducted by a strategy based on the off-rate selection [25,38] or a competition between the binders [25,28,39,40]. As we wanted to preserve the active conformation of Rho proteins all along the selection procedure, we chose an affinity selection to avoid long incubation time required in off-rate selection protocols [24]. This approach was successful since we obtained scFvs with affinities dramatically higher than the one of the scFvC1, while keeping the properties of being a conformational sensor.

The phylogenetic tree of small GTPases classifies into subfamilies among which the RhoA, RhoB and RhoC members shares a high homology in terms of secondary structure [5]. This homology assessed to more than 80% identity on the whole protein reaches more than 95% in the 100 amino-terminal residues that comprises the nucleotide binding loop and the switch I and II which are supposed to be implicated in the conformational recognition by effector proteins [3]. Despite this fact, molecular evolution of the scFvC1 permitted us to isolate scFvs able not only to discriminate Rho proteins in their active conformation but in addition to selectively bind to either RhoA/C or to RhoB. Strikingly, the differential of affinity of the scFvE3 for RhoB versus RhoA active conformation is quite modest (10 fold as determined by SPR experiments) but appeared to be sufficient to pull down specifically an activation of RhoB in cells stimulated by EGF. Rare studies have reported the isolation of conformation specific antibodies discriminating active conformations of small GTPases using phage display technology such as HRas [41], Rab6A [42], Rho [16], and our results confirm the efficiency of this fully in vitro strategy. Furthermore, as previously reported by Tanaka and Rabbitts this approach can constitute the first step in order to achieve the isolation of intrabodies when coupled to intracellular antibody capture technique [43].

To date we do not know the exact epitopes where these scFvs bind on Rho proteins and the mechanism that allow scFvE3 to discriminate RhoB from RhoA remains unknown without

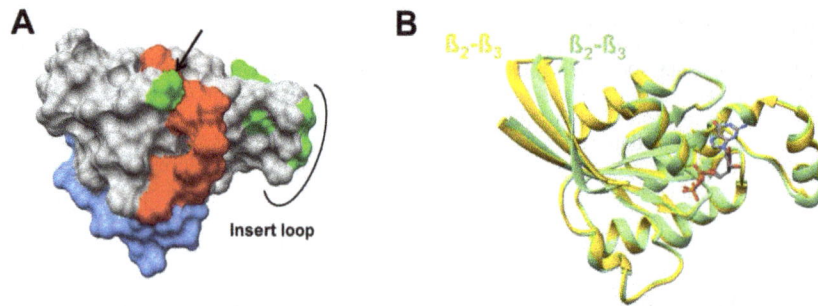

Figure 7. Structure comparison between RhoA and RhoB. A, The filling structure of active RhoAV14-GTPγS (Pdb code 1A2B) was created by UCSF-Chimera software [53]. Switch I & II are depicted in red and blue, respectively. The RhoA/RhoB differences in amino acids sequence are shown in green with the residue 29 indicated by an arrow. B, Structural divergence between RhoA and RhoB in the β2-β3 région. The *Matchmaker tool* in *UCSF-Chimera* (http://www.cgl.ucsf.edu/chimera) was used to generate the structural superposition. RhoA (Green, Pdb code 1DPF) & RhoB (Yellow, Pdb code 2FV8) structures are shown in ribbon model and the substrate GDP in stick. Both proteins are in the inactive form with only GDP in the active site.

performing co-crystallization studies. Nevertheless we assume that these conformational sensors may interact with residues near the switch regions as it has been shown for effector proteins *e.g.* mDia1 [44], PKN/PRK1 [45], ROCKI [46] and in a remarkable way for an anti-Hras-GTP intrabody [47]. Interestingly, this region is extremely homologous in secondary structure alignment between RhoA and RhoB, apart from the residues 10 and 29. Modeling these residue discripencies onto the filling structure of RhoA-GTP, since the GTP-bound form structure of RhoB has not been resolved until now, reveals that only position 29 in the switch I is exposed indeed (Figure 7A). Moreover 4 residues within the insert loop are as well exposed and might also explain the differential binding of the scFvs (Figure 7A). Apart from the switch domains, the ß2/ß3 region has been described to be involved in the specific binding of RhoA effectors [45,48]. As this region positioning does not change whatever the nucleotide bound in RhoA resolved structures, we superimposed available ribbon models of RhoA and RhoB inactive conformations. Strikingly, we observed a clear shift in this ß2/ß3 region at the protein surface (Figure 7B), that could be involved in the scFvs selectivities. The knowledge of the exact residues really implicated in the scFv binding would be of great importance to optimize affinity and selectivity of scFv by a targeted approach. This will be reached by co-crystalisation of Rho in the presence of their selective scFv. Nevertheless, we cannot exclude that the carboxy-terminal domain could be part of the binding site as it has been shown for the effector proteins selective of RhoB, MAP1A/LC2 [49] and p76[RBE] [50].

Rho GTPases become activated among several stimuli, then trigger signaling pathways that control many cellular processes, the deregulation of which may lead to disease such as cancer. The intracellular level of Rho GTP-bound form represents a criteria of choice to characterize the activity of these pathways and to understand physiopathological processes. To date, we use the ability of the Rho binding domain of Rhotekin to selectively bind to RhoA, RhoB, RhoC GTP-bound form to discriminate the ratio

between the GTP and GDP-bound form of Rho Proteins in cellular conditions in a semi-quantitative manner. This technique implies that activated Rho bind the RBD with the same affinity, which is in the range of 100nM [51]. However, the RBD recombinant domain is poorly stable and does not tolerate many tags or expression systems, remaining expressed as a GST fusion tool. Its relatively low affinity combined to its labile stability implies that the assay has to be done in critical scale condition and cannot be engineered to perform precise quantitation of Rho cellular activation [52]. Nevertheless, our results showed that scFvs could recognize with higher affinity than the RBD and with at least similar selectivity the GTP-bound Rho, some scFv being even more selective to a single Rho. Therefore our study opens up all the potential of scFv engineering tools to implement other recombinant format, with different tags or multimeric status, which will allow the establishment of reliable quantitation biosensors to address Rho activity biological function in vitro as well as in the cellular context [43]. Actually, deciphering intracellular signaling pathways requires sophisticated tools able to monitor not only the expression of proteins participating in the signal transduction but also the status of activation of switch proteins, witnesses of the implication of a particular pathway in a physiological phenomenon.

Acknowledgments

We thank Alan Hall for kindly providing plasmids encoding XPLN. We thank Stéphanie Cabantous and Jamòn De La Esquina for critical comments. We are grateful to Frederic Lopez from the CRCT protein core facility for helping in surface plasmon resonance measurements.

Author Contributions

Conceived and designed the experiments: PC JCF GF. Performed the experiments: PC CMG RG LL. Analyzed the data: PC JCF GF LL. Contributed reagents/materials/analysis tools: PC RG ILM CI LL. Wrote the paper: PC AO JCF GF LL.

References

1. Jaffe AB, Hall A (2005) Rho GTPases: biochemistry and biology. Annu Rev Cell Dev Biol 21: 247–269. doi:10.1146/annurev.cellbio.21.020604.150721.
2. Vega FM, Ridley AJ (2008) Rho GTPases in cancer cell biology. FEBS Lett 582: 2093–2101. doi:10.1016/j.febslet.2008.04.039.
3. Ihara K, Muraguchi S, Kato M, Shimizu T, Shirakawa M, et al. (1998) Crystal structure of human RhoA in a dominantly active form complexed with a GTP analogue. J Biol Chem 273: 9656–9666.

4. Cherfils J, Zeghouf M (2013) Regulation of Small GTPases by GEFs, GAPs, and GDIs. Physiol Rev 93: 269–309. doi:10.1152/physrev.00003.2012.
5. Boureux A, Vignal E, Faure S, Fort P (2007) Evolution of the Rho family of ras-like GTPases in eukaryotes. Mol Biol Evol 24: 203–216. doi:10.1093/molbev/msl145.
6. Wheeler AP, Ridley AJ (2004) Why three Rho proteins? RhoA, RhoB, RhoC, and cell motility. Exp Cell Res 301: 43–49. doi:10.1016/j.yexcr.2004.08.012.

7. Adamson P, Paterson HF, Hall A (1992) Intracellular localization of the P21rho proteins. J Cell Biol 119: 617–627.

8. Michaelson D, Silletti J, Murphy G, D'Eustachio P, Rush M, et al. (2001) Differential localization of Rho GTPases in live cells: regulation by hypervariable regions and RhoGDI binding. J Cell Biol 152: 111–126.

9. Gerald D, Adini I, Shechter S, Perruzzi C, Varnau J, et al. (2013) RhoB controls coordination of adult angiogenesis and lymphangiogenesis following injury by regulating VEZF1-mediated transcription. Nat Commun 4: 2824. doi:10.1038/ncomms3824.

10. Mazieres J, Antonia T, Daste G, Muro-Cacho C, Berchery D, et al. (2004) Loss of RhoB expression in human lung cancer progression. Clin Cancer Res Off J Am Assoc Cancer Res 10: 2742–2750.

11. Liu AX, Rane N, Liu JP, Prendergast GC (2001) RhoB is dispensable for mouse development, but it modifies susceptibility to tumor formation as well as cell adhesion and growth factor signaling in transformed cells. Mol Cell Biol 21: 6906–6912. doi:10.1128/MCB.21.20.6906-6912.2001.

12. Canguilhem B, Pradines A, Baudouin C, Boby C, Lajoie-Mazenc I, et al. (2005) RhoB protects human keratinocytes from UVB-induced apoptosis through epidermal growth factor receptor signaling. J Biol Chem 280: 43257–43263. doi:10.1074/jbc.M508650200.

13. Meyer N, Peyret-Lacombe A, Canguilhem B, Médale-Giamarchi C, Mamouni K, et al. (2014) RhoB Promotes Cancer Initiation by Protecting Keratinocytes from UVB-Induced Apoptosis but Limits Tumor Aggressiveness. J Invest Dermatol 134: 203–212. doi:10.1038/jid.2013.278.

14. Mamouni K, Cristini A, Guirouilh-Barbat J, Monferran S, Lemarié A, et al. (2014) RhoB Promotes γH2AX Dephosphorylation and DNA Double-Strand Break Repair. Mol Cell Biol 34: 3144–3155. doi:10.1128/MCB.01525-13.

15. Ren XD, Schwartz MA (2000) Determination of GTP loading on Rho. Methods Enzymol 325: 264–272.

16. Goffinet M, Chinestra P, Lajoie-Mazenc I, Medale-Giamarchi C, Favre G, et al. (2008) Identification of a GTP-bound Rho specific scFv molecular sensor by phage display selection. BMC Biotechnol 8: 34. doi:10.1186/1472-6750-8-34.

17. Chinestra P, Lajoie-Mazenc I, Faye J-C, Favre G (2012) Use of Phage Display for the Identification of Molecular Sensors Specific for Activated Rho. In: Rivero F, editor. Rho GTPases. New York, NY: Springer New York, Vol. 827. pp. 283–303. Available: http://www.springerlink.com/index/10.1007/978-1-61779-442-1_19. Accessed 2012 Apr 16.

18. Sheffield P, Garrard S, Derewenda Z (1999) Overcoming expression and purification problems of RhoGDI using a family of "parallel" expression vectors. Protein Expr Purif 15: 34–39. doi:10.1006/prep.1998.1003.

19. Friguet B, Chaffotte AF, Djavadi-Ohaniance L, Goldberg ME (1985) Measurements of the true affinity constant in solution of antigen-antibody complexes by enzyme-linked immunosorbent assay. J Immunol Methods 77: 305–319.

20. Martineau P (2010) Affinity Measurements by Competition ELISA. In: Kontermann R, Dübel S, editors. Antibody Engineering. Berlin, Heidelberg: Springer Berlin Heidelberg. pp. 657–665. Available: http://www.springerlink.com/index/10.1007/978-3-642-01144-3_41. Accessed 2012 Apr 12.

21. Gampel A, Mellor H (2002) Small interfering RNAs as a tool to assign Rho GTPase exchange-factor function in vivo. Biochem J 366: 393–398. doi:10.1042/BJ20020844.

22. Shimizu T, Ihara K, Maesaki R, Kuroda S, Kaibuchi K, et al. (2000) An open conformation of switch I revealed by the crystal structure of a Mg2+-free form of RHOA complexed with GDP. Implications for the GDP/GTP exchange mechanism. J Biol Chem 275: 18311–18317. doi:10.1074/jbc.M910274199.

23. Soundararajan M, Turnbull A, Fedorov O, Johansson C, Doyle DA (2008) RhoB can adopt a Mg2+ free conformation prior to GEF binding. Proteins 72: 498–505. doi:10.1002/prot.22017.

24. Thie H, Voedisch B, Dübel S, Hust M, Schirrmann T (2009) Affinity maturation by phage display. Methods Mol Biol Clifton NJ 525: 309–322, xv. doi:10.1007/978-1-59745-554-1_16.

25. Hawkins RE, Russell SJ, Winter G (1992) Selection of phage antibodies by binding affinity. Mimicking affinity maturation. J Mol Biol 226: 889–896.

26. Persson H, Wallmark H, Ljungars A, Hallborn J, Ohlin M (2008) In vitro evolution of an antibody fragment population to find high-affinity hapten binders. Protein Eng Des Sel PEDS 21: 485–493. doi:10.1093/protein/gzn024.

27. Daugherty PS, Chen G, Iverson BL, Georgiou G (2000) Quantitative analysis of the effect of the mutation frequency on the affinity maturation of single chain Fv antibodies. Proc Natl Acad Sci U S A 97: 2029–2034. doi:10.1073/pnas.030527597.

28. Juárez-González VR, Riaño-Umbarila L, Quintero-Hernández V, Olamendi-Portugal T, Ortiz-León M, et al. (2005) Directed evolution, phage display and combination of evolved mutants: a strategy to recover the neutralization properties of the scFv version of BCF2 a neutralizing monoclonal antibody specific to scorpion toxin Cn2. J Mol Biol 346: 1287–1297. doi:10.1016/j.jmb.2004.12.060.

29. Wright MJ, Deonarain MP (2007) Phage display of chelating recombinant antibody libraries. Mol Immunol 44: 2860–2869. doi:10.1016/j.molimm.2007.01.026.

30. Hoogenboom HR (1997) Designing and optimizing library selection strategies for generating high-affinity antibodies. Trends Biotechnol 15: 62–70. doi:10.1016/S0167-7799(97)84205-9.

31. Katayama M, Kawata M, Yoshida Y, Horiuchi H, Yamamoto T, et al. (1991) The posttranslationally modified C-terminal structure of bovine aortic smooth muscle rhoA p21. J Biol Chem 266: 12639–12645.

32. Baron R, Fourcade E, Lajoie-Mazenc I, Allal C, Couderc B, et al. (2000) RhoB prenylation is driven by the three carboxyl-terminal amino acids of the protein: evidenced in vivo by an anti-farnesyl cysteine antibody. Proc Natl Acad Sci U S A 97: 11626–11631. doi:10.1073/pnas.97.21.11626.

33. Wang D-A, Sebti SM (2005) Palmitoylated cysteine 192 is required for RhoB tumor-suppressive and apoptotic activities. J Biol Chem 280: 19243–19249. doi:10.1074/jbc.M411472200.

34. Lang P, Gesbert F, Delespine-Carmagnat M, Stancou R, Pouchelet M, et al. (1996) Protein kinase A phosphorylation of RhoA mediates the morphological and functional effects of cyclic AMP in cytotoxic lymphocytes. EMBO J 15: 510–519.

35. Tillement V, Lajoie-Mazenc I, Casanova A, Froment C, Penary M, et al. (2008) Phosphorylation of RhoB by CK1 impedes actin stress fiber organization and epidermal growth factor receptor stabilization. Exp Cell Res 314: 2811–2821. doi:10.1016/j.yexcr.2008.06.011.

36. Watanabe T, Ito Y, Yamada T, Hashimoto M, Sekine S, et al. (1994) The roles of the C-terminal domain and type III domains of chitinase A1 from Bacillus circulans WL-12 in chitin degradation. J Bacteriol 176: 4465.

37. Arthur WT, Ellerbroek SM, Der CJ, Burridge K, Wennerberg K (2002) XPLN, a guanine nucleotide exchange factor for RhoA and RhoB, but not RhoC. J Biol Chem 277: 42964–42972. doi:10.1074/jbc.M207401200.

38. Lu D, Shen J, Vil MD, Zhang H, Jimenez X, et al. (2003) Tailoring in vitro selection for a picomolar affinity human antibody directed against vascular endothelial growth factor receptor 2 for enhanced neutralizing activity. J Biol Chem 278: 43496–43507. doi:10.1074/jbc.M307742200.

39. Schier R, McCall A, Adams GP, Marshall KW, Merritt H, et al. (1996) Isolation of picomolar affinity anti-c-erbB-2 single-chain Fv by molecular evolution of the complementarity determining regions in the center of the antibody binding site. J Mol Biol 263: 551–567. doi:10.1006/jmbi.1996.0598.

40. Suzuki Y, Ito S, Otsuka K, Iwasawa E, Nakajima M, et al. (2005) Preparation of functional single-chain antibodies against bioactive gibberellins by utilizing randomly mutagenized phage-display libraries. Biosci Biotechnol Biochem 69: 610–619.

41. Horn IR, Wittinghofer A, de Bruïne AP, Hoogenboom HR (1999) Selection of phage-displayed fab antibodies on the active conformation of ras yields a high affinity conformation-specific antibody preventing the binding of c-Raf kinase to Ras. FEBS Lett 463: 115–120.

42. Nizak C, Monier S, del Nery E, Moutel S, Goud B, et al. (2003) Recombinant antibodies to the small GTPase Rab6 as conformation sensors. Science 300: 984–987. doi:10.1126/science.1083911.

43. Tanaka T, Rabbitts TH (2003) Intrabodies based on intracellular capture frameworks that bind the RAS protein with high affinity and impair oncogenic transformation. EMBO J 22: 1025–1035. doi:10.1093/emboj/cdg106.

44. Rose R, Weyand M, Lammers M, Ishizaki T, Ahmadian MR, et al. (2005) Structural and mechanistic insights into the interaction between Rho and mammalian Dia. Nature 435: 513–518. doi:10.1038/nature03604.

45. Maesaki R, Ihara K, Shimizu T, Kuroda S, Kaibuchi K, et al. (1999) The structural basis of Rho effector recognition revealed by the crystal structure of human RhoA complexed with the effector domain of PKN/PRK1. Mol Cell 4: 793–803.

46. Dvorsky R (2003) Structural Insights into the Interaction of ROCKI with the Switch Regions of RhoA. J Biol Chem 279: 7098–7104. doi:10.1074/jbc.M311911200.

47. Tanaka T, Williams RL, Rabbitts TH (2007) Tumour prevention by a single antibody domain targeting the interaction of signal transduction proteins with RAS. EMBO J 26: 3250–3259. doi:10.1038/sj.emboj.7601744.

48. Movilla N, Dosil M, Zheng Y, Bustelo XR (2001) How Vav proteins discriminate the GTPases Rac1 and RhoA from Cdc42. Oncogene 20: 8057–8065. doi:10.1038/sj.onc.1205000.

49. Lajoie-Mazenc I, Tovar D, Penary M, Lortal B, Allart S, et al. (2008) MAP1A light chain-2 interacts with GTP-RhoB to control epidermal growth factor (EGF)-dependent EGF receptor signaling. J Biol Chem 283: 4155–4164. doi:10.1074/jbc.M709639200.

50. Mircescu H, Steuve S, Savonet V, Degraef C, Mellor H, et al. (2002) Identification and characterization of a novel activated RhoB binding protein containing a PDZ domain whose expression is specifically modulated in thyroid cells by cAMP. Eur J Biochem FEBS 269: 6241–6249.

51. Blumenstein L, Ahmadian MR (2004) Models of the Cooperative Mechanism for Rho Effector Recognition: IMPLICATIONS FOR RhoA-MEDIATED EFFECTOR ACTIVATION. J Biol Chem 279: 53419–53426. doi:10.1074/jbc.M409551200.

52. Guilluy C, Dubash AD, García-Mata R (2011) Analysis of RhoA and Rho GEF activity in whole cells and the cell nucleus. Nat Protoc 6: 2050–2060. doi:10.1038/nprot.2011.411.

53. Pettersen EF, Goddard TD, Huang CC, Couch GS, Greenblatt DM, et al. (2004) UCSF Chimera–a visualization system for exploratory research and analysis. J Comput Chem 25: 1605–1612. doi:10.1002/jcc.20084.

9

Carrier of Wingless (Cow), a Secreted Heparan Sulfate Proteoglycan, Promotes Extracellular Transport of Wingless

Yung-Heng Chang[1,2], Yi Henry Sun[1,2]*

1 Graduate Institute of Life Sciences, National Defense Medical Center, Taipei, Taiwan, Republic of China, 2 Institute of Molecular Biology, Academia Sinica, Taipei, Taiwan, Republic of China

Abstract

Morphogens are signaling molecules that regulate growth and patterning during development by forming a gradient and activating different target genes at different concentrations. The extracellular distribution of morphogens is tightly regulated, with the *Drosophila* morphogen Wingless (Wg) relying on Dally-like (Dlp) and transcytosis for its distribution. However, in the absence of Dlp or endocytic activity, Wg can still move across cells along the apical (Ap) surface. We identified a novel secreted heparan sulfate proteoglycan (HSPG) that binds to Wg and promotes its extracellular distribution by increasing Wg mobility, which was thus named Carrier of Wg (Cow). Cow promotes the Ap transport of Wg, independent of Dlp and endocytosis, and this function addresses a previous gap in the understanding of Wg movement. This is the first example of a diffusible HSPG acting as a carrier to promote the extracellular movement of a morphogen.

Editor: Amit Singh, University of Dayton, United States of America

Funding: YHS is supported by grants from Academia Sinica, Taiwan, Republic of China (AS-91-IMB-4PP, AS-95-TP-B04; http://www.sinica.edu.tw/index.shtml), and National Science Council, Taiwan, Republic of China (NSC 96-2321-B-001-002, NSC 97-2321-B-001-002, NSC 98-2321-B-001-034, NSC 99-2321-B-001-016, NSC 100-2321-B-001-012; http://www.most.gov.tw/mp.aspx?mp = 7). The funders had no role in study design, data collection and analysis, decision to publish, or preparation of the manuscript.

Competing Interests: The authors have declared that no competing interests exist.

* Email: mbyhsun@gate.sinica.edu.tw

Introduction

Morphogens are signaling molecules that can be distributed in a developing tissue along a concentration gradient and affect development in a concentration-dependent manner. The formation and interpretation of the gradient are regulated at multiple levels.

The *Drosophila* morphogen Wingless (Wg) is one of the founding members of the Wnt family of signaling molecules. In *Drosophila* embryo and imaginal disc development, Wg has been shown to act as a long-range morphogen [1], [2], [3], [4]. In the best-studied wing disc, Wg is expressed in several rows of cells at the dorsoventral (D-V) boundary in the prospective wing pouch region [3]. Wg can be secreted from producing cells or localized extracellularly to form a concentration gradient to regulate target genes at different levels [5], [6]. Although Wg is secreted from the apical (Ap) surface of its producing cells [7], [8], extracellular Wg (exWg) is localized primarily on the basolateral (Ba) surface [5]. ExWg can be detected within a few rows of cells away from its producing cells at the Ap surface but spreads more than 20 cells away at the lateral surface [7]. These results suggest that the long-range movement of exWg occurs on the Ba surface. However, the mechanisms by which exWg moves short distances along the Ap surface and longer distances along the Ba surface remain unclear.

In receiving cells, Wg can also be found in puncta representing internalized Wg. The internalization of Wg is dependent on endocytosis and occurs at both Ap and Ba surfaces [5], [9], [10].

Whereas the secretion and degradation of Wg are dependent on dynamin, the movement or distribution of exWg is independent of endocytosis [5], [7], [9], [10], [11].

Wg distribution is affected by heparan sulfate proteoglycans (HSPGs), which are proteins modified by heparan sulfate (HS) glycosaminoglycan (GAG) chain attachments. Enzymes for GAG and HS synthesis, such as Sulfateless (Sfl) and Brother of tout-velu (Botv), are required for exWg distribution [12], [13], [14]. These results suggest that the exWg movement requires HSPGs. Within large *sfl* and *botv* mutant clones, although exWg is reduced, there is Wg accumulation within and behind the clone, suggesting that some HSPG from neighboring wild-type (WT) cells can act non-autonomously [12], [14], [15]. Because the two HSPGs known to affect Wg signaling, Dally and Dally-like (Dlp) (see below), are membrane-anchored, an unidentified diffusible HSPG is predicted to serve this role.

Dally and Dlp belong to the glypican family of HSPGs, and both can bind to Wg and are involved in regulating Wg distribution and signaling [14], [16]. The movement of Wg from the Ap to Ba surface was reported to be dependent on Dlp through transcytosis [7], although this finding was contradicted by another study [17]. Moreover, this vertical intracellular translocation of Wg does not explain the lateral intercellular spread of exWg. It has been proposed that Wg bound to Dlp can be transferred to adjacent Wg receptors, depending on the ratio of Dlp to DFz2

[16], [17], [18], although whether this mechanism transfers Wg to adjacent cells has not yet been demonstrated.

Extracellular movement of Wg at or near producing cells likely occurs independent of membrane-anchored Dlp and DFz2, as these levels are low in the Wg-producing cells at the D-V border [14], [15], [16], [19], [20]. Furthermore, *dlp*-null clones do not affect the exWg level in this region [14], which leaves open the question of what factor is responsible for moving Wg from its source to adjacent Dlp-expressing cells. Dally is present at high levels at the D-V border, but it plays only a minor role that is partially redundant with Dlp [5], [14], [16]. Even in a *dally dlp* double mutant clone, exWg is detected away from the producing cells [14]. The extracellular hydrolase Notum/Wingfull can modify Dlp to reduce its ability to bind and stabilize Wg, thereby reducing, rather than promoting, the range of Wg distribution [21], [22]. Therefore, some unidentified factor must be responsible for moving Wg away from its source.

In this study, we identified Carrier of Wg (Cow) as a novel secreted HSPG. Our results showed that Cow can bind to exWg to increase its rate of movement and stability. The identification of Cow answers four previously unknown aspects of Wg gradient formation. First, Cow is localized primarily at the Ap surface and is responsible for the Ap movement of Wg. Second, Cow is a diffusible HSPG, which can explain the non-autonomous rescue of Wg movement in clones defective for HS synthesis. Third, Cow is present at the D-V border and is responsible for moving Wg away from its source to interact with Dlp and receptors. Fourth, diffusible Cow can mediate the transfer of Wg to adjacent cells, a role not satisfactorily explained by membrane-anchored Dlp.

Materials and Methods

Transgenes

Human *Testican-2* cDNA was obtained from the Human Unidentified Gene-Encoded (HUGE) Large Proteins Database (Kazusa DNA Research Institute) and cloned into the *pUAST-flag* vector with a Flag tag at the N-terminal [23] to generate *UAS-Testican-2*. The *cow* cDNA was amplified by RT-PCR and cloned into *pUAST-flag* to generate *UAS-Cow*. The protein product does not contain the Flag tag because it is cleaved at the signal peptide. For Flag-tagged Cow, the Flag tag "DYKDDDDK" was inserted after the signal peptide of Cow to generate *UAS-SP-Flag-Cow*. For *SP-EGFP-Cow*, the EGFP replaced the Flag tag in *SP-Flag-Cow*. For *Cow-GPI*, the Dally GPI domain (aa577–626) was added to the C-terminus of Cow. The two putative GAG sites SG1 (I**SG**Y) and SG2 (**NSG**N) were mutated to I**AA**Y and **NAA**N, respectively, to produce *SP-Flag-Cow-mSG1+2*. The *Cow-miRNA* constructs were designed based on the method of Chen et al. [24]. The targeting sites of *Cow-miRNA-1* and *Cow-miRNA-2* were TTAGCATAGTTGCTGCGTAAGA in the 5′-untranslated region (UTR) and CCACAAGAATCGTGATGAGATA in the N region, respectively (see Figure S1B). *SP-Flag-Cow*, *SP-EGFP-Cow*, *SP-Flag-Cow-mSG1+2*, *Cow-miRNA-1* and *Cow-miRNA-2* were cloned into the *pUAST* vector [25]. The *Cow-dsRNA* construct was made by cloning the sense and anti-sense sequences of the Cow TY-C region into the *pWIZ* vector [26] at *Bgl*II/*Xho*I and *Nhe*I/*Xba*I sites, respectively. The detailed construction process is available upon request.

RT-PCR

The primer sets for RT-PCR and the conditions for semi-quantitative analysis are available upon request.

Fly stocks

UAS-GFP-wg [27], *UAS-DFz2* and *UAS-DFz2N* [28], *UAS-Dlp-HA* [16], *wg{KO; Gal4}* [29], *wg{KO; wg-HA}* [30], *wg^{cx4}*, *sfl^{03844}*, *Df(3R)Exel6193*, *Df(3R)BSC527*, *Df(3R)BSC619*, *Mi{ET1}CG13830^{MB00767}*, *neur-lacZ* (*neur^{A101}*), *UAS-Shi^{ts}*, *UAS-lacZ*, *UAS-myrRFP*, *UAS-FLP*, *tub-p-Gal4*, *ap-Gal4*, *en-Gal4*, *hh-Gal4*, *ptc-Gal4*, *MS1096-Gal4*, and *nub-Gal4* were obtained from the Bloomington Drosophila Stock Center. The *cow^{5Δ}* allele was generated by imprecise excision of the *Minos* transposable element *Mi{ET1}CG13830^{MB00767}* from the 3′-UTR of *cow* (see Figure S1A).

Immunostaining and *in situ* hybridization of the embryo and wing discs

Immunostaining and *in-situ* hybridization of the embryo and wing discs were performed following standard protocols [31], [32]. The primary antibodies included guinea pig anti-Sens (1:1,000) [33], rat anti-Dll (1:250) [3], goat anti-Dll (1:100; Santa Cruz Biotechnology), rabbit anti-GFP (1:250; Invitrogen), mouse anti-HA (1:200; Abcam), mouse anti-β-Galactosidase (1:500; Promega), rabbit anti-β-Galactosidase (1:1,000; Cappel), mouse anti-Flag (1:500; Sigma), mouse anti-Cut (1:150), mouse anti-Wg (1:200), and mouse anti-Ptc (1:250), which were obtained from the Developmental Studies Hybridoma Bank. A polyclonal rabbit anti-Cow antibody was raised against the synthesized peptide Ac-YLEEEAKRRVNQQDNQDSQDC-amide (aa112–131) by Quality Controlled Biochemicals (Hopkinton, MA, USA). The extracellular staining protocol was performed as previously described [5], using 3× standard antibody concentrations. Secondary antibodies conjugated with FITC, Cy3 and Cy5 were purchased from Jackson ImmunoResearch. Some samples were co-stained with DAPI (1:1,000; Sigma), FITC-Phalloidin (1:200; Invitrogen), Texas Red-X-Phalloidin (1:200; Invitrogen) and Alexa Fluor 680-Phalloidin (1:200; Invitrogen). For comparative purposes, wing discs were dissected, fixed and stained in parallel and imaged under identical conditions. Images were collected with Zeiss LSM510 META and LSM780 confocal microscopes.

Embryo cuticle and adult wing preparation

Embryo cuticle preparation was performed according to a published protocol [16], and samples were mounted in Hoyer's solution and photographed under dark-field microscopy. To examine the chemosensory bristle number, adult wings were mounted in Hoyer's solution and photographed under differential interference contrast (DIC) and phase-contrast microscopy.

Generation of germline and mutant clones

Generation of the germline clone was performed according to the described protocol [34]. *hs-FLP/Y; FRT82B ovo^{D1}/TM3, Sb* males were crossed to *hs-FLP; FRT82B cow^{5Δ}/TM3, Sb* virgins. Progeny at the third instar were heat shocked at 37°C for 1 h once on day 5 and once on day 6. The *hs-FLP; FRT82B cow^{5Δ}/FRT82B ovo^{D1}* females were crossed to *FRT82B cow^{5Δ}/TM3, Ser, twi-GFP* males to generate embryos that were maternal and zygotic *cow^{5Δ}* mutants. To generate the *cow^{5Δ}* mutant clone in wing disc and adult wing, the *ap-Gal4/UAS-FLP; FRT82B cow^{5Δ}/FRT82B ubi-GFP Minute* was used.

Immunoprecipitation and western blotting

The Cow protein levels in WT and *cow^{5Δ}* homozygous embryos were analyzed by western blot. *cow^{5Δ}* was balanced with the *twist-GFP*, and the non-GFP homozygous *cow^{5Δ}* embryos were selected by detecting the level of Cow protein. WT and *sfl^{03844}* homozygous embryos were selected and analyzed for HS modification of Cow protein. sfl^{03844} was also balanced with $twist$-GFP, and non-GFP sfl^{03844} homozygous embryos were collected and analyzed for the HS modification state of Cow. Coimmunoprecipitation of Wg and Cow was performed following a published protocol [20]. To address physiological interactions between GFP-Wg and SP-Flag-Cow, GFP-Wg and SP-Flag-Cow were coexpressed in embryos using tub-Gal4 and then collected for immunoprecipitation assays. To examine endogenous Wg-HA and Cow protein interactions, protein extracts containing endogenous Wg-HA and Cow proteins were collected from WT, wg{KO; Gal4} and wg{KO; wg-HA} at the larval stage and then analyzed using the immunoprecipitation protocol. The primary antibodies used included mouse anti-HA (Roche), mouse anti-Flag (Sigma) and mouse anti-α-tubulin (Sigma). Light chain-specific secondary antibodies included goat anti-mouse and goat anti-rabbit conjugated with horseradish peroxidase (HRP) (Jackson ImmunoResearch).

Extracellular binding assays for Cow with Wg and Testican-2 with Wnt5a

To test whether secreted Cow and Wg could interact in the extracellular space, the empty UAS vector UAS-HA-Wg [22] and UAS-SP-Flag-Cow were separately co-transfected with the actin-Gal4 plasmid in S2 cells. After 48 h, the culture supernatants were separately collected. Five hundred micrograms of protein from each supernatant was mixed in different combinations and analyzed by immunoprecipitation as described above. In the binding assay using secreted Testican-2 with Wnt5a, the empty pAC vector, pAC-SP-Flag-Testican-2 and pAC-Wnt5a-HA [35] (with the Wnt5a in the pcDNA moved to the pAC vector) were separately transfected into S2 cells and tested as described above.

Wg protection assay

S2 cells were transfected with the HA-Wg vector for 48 h. The HA-Wg conditioned medium was added to replace the medium of empty or Cow-dsRNA vector-transfected cells. The HA-Wg content from the culture supernatant at different time points was measured by western blotting.

Intensity plot for exWg

Staining for exWg in wing discs was performed according to the protocol of Strigini and Cohen (2000) [5]. Each plotting line was derived from a region of 600×650 pixels in the center of the wing pouch region to minimize the effect of the curvature of the D-V border. There are 600 single-pixel lines perpendicular to the D-V border; each 10 adjacent lines were averaged, and the 60 averaged lines were averaged to produce the intensity plot from one disc. The results from three discs were then averaged to obtain the intensity plot for each condition.

Fluorescence recovery after photobleaching (FRAP)

The dpp>GFP-Wg+lacZ and dpp>GFP-Wg+Cow third instar wing discs were analyzed following the published procedure [6]. The time-lapse images were collected using a Zeiss LSM710 confocal microscope.

Results

Cow is a secreted HSPG

We identified the Drosophila CG13830 gene (Figure S1A) and its human homolog Testican-2 in a gain-of-function screen. We named it carrier of wg (cow) for reasons to be described below.

Cow belongs to the testican family of secreted HSPGs with a signal peptide (SP) and five conserved domains (Figure S1B) [36]. We generated a mutation, $cow^{5\Delta}$, with a deletion of the 3′UTR (Figure S1A) that is protein-null (Figure S1C–D). We also generated UAS-Cow-dsRNA and two UAS-Cow-miRNA constructs (Figure S1B; see Materials and Methods) that effectively reduced the Cow level (Figure S1D–E). The expression patterns of cow at the RNA and protein levels throughout development are presented in Figure S1. In the wing disc, Cow is detected throughout the disc and is primarily localized to the Ap surface (Figure 1A).

We next tested whether Cow can be secreted, reminiscent of its human homolog Testican-2 [37]. We generated an EGFP-Cow fusion construct (SP-EGFP-Cow) and a construct of Cow with a truncated signal peptide (Flag-tSP-Cow). Both were expressed by the ptc-Gal4 (abbreviated as ptc>myrRFP+SP-EGFP-Cow +Flag-tSP-Cow) at the anterior-posterior (A-P) boundary of the wing disc (Figure 1B). Both were detectable using the anti-Cow antibody (Figure 1C–D). The GFP signal was broader than the Ptc expression region, and it appeared in puncta on both the Ap and Ba surfaces (Figure 1B–C). However, Flag-tSP-Cow was restricted to ptc-Gal4-expressing cells (Figure 1B and 1D). Extracellular staining using anti-GFP was performed on ptc>myrRFP+ SP-EGFP-Cow discs to detect EGFP-Cow in the extracellular space. The GFP fluorescent signal could be detected in the ptc-expressing region, and the extracellular anti-GFP signal was broader than the ptc domain (Figure 1E). These results suggest that Cow can be secreted and dispersed away from producing cells and that this secretion requires its signal peptide. When SP-Flag-Cow was transfected into S2 cells, an anti-Flag antibody detected bands at 100 kDa and 75 kDa in the culture supernatant (Figure S1F). These results clearly demonstrated that Cow was secreted.

We then tested whether Cow is HS-modified. Cow has a predicted molecular weight of 72 kDa, and an anti-Cow antibody detected two bands at approximately 100 kDa and 75 kDa in embryos, wing discs, adults and S2 cells (Figure S1C). Both bands were lost in $cow^{5\Delta}$ mutant embryos and were enhanced when Cow was overexpressed (Figure S1C–D). These results demonstrated the specificity of the anti-Cow antibody. The 100-kDa band was not detected in the embryonic extract of the sfl homozygous mutant (Figure 1F), suggesting that it was modified by HS. When the two putative GAG sites in Cow were mutated, the 100-kDa band was significantly reduced (Figure S1F), supporting the conclusion that this was the HS-modified form. In comparison, the 75-kDa band likely represented unmodified Cow. These results demonstrate that Cow, like its human homolog Testican-2, is HS-modified and that both HS-modified and unmodified Cow can be secreted.

Genetic relationship between cow and wg

Homozygous $cow^{5\Delta}$ mutants are mostly embryonic lethal; 50% die before cuticle formation, and 9.5% of $cow^{5\Delta}$ embryos showed a weak denticle belt fusion phenotype (Figure 1H). Expressing Cow-miRNA-1 or Cow-dsRNA in the embryo also produced similar phenotypes (Figure 1J–K). Similar embryonic denticle belt fusion phenotypes have been reported in many mutants, including mutations in the Hedgehog (Hh) and Wg signaling pathways [38]. Because Cow is a HSPG, we investigated its relationship with the morphogens Hh and Wg. When the wg dosage was reduced, the $cow^{5\Delta}$ phenotype became very strong (Figure 1L), similar to the wg^{cx4} homozygous mutant [38]. The hh gene is also deleted in two deficiencies with deleted cow (Figure S1A), but the combination of hh deletion and cow mutation showed no enhancement of the denticle fusion phenotype (Figure 1I and unpublished data). Deletion of one copy of the Hh coreceptor gene smoothened

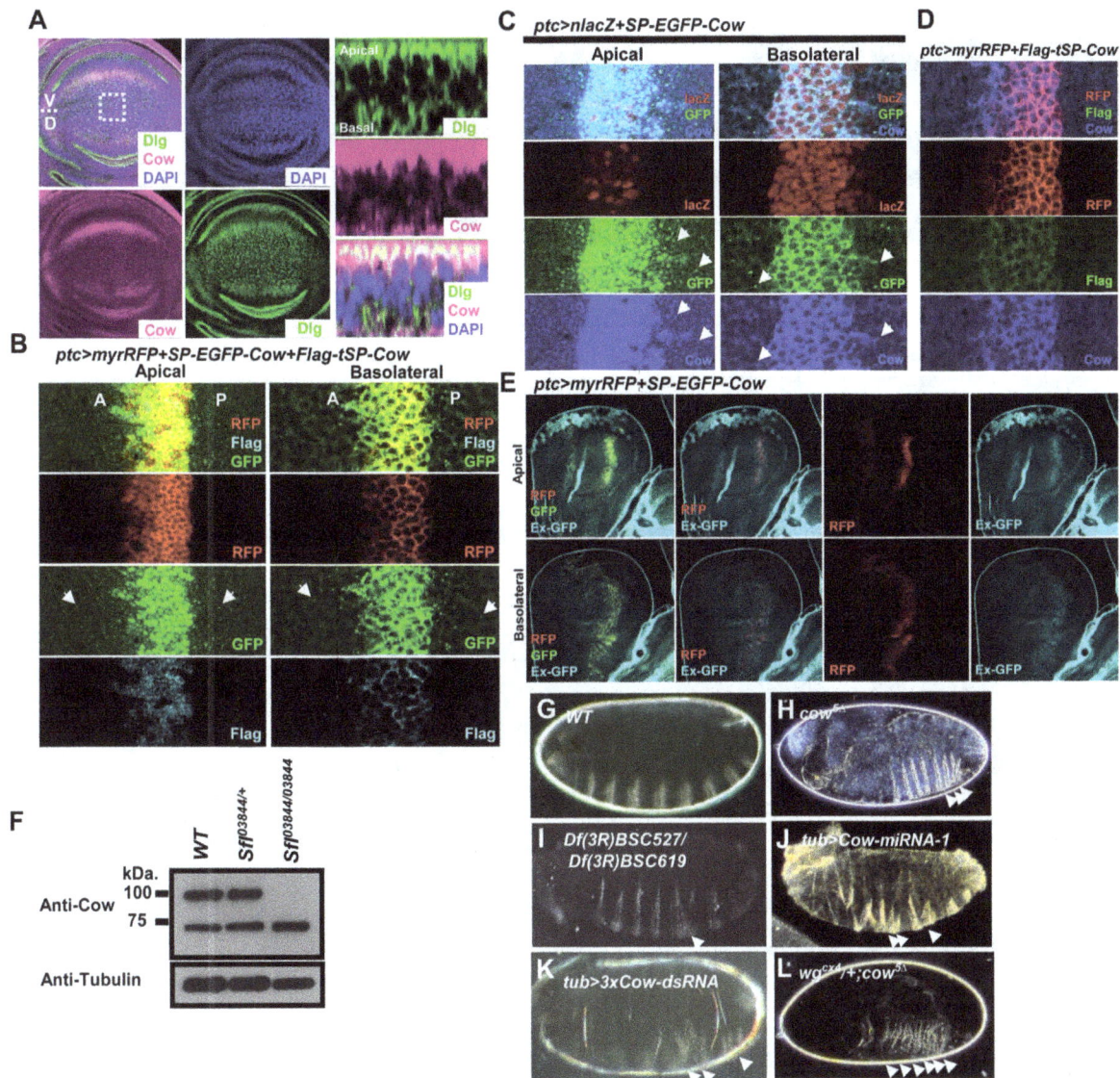

Figure 1. Cow is a secreted HSPG and genetically interacts with *wg*. (A) Cow protein (stained with anti-Cow, magenta) is detected throughout the wing disc. Dlg, green; DAPI, blue. Z-sections show that Cow is located primarily at the Ap surface but is also present at the Ba surface. In this and all subsequent figures on wing discs, the V side is located at the top, and the P side is at the right. (B) *SP-EGFP-Cow* and *Flag-tSP-Cow* expression were induced along the A-P border using *ptc-Gal4* (in *ptc>myrRFP+SP-EGFP-Cow+Flag-tSP-Cow*). (C) *SP-EGFP-Cow* was induced by *ptc-Gal4* (*ptc>nlacZ+SP-EGFP-Cow*) and detected with an anti-Cow antibody. The *ptc*-expressing cells are labeled with *nlacZ* (red). The GFP fusion Cow protein, SP-EGFP-Cow (GFP, green), could be detected with an anti-Cow antibody (blue). The distribution of both signals was broader than the *ptc*⁺ cells and showed a punctate signal. (D) Flag-tagged Cow with a truncated signal peptide (*Flag-tSP-Cow*) was expressed in the *ptc* (red) expression domain (in *ptc>myrRFP+Flag-tSP-Cow*). A Flag signal (green) was detected only in the Ptc domain, indicating that the signal peptide was required for its secretion. Anti-Cow antibody (blue) detected the Flag-tSP-Cow in the Ptc domain as well as the endogenous Cow present throughout the disc. In (C) and (D), the level of *ptc-Gal4*-driven Cow expression was much higher than that of endogenous Cow. (E) Extracellular GFP staining of *ptc>myrRFP+SP-EGFP-Cow* was demonstrated in the haltere; the extracellular GFP staining signal (cyan) can be detected at the Ap and Ba surfaces of the haltere. (F) The WT, *sfl*⁰³⁸⁴⁴/*twi-GFP* and *sfl*⁰³⁸⁴⁴ homozygous embryo protein extracts were probed using anti-Cow antibody. Tubulin was used as the loading control. (G–L) Embryonic cuticle phenotypes of (G) wild-type (WT), (H) *cow*⁵ᐃ, (I) *Df(3R)BSC527/Df(3R)BSC619*, (J) *tub>Cow-miRNA-1*, (K) *tub>3xCow-dsRNA*, and (L) *wg*ᶜˣ⁴/+; *cow*⁵ᐃ.

(*smo*) did not affect the *cow*⁵ᐃ embryonic phenotype (unpublished data). These results showed that *cow* has a strong genetic interaction with *wg* but not with Hh signaling.

When Cow was knocked down in the developing wing, driven by *MS1096-Gal4*, which is expressed in the entire wing pouch region in the wing disc [39], [40], ectopic chemosensory bristles developed along or near the wing margin, on both the anterior (A) and posterior (P) sides (Figure S2B–B'). In contrast, ubiquitous

overexpression of Cow or human Testican-2 caused the loss of chemosensory bristles along the wing margin (Figure S2C–C', D–D' and E) and extra wing vein tissue (Figure S2C–D). The extra wing vein phenotype has not been reported for Wg signaling, suggesting that Cow has additional functions beyond influencing Wg. The cell fate of wing margin chemosensory bristles is determined by Wg signaling [1], [41], and ectopic chemosensory bristles were also observed upon overexpression of the Wg

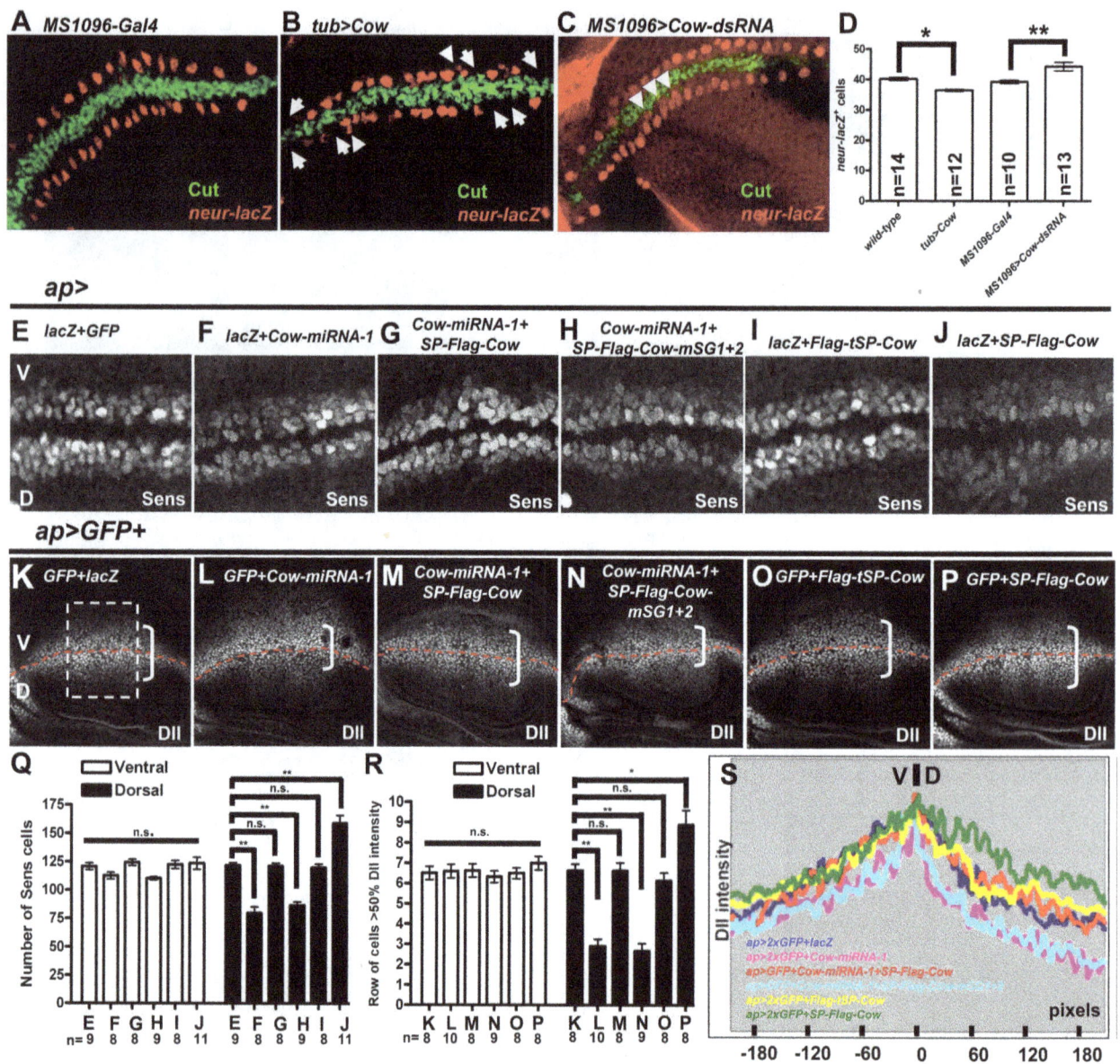

Figure 2. Cow affects the expression of Wg target genes in a biphasic pattern. (A–C) Cut (green) and *neur-lacZ* (red) expression along the D-V border of late third instar wing discs of (A) *MS1096-Gal4* (representing WT), (B) *tub>Cow*, and (C) *MS1096>Cow-dsRNA. neur-lacZ* cells form a single row on each side of the D-V border in WT. Loss of *neur-lacZ* cells in their normal location is indicated by arrows. *neur-lacZ* cells located outside of their normal location are indicated by arrowheads. (D) Summary of the numbers of *neur-lacZ* positive cells in A–C. (E–J) Sens+ cells (white) in the V and D sides of wing discs from the indicated genotypes. The numbers of Sens+ cells in (E–J) are summarized in (Q). Dll (white) expression in wing discs of the indicated genotypes. To eliminate dosage variation, the number of *UAS* constructs was equal in all groups. Data for Dll are presented as merged images of several Z sections. For Dll intensity, we chose a rectangle (shown in K) with its center localized on the intersection of the A/P and D/V axes. The results are summarized in (R) as the number of rows of cells at 50% Dll intensity and in (S) as plots of Dll intensity measured from the rectangle along the D-V axis. In this and all subsequent figures, n.s., not statistically significant, *, p<0.01, **, p<0.001.

receptor DFz2 (Figure S2H; [28], [42]). Thus, for the wing margin chemosensory bristles, the loss of function (LOF) of *cow* produces a phenotype similar to enhanced Wg signaling, whereas gain of function (GOF) of Cow gives a phenotype similar to loss of Wg. This is in contrast to the embryonic segmentation phenotype, where loss of *cow* and *wg* caused similar denticle belt fusion phenotypes. Consistent with the lack of genetic interaction with the Hh pathway in the embryonic phenotype, neither GOF nor LOF of *cow* caused changes in the wing L3–L4 intervein area, which is regulated by Hh signaling and by Dlp [7], [43].

This genetic interaction suggested that Cow may affect Wg signaling. When *cow*[5Δ] mutant clones were generated in the wing using *ap>FLP*, 89.4% of the *cow*[5Δ] mutants died before the larval stage, and 9.3% died at the pupal stage. The escapers (1.3%) all showed a wing-loss phenotype, (Figure S2F), which is similar to the *wg*[1] phenotype. We then tested the epistatic relationship between *cow* and the Wg receptor DFz2. Expression of dominant-negative DFz2 (DFz2N), driven by *MS1096-Gal4*, caused the loss or strong reduction of wings (Figure S2G; [28]), whereas coexpression of Cow, Cow-dsRNA or Cow-miRNA with DFz2N did not affect the *MS1096>DFz2N* wing loss (Figure S2G) or the *MS1096>DFz2*

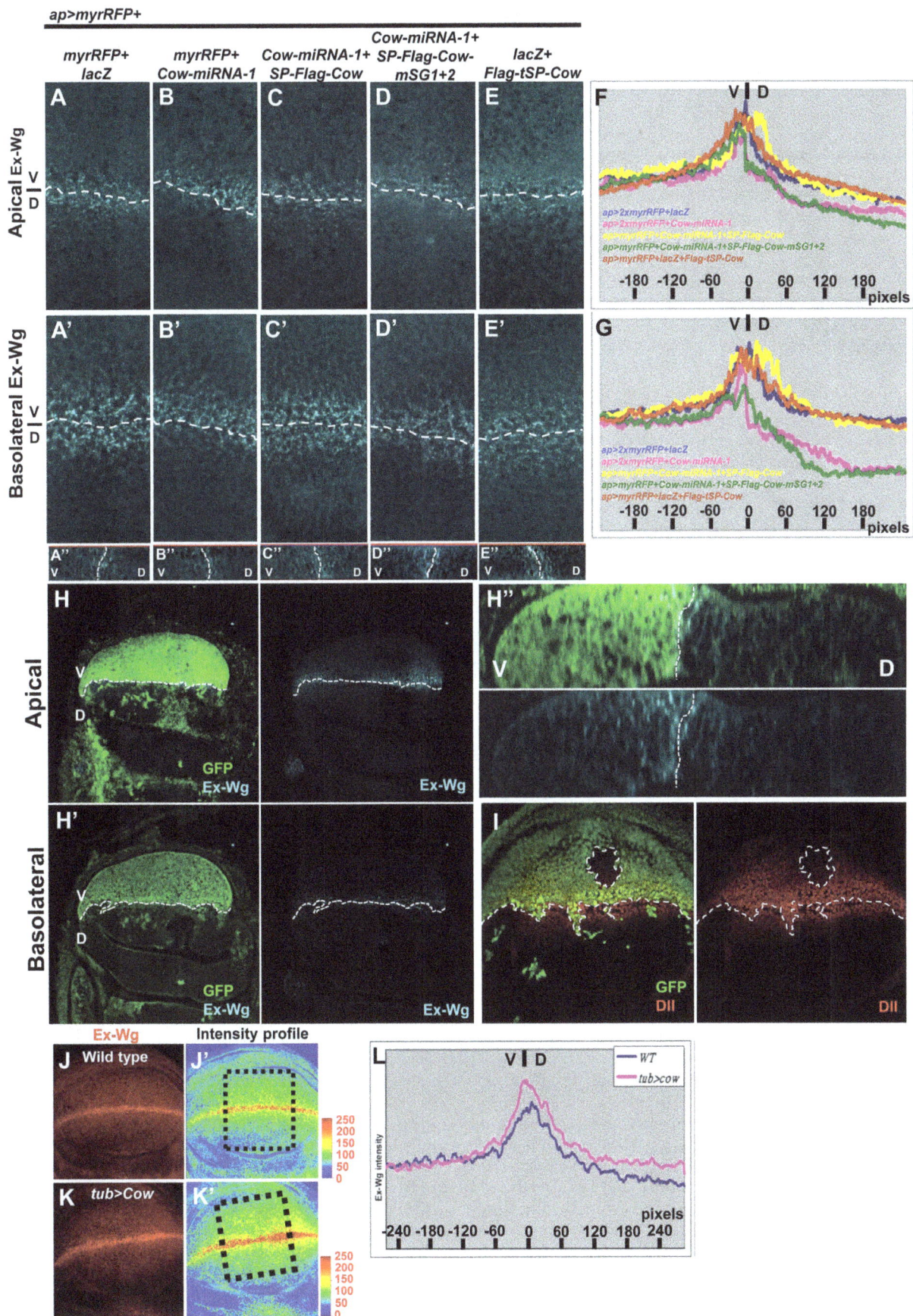

Figure 3. Cow promotes the extracellular distribution of Wg. (A–E) Staining of exWg (cyan) in wing discs of the indicated genotypes. (A″, B″, C″, D″ and E″) Z sections of each combination are shown. The number of *UAS* constructs was equalized to eliminate dosage differences. The exWg intensity plots (see Materials and Methods) for each condition at the Ap (F) and Ba surfaces (G). The exWg distribution (H–H″) and Dll range (I) were examined in the *cow⁵ᐟᵈ* mutant clone. Staining of exWg (red) in (J) WT and (K) *tub>Cow*. The intensity of the exWg signal in J and K is color-coded in J′ and K′, respectively, and intensity plots are shown in (L).

ectopic bristles (Figure S2H). These results suggest that Cow affected Wg signaling at a step upstream of the receptor DFz2, possibly at Wg itself. When Cow was ubiquitously overexpressed, the expression level of *wg-lacZ* and the number of rows of *wg-lacZ⁺* cells at the D-V border in the wing disc were not changed (unpublished data). Semi-quantitative RT-PCR also showed that the *wg* RNA level in late third instar wing discs of *tub>Cow* was not changed (unpublished data). These results demonstrate that the effect of Cow on Wg signaling is not at the level of transcription, but most likely at the level of Wg protein.

Cow affected the expression of Wg target genes in a biphasic pattern

We next investigated whether *cow* affects Wg signaling by examining the expression of Wg target genes. In the wing disc, Wg is expressed in a stripe of a few rows of cells at the D-V boundary and at the boundary of the wing pouch region [41], [44], [45]. The *wg⁺* cells at the D-V border also express Cut [41]. We examined the expression of several known Wg target genes in the wing disc: *neuralized (neur)*, *senseless (sens)*, and *Distalless (Dll)* [1], [3], [46]. Although both *neur* and *sens* are viewed as high-threshold Wg target genes, they appear to require different levels of Wg for their activation, and *sens* expression is broader than *neur* expression (see below). In particular, *neur* is only activated within and immediately adjacent to the ectopic Wg-expressing clone [1], whereas *sens* can be activated several cells away [46]. In addition, *sens* expression occurs earlier than *neur* in the same cell [33]. All of these findings suggest that *sens* responds to a lower level of Wg than *neur*. Therefore, we suggest that *neur* is a short-range target and that *sens* is an intermediate-range target.

Furthermore, *neur* expression, as shown by *neur-lacZ*, flanked the D-V border in Wg/Cut⁺ cells (Figure 2A; [1]). In *tub>Cow*, there was a loss of *neur⁺* cells in the normal domain (arrow in Figure 2B, compare with Figure 2A), accompanied by a few additional *neur⁺* cells just outside its normal domain (arrowhead in Figure 2B). When Cow was knocked down in the wing pouch area, there were extra *neur⁺* cells, and some of these extra *neur⁺* cells may be within the Cut expression domain (Figure 2C). The total number of *neur⁺* cells decreased when Cow was overexpressed and increased when Cow was knocked down (Figure 2D).

Sens is expressed in a few rows of cells along the D-V boundary (Figure 2E and S3A; [33], [46]). The number of Sens⁺ cells flanking the D-V boundary increased when either DFz2 or Cow was overexpressed (Figure S3B–C and E), whereas when Cow was knocked down, the number of Sens⁺ cells decreased (Figure S3D–E). The numbers of Sens⁺ cells were equal in the V and D sides in WT discs (Figure 2E and Q). When Cow was knocked down in the D side, driven by *ap-Gal4*, the number of Sens⁺ cells was reduced in the D compared to V side (Figure 2F and Q). This Sens⁺ reduction could be rescued by coexpression of SP-Flag-Cow (Figure 2G and Q). The Sens⁺ cells in the D side increased when SP-Flag-Cow was overexpressed (Figure 2J and Q), and the signal peptide-truncated Cow, Flag-tSP-Cow, did not show any effect on Sens⁺ cell number (Figure 2I and Q), consistent with the previous conclusion that Cow functions as a secreted protein.

Dll is expressed in a broad domain in the wing disc and is equally expressed in the D and V sides of the wing disc (Figure 2K

and R–S). When Cow was knocked down in the D side, the range of Dll was reduced in this side (Figure 2L and R–S), and this reduction could be rescued by co-expression of *SP-Flag-Cow* (Figure 2M and R–S). Overexpression of Cow in the D side caused expansion of the Dll range (Figure 2P and R–S), whereas expression of Flag-tSP-Cow did not affect the Dll range (Figure 2O and R–S), confirming that Cow acts as a secreted form.

In summary, *cow* showed a biphasic effect on Wg target genes in the wing disc. For the short-range target *neur*, *cow* had the opposite effect of Wg signaling. For the intermediate-range target Sens and the long-range target Dll, *cow* had the same effect as Wg. Similar biphasic effects have been reported for Dlp [7], [16], [17], [18], [19], [20].

HS-modified Cow promoted exWg movement and stability

The apparent contrast between wing and embryo phenotypes and the biphasic activity of target genes can be explained by a simple unifying model in which Cow promotes Wg movement, thereby reducing the Wg level near its source and broadening its distribution. We used an immunostaining protocol to measure the extracellular distribution of Wg produced from the D-V boundary in the wing disc [5] and tested whether Wg distribution is promoted by Cow. In the control wing disc, exWg showed a similar distribution in the D and V sides, in both Ap and Ba regions (Figure 3A–A″ and F–G). When Cow was knocked down in the D side, the exWg range in D was reduced in both the Ap and Ba domains (Figure 3B–B″ and F–G). This reduction in the range of exWg could be rescued by co-expressing *SP-Flag-Cow* (Figure 3C–C″ and F–G), indicating that the effect was specific to Cow and ruling out off-target effects of the *Cow-miRNA*. We also generated a *cow⁵ᐟᵈ* mutant clone in D using *ap>FLP*, in which the exWg level is significantly reduced in both the Ap and Ba domains (Figure 3H–H″), and the Dll range was also reduced (Figure 3I).

We then tested the requirement for HS modification of Cow. Cow with mutations in the putative GAG sites failed to rescue the reduction of range of exWg (Figure 3D–D″ and F–G) and Sens and Dll (Figure 2H, N, and Q–S) caused by *Cow-miRNA-1*. These results suggest that the HS modification is required for Cow to promote Wg movement.

When Cow was ubiquitously overexpressed, the exWg distribution along the D-V axis was broader than that in WT cells (Figure 3J–K), which were stained and imaged in parallel under the same conditions. This difference was best shown by color-coding (compare Figure 3J′ and K′) and intensity plots (Figure 3L). The Flag-tSP-Cow did not change the range of exWg (Figure 3E–E″ and F–G), suggesting that this effect was mediated by secreted Cow.

One possible explanation for this increased range of distribution is that Cow may stabilize exWg. Therefore, we measured the total exWg signal over the entire wing disc by confocal microscopy. *tub>Cow* wing discs were 5.8% smaller than WT discs, and when normalized for wing disc size, *tub>Cow* produced a 12.6% increase in exWg (n = 16 for WT and n = 12 for *tub>Cow*). This result suggested that Cow could increase the stability of exWg because Cow did not change *wg* transcription and acted as a

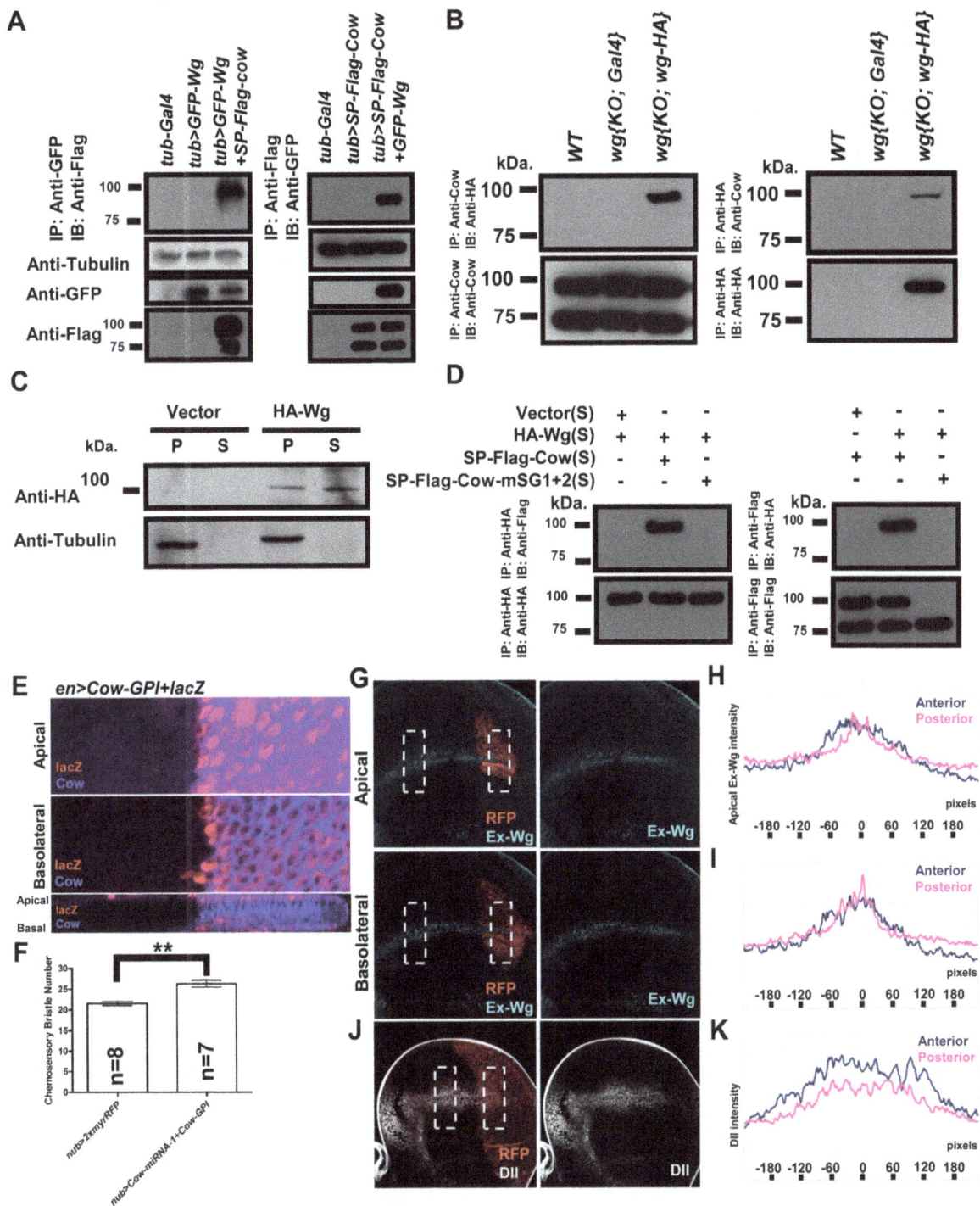

Figure 4. HS-modified Cow can bind to Wg. (A) The embryo extracts were immunoprecipitated with anti-GFP and then probed with anti-Flag on western blot. When GFP-Wg and SP-Flag-Cow were coexpressed (in *tub>GFP-Wg+SP-Flag-Cow*), a broad band at approximately 100 kDa was detected. This band was not detected in extracts from WT (*tub-Gal4*) or *tub>GFP-Wg* embryos. In the reverse experiment, embryo extracts were immunoprecipitated with anti-Flag and then stained with anti-GFP. The GFP signal was detected in *tub>SP-Flag-Cow+GFP-Wg* but not in *tub-Gal4* or *tub>SP-Flag-Cow*. (B) The larval extracts of *WT*, *wg{KO; Gal4}* and *wg{KO; wg-HA}* were immunoprecipitated with anti-HA or anti-Cow and then probed with anti-Cow and anti-HA antibodies, respectively. (C) S2 cells were separately co-transfected with *actin-Gal4* plasmid and empty *UAS* vector, or *UAS-HA-Wg*. After 48 h, the cell pellet (P) and supernatant (S) were probed using western blotting. HA-Wg could be detected by anti-HA as a band at approximately 100 kDa, consistent with a previous report on HA-Wg [22]. (D) HA-Wg, SP-Flag-Cow and SP-Flag-Cow-mSG1+2 were separately expressed in S2 cells, and then the culture supernatants were mixed in different combinations, immunoprecipitated and probed in western blots. When anti-HA was used to immunoprecipitate, anti-Flag detected Flag-Cow at approximately 100 kDa but did not detect in Flag-Cow-mSG1+2. These results indicate that only the HS-modified Cow can interact stably with Wg. The Anti-HA blot showed that HA-Wg was expressed at similar levels in the three samples. Using anti-Flag for immunoprecipitation, anti-HA detected HA-Wg as the expected 100-kDa band. Anti-Flag blotting showed that both the 75-kDa and 100-kDa bands of Cow were present in the supernatant. (E) Cow-GPI was expressed in the P compartment of the wing disc (in *en>Cow-GPI+lacZ*). Anti-Cow (blue) detected Cow in the P compartment. The signal at the Ap surface was much stronger than that at the Ba surface.

The induced expression level was also much higher than the endogenous Cow level, which was below the detection threshold under these conditions. There was no Cow signal outside the cells where it was expressed (marked by *lacZ*, red), indicating that the GPI-anchored Cow remained within the cells where it was produced. (F) Chemosensory bristle numbers from *nub-Gal4>2xmyrRFP* and *nub>Cow-miRNA-1+Cow-GPI* are summarized. Ectopic expression of Cow-GPI increased the chemosensory bristle numbers on the adult wing margin. With expression of Cow-GPI (in *tub>Cow-GPI*), the wing margin chemosensory bristles increased from 22.60±0.31 (*WT*) to 24.15±0.48 (*tub>Cow-GPI*), representing a 6.9% increase. When Cow-GPI was expressed in combination with Cow knockdown (*nub>Cow-miRNA-1+Cow-GPI*), the chemosensory bristle number increased from 21.57±0.48 in controls (*nub>2xmyrRFP*) to 26.29±0.84 (*nub>Cow-miRNA-1+Cow-GPI*), representing a 21.9% increase. (G) Most of the *hh>myrRFP+Cow-miRNA-1+Cow-GPI* individuals died before the larval stage, and the remaining 3.8% escapers were dissected for exWg staining. Ap and Ba exWg staining of *hh>myrRFP+Cow-miRNA-1+Cow-GPI*. A box of 120×240 pixels was chosen at a symmetrical position in the A and P sides of the Ap and Ba surfaces. There are 120 single-pixel lines perpendicular to the D-V border. The intensity values of ten adjacent lines were averaged to produce 12 groups from one disc. The results from two discs were then averaged to obtain the intensity plots (H and I). The results from the P side (red) and A side (blue) were compared. Compared with the A side, the exWg in the P side showed a narrower distribution. (J) The level and range of distribution of Dll were similarly analyzed in *hh>myrRFP+Cow-miRNA-1+Cow-GPI*. The P side showed a narrower range of expression of Dll than the A side. (K) The Dll plot of *hh>myrRFP+Cow-miRNA-1+Cow-GPI*. For (G), **, p<0.001.

secreted form. We also tested whether Cow stabilized Wg in the extracellular environment. Secreted HA-Wg from transfected S2 cells was added to S2 cells that were transfected with *Cow-dsRNA* vectors (Figure S4A–C). The level of HA-Wg in the supernatant was decreased by 47% in the *Cow-dsRNA*-transfected S2 cells after 48 h in culture, whereas this level was decreased by only 7% in control S2 cells (Figure S4A and C). These results strongly support a model in which Cow protein is required for exWg stability.

HS-modified Cow can bind to Wg extracellularly

We then tested whether Cow can physically interact with Wg by coexpressing epitope-tagged forms of Cow and Wg (*tub>GFP-Wg+SP-Flag-Cow*). Coimmunoprecipitation from embryo extract showed that HS-modified Cow interacted with Wg (Figure 4A). We further tested whether Cow and Wg could form a protein complex at endogenous levels. In the *wg{KO; wg-HA}*, the *wg* was replaced with *wg-HA* in the endogenous genomic locus [29], [30]. Coimmunoprecipitation from larval extract again showed that only the HS-modified Cow formed a protein complex with Wg-HA (Figure 4B).

When HA-Wg and SP-Flag-Cow were separately expressed in S2 cells and the supernatants (Figure 4C) were mixed, coimmunoprecipitation showed that Cow and Wg interacted after their secretion, and the interaction with Wg involved only the HS-modified form of Cow. This result was further supported by the finding that the interaction with Wg was lost when the GAG sites in Cow were mutated (Figure 4D). The same extracellular binding assay also showed that human Testican-2 could bind Wnt5a extracellularly (Figure S4D–G), suggesting that the Testicans may have an evolutionarily conserved role in regulating Wnt signaling.

We then tested whether the binding between Cow and Wg occurs in vivo. We generated a GPI-anchored Cow (Cow-GPI) to test whether membrane-anchored Cow would restrict exWg movement. Expression of Cow-GPI in the P domain showed that Cow-GPI was indeed restricted to P in both the Ap and Ba surface (Figure 4E) and that it caused an increase in chemosensory bristles (Figure 4F). Although the *en>Cow-GPI* expression level was much higher than that of endogenous Cow (Figure 4E), endogenous Cow was knocked down to observe the full effect of Cow-GPI on Wg. The exWg range became narrower in the P region compared to the A region, whereas the exWg level at the D-V border became higher at the Ap (Figure 4G–H) and Ba (Figure 4G and I) surfaces. Similarly, the Dll range became narrower in the P region (Figure 4J–K). When *Cow-GPI* expression was combined with Cow knockdown in the wing pouch, there was a significant increase in chemosensory bristles along the wing margin (Figure 4F). These results are consistent with the expectation that membrane-anchored Cow would bind

Wg and restrict its movement, thereby reducing its range but retaining more Wg at its source, thus providing in vivo evidence that Cow can bind to Wg and affect its distribution.

Cow enhanced the rate of extracellular movement of Wg

One possible mechanism by which Cow might promote Wg movement is through binding to Wg and enhancing its rate of movement. We transiently expressed GFP-Wg in the presence or absence of coexpressed Cow, and we followed the movement of Wg over time. We combined *ap-Gal4* or *en-Gal4* with *tub-Gal80^{ts}* and shifted the temperature from 17°C to 30°C during the third instar to induce GFP or GFP-Wg expression. The *ap-Gal4* and *en-Gal4* allowed us to observe the movement of Wg and Cow proteins along the D-V axis and A-P axis, respectively (Figure 5A–E, N–R, and S5A). The distances of GFP-Wg puncta signals from the D-V and A-P borders at 12 and 20 h after the temperature shift (Figure 5B–E, O–R and Table) were used to estimate the apparent rate of movement of GFP-Wg during this interval. EGFP-Cow moved 1.46- to 2.17-fold faster than GFP-Wg (Figure 5Table), and when Cow was co-expressed, the apparent mobility of GFP-Wg increased by 47% at Ap and 56% at Ba surfaces along the D-V axis and by 40% at the Ap surface and 44% at the Ba surface along the A-P axis (Figure 5, F–I, S–V and Table). After a transient 12-h induction, Wg showed similar mobility (Figure 5, Table). Thus, we propose that by binding to Wg, Cow serves as a carrier to enhance the speed of Wg movement; as a result of this role, we named the protein Carrier of Wg (Cow).

Strigini and Cohen (2000) used a different method to measure the rate of Wg movement; these authors shifted *shi^{ts}* mutants to a non-permissive temperature to deplete exWg and internalized Wg in the receiving cells. Shifting back to a permissive temperature allowed the reinitiation of Wg secretion and the estimation of its rate of movement. We modified this approach by expressing Shi^{ts} in the wing disc, with or without Cow knockdown, and applied the same temperature-shift regime. The Wg distribution was examined immediately or 60 min after shifting back to the permissive temperature. The rates of Wg movement in *nub>Shi^{ts}+lacZ* were 5.04 μm/h at the Ap surface and 10.15 μm/h at the Ba surface during this interval (Figure S5B–E and J). With Cow knockdown, the Wg rates decreased to 0.93 μm/h at the Ap surface and 2.12 μm/h at the Ba surface (Figure S5F–J). This result showed that Cow is required for enhancing the rate of Wg movement. In addition, the Wg level was lower in the *nub>Shi^{ts}+Cow-miRNA-1* discs than in WT, consistent with the above in vitro result (Figure 3 and S4A–C) showing that Cow also stabilized Wg.

To further investigate whether Cow enhanced the movement of Wg, FRAP was used to monitor the kinetics of GFP-Wg spreading [6]. At both the A and P regions, the GFP-Wg recovery was faster

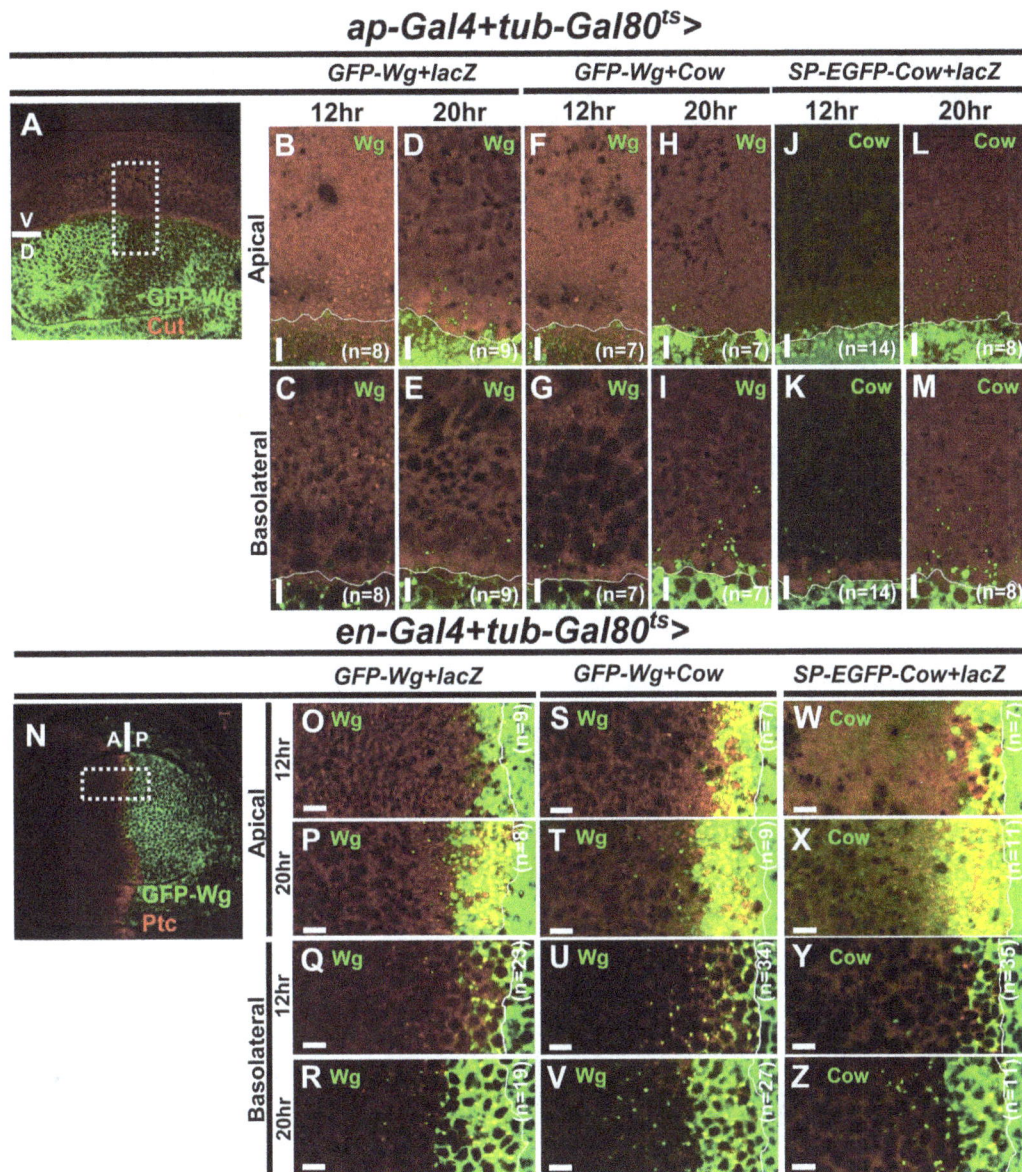

Continuous expression				
	Velocity (μ m/hr)	GFP-Wg	GFP-Wg + Cow	EGFP-Cow
Apical	D→V	2.22	3.27	3.54
	P→A	1.19	1.67	1.74
Basolateral	D→V	1.31	2.04	2.30
	P→A	0.71	1.02	1.54

Pulse-chase expression				
	Velocity (μ m/hr)	GFP-Wg	GFP-Wg + Cow	EGFP-Cow
Apical	D→V	1.90	3.11	3.19
	P→A	1.28	1.88	2.04
Basolateral	D→V	1.20	2.19	2.26
	P→A	0.71	1.25	1.57

Figure 5. Cow enhances the rate of extracellular movement of Wg. For continuous expression, the *ap-Gal4+tub-Gal80*[ts] (A-M) or *en-Gal4+tub-Gal80*[ts] (N-Z) larvae were shifted from 17°C to 30°C to induce UAS-transgene expression. The boxed areas in A and N were enlarged and shown in B-M and O-Z. The entire wing disc was divided into 10 equal sectors parallel to the D-V axis in *ap-Gal4+tub-Gal80*[ts] and to the A-P axis in *en-Gal4+tub-Gal80*[ts]. The distances between the farthest signal puncta (arrowhead) and the D-V or A-P border for each sector were averaged to obtain the travel distance. The results from multiple discs were then averaged for each time point. The differences in distance traveled from 12 h to 20 h were used to calculate the velocity of movement and are summarized in the table. For pulse-chase expression, the *ap-Gal4+tub-Gal80*[ts] and *en-Gal4+tub-Gal80*[ts] flies were shifted to 30°C for 12 h to induce GFP-Wg expression and then shifted back to 17°C to turn off GFP-Wg expression. The GFP-Wg movement during the 8 h following the end of the induction was monitored. The calculated velocities are summarized in the table.

in *dpp>GFP-Wg+Cow* than in *dpp>GFP-Wg+lacZ* (Figure 6A–D), strongly confirming that Cow can enhance GFP-Wg movement. Furthermore, GFP-Wg showed more rapid recovery in the P region than in the A region (Figure 6E). Although Kicheva et al. (2007) [6] calculated Wg mobility based on a non-directional model, they performed FRAP only in the P compartment. Our experimental results showed that Wg mobility varied between directions, suggesting an additional level of regulation.

Next, we asked whether this enhancement of Wg speed by Cow is relevant during development. To this end, we monitored the Wg distribution, Dll pattern and wing pouch size in Cow GOF and LOF mutants during wing disc development. During the period of 84–112 h after egg laying (AEL), the disc growth resulted in the

wing pouch margin moving away from the D/V boundary at 0.98 μm/h for the D margin and 1.50 μm/h for the V margin (Figure S5K–P). These results were within the same range as our estimate of Wg movement. Thus, the enhancement of Wg mobility by Cow matches the requirement for the establishment of a Wg gradient during this short developmental time period.

Cow promoted Wg Ap transport independent of endocytosis

We next tested whether Cow-mediated Wg transport occurs at the Ap or Ba surface, as well as whether it is dependent on endocytosis. Endocytosis was blocked in the P region by expressing Shi[ts] and shifting the temperature from 17°C to 32°C for 6 h.

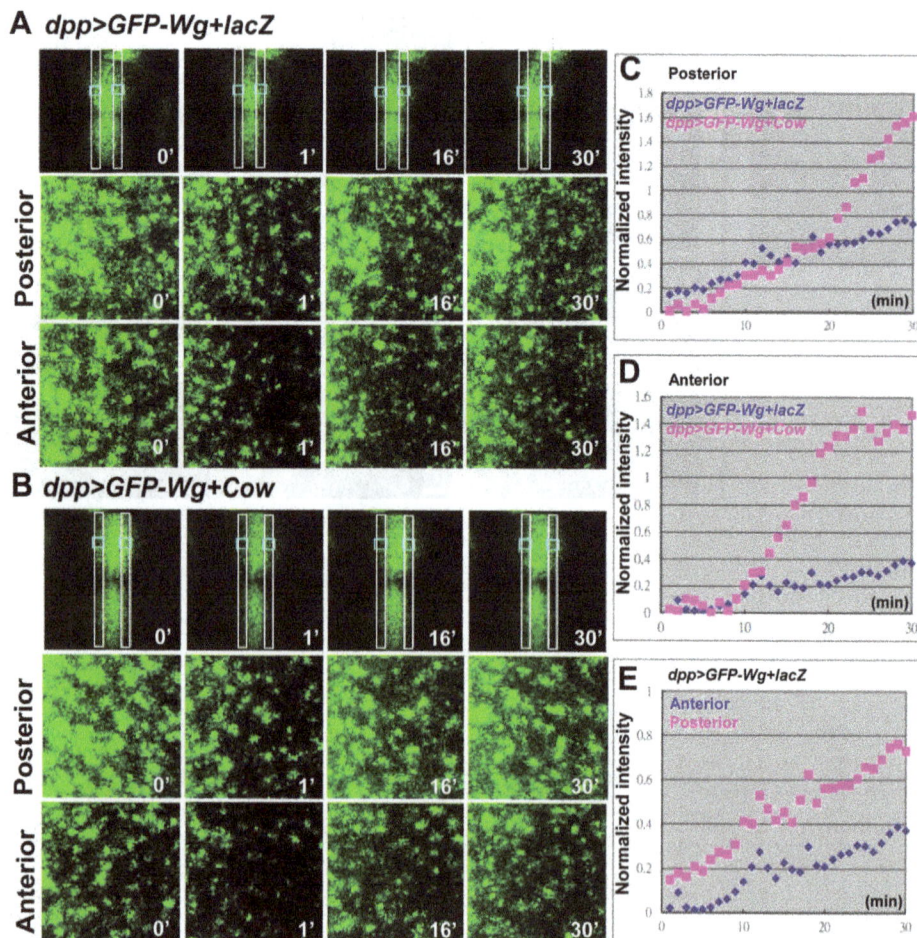

Figure 6. Cow enhances GFP-Wg recovery in FRAP. GFP-Wg was ectopically expressed in the *dpp* domain without (*dpp>GFP-Wg+lacZ*) (A) or with Cow (*dpp>GFP-Wg+Cow*) (B) coexpression. GFP-Wg recovery was analyzed in the A and P compartments using FRAP time-lapse images. The FRAP images were recorded before photobleaching (0 min) and after photobleaching (1, 16 and 30 min) at A and P. Normalized intensity of GFP-Wg recovery from *dpp>GFP-Wg+lacZ* and *dpp>GFP-Wg+Cow* at P (C) and A (D). The recovery in *dpp>GFP-Wg+Cow* overshoots at 20–30 min and returns to the original intensity at approximately 1–2 h. (E) The recovery curve of GFP-Wg showed faster recovery in P than in A.

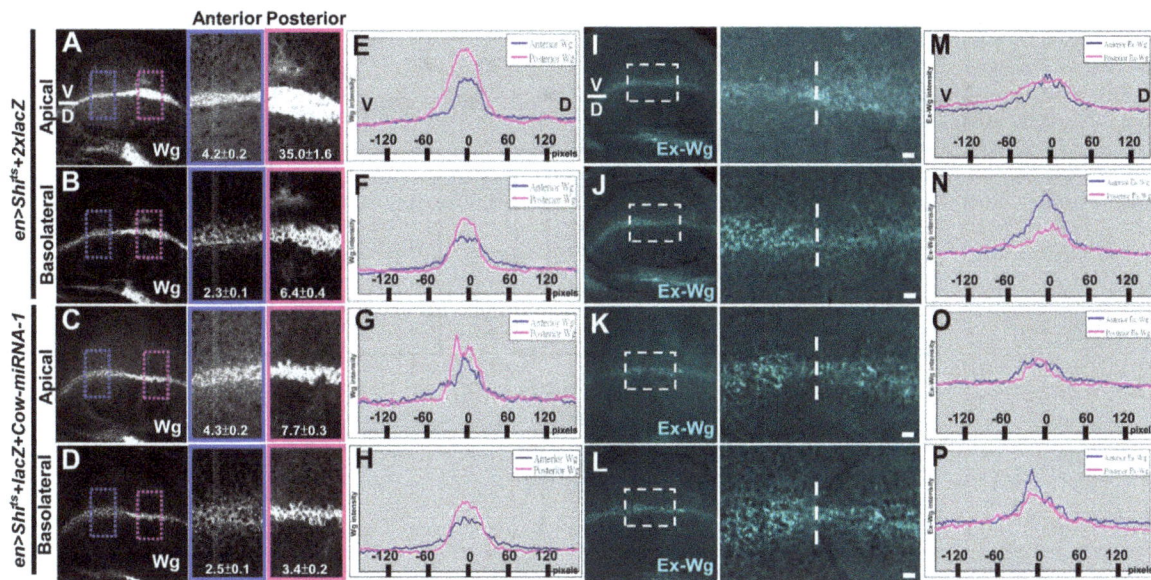

Figure 7. Cow promotes the Ap transport of Wg. (A–B) en>Shi^ts^+2xlacZ (n = 14) and (C–D) en>Shi^ts^+lacZ+Cow-miRNA-1 (n = 11) larvae raised at 17°C were shifted to 32°C for 6 h at late third instar and were examined immediately using conventional Wg staining. Because the signal intensity varies among discs, the A compartment served as an internal control for each disc. For intensity plots, the intensity in each disc was normalized to the background and maximum signal intensity. (E–H) The Wg signal intensity in each boxed region (A–D) was transformed into an intensity plot. P, red; A, blue. (I–J) The exWg staining in the Ap (I) and Ba (J) surfaces of en>Shi^ts^+2xlacZ (n = 21). (K–L) The exWg staining in the Ap (K) and Ba (L) surfaces of en>Shi^ts^+lacZ+Cow-miRNA-1 (n = 13). The expansion of Ap exWg in (I) was blocked in (K) (100%). The Ba exWg was similar in (J) and (L) (100%).

Because there is variation between discs, the A compartment in the same disc can be used as an internal control for quantitative comparison with the P compartment. Total Wg accumulation was higher in the P region (8.3X in Ap, 2.8X in Ba) than A region, and its distribution was also wider (Figure 7A–B and E–F). The expression of exWg in the P region was higher and broader at the Ap surface but lower and narrower at the Ba surface (Figure 7I–J and M–N). These results are consistent with Ap transport being independent of endocytosis and with transcytosis being responsible for the movement of Wg to the Ba surface [5], [7], [9]. Because Shi is also required for Wg secretion, the effect may also be due to a block in Wg secretion. However, when Wg secretion was blocked in evi^2 mutant clones, exWg was strongly reduced [47]. Thus, the wider distribution of exWg and Ap accumulation was not due to a block of secretion by Shi^ts^. Moreover, the higher level of Wg in the endocytic-deficient P compartment is consistent with endocytic Wg being targeted for degradation [5].

When Cow was knocked down in the P region together with endocytosis blockage, the accumulation and expansion of Ap Wg were significantly reduced (Figure 7C and G), and a similar effect was found for exWg (Figure 7K and O). The range of Wg was broader in en>Shi^ts^ cells (P>A in Figure 7A) but was reduced in en>Shi^ts^+Cow-miRNA-1 cells (A>P in Figure 7C), and the range of exWg was wider in en>Shi^ts^ cells (P>A in Figure 7I) and reduced in en>Shi^ts^+Cow-miRNA-1 cells (A>P in Figure 7K). These results suggest that the Ap transport of Wg depends on Cow. The Ba distribution of Wg was similar to that observed in en>Shi^ts^+2xlacZ (Figure 7D, H, L and P), suggesting either that Cow does not play a major role in the transcytosis of Wg through the Ba surface or that Cow activity is dependent on Shi function. Furthermore, the accumulation of Wg was reduced on both the Ap and Ba surfaces (Figure 7C–D and G–H), supporting the idea that Cow stabilizes exWg. Together, these results show that Cow is required primarily for exWg transport at the Ap surface and is not

dependent on endocytosis. At 84 h AEL, Wg appeared primarily at the Ap surface around the D/V border (unpublished data), supporting the importance of Wg Ap transport for establishment of the Wg gradient.

Discussion

Cow serves as a carrier of Wg to promote Wg extracellular movement

In this study, we showed that Carrier of Wg (Cow) is a secreted HSPG that can physically interact with Wg (Figure 4). The binding to Wg is dependent on the HS modification on Cow, and it can occur after both proteins are secreted (Figure 4). We also measured the apparent rates of Cow and Wg movement (Figure 5), and our results showed that (a) Cow moved faster than Wg, (b) overexpression of Cow enhanced the rate of Wg movement, and (c) knockdown of Cow reduced the rate of Wg movement. Thus, we suggest that Cow serves as a carrier of Wg to enhance the rate of its extracellular movement. This enhancement of Wg movement by Cow is important for the establishment of the Wg gradient during development. Moreover, the role of Cow as a carrier for a morphogen is unique among HSPGs in that Cow is a secreted HSPG, whereas the previously studied syndecans and glypicans are membrane-bound.

The measurements of Wg and Cow mobility were performed using endogenous Wg and Cow. Therefore, the true mobility of Wg and Cow, without the presence of the other, has not been determined. Because the cow mutant phenotype is dominantly affected by reducing the wg dosage (Figure 1) and because overexpression affected Wg signaling (Figure 2, 3 and S3), we expect that neither is present in large excess over the other. The ptc-Gal4-driven expression of Cow was also much higher than the endogenous level of Cow (Figure 1C–D). Therefore, we expect that EGFP-Cow would be expressed much more highly than

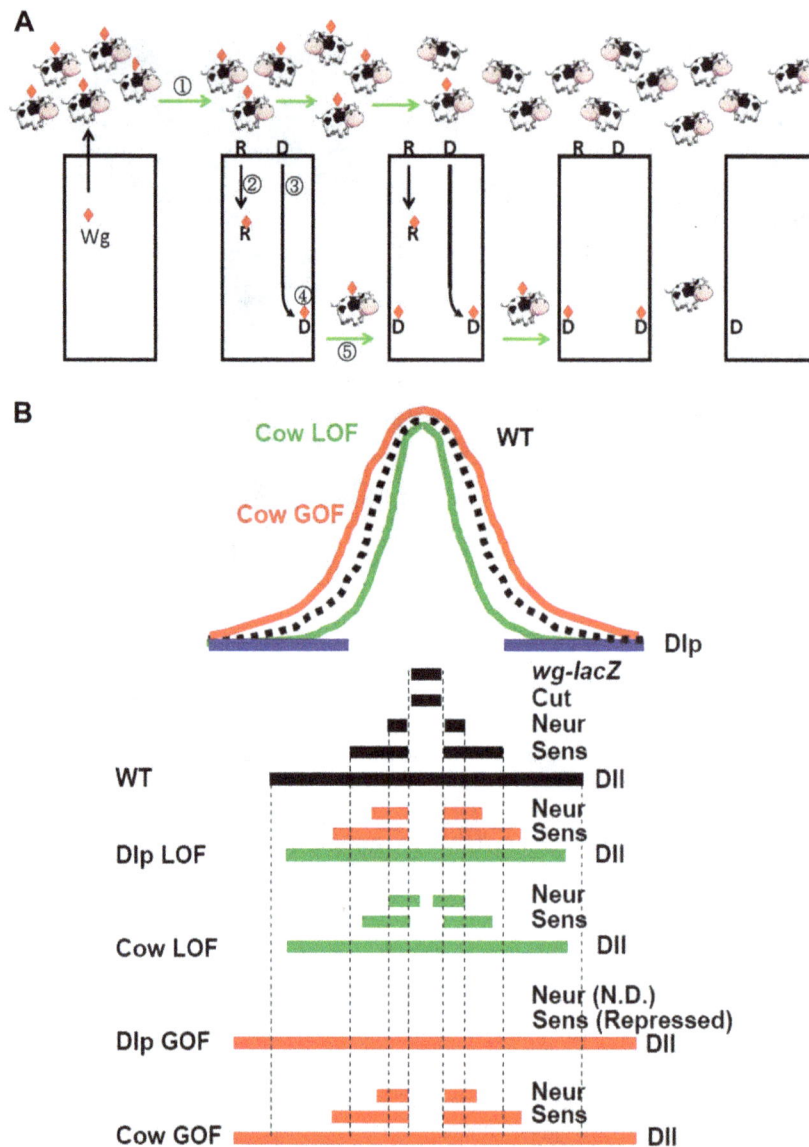

Figure 8. Roles of Cow in formation of the Wg gradient. (A) Our model for shaping the Wg gradient. Wg (red diamond) is produced by the D-V border cells (leftmost cell) and is secreted at the Ap side. This Wg can then be spread by several pathways: (1) ExWg transport on the Ap surface is independent of Dlp and endocytosis. Our results suggest that Cow binds to Wg and enhances its mobility on the Ap surface. (2) Wg can be internalized by receptor (**R**)-mediated endocytosis, which targets it for lysosomal degradation [15]. (3) Wg can also be internalized by Dlp (**D**)-mediated transcytosis, an intracellular process that translocates Wg from the Ap to the Ba surface. Cow does not play an essential role in the two Wg internalization processes because Wg internalization can still occur when Cow is knocked down. (4) Wg is then released to the outside of the cell, presumably by exocytosis at the Ba surface. (5) The Ba exWg movement across the intercellular gap is presumably facilitated by binding to Cow. In summary, Cow participates in steps 1 and 5 described above to facilitate the extracellular distribution of Wg. (B) Summary of Cow and Dlp LOF and GOF effects on Wg distribution and Wg target gene expression. The expression ranges of Dlp, *wg-lacZ*, Cut, *neur-lacZ*, Sens and Dll along the D-V axis are shown. Red, expansion of target gene; green, reduction of target gene.

endogenous Wg, and this measurement represents the mobility of free EGFP-Cow. The large excess of overexpressed Cow likely enabled the measurement of its lateral mobility independent of endocytosis.

We used two methods to measure the mobility of Wg. The first was to use Gal4/Gal80^{ts} and temperature shifting to transiently induce Wg expression and then measure Wg distribution within an 8-h period. The estimated rates were 2.22 μm/h and 1.31 μm/h at the Ap and Ba surfaces, respectively, for the D-V axis. The second is a modification of the method used by Strigini and Cohen (2000) [5]. The estimated rate of Wg movement was 5.04 μm/h at

the Ap surface and 10.15 μm/h at the Ba surface. In contrast to the first method, which addresses newly synthesized Wg, the second method addresses the release of intracellularly accumulated Wg, which is therefore driven by a higher concentration gradient and thus shows higher mobility. However, the second method indicated much slower rates than the calculated rate of 50 μm/ 30 min reported by Strigini and Cohen (2000) [5]. One important difference is that Strigini and Cohen (2000) used *shi^{ts}* mutant discs [5], whereas we expressed Shi^{ts} in the wing disc using *nub-Gal4*. Moreover, this difference was likely caused by the incomplete blocking of Wg secretion by the dominant-negative Shi^{ts}, as evidenced by the exWg distribution in en>Shi^{ts}.

Several extracellular molecules have been reported to influence Wg trafficking, including secreted Wingless-interacting molecule (Swim) and Lipophorin [48], [49], which are involved in Wg long-range travelling; exosomes [50], which do not affect Wg gradient formation [51]; and Secreted Frizzled-Related Proteins (SFRPs) [52], [53], which have no homolog in Drosophila. Our study shows that Cow is required for the short-range transport of Wg. Because Wg is lipid-modified [54], its diffusion may be hindered by hydrophobic interactions with cell membranes. Its interaction with Cow may also help to reduce its interaction with the cell membrane, thereby accelerating its movement.

We note that because Cow is a secreted protein and appears uniformly in the wing disc, it cannot provide directionality to Wg transport. Instead, it simply enhances the mobility of Wg and allows the Wg gradient to be established more quickly.

The role of Cow in the formation of the exWg gradient

We propose that Cow plays two distinct roles in the formation of the exWg gradient (Figure 8A). (1) Cow is responsible for the Ap transport of exWg. This process is independent of endocytosis and is especially important because it is responsible for moving Wg out from its producing cells, which express only low levels of DFz2 and Dlp. (2) Cow is responsible for the intercellular transfer of Wg at the Ba surface. Because Cow is diffusible, it is expected to be more efficient to carry Cow over the intercellular space than to exchange Wg between membrane-bound Dlp or other receptors on adjacent cells. In addition, Cow also slightly increased exWg stability, perhaps by binding to Wg, or by diverting Wg away from endocytosis and degradation.

Because Wg can be bound by its receptor DFz2, by Dlp and Dally, and by Cow, these factors may compete for binding to Wg. It has also been proposed that the relative levels of Wg, DFz2 and Dlp can affect the morphogen activity gradient [17]. Our study adds another potential binding partner to this process. The relative levels of Wg, DFz2, Dlp and Cow likely determine not only the shape of the Wg gradient but also its relative distribution on the Ap versus Ba surfaces.

The biphasic activity of Cow can be explained by its effect on Wg mobility (Figure 8B). Cow knockdown reduced Wg mobility, causing Wg to spread less and accumulate near the Wg-producing region. For short-range targets, the effect was similar to Wg GOF, whereas for intermediate- and long-range targets, the effect was similar to Wg LOF. This model can also explain the apparent contrast between the wing and embryo phenotypes in Cow knockdown. In the embryo, Wg specifies the naked cuticle fate over 5 rows of cells in the anterior segments. Therefore, the naked cuticle fate can be viewed as a long-range target of Wg, in which Wg LOF loses the naked cuticle fate and produces a denticle fusion phenotype. Cow knockdown also produces a phenotype similar to Wg LOF. In the wing, the chemosensory bristles at the wing margin are controlled by the short-range target neur [55], and Cow knockdown caused an increase in chemosensory bristles, similar to the effect of Wg GOF.

Gradient vs. cellular memory

Recently, it was shown that replacement of endogenous Wg with a membrane-tethered Wg is sufficient for wing development with normal patterning [30]. It has also been suggested that early Wg expression is coupled to cellular memory of target gene expression and that the spreading of Wg is therefore dispensable for patterning. However, this hypothesis does not readily explain how different Wg target genes are expressed at different ranges from the Wg source,

which can be explained by the Wg gradient model and is supported by previous studies [1], [5], [7], [9], [14], [15], [16], [17], [18], [19], [20], [21], [22], [42]. In addition, we found that Dll expression is activated only after 84 h AEL (Figure S5), which is past the early Wg expression phase suggested for cellular memory [30]. The contradiction between the two modes of Wg patterning mechanisms requires further study for clarification [56].

Conserved function and medical implication

The mammalian testicans can regulate neurite outgrowth [57] and proteases activity [58], [59], [60], [61], [62]. However, the role of testicans in regulating signaling pathways has not been studied. Our study on the fly testican Cow is the first to demonstrate a role for the testican family in morphogen signaling as a diffusible HSPG. In addition, we showed that human Testican-2 could bind to Wnt5a extracellularly, suggesting that the testican family may have a general role in regulating Wnt distribution and thus Wnt signaling.

Misregulation of Wnt signaling is well known to contribute to human diseases, including cancer [63]. Accordingly, components of the Wnt signaling pathway have been developed as therapeutic targets for cancer [64]. The Reggie protein, which affects Wnt secretion and spreading, is also associated with various types of cancer [65]. Our original identification of Cow was the result of an overexpression screen of genes with elevated expression in human hepatocellular carcinoma [66]; thus, Cow may be involved in oncogenesis. Moreover, our finding of the novel and conserved role of the testican protein family in binding to Wnt ligands may reveal their involvement in human diseases.

Supporting Information

Figure S1 Characterization of cow. (A) The cow locus and deletion mutants are shown. Mi{ET1}CG13830^{MB00767} is a Minos transposon inserted into the 3'-UTR of cow. Using imprecise excision, we generated a mutation, cow$^{5\Delta}$, which has a deletion beginning at 396 bp downstream of the cow open reading frame (ORF) and extending 9,119 bp downstream. The coding region of cow was not affected, but part of the 3'-UTR, including two putative polyadenylation signals, was deleted. RT-PCR for the cow coding region showed that the cow transcript was present in the cow$^{5\Delta}$ mutant embryo at levels comparable to that in WT (unpublished data). However, Cow protein, as detected in western blot using an anti-Cow antibody we generated, was undetectable in cow$^{5\Delta}$ embryo extracts (Figure S1C–D). Thus, cow$^{5\Delta}$ is a protein null mutant. The downstream CG17111 gene was also deleted in cow$^{5\Delta}$. RT-PCR confirmed that cow$^{5\Delta}$ embryos contained no CG17111 transcript, whereas the CG6697 further downstream was not affected (unpublished data). cow$^{5\Delta}$ homozygotes are lethal; 65.5% die at the embryonic stage and the rest die at the first instar. A combination of two deficiencies, Df(3R)BSC527/Df(3R)BSC619, which deleted the entire coding region of cow, and cow$^{5\Delta}$/Df(3R)Exel6193 also showed similar early lethality and denticle belt fusion phenotypes (Figure 1I and unpublished data). These results suggest that cow$^{5\Delta}$ is a functionally null mutation. When Cow was knocked down in tub>Cow-miRNA-1 (Figure 1J) and tub>3xCow-dsRNA (Figure 1K), the cuticle phenotype was much stronger than that of cow$^{5\Delta}$, although the protein level was only reduced in these knockdowns (Figure S1D). It is possible that the knockdown was not complete in the early stages, as shown by the residual Cow level, thus allowing the embryos to develop past the early phase of cow requirement and allowing the full strength of the cow phenotype to be observed. In the cow$^{5\Delta}$, Df(3R)BSC527/

Df(3R)BSC619 and *cow⁵ᴬ/Df(3R)Exel6193* mutants, only the weaker phenotype was observed. Expressing *cow*, using *UAS-Cow* driven by *da-Gal4* (abbreviated as *da>Cow*), in the *cow⁵ᴬ* mutant nearly completely rescued the lethality, and the adults showed no apparent phenotype. These results demonstrate that *cow⁵ᴬ* lethality was due to the loss of Cow. The embryonic lethality of *tub>Cow-miRNA-1* could also be rescued to adulthood by coexpression of a *cow* transgene without the 5′-UTR target for *Cow-miRNA-1* (Figure S1B). The efficient rescue excluded off-target effects of the miRNA. In situ hybridization showed that *cow* was expressed relatively uniformly in the early embryo beginning at the cellular blastoderm and that it developed a segmentally repeated pattern at stage 13 (unpublished data). There was also evidence of maternal contribution, as *cow* embryos devoid of maternal contribution generated by germline clones died before cellularization (unpublished data). Cow expression was undetectable in *Df(3R)BSC527* (unpublished data), which deleted nearly the entire *cow* gene. Cow is expressed throughout the imaginal discs (unpublished data). RT-PCR showed that the *cow* transcript level is highest in the embryo, low in larva, and increased in pupa and adult (unpublished data). RNA-Seq and RNA tiling microarray data [67] show that the *cow* transcript is expressed at low levels in early embryonic stages (0–12 h), increases after 12 h, peaks at 16–18 h and is maintained until the end of embryonic stage. It then gradually decreases during larval stages and increases to high levels during pupal and adult stages. These results are consistent with our findings. Cow is also expressed at high levels in the larval and adult CNS [68]. (B) The Cow protein contains five domains conserved in the testican family: signal peptide (SP, amino acids 1–35), N region (N, 36–104), follistatin-like domain (FS, 168–274), extracellular Ca^{2+} binding EF-hand motif (EC, 428–531), thyroglobulin-like domain (TY, 535–594) with a CWCV motif, and C region (602–625). There are two GAG attachment sites in the FS domain. (C) In western blot, an anti-Cow antibody detected two bands at approximately 100 kDa and 75 kDa, respectively, in WT (*tub-Gal4*) embryos (E), wing disc (W), adults (A) and *act>GFP* S2 cells. Both bands were enhanced when SP-Flag-Cow was expressed in flies (*tub>SP-Flag-Cow*) or in S2 cells (*act>SP-Flag-Cow*). *cow⁵ᴬ* homozygous embryos showed no detectable Cow protein. Tubulin served as a loading control. (D) Knockdown of Cow by the ubiquitous expression of either *Cow-miRNA-1* or *Cow-dsRNA* reduced the level of Cow protein. Tubulin served as a loading control. *tub>Cow-miRNA-1* reduced the level of Cow, but *tub>3xCow-dsRNA* produced a much stronger reduction of the Cow level. These mutants were lethal, with over 90% death at the embryo stage and the remaining individuals dying at the first instar. The embryo cuticle phenotype is similar to that of the *cow* mutant. (E) When Cow was knocked down by *ap-Gal4* (*ap>GFP+Cow-miRNA-1*) in the D side of the wing disc, endogenous Cow was not detected with the anti-Cow antibody (cyan) in D GFP⁺ cells (green) when compared to the V control side. This result demonstrates the specificity of the Cow antibody and the efficiency of the RNA interference constructs. This knockdown could be observed at both the Ap and Ba surfaces. Less Cow protein was detected at the V side near the D-V border, suggesting that Cow may move toward the D cells from the V cells near the border. (F) Anti-Flag detected bands at both 100 kDa and 75 kDa of Flag-Cow in the cell pellet and supernatant of S2 cells transfected with *UAS-SP-Flag-Cow*. When the two putative glycosylation sites were mutated (in Flag-Cow-mSG1+2), the 100-kDa band was dramatically reduced and the 75-kDa band was enhanced. This suggested that the 100-kDa band represents the HS-modified form, whereas the 75-kDa band

represents the unmodified form. Anti-tubulin was used to demonstrate the purity of the supernatant.

Figure S2 Genetic interaction of *cow* and *wg*. (A–D) Adult wing of (A) *tub-Gal4* (representing WT), (B) *MS1096>Cow-dsRNA*, (C) *tub>Cow*, and (D) *tub>Testican-2*. The phenotypes of *MS1096>Cow-miRNA-1* (unpublished data), *MS1096>Cow-miRNA-2* (unpublished data) and *MS1096>Cow-dsRNA* were similar. There were ectopic chemosensory bristles along or near the A and P margins (arrowheads in B′) with 60–85% penetrance. (C–D) Overexpression of Cow or human Testican-2 produced extra vein tissue (arrows) and loss of chemosensory bristles at the wing margin (arrowheads in C′ and D′ compared with WT in A′). The results are summarized in E. (E) Chemosensory bristle numbers from *tub-Gal4*, *tub>Cow* and *tub>Testican-2*. Overexpression of Cow or Testican-2 reduced the numbers of chemosensory bristles on the adult wing margin. The differences in comparison with the control, *tub>Gal4*, were statistically significant (**: $p<0.001$). (F) The adult phenotype of the *cow⁵ᴬ* mutant clone (*ap-Gal4/UAS-FLP; FRT82B ubi-GFP Minute/FRT82B cow⁵ᴬ*) that caused the wing-loss phenotype. (G) Adult phenotypes of *MS1096>DFz2N*, *MS1096>DFz2N+2XCow*, *MS1096>DFz2N+2XCow-dsRNA* and *MS1096>DFz2N+Cow-miRNA-*, which each led to wing loss. (H) Adult wing phenotype of *MS1096>DFz2*, *MS1096>DFz2+2XCow*, *MS1096>DFz2+2X Cow-dsRNA* and *MS1096>DFz2+Cow-miRNA-1*.

Figure S3 Cow affects the expression of Wg target genes. (A–E) Sens expression (green) along the D-V border of the wing disc in (A) WT, (B) *MS1096>DFz2*, (C) *MS1096>2xCow*, and (D) *MS1096>2xCow-dsRNA*. The number of Sens⁺ cells in each group is summarized in (E). The results in B-D are significantly different from that of the WT (A).

Figure S4 HS-modified Cow can stabilize Wg. (A–C) S2 cells were transfected with the HA-Wg construct. Conditioned medium with secreted HA-Wg was then added back to new S2 cells transfected with control or *Cow-dsRNA* vectors. The HA-Wg and Cow in the culture supernatant were then examined by western blot. (A and C) HA-Wg was maintained at similar levels in the supernatant of empty control vector cells over 48 h (95% at 12 h, 94.5% at 24 h, 93% at 48 h relative to 100% at 0 h). When Cow was knocked down with the *Cow-dsRNA* vector, the HA-Wg level in the supernatant gradually decreased (88.5% at 12 h, 72% at 24 h, 53% at 48 h relative to 100% at 0 h). (A and B) Cow protein level decreased gradually in the supernatant after *Cow-dsRNA* transfection (75 kDa: 94% at 12 h, 81.5% at 24 h, 59.5% at 48 h relative to 100% at 0 h, 100 kDa: 96% at 12 h, 84% at 24 h, 59% at 48 h relative to 100% at 0 h), but Cow was maintained at the same level in cells transfected with the empty control vector (75 kDa: 94.5% at 12 h, 99% at 24 h, 101.5% at 48 h relative to 100% at 0 h, 100 kDa: 99.5% at 12 h, 99% at 24 h, 99% at 48 h relative to 100% at 0 h) (n = 3). Comparison of 75 kDa intensity at different time of dsRNA treatment with control (time 0): *, $p<0.01$, **, $p<0.001$, ***, $p<0.0001$. Comparison of 100 kDa intensity at different time of dsRNA treatment with control (time 0): ##, $p<0.001$, ###, $p<0.0001$. (D) S2 cells were transfected with empty control or Wnt5a-HA vectors. Wnt5a was detected in the cell pellet and culture supernatant with an anti-HA antibody 48 h after transfection. (E) S2 cells were transfected with empty control or SP-Flag-Testican-2 vectors. SP-Flag-Testican-2 was detected in the cell pellet and culture supernatant with an anti-Flag antibody 48 h after transfection. Anti-tubulin showed that the supernatant was not contaminated with cells. (F–G) Culture supernatants from S2 cells transfected with control vector, Wnt5a and SP-Flag-Testican-2, respectively, were mixed in different combinations. (F) When anti-HA was used for immunoprecipitation, anti-Flag detected SP-Flag-Testican-2. (G) When anti-Flag was used for immunoprecipitation, anti-HA detected Wnt5a-HA. For (C), n.s., not statistically significant, **, p<0.001, ***, p<0.0001.

Figure S5 Cow enhances the rate of extracellular movement of Wg. (A) In monitoring the speed of Wg movement, we used en-$Gal4$+tub-$Gal80^{ts}$ to drive UAS-GFP expression in an inducible fashion. The en-$Gal4$+tub-$Gal80^{ts}$> GFP larvae were raised at 17°C and shifted to 30°C to inactivate Gal80^{ts}, thereby inducing GFP expression. The GFP signal in the P compartment became weakly detectable starting at 3 h, was strong at 6 h, and reached full expression at 16 h after the temperature shift. Similar kinetics were observed for UAS-GFP driven by the ap-$Gal4$ (unpublished data). (B–J) The rate of movement of endogenous Wg was estimated by blocking endocytosis (expressing Shi^{ts} and shifting temperature from 17°C to the non-permissive 32°C for 3 h for late third instar wing disc) and then relieving the endocytic block (by shifting back to 17°C) to initiate Wg distribution. Wg distribution was examined immediately or 60 min after the shift back to 17°C. nub-$Gal4$ was used to drive expression in the wing pouch area of the wing disc. (B–E) The Wg distribution ranges in the Ap (B and D) and Ba (C and E) surfaces of nub>Shi^{ts}+$lacZ$ at 0 min (B–C) and 60 min (D–E). Significant numbers of Wg puncta were still observed at the Ap surface (B), but not at the Ba surface (C), away from its D-V source immediately after the return to the permissive 17°C. The distance of Wg puncta farthest from the Wg source was averaged to generate the Wg distribution range. (F–I) With knockdown of

Cow, the Wg distribution at the Ap (F and H) and Ba (G and I) surfaces of nub>Shi^{ts}+Cow-$miRNA$-1 was scored at 0 min (F–G) and 60 min (H–I). (J) The apparent rates of movement for endogenous Wg at the Ap and Ba surfaces are summarized for nub>Shi^{ts}+$lacZ$ and nub>Shi^{ts}+Cow-$miRNA$-1. (K–O) The Wg (white), Dll (red) and wing pouch size were measured at the indicated times 2 h after egg collection and cultured at 25°C (K–K''', n = 11) for 70–72 h, (L–L''', n = 7) 76–78 h, (M–M''', n = 14) 82–84 h, (N–N''', n = 11) 94–96 h and (O–O''', n = 10) 110–112 h in nub>GFP (representing WT). (P) The distances between the wing pouch margins and the D-V boundary were measured to calculate the rate of growth. White bar, 20 μm. In WT, Wg was widely detected at low levels in the wing disc at 70–72 h AEL and gradually became high at the D/V boundary between 76–78 h and 82–84 h AEL (K-M; Alexandre et al., 2014). In addition, Wg spread from the D/V source toward the border of the wing pouch (Figure S5, N–O). Dll was undetectable until 84 h AEL, and its range expanded from 96 h to 112 h AEL (K''–O''). Therefore, we focused our analysis on the period from 84–112 h AEL.

Acknowledgments

We thank Stephen M. Cohen, Konrad Basler, Jean-Paul Vincent, Hugo J. Bellen, the Kazusa DNA Research Institute, Bloomington Stock Center and the VDRC Stock Center for providing fly stocks and reagents. We are grateful to Chun-lan Hsu and Yu-Chi Yang for preparing fly food and maintaining fly stocks, to Chiou-Yang Tang for making transgenic flies, to Sue-Ping Lee and the IMB Imaging Core for help in confocal microscopy.

Author Contributions

Conceived and designed the experiments: YHC YHS. Performed the experiments: YHC. Analyzed the data: YHC YHS. Contributed reagents/materials/analysis tools: YHC. Wrote the paper: YHC YHS.

References

1. Zecca M, Basler K, Struhl G (1996) Direct and long-range action of a wingless morphogen gradient. Cell 87: 833–844.
2. Lawrence PA, Sanson B, Vincent JP (1996) Compartments, wingless and engrailed: patterning the ventral epidermis of Drosophila embryos. Development 122: 4095–4103.
3. Neumann CJ, Cohen SM (1997) Long-range action of Wingless organizes the dorsal-ventral axis of the Drosophila wing. Development 124: 871–880.
4. Sanson B, Alexandre C, Fascetti N, Vincent JP (1999) Engrailed and hedgehog make the range of Wingless asymmetric in Drosophila embryos. Cell 98: 207–216.
5. Strigini M, Cohen SM (2000) Wingless gradient formation in the Drosophila wing. Curr Biol 10: 293–300.
6. Kicheva A, Pantazis P, Bollenbach T, Kalaidzidis Y, Bittig T, et al. (2007) Kinetics of morphogen gradient formation. Science 315: 521–525.
7. Gallet A, Staccini-Lavenant L, Therond PP (2008) Cellular trafficking of the glypican Dally-like is required for full-strength Hedgehog signaling and wingless transcytosis. Dev Cell 14: 712–725.
8. Bartscherer K, Boutros M (2008) Regulation of Wnt protein secretion and its role in gradient formation. EMBO Rep 9: 977–982.
9. Marois E, Mahmoud A, Eaton S (2006) The endocytic pathway and formation of the Wingless morphogen gradient. Development 133: 307–317.
10. Rives AF, Rochlin KM, Wehrli M, Schwartz SL, DiNardo S (2006) Endocytic trafficking of Wingless and its receptors, Arrow and DFrizzled-2, in the Drosophila wing. Dev Biol 293: 268–283.
11. Seto ES, Bellen HJ (2006) Internalization is required for proper Wingless signaling in Drosophila melanogaster. J Cell Biol 173: 95–106.
12. Baeg GH, Lin XH, Khare N, Baumgartner S, Perrimon N (2001) Heparan sulfate proteoglycans are critical for the organization of the extracellular distribution of Wingless. Development 128: 87–94.
13. Takei Y, Ozawa Y, Sato M, Watanabe A, Tabata T (2004) Three Drosophila EXT genes shape morphogen gradients through synthesis of heparan sulfate proteoglycans. Development 131: 73–82.
14. Han C, Yan D, Belenkaya TY, Lin XJ (2005) Drosophila glypicans Dally and Dally-like shape the extracellular Wingless morphogen gradient in the wing disc. Development 132: 667–679.
15. Baeg GH, Selva EM, Goodman RM, Dasgupta R, Perrimon N (2004) The Wingless morphogen gradient is established by the cooperative action of Frizzled and Heparan Sulfate Proteoglycan receptors. Dev Biol 276: 89–100.
16. Franch-Marro X, Marchand O, Piddini E, Ricardo S, Alexandre C, et al. (2005) Glypicans shunt the Wingless signal between local signalling and further transport. Development 132: 659–666.
17. Yan D, Wu Y, Feng Y, Lin SC, Lin X (2009) The core protein of glypican Dally-like determines its biphasic activity in wingless morphogen signaling. Dev Cell 17: 470–481.
18. Hufnagel L, Kreuger J, Cohen SM, Shraiman BI (2006) On the role of glypicans in the process of morphogen gradient formation. Dev Biol 300: 512–522.
19. Kirkpatrick CA, Dimitroff BD, Rawson JM, Selleck SB (2004) Spatial regulation of wingless morphogen distribution and signaling by dally-like protein. Dev Cell 7: 513–523.
20. Kreuger J, Perez L, Giraldez AJ, Cohen SM (2004) Opposing activities of dally-like glypican at high and low levels of wingless morphogen activity. Dev Cell 7: 503–512.
21. Giraldez AJ, Copley RR, Cohen SM (2002) HSPG modification by the secreted enzyme Notum shapes the Wingless morphogen gradient. Dev Cell 2: 667–676.
22. Gerlitz O, Basler K (2002) Wingful, an extracellular feedback inhibitor of Wingless. Genes Dev 16: 1055–1059.
23. Yao JG, Sun YH (2005) Eyg and Ey Pax proteins act by distinct transcriptional mechanisms in Drosophila development. EMBO J 24: 2602–2612.
24. Chen CH, Huang HX, Ward CM, Su JT, Schaeffer LV, et al. (2007) A synthetic maternal-effect selfish genetic element drives population replacement in Drosophila. Science 316: 597–600.
25. Brand AH, Perrimon N (1993) Targeted Gene-Expression As A Means of Altering Cell Fates and Generating Dominant Phenotypes. Development 118: 401–415.
26. Lee YS, Carthew RW (2003) Making a better RNAi vector for Drosophila: use of intron spacers. Methods 30: 322–329.
27. Pfeiffer S, Ricardo S, Manneville JB, Alexandre C, Vincent JP (2002) Producing cells retain and recycle Wingless in Drosophila embryos. Curr Biol 12: 957–962.
28. Zhang J, Carthew RW (1998) Interactions between Wingless and DFz2 during Drosophila wing development. Development 125: 3075–3085.

29. Baena-Lopez LA, Alexandre C, Mitchell A, Pasakarnis L, Vincent JP (2013) Accelerated homologous recombination and subsequent genome modification in Drosophila. Development 140: 4818–4825.

30. Alexandre C, Baena-Lopez A, Vincent JP (2014) Patterning and growth control by membrane-tethered Wingless. Nature 505: 180–185.

31. Vincent JP, O'Farrell PH (1992) The state of engrailed expression is not clonally transmitted during early Drosophila development. Cell 68: 923–931.

32. Jang CC, Chao JL, Jones N, Yao LC, Bessarab DA, et al. (2003) Two Pax genes, eye gone and eyeless, act cooperatively in promoting Drosophila eye development. Development 130: 2939–2951.

33. Nolo R, Abbott LA, Bellen HJ (2000) Senseless, a Zn finger transcription factor, is necessary and sufficient for sensory organ development in Drosophila. Cell 102: 349–362.

34. Chou TB, Perrimon N (1996) The autosomal FLP-DFS technique for generating germline mosaics in Drosophila melanogaster. Genetics 144: 1673–1679.

35. Banziger C, Soldini D, Schutt C, Zipperlen P, Hausmann G, et al. (2006) Wntless, a conserved membrane protein dedicated to the secretion of Wnt proteins from signaling cells. Cell 125: 509–522.

36. Hartmann U, Maurer P (2001) Proteoglycans in the nervous system - the quest for functional roles in vivo. Matrix Biol 20: 23–35.

37. BaSalamah MA, Marr HS, Duncan AW, Edgell CJ (2001) Testican in human blood. Biochem Biophys Res Commun 283: 1083–1090.

38. Bejsovec A, Wieschaus E (1993) Segment polarity gene interactions modulate epidermal patterning in Drosophila embryos. Development 119: 501–517.

39. Capdevila J, Guerrero I (1994) Targeted expression of the signaling molecule decapentaplegic induces pattern duplications and growth alterations in Drosophila wings. EMBO J 13: 4459–4468.

40. Neumann CJ, Cohen SM (1996) A hierarchy of cross-regulation involving Notch, wingless, vestigial and cut organizes the dorsal/ventral axis of the Drosophila wing. Development 122: 3477–3485.

41. Couso JP, Bishop SA, Arias AM (1994) The Wingless Signaling Pathway and the Patterning of the Wing Margin in Drosophila. Development 120: 621–636.

42. Cadigan KM, Fish MP, Rulifson EJ, Nusse R (1998) Wingless repression of Drosophila frizzled 2 expression shapes the wingless morphogen gradient in the wing. Cell 93: 767–777.

43. Crozatier M, Glise B, Vincent A (2004) Patterns in evolution: veins of the Drosophila wing. Trends Genet 20: 498–505.

44. Baker NE (1988) Transcription of the segment-polarity gene wingless in the imaginal discs of Drosophila, and the phenotype of a pupal-lethal wg mutation. Development 102: 489–497.

45. Couso JP, Bate M, Martinez-Arias A (1993) A wingless-dependent polar coordinate system in Drosophila imaginal discs. Science 259: 484–489.

46. Jafar-Nejad H, Tien AC, Acar M, Bellen HJ (2006) Senseless and Daughterless confer neuronal identity to epithelial cells in the Drosophila wing margin. Development 133: 1683–1692.

47. Bartscherer K, Pelte N, Ingelfinger D, Boutros M (2006) Secretion of Wnt ligands requires Evi, a conserved transmembrane protein. Cell 125: 523–533.

48. Panakova D, Sprong H, Marois E, Thiele C, Eaton S (2005) Lipoprotein particles are required for Hedgehog and Wingless signalling. Nature 435: 58–65.

49. Mulligan KA, Fuerer C, Ching W, Fish M, Willert K, et al. (2012) Secreted Wingless-interacting molecule (Swim) promotes long-range signaling by maintaining Wingless solubility. Proc Natl Acad Sci USA 109: 370–377.

50. Gross JC, Chaudhary V, Bartscherer K, Boutros M (2012) Active Wnt proteins are secreted on exosomes. Nat Cell Biol 14: 1036–1045.

51. Beckett K, Monier S, Palmer L, Alexandre C, Green H, et al. (2013) Drosophila S2 cells secrete wingless on exosome-like vesicles but the wingless gradient forms independently of exosomes. Traffic 14: 82–96.

52. Bovolenta P, Esteve P, Ruiz JM, Cisneros E, Lopez-Rios J (2008) Beyond Wnt inhibition: new functions of secreted Frizzled-related proteins in development and disease. J Cell Sci 121: 737–746.

53. Esteve P, Sandonis A, Ibanez C, Shimono A, Guerrero I, et al. (2011) Secreted frizzled-related proteins are required for Wnt/beta-catenin signalling activation in the vertebrate optic cup. Development 138: 4179–4184.

54. Buechling T, Boutros M (2011) Wnt signaling signaling at and above the receptor level. Curr Top Dev Biol 97: 21–53.

55. Yeh E, Zhou L, Rudzik N, Boulianne GL (2000) Neuralized functions cell autonomously to regulate Drosophila sense organ development. The EMBO journal 19: 4827–4837.

56. Morata G, Struhl G (2014) Developmental biology: Tethered wings. Nature 505: 162–163.

57. Schnepp A, Lindgren PK, Hulsmann H, Kroger S, Paulsson M, et al. (2005) Mouse testican-2 - Expression, glycosylation, and effects on neurite outgrowth. J Biol Chem 280: 11274–11280.

58. Nakada M, Yamada A, Takino T, Miyamori H, Takahashi T, et al. (2001) Suppression of membrane-type 1 matrix metalloproteinase (MMP)-mediated MMP-2 activation and tumor invasion by testican 3 and its splicing variant gene product, N-Tes. Cancer Res 61: 8896–8902.

59. Nakada M, Miyamori H, Yamashita J, Sato H (2003) Testican 2 abrogates inhibition of membrane-type matrix metalloproteinases by other testican family proteins. Cancer Res 63: 3364–3369.

60. Hausser HJ, Decking R, Brenner RE (2004) Testican-1, an inhibitor of pro-MMP-2 activation, is expressed in cartilage. Osteoarthritis Cartilage 12: 870–877.

61. Edgell CJ, BaSalamah MA, Marr HS (2004) Testican-1: a differentially expressed proteoglycan with protease inhibiting activities. Int Rev Cytol 236: 101–122.

62. Berger EA, McClellan SA, Barrett RP, Hazlett LD (2011) Testican-1 promotes resistance against Pseudomonas aeruginosa-induced keratitis through regulation of MMP-2 expression and activation. Invest Ophthalmol Vis Sci 52: 5339–5346.

63. Clevers H, Nusse R (2012) Wnt/beta-catenin signaling and disease. Cell 149: 1192–1205.

64. Anastas JN, Moon RT (2013) WNT signalling pathways as therapeutic targets in cancer. Nat Rev Cancer 13: 11–26.

65. Solis GP, Luchtenborg AM, Katanaev VL (2013) Wnt secretion and gradient formation. Int J Mol Sci 14: 5130–5145.

66. Tsou AP, Chuang YC, Su JY, Yang CW, Liao YL, et al. (2003) Overexpression of a novel imprinted gene, PEG10, in human hepatocellular carcinoma and in regenerating mouse livers. J Biomed Sci 10: 625–635.

67. Graveley BR, Brooks AN, Carlson JW, Duff MO, Landolin JM, et al. (2011) The developmental transcriptome of Drosophila melanogaster. Nature 471: 473–479.

68. Chintapalli VR, Wang J, Dow JA (2007) Using FlyAtlas to identify better Drosophila melanogaster models of human disease. Nat Genet 39: 715–720.

10

Clonal Architectures and Driver Mutations in Metastatic Melanomas

Li Ding[1,2,3,4,9], Minjung Kim[5,9], Krishna L. Kanchi[1,9], Nathan D. Dees[1], Charles Lu[1], Malachi Griffith[1,3], David Fenstermacher[5], Hyeran Sung[5], Christopher A. Miller[1], Brian Goetz[6], Michael C. Wendl[1], Obi Griffith[1,2], Lynn A. Cornelius[2,6], Gerald P. Linette[2,4], Joshua F. McMichael[1], Vernon K. Sondak[5], Ryan C. Fields[4,6], Timothy J. Ley[1,2,4], James J. Mulé[5], Richard K. Wilson[1,3,4], Jeffrey S. Weber[5]*

1 The Genome Institute, Washington University in St. Louis, St. Louis, Missouri, United States of America, **2** Department of Medicine, Washington University in St. Louis, St. Louis, Missouri, United States of America, **3** Department of Genetics, Washington University in St. Louis, St. Louis, Missouri, United States of America, **4** Siteman Cancer Center, Washington University in St. Louis, St. Louis, Missouri, United States of America, **5** Donald A. Adam Comprehensive Melanoma Research Center, Moffitt Cancer Center, Tampa, Florida, United States of America, **6** Department of Surgery, Washington University in St. Louis, St. Louis, Missouri, United States of America

Abstract

To reveal the clonal architecture of melanoma and associated driver mutations, whole genome sequencing (WGS) and targeted extension sequencing were used to characterize 124 melanoma cases. Significantly mutated gene analysis using 13 WGS cases and 15 additional paired extension cases identified known melanoma genes such as BRAF, NRAS, and CDKN2A, as well as a novel gene EPHA3, previously implicated in other cancer types. Extension studies using tumors from another 96 patients discovered a large number of truncation mutations in tumor suppressors (TP53 and RB1), protein phosphatases (e.g., PTEN, PTPRB, PTPRD, and PTPRT), as well as chromatin remodeling genes (e.g., ASXL3, MLL2, and ARID2). Deep sequencing of mutations revealed subclones in the majority of metastatic tumors from 13 WGS cases. Validated mutations from 12 out of 13 WGS patients exhibited a predominant UV signature characterized by a high frequency of C->T transitions occurring at the 3' base of dipyrimidine sequences while one patient (MEL9) with a hypermutator phenotype lacked this signature. Strikingly, a subclonal mutation signature analysis revealed that the founding clone in MEL9 exhibited UV signature but the secondary clone did not, suggesting different mutational mechanisms for two clonal populations from the same tumor. Further analysis of four metastases from different geographic locations in 2 melanoma cases revealed phylogenetic relationships and highlighted the genetic alterations responsible for differential drug resistance among metastatic tumors. Our study suggests that clonal evaluation is crucial for understanding tumor etiology and drug resistance in melanoma.

Editor: Patrick Tan, Duke-National University of Singapore Graduate Medical School, Singapore

Funding: This work was supported by the Donald A. Adam Comprehensive Melanoma Research Center. The authors gratefully thank the Our Mark on Melanoma Inc. Foundation and the Come Out Swinging Inc. Foundation for additional funding. The funders had no role in study design, data collection and analysis, decision to publish, or preparation of the manuscript.

Competing Interests: The authors have declared that no competing interests exist.

* Email: Jeffrey.weber@moffitt.org

꜉ These authors contributed equally to this work.

Introduction

The incidence of invasive melanoma in the United States in 2013 is estimated to be 76,690, with approximately 9,480 deaths [1]. While the death rate due to melanoma is relatively low, and many tumors are found at an early stage when they can be completely resected and cured, the development of metastatic disease is a harbinger of poor outcome. Though four new drugs that prolong overall or progression-free survival were recently approved for stage IV disease [2,3], the median survival for metastatic melanoma remains poor. Work in the last decade has identified a number of common and/or driver mutations in melanoma and helped to elucidate the pathways determining melanoma oncogenesis, proliferation, and metastasis. These discoveries have led to the development of inhibitors for BRAF and KIT (C-Kit) signaling, some of which have shown benefit in melanoma patients [4,5].

The driver mutations in BRAF and NRAS that have been identified cannot fully explain melanoma oncogenesis, as these same mutations have been found at similar rates in benign nevi, or moles [6,7]. These common, benign skin lesions infrequently undergo malignant transformation into melanoma, but invariably remain in their growth-arrested state. BRAF V600E-induced oncogenic senescence has been implicated in melanoma cell cycle arrest [8], together with loss of function of genes including TP53, NF2, and IRF1 [9], CDKN2A (INK4A/ARF) and CDK4 [10]. In fact, loss of senescence is an important process in RAS- and RAF-induced transformation, implying that additional but still unknown genomic changes must be involved in transformation to melanoma. The mutational variants identified by WGS can provide insight into this process.

The first melanoma genome sequenced was derived from an established cell line [11] and it showed a large number of SNVs, most corresponding to a UV signature of C->T transitions [12].

Recently, several exome-based studies have been conducted to identify genes driving melanoma development. Directed sequencing has demonstrated *ERBB4* mutations in melanomas [13] and copy number analysis has revealed amplification of the histone methyltransferase gene *SETDB1* [14]. Frequent somatic mutations in *MAP3K5* and *MAP3K9* were identified in metastatic melanomas by exome sequencing [15]. A preliminary study of matched whole exome sequencing of melanomas and matched normals indicates that *GRIN2A* was frequently mutated in 14 specimens, and subsequent analysis of over 100 samples showed it to be mutated in 33% of cases [16]. *TRRAP* and *GRM3* were also found to be mutated in a small proportion of tumors, indicating that the glutamate pathway might play an important role in melanoma development and progression [17]. Another exome sequencing study described six novel melanoma genes: *PPP6C*, *RAC1*, *SNX31*, *TACC1*, *STK19*, and *ARID2* [*18*]. Recently, *PREX2*, a phosphatidylinositol-3,4,5-trisphosphate-dependent Rac exchange factor 2, was found to be mutated in 14% of 107 melanoma cases [19]. Whole genome sequencing of an acral melanoma primary and lymph nodal metastasis showed 40 SNVs in the primary, of which 39 were also present in the metastasis [20].

Our analysis of 15 metastatic melanoma tumors from 13 patients shows very high numbers of non-synonymous SNVs and it constitutes a useful catalogue of copy number alterations, insertions, deletions and translocations within those genomes. *EPHA3* (ephrin type-A receptor 3) was found to be significantly mutated in 28 melanoma cases. Extension analysis using 97 tumors from 96 patients revealed a number of truncation mutations in well-known tumor suppressors, protein phosphatases, as well as genes involved in chromatin remodeling. Notably, the majority of these truncation mutations co-occur with *BRAF* and *NRAS* mutations, suggesting a potential cooperating role during the progression of melanoma. We also performed deep sequencing of somatic mutations that uncovered the clonal structures of melanomas, helped to dissect diverse mutational mechanisms in subclones, and further established the initiation roles of *BRAF* and *NRAS* mutations in melanoma. More importantly, our comparative analysis of 4 metastases from different geographical locations in 2 cases revealed a clonal evolution path and underscored the genetic alterations responsible for drug resistance.

Results

Genomic analysis of the whole genome sequencing data of 15 tumors from 13 melanoma cases

All 13 WGS patients, from whom 15 tumors and 13 sets of normal PBMC were included in this study, had stage IV melanoma. The metastatic samples were from diverse locations including lung, chest wall, brain, lymph node, stomach, small intestine, and adrenal gland (**Figure 1**). Using an Illumina paired-end sequencing strategy, tumor and normal genomes were sequenced to at least 29.5X-fold and 35X-fold haploid coverage, respectively, with corresponding diploid coverage of 98.84% or better based on concordance with SNP array data (**Table S1 in File S1**). Candidate somatic changes were predicted using multiple algorithms [21–25] and selected for hybridization capture-based validation (Supplementary Materials and Methods). We included capture probes corresponding to all putative somatic single nucleotide variants (SNVs) and small insertions/deletions (indels) that overlap with coding exons, splice sites, and RNA genes (tier 1), a number of high-confidence SNVs and indels in non-coding conserved or regulatory regions (tier 2), and non-repetitive regions of the human genome (tier 3). In addition, we included predicted somatic structural variants (SVs) genome-wide for validation (Methods). Analysis of the high depth sequencing data resulting from the captured target DNAs of 15 tumor and 13 normal samples (**Table S2, S3** in File S1 and Supplementary Materials and Methods in File S1) confirmed 17,361 tier 1 point mutations, with a validation rate of 93.6%, 84 tier 1 indels, and 411 somatic SVs (**Tables S4 and S5 in File S2**). Seven of 15 tumors had over 1,000 tier 1 SNVs. This is among the highest mutation frequency of known cancers. For comparison, AMLs have a median of 13 tier 1 changes per genome [cite TCGA AML] [26] and metastatic breast cancers have been reported with between 32 (lobular) [27] and 50 (basal-like) [28]. C->T transitions were predominant in all 15 tumors, consistent with a UV damage signature (**Figure 1**). Notably, MEL9, with 6,795 validated tier 1 point mutations (7.7-fold times the average number from the other 12 cases), exhibited the highest C->T transition rate among 15 tumors. The patterns of point mutations were very similar for the paired metachronous tumors from patients 5 and 13 (**Figure 1 and Table S6a in File S2**). Notably, we identified 443 tier 1 dinucleotide mutations in 13 WGS cases and among them, an average of 74% (ranging from 68% to 78%) are CC->TT changes, consistent with previous reports [11]. The ratio between dinucleotide and point mutations in Tier 1 ranges from 0.76% to 6.28% while the ratio in Tiers 1–3 ranges from 0.46% to 2.43%, consistent with the higher GC content in the coding sequences. (**Table S7 in File S1**)

We validated 84 coding small indels (65 deletions and 19 insertions) ranging from 1 to 37 base pairs (bp) in length, including a complex frameshift indel (4 bp deletion and 2 bp insertion) in *TP53* and 1 bp deletions in *STAG2* and *CDKN2A* (**Table S5 in File S2**). We also identified a total of 411 validated rearrangements for the 13 cases (range 10–87), including 240 deletions, 95 inversions, and 69 translocations (**Table S5 in File S2**). Across all 15 tumors, there was a median of 11 chromosomal rearrangements disrupting protein-coding regions per tumor. Recurrent deletions in tumor suppressor *FHIT*, the fragile histidine triad gene, were identified in MEL5 and MEL10, while large deletions interrupting *CDKN2A* were identified in 5 tumors from 4 patients (MEL2, 5, 8, 9, and 10). Further, one 1 bp deletion and one nonsense mutation in *CDKN2A* were identified in MEL3 and MEL11, respectively. A deletion involving *CTNNA2* was also detected in MEL10. A focal amplification of *CCND1* was observed in MEL1 and MEL4, which resulted in increased expression levels, according to DNA microarray analysis from those tumors (data not shown). SVs and copy number alterations for each of the 15 sequenced tumors are shown in **Figure S1** in File S1, providing a comprehensive view of genetic aberrations in melanoma metastases.

Previous studies showed that *TERT* promoter mutations are frequent in familial and sporadic melanoma [29,30] and other cancer types [31]. We identified somatic mutations within the promoter of *TERT* in 9 of 13 (68%) melanoma cases. Four distinct somatic base substitutions were observed G>A, 101–146 bases upstream of the *TERT* transcription start site (TSS). Three are point mutations (C205T (MEL8), C228T (MEL4, 5, 10, 11), and C250T (MEL1, 8, 9, 12)) and one is a dinucleotide mutation (C242T, C243T (MEL13)) (**Table S6b in File S2**). Sequence coverage levels achieved around the *TERT* promoter region (500bp upstream from TSS) were low (~7.4X on average) due to the sequence context (high GC) and repeat content. It is therefore possible that the prevalence of *TERT* promoter mutation could be higher than 68% in our sample set.

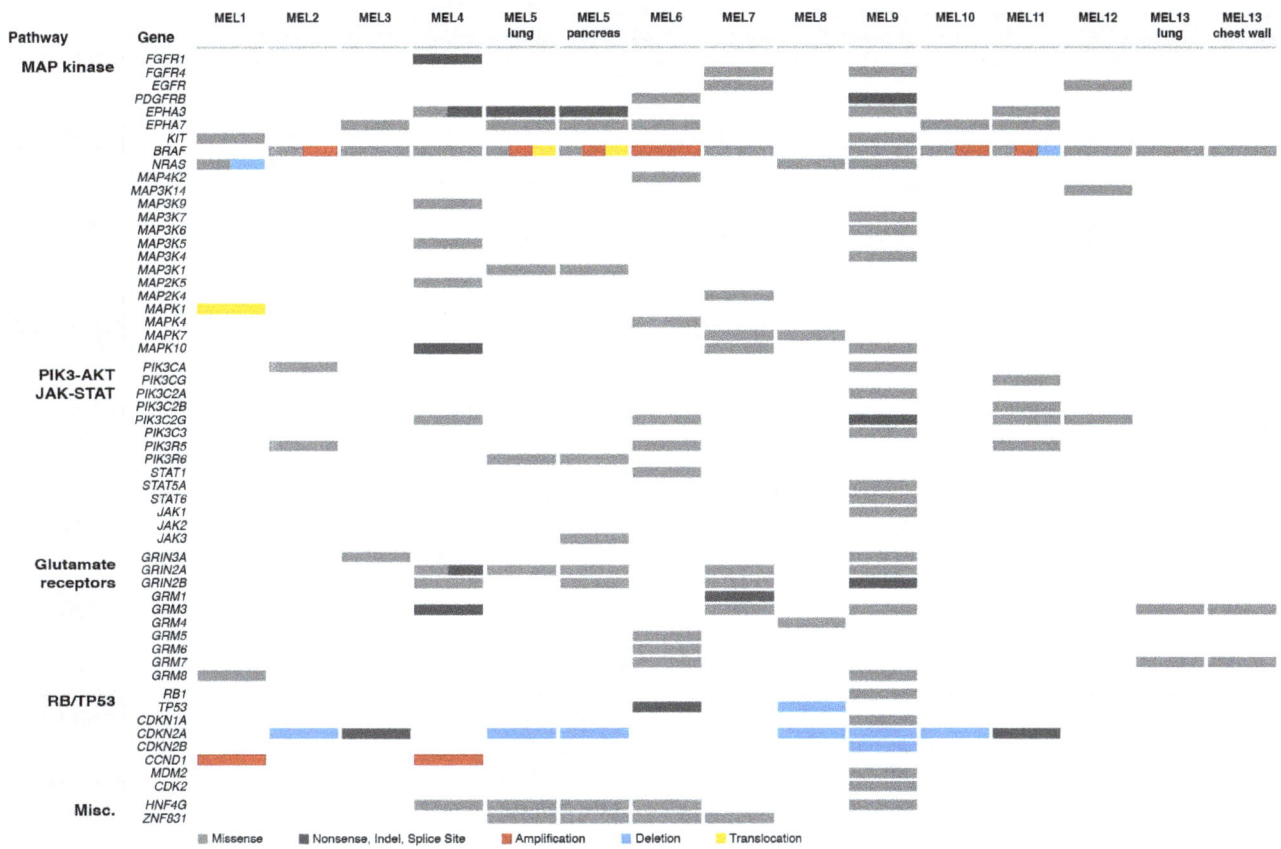

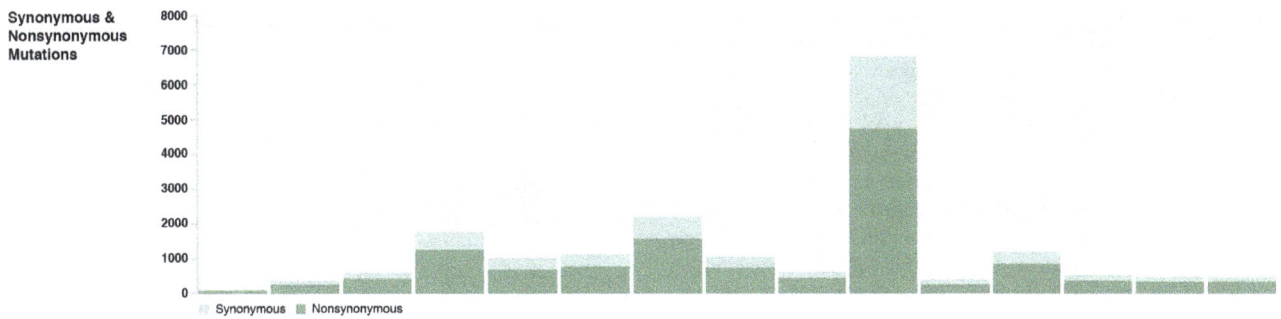

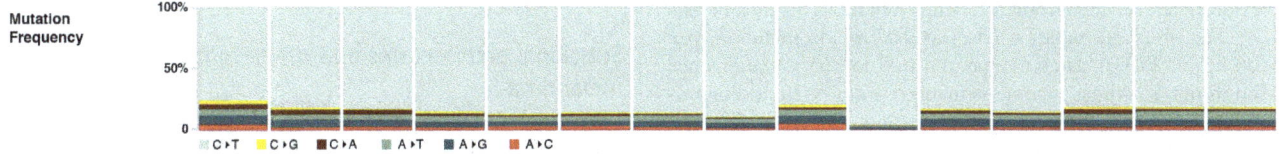

Figure 1. Mutation pattern, spectrum, and clinical features in 15 metastases from 13 WGS melanoma cases. Mutations found in genes from MAP kinase, PI3K-AKT, RB/TP53 pathways and glutamate receptors are shown. Copy number alterations and structural variants found in *BRAF*, *NRAS*, *TP53*, *CDKN2A/2B*, and *CCND1* are also displayed. The numbers and frequencies of tier 1 transition and transversion events identified in all 15 tumors are shown.

Significantly mutated genes (SMG) and pathways in melanoma

After our initial discovery using 13 WGS cases followed by the validation analysis described above, we performed further extension screening in 15 melanoma cases (25 metastatic tumors and matched normal tissue; 6 cases with multiple metastases). 1,209 genes were chosen for screening based on our initial WGS results and mutations and genes reported in several recent genomic studies of melanoma (**Tables S8a and S8b in File S2**) [13,15–18]. Using mutations identified in all 28 cases, we performed MuSiC [32] analysis to discover genes displaying significantly higher mutation rates than expected based on the background mutation rate. A small group of genes was identified as significant after applying a 5% false discovery rate threshold (**Table 1**). This group included *BRAF* and *NRAS*, which were found to be mutated in 18 and 4 patients of 28, respectively (**Figure 2**). MEL9, the adrenal gland metastasis that was hypermutated, harbored mutations in both *BRAF* (H574L) and *NRAS* (Q61R); these two mutations were found to be present in the same variant allele frequency cluster (see Subclonal architecture in melanoma below). Meanwhile, mutations were not detected in either *BRAF* or *NRAS* in MEL6, a lung metastasis, and also four other tumors from the latter discovery group of 15 cases. The SMG list includes other genes known to be potentially involved in cancer. For instance, protein tyrosine kinase *EPHA3*, known mostly for its role in lung cancer [33], had 7 missense and 3 nonsense mutations, and tumor suppressor *CDKN2A* harbored one splice site, one nonsense, and one frame-shift indel, respectively (**Figure 2**). Mutations in a wide variety of protein families were seen in this study, including a large number of non-synonymous mutations in protein tyrosine phosphatases (e.g., *PTPRB*, *PTPRT*, and *PTPN13*), and protein tyrosine kinases (e.g., *EPHA7*, *EPHA3*, *KIT*, *FGFR4*, *FGFR1*, and *ROS1*). Of note, 8 missense and 1 nonsense mutations were found in *ASXL3*, a member of the polycomb group. The existence of mutations in glutamate receptors was described in prior exome sequencing studies [16], and our data not only confirmed that *GRIN2A* was mutated in melanoma (5 out of 28 cases) but also showed that *GRIN2B* was recurrently mutated (**Figure 2**). In addition, a number of mutations have also been found in other metabotropic glutamate receptors, such as *GRM1* and *GRM3-8*. Specifically, out of 23 nonsynonymous mutations from GRM genes, one nonsense and four missense mutations were from *GRM3*, previously shown to harbor activating mutations in melanomas [17]. The observed mutation rate was 0.22 to 143 mutations per Mbp in the TCGA dataset compared to 3 to 155 mutations per Mbp in our 15 whole genome sequenced samples. In addition to the similar distribution of mutation rates, we also observed recurrent single nucleotide variants including S225F and G394E in EPHA7 and G114E and R136* in EPHA3 from both datasets. The Comparison of the number of mutations in significant genes between this study and TCGA report [34] is shown in Table S15 in File S1.

Melanomas harbor a number of aberrantly regulated signaling pathways, including INK4A-CDK4/6-RB, ARF-TP53-MDM2, RAS-RAF-MAPK, PTEN-PI3K-AKT, and aMSH-MC1R-cAMP-MITF, all of which may be altered via genetic, genomic, or epigenetic mechanisms [35,36]. Mutations and rearrangements

were identified in *BRAF*, *NRAS*, and several MAP kinases including *MAP3K1*, *MAP4K2*, and *MAP3K14* in the MAPK signaling pathways; 24/28 patients had at least one mutation in this pathway. In addition, 18 patients harbored somatic alterations in genes affecting the RB/TP53 pathway, including *CDKN2A*, *CCND1*, *MDM2*, and *CDK2* (**Figure 1 and Tables S9a, S9b, and S9c in File S2**). *GRM3* and *GRIN2A/2B* alterations were also frequently observed in the tumors sequenced herein, suggesting the importance of glutamate mediated transduction in melanoma (**Figures 1 and 2**).

Recurrence analysis using 96 additional melanoma cases

In addition to the 15 paired targeted samples, we also screened 1,209 genes using an extension set of 97 melanoma tumor samples from 96 patients. Since matched normals are not available for these 96 patients, we focused our analysis on known mutations and truncation mutations. We identified 1,716 recurrent nonsynonymous mutations found previously in our paired discovery samples, in the COSMIC (Catalog of Somatic Mutations in Cancer) database [37], or in recent melanoma studies (**Table S10 in File S2**) [13,15–18], as well as 1,287 truncation (nonsense, splice site, and frame-shift) variants in 616 genes (**Table S11 in File S2**). *BRAF* and *NRAS* mutations were identified in 60 and 19 patients, respectively. Additionally, three patients harbored mutations in both *BRAF* and *NRAS*; 10 patients exhibited dinucleotide polymorphisms (DNPs) in *BRAF*, and one DNP was identified in *NRAS*. Further analysis revealed a number of truncation mutations in well-established tumor suppressors (*TP53* and *RB1*), protein phosphatase genes (e.g., *PTEN*, *PTPRB*, *PTPRD*, *PTPRN2*, *PTPRT*, and *PPP1R3A*), and chromatin remodeling genes (e.g., *ASXL3*, *MLL2*, and *ARID2*) (**Figure S2 in File S1**). In addition, three truncation variants (2 nonsense and 1 splice site) were found in the *PREX2* gene, consistent with previous findings (**Figure S2** in File S1) [19]. Our analysis using MuSiC [32] identified *ASXL3* and *PTPRT* as harboring co-occurring truncation mutations in our 96-patient extension set (P = 0.002). All three patients containing a *PREX2* truncation also had an *NRAS* mutation (P = 0.008 for the co-occurence); none of the three had a *BRAF* mutation (P = 0.062 for the mutual exclusion between *PREX2* and *BRAF*). *ASXL3* truncations also co-occurred with *NRAS* (P = 0.044). Interestingly, tumor suppressors *PTEN* and *RB1* co-occurred in our extension dataset (P = 0.019), harboring six and four truncation events, respectively.

Subclonal architectures and driver mutations in melanoma

We took hundreds of validated somatic mutations with read depths of hundreds to thousands from capture validation and applied the SciClone algorithm (https://github.com/genome/sciclone) to cluster mutations with similar allelic fractions. These clusters are indicative of distinct subclonal populations of tumor cells. Multiple subclones were observed in the majority of 15 WGS tumors (**Table 2**). Due to the high mutation rate and complex copy number landscape in melanoma, the boundaries of some clusters could not be clearly separated using genome-wide data. We then selected "stable" genomic regions based on LOH and CNV analyses using VarScan 2 [38] and used somatic mutations

Table 1. Significantly mutated genes identified in 28 paired melanoma cases.

Gene	Indels	SNVs	Total Mutations	Mutated Cases	Mutation Frequency	Covered bps	Mutation per Mbp	P-value	FDR
BRAF	0	22	22	18	0.6429	62953	349.47	0	0
NRAS	0	4	4	4	0.1429	16408	243.78	2.1749E−07	0.00059482
CDKN2A	1	3	4	4	0.1429	16864	237.19	4.4204E−06	0.00743995
EPHA3	0	10	10	7	0.25	89708	111.47	2.1443E−05	0.02606568

from these regions for plotting (**Figure 3 and Table S12a in File S2**). In MEL1, two distinct clusters at 36.7% and 21.7% Variant Allele Frequency (VAF)s were identified. The majority of mutations were from the 21.7% VAF cluster. MEL8 displayed a similar pattern as MEL1, with one cluster at 37.8% VAF and another cluster at 23.4%. The hypermutated MEL9 tumor has the founding cluster at 46.8% VAF and the secondary but dominant cluster centered at 19.8%, suggesting that a massive mutation expansion took place in the 17.6% VAF cluster. Likewise, MEL10 had two clusters, centered at 30.6% and 19.5% VAFs respectively (**Figure 3**). These estimates of tumor heterogeneity represent a lower bound, and it is possible that additional subclone(s) were present in these samples but not detected. Our results demonstrate that melanoma is a disease characterized by significant intra-tumor heterogeneity.

By associating mutations with specific subclonal populations, we can infer the relative order in which these mutations were acquired. 12 of 13 NRAS and BRAF mutations in the cohort are clonal (or lie in copy number amplified regions), suggesting that these mutations are involved in melanoma initiation (**Figure 3 and Table 2**). In MEL1, our analysis showed that CTNNB1 T41I missense (a highly recurrent site) maps to the founding clone while KIT P157S (a novel mutation) resides in a subclone, suggesting the former was an early event, while the latter may have contributed to progression. Similarly, NRAS Q61K is present in all cells of MEL8, suggesting an initiation role, while the ARID2 S1382F missense mutation is subclonal (**Figure 3**). MEL9 is characterized by both NRAS and BRAF mutations in founding clone, along with a CTNNB1 P16S mutation that arose later in the evolution of the tumor.

Distinct mutational signatures in founding and secondary clones of a hypermutated sample (MEL9)

It has been shown that DNA damage caused by UV light often leads to the formation of covalent links between two adjacent pyrimidine residues [11]. As a result, C->T mutations in melanoma samples often occur at dipyrimidine sequences. Our analysis of 13 WGS melanoma cases showed that 12 cases had greater than 89.9% of C->T mutations occurring at the 3′ base of a pyrimidine dinucleotide, supporting previous findings [11]. However, MEL9, the hypermutated tumor, lacks this signature and has only 59.5% C->T occurring at the 3′ base of a dipyrimidine, comparable to 53% expected by chance (**Figure 4**). We reasoned that the UV signature in MEL9 might be masked by a large number of subsequent mutations arising from some other mechanism. One candidate was that these mutations were the result of a DNA repair defect (e.g., S418F and G1134R in MSH6, G2569S in BRCA2, or G648E in ERCC6). To test this hypothesis, we independently analyzed mutations from the founding and subclonal populations described above. (**Figure 4**). Strikingly, these two subclones in MEL9 exhibited two very distinct phenotypes. The founding clone exhibited a classic UV-damage phenotype with an abundance of C->T transitions and a disproportionately higher number of pyrimidine bases preceding the mutated cytosine bases (Proportion test $P = 1.60 \times 10^{-10}$). (**Figure 4**). In contrast, the subclone exhibited a typical pyrimidine base frequency preceding the mutated C base (59.5%, $P = 0.17$); interestingly this subclone had a significantly higher frequency of pyrimidine bases following the mutated C base ($P = 1.72 \times 10^{-46}$), consistent with findings in another hypermutated melanoma reported by Berger et al. [19]. Our hypothesis is that UV-driven mutations in the originating, founding clone of MEL9 damaged a DNA-repair gene and spurred a massive deficit in DNA repair. The resulting large number of mutations,

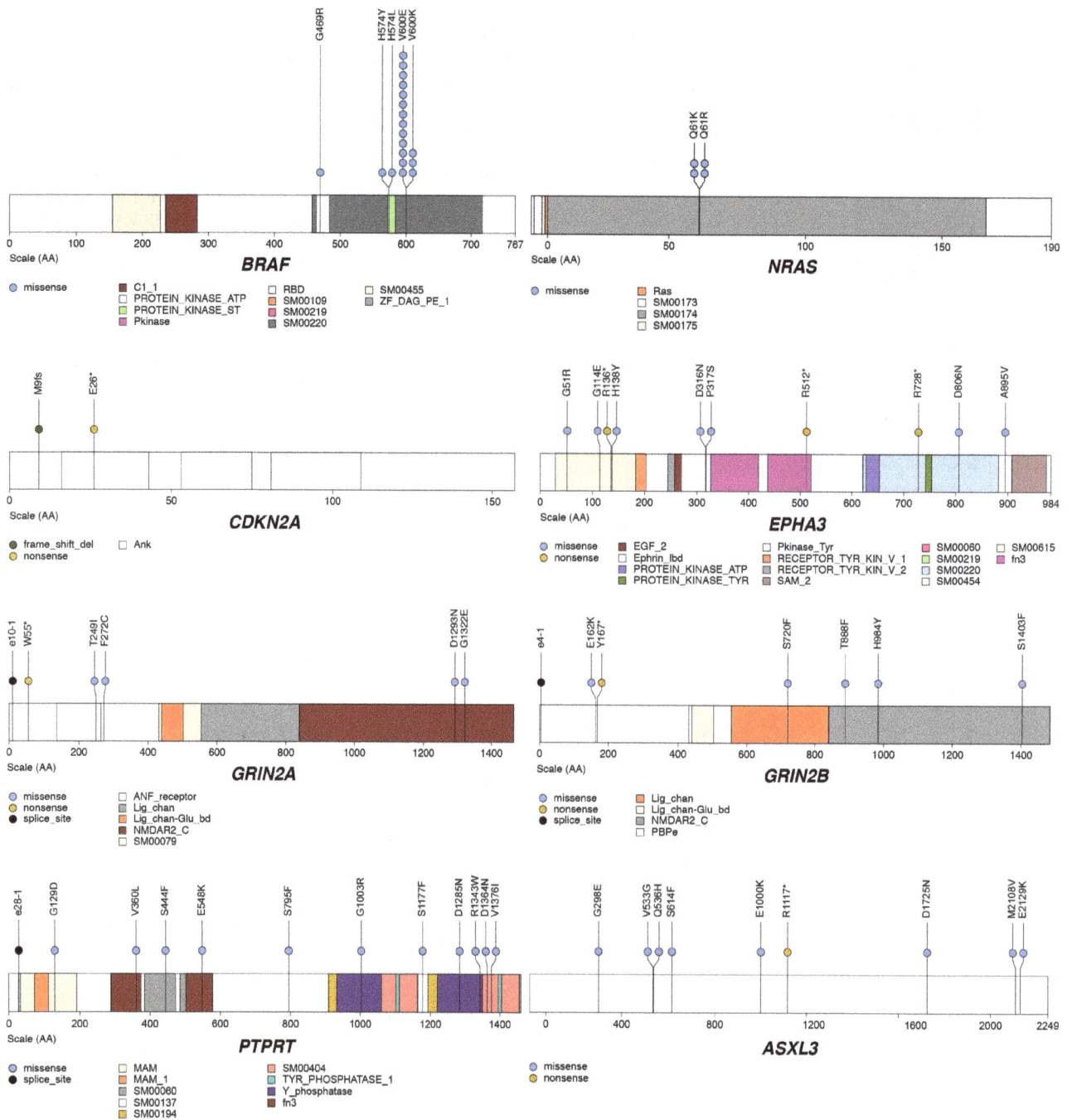

Figure 2. Mutation distribution in *BRAF, NRAS, CDKN2A, EPHA3, GRIN2A, GRIN2B, PTPRT,* **and** *ASXL3.* The locations of conserved protein domains are highlighted. Each nonsynonymous substitution, splice site mutation, or indel is designated with a circle at the representative protein position with color to indicate the translational effects of the mutation.

occurring later than the UV damage, make up the lower-VAF subclone. The mutation context observed in the secondary clone of MEL9 does not match the patterns expected from defects in *MSH2* and MSH6, and it may be attributable to another repair pathway. As a control, we also dissected the mutation spectrum in the founding clone and subclone of MEL10. The subclone for MEL10 also has a larger number of mutations, (**Figure 4**) but both show a typical UV-signature, with a significant number of Cytosine and Thymine bases preceding C->T transition sites (P-

value for founding clone $=4.01\times10^{-21}$, P-value for secondary clone $=3.83\times10^{-65}$).

Clonal and phylogenetic relationship among metastatic tumors from different sites

Among the 28 paired cases, 8 (2 WGS and 6 extension cases) had multiple metastasis samples, allowing the examination of relationships of different tumors from the same individual. First, we investigated the two WGS cases (MEL5 and MEL13) with two metastasis samples each. The rearrangement and copy number

Table 2. Clonal numbers and variant allele frequencies for driver mutations in 15 WGS metastatic tumors.

Sample	BRAF VAF	NRAS VAF	Focal Amplification of BRAF and/or NRAS	CN for BRAF	Number of subclones	Founding cluster frequency	Subclone cluster frequency
MEL1	NA	(Q61R) 26.89%	No	2	2	36.69	21.74
MEL2	56.88%	NA	Yes	3.78	2	46.26	25.31
MEL3	32.31%	NA	No	2	2	29.02	19.9
MEL4	28.35%	NA	No	2	2	23.53	12.55
MEL5 (lung metastasis)	(V600K)67%(WGS -122 reads)	NA	Yes	5.26	2	22.24	14.32
MEL5 (pancreas metastasis)	(V600K)99%(WGS- 181 reads)	NA	Yes	9.67	2	39.41	45.6
MEL6	NA	NA	Yes	2.59	1	26.33	NA
MEL7	32.34%	NA	No	2	1	28.08	NA
MEL8	NA	(Q61K)34.58%	No	1.3	2	37.77	23.37
MEL9	43.30%	58.17%	No	2	2	46.8	19.81
MEL10	(V600E) 25.31%	NA	Yes	2.97	2	30.62	19.47
MEL11	70%(WGS-54 reads)	NA	Yes	2.64	2	41.07	23.95
MEL12	58.44%	NA	No	2	2	32.85	19.27
MEL13 (lung metastasis)	25.36%	NA	No	2	1	30.2	NA
MEL13 (chestwall metastasis)	32.25%	NA	No	2	1	32.38	NA

MEL1

MEL8

MEL9

MEL10

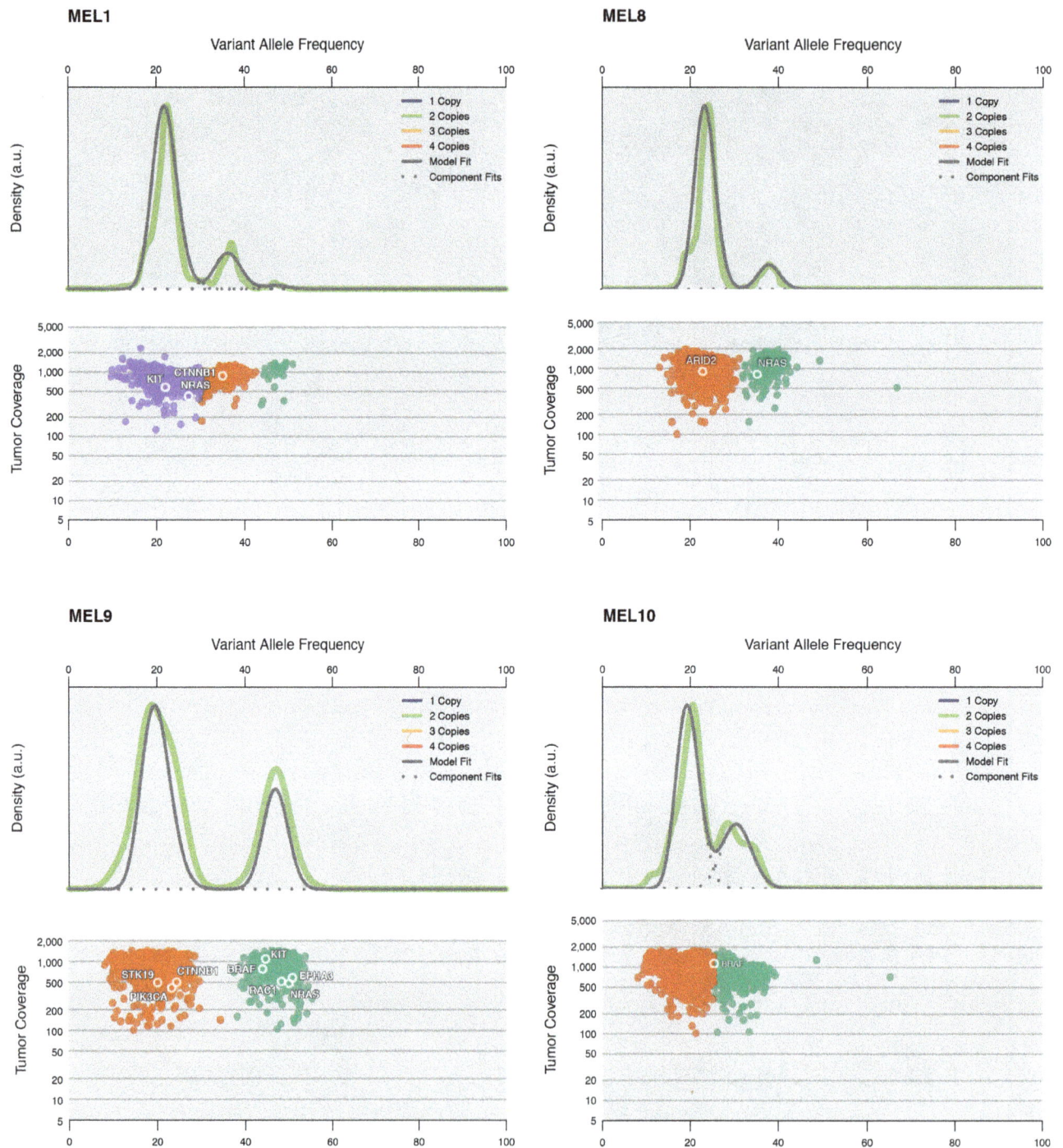

Figure 3. Overview of subclonal landscape in melanoma (MEL1, 8, 9, and 10) and their associated driver mutations. Two plots are shown for each case: kernel density (top), followed by the plot of tumor variant allele frequency by sequence depth for sites from selected copy number neutral regions (see Methods). Data shown are from chromosomes 1, 3, 4, 6, 10, and 13 for MEL1, from chromosomes 1, 2, 5, 12, and 20 for MEL8, from chromosomes 1, 3, 4, 6, 7, 12, and 13 for MEL9, and from chromosomes 6, 7, 10, 13, and 15 for MEL10. The data show evidence of two clusters in MEL1, MEL8, MEL9 and MEL10 with the majority of mutations from the lower allele frequency clusters. Mutations detected in significantly mutated genes in this study and genes implicated in Hodis *et al.* [18] were labeled.

patterns were almost identical for the MEL13 paired metachronous tumors from chest wall and lung (**Figure 5**). In MEL5, a significant number of inversions on chromosome 3 were found to be present in the pancreatic metastasis but not the lung metastasis (**Figure 5**). Clonality analysis using point mutations from selected copy number neutral regions revealed at least two clusters each in

the lung (22.2% and 14.3%) and pancreas (45.6% and 39.4%) metastasis from MEL5 (**Figure 6 and Table S12b–c in File S2**). The MEL5 lung metastasis has two distinct mutation clusters, while the pancreas metastasis harbors two clusters with overlapping boundaries. A comparison of genome-wide tier 1 mutations in the pancreas versus lung metastases (**Figure 6 and Table**

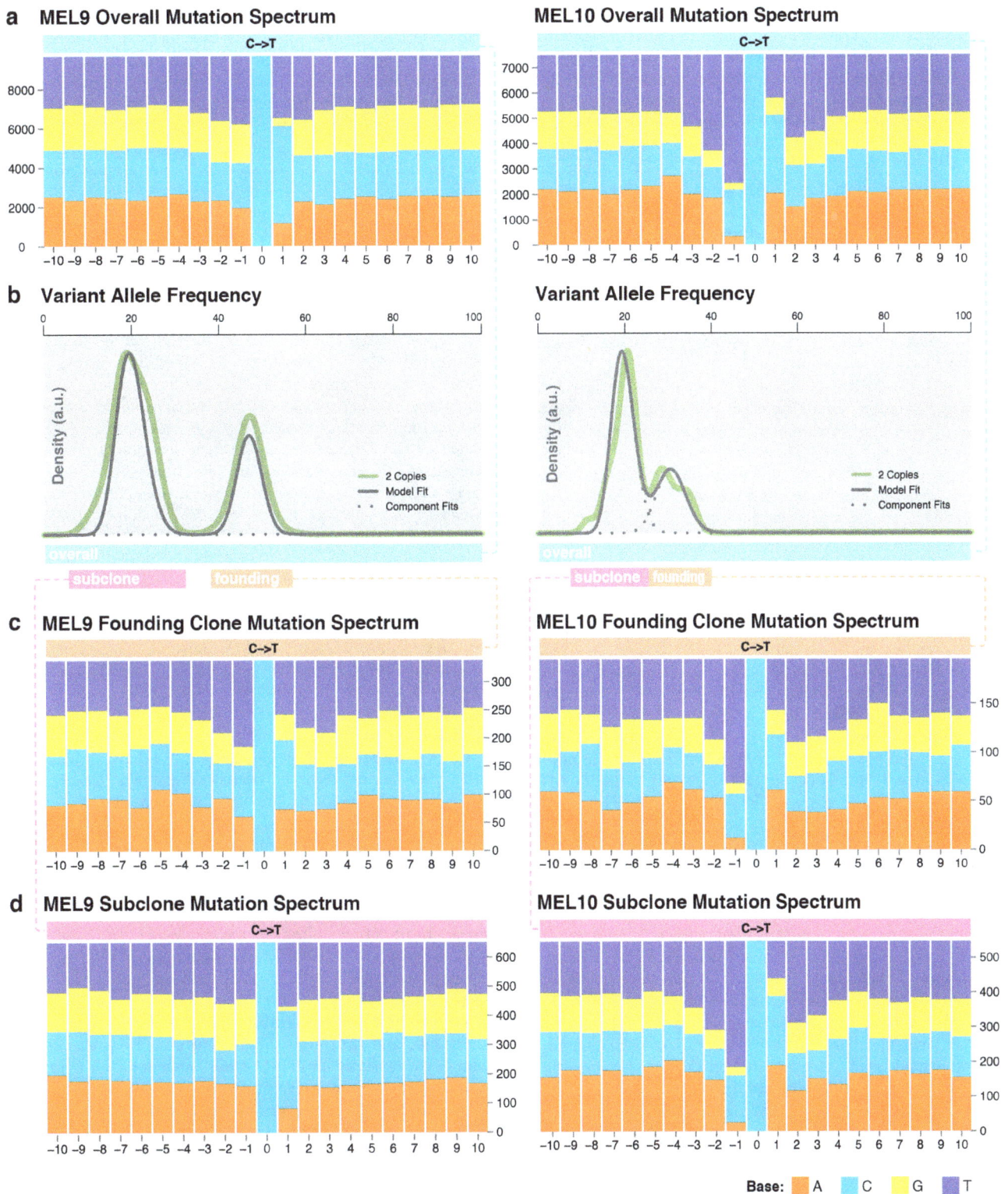

Figure 4. Dissecting mutational mechanisms using subclonal mutations. (a) Overall sequence context surrounding C->T transitions in MEL9 and MEL10. (b) Density plots showing the founding clone and subclone in MEL9 and MEL10. (c) Mutation context analysis of the founding clones detected a UV mutation signature in both MEL9 and MEL10. (d) Mutation context analysis of secondary clones detected a UV signature in MEL10 but not MEL9.

S12f in File S2) shows that greater than 99% of the tier 1 mutations (1127/1139) are shared between these two samples, indicating they likely emerged from the same progenitor clone in the primary tumor. Many MEL5 mutations appear to be enriched in the pancreas sample. Additionally, a wider range of VAFs present in the pancreas sample indicates that numerous copy number altering events occurred after the initial development of the pancreas metastasis. Both the lung and chest wall metastatic tumors in MEL13 have a very similar clonal pattern (**Figure 6 and Table S12d–e in File S2**), with one dominant higher VAF peak (24.9% and 32.4%, respectively) shouldered by less distinct clusters of mutations. The slight difference in the peak VAFs of the dominant clusters in these two metastases suggests that the chest wall biological specimen has a higher purity (64.8% tumor) than that of the lung metastasis (49.8% tumor). Consistent with the kernel density plots (**Figure 6**), an analysis of VAFs of genome-wide tier 1 mutations in MEL13 (**Figure 6 and Table S12g in File S2**) shows similar clonal architecture in the two metastasis samples, as most occur at comparable VAFs. Again, over 99% of the mutations are shared between the two metastases (503 of 505) in MEL13, with only 2 mutations being sample specific. This result suggests that both MEL13 metastases are derived from the same clone/subclone in the primary tumor.

Two cases (MEL167 and MEL174) with 4 synchronous metastatic samples were sequenced with a targeted extension panel. MEL167 has three tumor samples from small bowel and one from a lymph node. The relative locations of small bowel tumors 1, 2, and 3 from MEL167 are shown in **Figure 7**. Copy number analysis using VarScan 2 shows that small bowel tumors 2 and 3 share amplifications on chromosomes 7q, 13q, and 20, consistent with their close proximity in location. The lymph node tumor shared amplifications on chromosomes 2 and 3 with small bowel tumor 2 while small bowel tumor 1 showed no major amplifications genome-wide (**Figure 7**). Phylip-based phylogenetics analysis (Supplementary Materials and Methods in File S1) using mutations and their purity-corrected frequencies (**Table S13 In File S1 and Figure S3** in File S1) recapitulated the copy number analysis-based findings and identified small bowel tumors 2 and 3 as closely related and most divergent from the normal. The lymph node and small bowel tumor 1 are more similar to each other and appear to be less divergent from the the normal tissue (**Figure 7**). Mutations in *TP53* (H179Y), *NRAS* (Q61K), *ATR* (G2120C), and *EPHA3* (P317S) are shared by all four tumors, suggesting they are likely founding mutations from the primary tumor (**Figure 7 and Table S14 in File S2**). On the other hand, *PREX2* (P614S) and *ZNF831* (P1639S) are only present in some tumors.

We also analyzed a quartet of synchronous tumors from MEL174: two from the liver (liver tumors 1 and 2), one omentum tumor, and one portal nodule tumor (**Figure 8**). Liver tumor 2 shows some similarities to the portal nodule tumor, with both having major amplifications on chromosomes 7 and 8, while liver tumor 1 and omentum tumor show a "quieter" overall copy number landscape. Copy number alterations and mutation-based phylogenetic largely agree on the relationship between these tumors (**Table S13** in File S1 **and Figure S3 in File S1**). Liver tumor 1 and the omentum tumor responded to treatment with the BRAF inhibitor vemurafinib, while the others continued growing/ progressing during treatment. We found that both *BRAF* (V600E) and *CTNNA2* (R755*) have much higher variant/mutant allele frequencies in the non-reponsive portal nodule (*BRAF*: 65.2%; *CTNNA2*: 15.6%) and liver tumor 2 (33.3%; 16.1%) than in the omentum nodule (8.2%; 1.1%) and liver tumor 1(14.3%; 2.8%). The *CTNNA2* mutation is almost undetectable in the omentum nodule with only one read, out of 94 total reads, supporting the mutant allele. A *MAP2K1* (E203K) mutation is only detected in liver tumor 2 despite the high coverage (>40X) in all 4 metastases (**Table S14 in File S2**). After applying purity-based VAF adjustments, the portal nodule has the highest adjusted VAF at 72.47%, followed by liver tumor 2, liver tumor 1, omentum nodule

at 55.55%, 47.63%, and 41% respectively. Our data indicate that sample-specific genetic alterations and variable frequencies of mutations might contribute to differential treatment responses among metastasis samples from the same patients, consistent with a previous report [39] (**Figure 8**). Moreover, our analysis shows that all four metastatic tumors from both patients were derived from the same primary tumor but their patterns of mutations diverged considerably during evolution and metastatic growth.

Discussion

This study represents a comprehensive whole genome and targeted sequencing analysis of 124 human melanoma cases, including 13 WGS, 15 paired targeted, and 96 unpaired targeted extension cases. Besides the expected coding mutations in *BRAF*, *NRAS*, *CDKN2A*, and other genes detected, significant numbers of recurrent copy number variants and structural rearrangements found in this analysis of 13 WGS cases suggest that they may be important initiating and metastatic events in melanoma. Examples include the amplification of *CCND1* on chromosome 11 in two patients, and *CDKN2A* deletions, which were observed in 5 of 13 patients, confirming previous data [40,41]. Our *in silico* significance and proximity analysis of 28 paired cases (13 WGS and 15 targeted cases) identified known (e.g., *BRAF*, *NRAS*, *CDKN2A*, and *GRIN2A*) and novel (e.g., *EPHA3*, *GRIN2B*, *and ASXL3*) genes involved in melanoma (**Table 1 and Figure 2**). Moreover, our extension analysis of 96 patients using targeted sequencing of 1,209 genes, revealed a number of truncation mutations in *TP53*, *RB1*, *PTEN*, *PTPRD*, *ARID2*, *ASXL3*, and other genes co-occuring with *BRAF* and *NRAS* mutations, suggesting their cooperating roles during the progression of melanoma. These alterations in other pathways, such as the PI3K pathway, may partially explain the resistance of *BRAF*-mutant melanoma to RAF inhibitors.

Previous melanoma whole genome [11,19,42] and whole exome [18,43,44] studies have uncovered a complex landscape of melanoma genomes including high mutation rate, a complex copy number landscape, predominance of UV-related C>T transitions, and frequent genetic alterations in well-known drivers of melanomagenesis such as *BRAF*, *NRAS*, *TP53*, *CDKN2A*, and *PTEN*, which was corroborated in our study.

As described, all 13 whole genome analyzed cases harbored mutations or copy number alterations in either *NRAS* or *BRAF* (with 1 tumor having both). Their presence largely in the founding clone, as determined by VAF, is evidence of the contribution of these genomic changes to melanoma initiation. Interestingly, WGS of these tumors identified the mutations, copy number alterations, and structural variants that occurred concomitantly with these previously identified "driver" mutations. These findings provide a mutational profile of melanoma metastasis against the background of an initiating oncogenic event and point to genomic changes that may be implicated in the loss of senescence and oncogenic transformation in melanoma. In addition, the predominance of C->T transitions provides impressive evidence of UV-related DNA damage in WGS metastatic tumors and suggests that many of the potential driver SNVs were present long before the spread of metastatic disease occurred. The sheer number of SNVs in these tumors is striking, with one tumor possessing over 6,000 tier 1 mutations (~180 mutations per Mbp), and the average of greater than 1,300 tier 1 SNVs per tumor constitutes a large number by any metric.

In this study, we focused on subclone structures and intra-tumor heterogeneity of melanoma, which was not addressed in previous studies. Our results for 15 metastatic melanoma tumors demon-

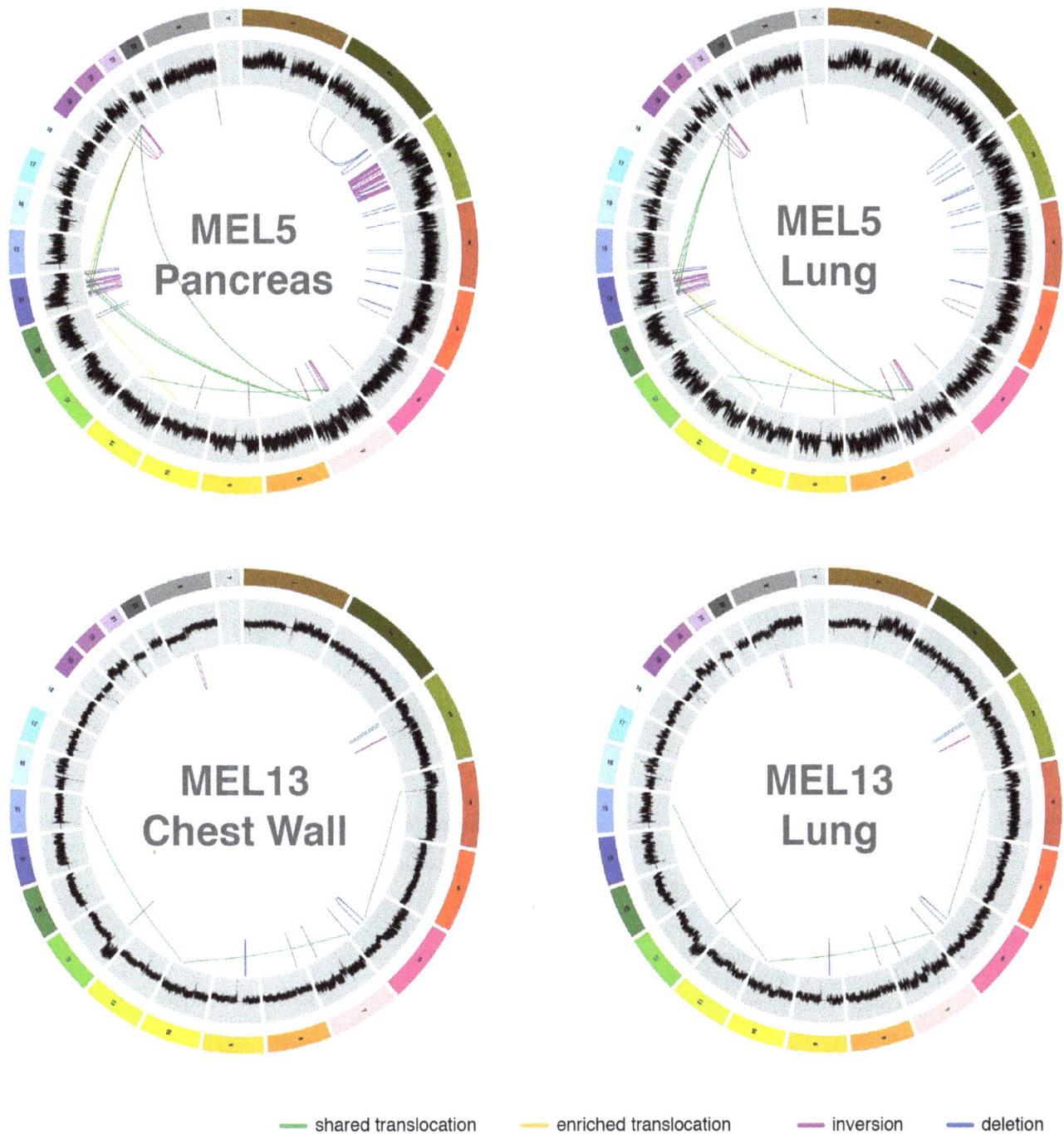

Figure 5. Comparison of Circos plots of the metastatic samples from two tissue sites of the same individuals (MEL5 and MEL13). In MEL5, pancreas tumor specific structural variants (inversions and deletions) are clustered on chromosomes 2 and 5, and pancreas or lung enriched rearrangements are drawn in yellow. In MEL13, highly similar copy number and structural variant patterns between lung and chest wall metastases are shown. No purity-based copy number corrections were used for plotting copy number.

strate that melanoma is mainly a multi-clonal disease which harbors diverse numbers of clonal populations with various frequencies and densities. In contrast to point mutations, which are likely the result of UV damage, the rate of structural variation in this study was similar to that previously described for most malignancies. Whole genome sequencing data herein indicate that complex rearrangements may generate important gain- and loss-of-function driver events in melanoma oncogenesis. Moreover,

many rearrangements may occur preferentially in genes that are spatially localized together within transcriptional or chromatin compartments, perhaps initiated by DNA strand breaks and erroneous repair.

Subclone-specific mutation signature analysis in a hypermutator sample (MEL9) revealed that the founding clone and the subclone displayed a distinct mutation context, suggesting different mutational mechanisms for two subclones from the same tumor. The

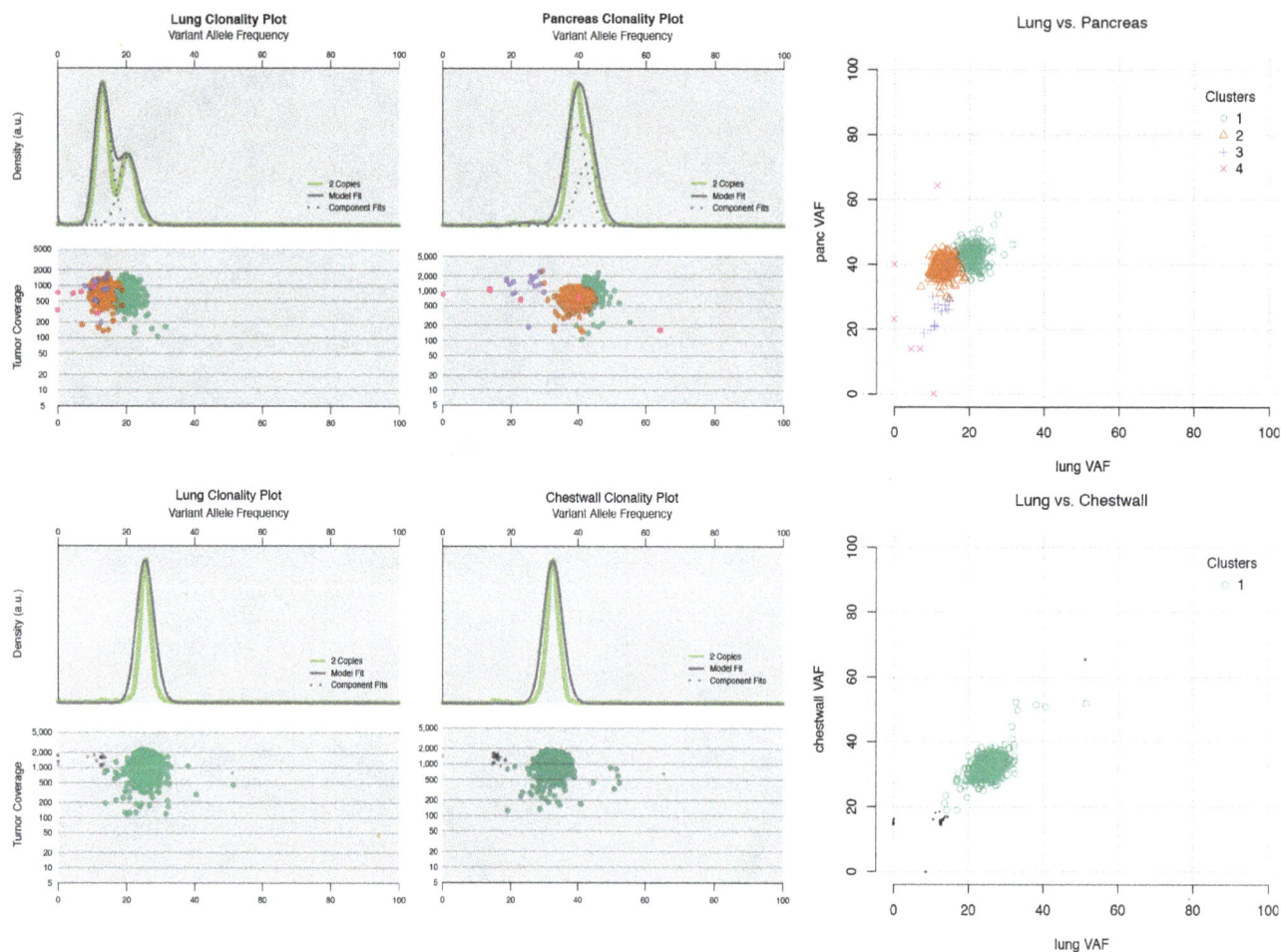

Figure 6. Comparison of clonality patterns of metastatic samples from two tissue sites of the same individuals (MEL5 and MEL13). Kernel density and variant allele frequency by sequence depth plots for each metastasis in MEL5 and MEL13. Data shown are from chromosomes 17, 18, and 21 for MEL5, and from chromosomes 3, 7, and 14 for MEL13. The plots indicate multiple clones in MEL5 with enrichment from lung to pancreas in MEL5, and nearly identical clonal pattern in both metastases in MEL13.

vast majority of mutations and re-arrangements were shared between the two metastatic samples for both cases with metachronous tumors, suggesting they are derived from the same subclone(s) of the primary tumor. A significant number of pancreas-specific inversions were identified in MEL5, though their potential role in the progression of the disease is unclear.

Finally, in studying the clonal architecture of all WGS cases, we confirmed that clonal heterogeneity is a common phenomenon in melanoma and driver initiation and progression mutations are all required for the development of melanoma. This suggests that knowing the clonal architecture of each patient's tumor will be essential for understanding the evolutionary history of each tumor and for formulating optimized treatment options. Importantly, our phylogenetic study of two patients each with 4 different synchronous metastases clearly revealed the complex relationships among these tumors derived from the originating primary tumor and suggest that melanoma therapy is a moving target and constant monitoring of tumor genomes may be required to develop an effective treatment plan as evidenced by the differential response to Vemurafenib of 4 synchronous metastases from MEL174.

Methods

Illumina library construction and sequencing

The procedure described by Mardis *et al* [45] was followed for library construction and sequencing. Briefly, Illumina DNA sequencing was used to generate between 117 and 286 million base pairs of sequence data for each of the 15 metastatic tumors and 13 matched normal samples, with haploid coverages ranging from 29.51 to 63.22 (Table S1 in File S1). Comparison of heterozygous SNPs detected in the whole genome sequencing (WGS) data with SNPs array genotypes confirmed bi-allelic detection of between 98.7 and 99.7% of the heterozygous array SNPs in the 13 cases. Detailed coverage statistics for all cases are included in Table S1 in File S1.

Mutation detection pipeline

For each sample, reads were aligned using BWA 0.5.9 (http://sourceforge.net/projects/bio-bwa/) on a per-lane basis, merged into a single bam file, and duplicate reads were removed using Picard 1.29 (http://picard.sourceforge.net). Sample variants were called using Samtools (svn rev 599) [23]. Somatic single nucleotide variants were detected using SomaticSniper [25]. High quality somatic predictions were defined as those sites with a SomaticSni-

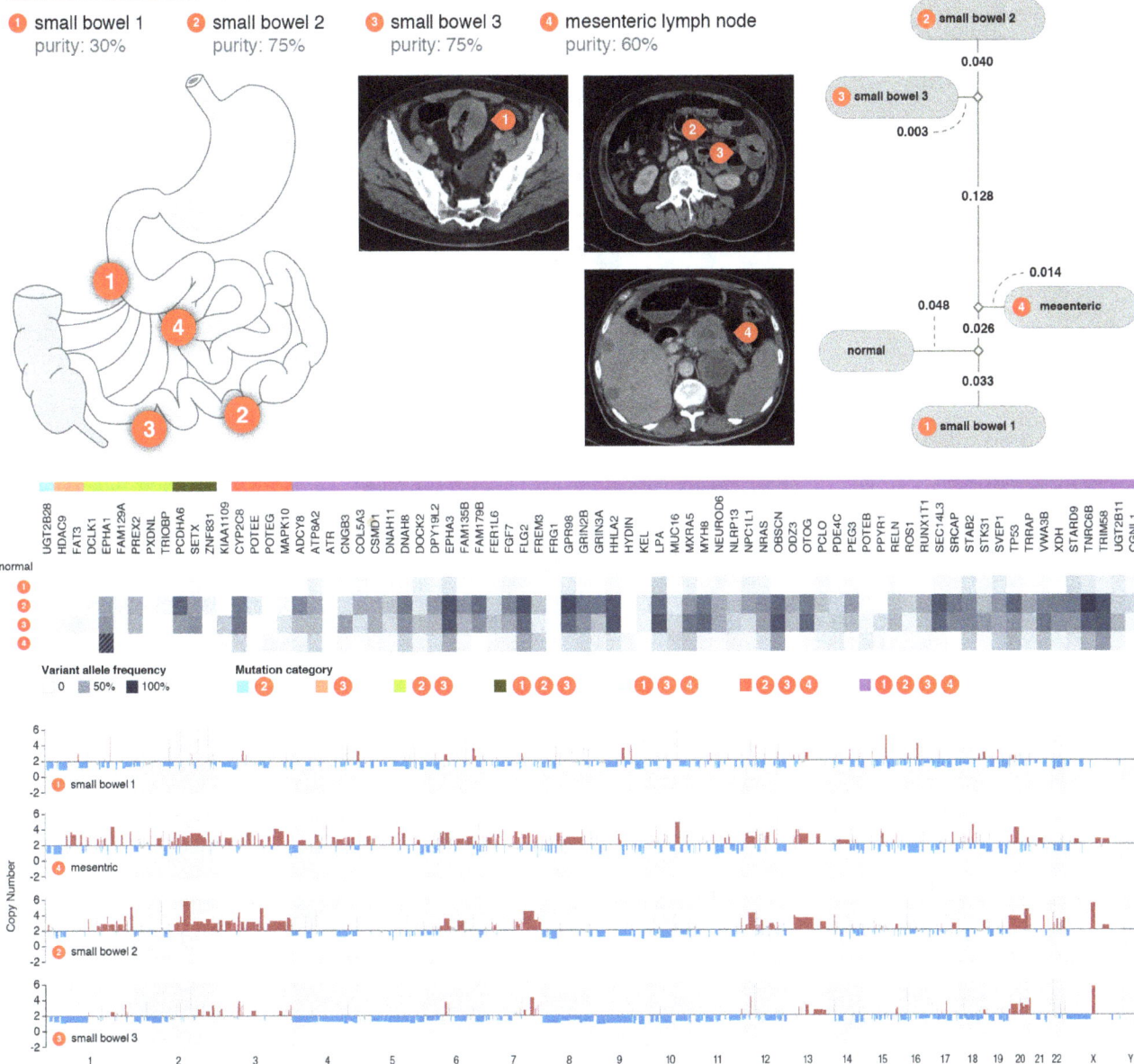

Figure 7. Phylogenetic and mutational relationships among four metastatic samples from different sites of the same individuals (MEL 167). Geographic locations and CT scans of metastasis samples in MEL167 with three tumor samples from small bowel (mass 1, mass 2, and mass 3) and one from mensenteric lymph node. Phylogenetic relationships, mutation patterns, and copy number landscape in all four tumors were shown. Purity based VAF corrections were applied prior to phylogenetic analysis.

per somatic score greater than 40 and an average mapping quality greater than 40. Indels in all samples were called using a combination of Pindel [22] and GATK [21]. Somatic variants were grouped into tiers based on genome annotation as described previously [45]. Y chromosome variants were filtered for all female patients.

Structural variant detection

Structural variants (SVs) in all samples were predicted by BreakDancer [24] and SquareDancer (https://github.com/genome/gms-core/blob/master/lib/perl/Genome/Model/Tools/Sv/SquareDancer.pl). All SV predictions were filtered using TIGRA (Chen et al., in preparation) to identify assembled breakpoints in SV flanking reads. The same procedure as described in Ding et al. [26] for selecting somatic SVs was used.

6491 structural variants from the 13 WGS cases were sent for capture validation

Validation of structural variants

All BWA-aligned capture reads and their mates that map within 1000 bp of the structural variant breakpoints were realigned by CrossMatch (version 1.080721) to the assembled SV contigs and to the reference. The threshold for an acceptable alignment is < = 1 unaligned base at either end, < = 1% substitutions, < = 1% indels

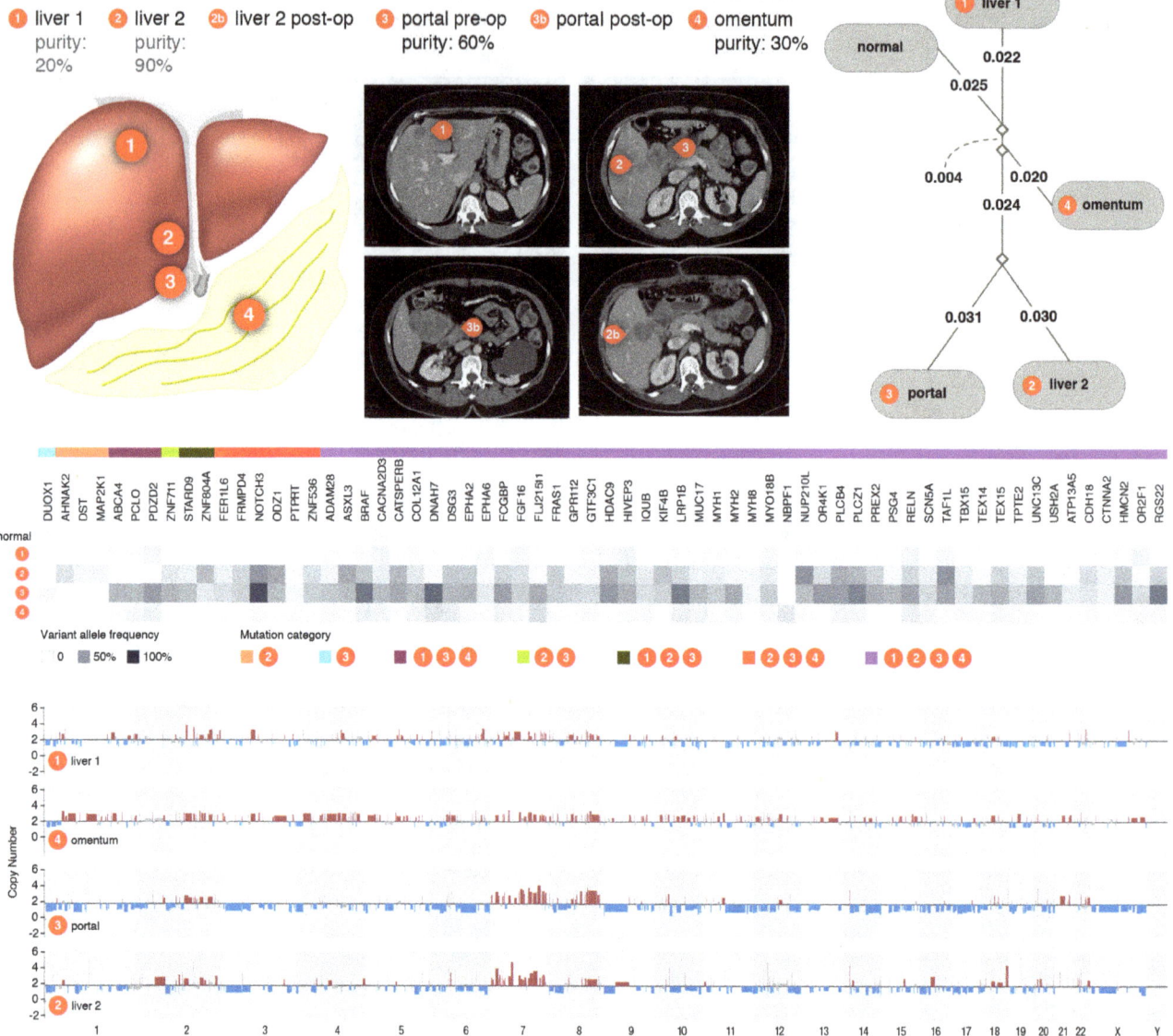

Figure 8. Phylogenetic and mutational relationships among four metastatic samples from different sites of the same individuals (MEL 174). Geographic locations and CT scans of metastasis samples in MEL 174 with 2 samples from liver (liver tumor 1, liver tumor 2), one from omentum, and one from portal nodule. Phylogenetic relationships among 4 metastatic tumors were shown. Purity based VAF corrections were applied prior to phylogenetic analysis. Increased variant allele frequency of BRAF (V600E) in liver 2 and portal module, both tumors showed resistance to vemurafinib treatment. BRAF amplifications were also shown in liver 2 and portal module tumors.

and a CrossMatch score $>= 50$. An SV-supporting read is required to span the breakpoint on the SV contig, align to 10 bases of flanking on each side of the breakpoint, and have no alignment to the reference above minimum alignment criteria. SV-supporting reads were tabulated in the tumor and normal sample separately, and a Fisher's Exact test was applied to these counts to determine the somatic status of each variant. The same method for determining SV-supporting reads was applied to the WGS alignment data for those calls deemed somatic via all other criteria. Variants with any SV-supporting reads in the normal WGS sample were filtered out as potential germline variants or alignment artifacts. An additional filter was put in place to filter ALU sequences and the remaining high confidence SV events

were manually reviewed based on BWA mapping of supporting capture validation data to the assembled SV contigs spanning the breakpoint. 411 structural variants from the 13 WGS cases passed the final manual review and filtering.

Kernel density analysis for identifying clusters and estimating allele frequencies for each tumor

Tumor clonality estimates were determined using the mutation allele frequencies from sites with deep coverage from capture validation data. To minimize the effect of coverage on allele frequency estimations, only mutations with $>100x$ coverage in both the normal and tumor validation data were included in this analysis. Efforts were made to exclude somatic SNVs from regions containing copy number alterations identified in WGS data. Varscan 2 was utilized on whole-genome sequencing data to eliminate all LOH SNV calls. For each chromosome, the variant allele frequencies were plotted from both the tumor and normal samples at sites where the normal sample's variant allele frequency fell between 40% and 60%. Chromosomes where variants frequently exhibited highly variable variant allele frequencies in the tumor sample were excluded from the clonality analysis. Thus only a few diploid chromosomes were chosen to represent each sample for this analysis. The remaining SNVs were further segregated according to their segmented copy number states as predicted by cnvHMM (states of copy number equal to 2, 3, or 4), and each copy number state was analyzed individually. For each copy number state in the tumor, a kernel density estimate (KDE) plot was drawn for tumor variant allele frequencies using the density function in R. A customized R function evaluated each KDE plot to determine the number of significant peaks in variants existing in the copy number neutral, or diploid regions. These clusters thus identified served as an estimation of the number and relative composition of clones and subclones present in each tumor. Only copy number neutral data is presented in Figure 6.

Significantly mutated gene analysis

We used the SMG test in the MuSiC suite[32], using the particular options available for accommodating the large numbers of somatic mutations discovered in some melanoma cases. The SMG test identifies genes that have significantly higher somatic and germline mutation rates than background. To account for hypermutated samples, the ability to sub-group cases based upon similarities in their overall mutation counts was utilized, and P-values were calculated for mutated genes in each sub-group independently. All P-values were combined using the same methods as described in MuSiC suite[32]. For the purposes of this analysis, MEL4, MEL6, and MEL9 was considered as a hypermutant, with all other samples being placed in a separate sub-group for the SMG analysis. For the analysis of significantly mutated genes, genes not typically expressed in melanoma tumor samples were filtered if they had an average RPKM≤0.5. For the RNA-seq based gene expression analysis, we used the Pancan12 per-sample log2-RSEM matrix from doi:10.7303/syn1734155.1. A gene qualified as expressed if it had at least 3 reads in at least 70% of samples. For every gene, the average per-sample RNA-seq by Expectation Maximization (RSEM) value was calculated across TCGA SKCM samples from the same tumor-type.

Proximity analysis

Validated somatic mutations were identified that clustered within specific protein regions across multiple individuals. This was accomplished by querying the distance between amino acids for every pair of mutations on a given transcript of a mutated gene within the sample set, and then determined which mutations fell within "close" proximity, where "close" was defined to be within a limit of 10 amino acids. We used the MuSiC suite[32] for the proximity analysis.

Supporting Information

File S1 Supplementary Materials and Methods, along with Supplementary Figures S1 to S4 and Supplementary Tables S1 to S3, S7, S13 and S15 and references cited in the Supplementary materials and method section. Figure S1 in File S1: Copy number patterns and structural variants identified in all 15 tumors sequenced. **Figure S2 in File S1**: Extension heatmap of recurrent *BRAF* and *NRAS* mutations

as well as truncation mutations in selected tumors suppressor, protein phosphatase and chromatin remodeling genes. The mutual exclusion and co-occurrence of common recurrent drivers and truncation mutations are shown in the 96 extension cases. Three melanoma cases harbored both *BRAF* and *NRAS* mutations. **Figure S3 in File S1**: Density plots of variants to estimate the purity of the four metastases for each patient MEL167 and MEL174. Both the density plots and the Scatter plots were used for purity estimation of the metastases samples. **Figure S4 in File S1**: Flowchart showing the detailed overview of the analysis steps and the pipeline used for the analysis. **Table S1 in File S1**: WGS haploid coverage and SNP array concordance. Haploid and diploid coverage estimates are given for 15 whole-genome sequenced samples. Haploid coverage is calculated as the amount of non-redundant mapped read bases divided by the haploid size of the human genome. Diploid coverage is estimated from the fraction of heterozygous SNPs from high-density SNP array data that were present in SAMtools raw (unfiltered) or filtered SNP calls. **Table S2 in File S1**: Capture validation coverage. Custom capture validation coverage of putative somatic mutations is reported for the 13 cases in which such data were generated. Shown are the fraction of bases targeted that were covered >1x, >10x, and >20x in each sample. **Table S3 in File S1**: Tier 1–3 somatic SNVs predicted and validation rate. Numbers of validated somatic SNVs in tiers 1, 2, and 3 are shown for the 13 cases having both whole genome sequence data and custom capture validation. **Table S7 in File S1**: Dinucleotide polymorphisms (DNP) in 13 WGS cases. **Table S13 in File S1**: Purity estimation of the multi-metastases samples MEL167 and MEL174 using the density plots. **Table S15 in FileS1**: Comparison of the number of Tier1 SNVs in the TCGA melanoma dataset to the number of SNVs in the WGS dataset.

File S2 The Zipped file contains the Supplementary Tables S4, S5, S6, S8, S9, S10, S11, S12, S14. Table S4 in File S2: Validated somatic point mutations and indels: See separate.xlsx file. **Table S5 in File S2**: Validated somatic structural variants: See separate.xlsx file. Validated structural variants in 15 whole genome sequenced samples are listed with Patient ID, chromosomal positions of each breakpoint (A, B), the type of event (DEL deletion, CTX translocation, INV inversion, INS insertion, ITX tandem duplication), and the size of event in bp. If a breakpoint is within a gene, the gene and transcript name are given with the direction of transcription and transcript substructure where the breakpoint is found (intron numbering is relative to the first translat ed exon). Genes completely deleted are listed in the final column. **Table S6 in File S2**: (a) Point mutations, indels, structural variations, and copy number variations presented in Figure 1. (b) TERT promoter mutations identified in 13 WGS cases: See separate.xlsx file. **Table S8 in File S2**: (a)Average haploid coverage across each targeted gene in the extension experiment. (b) Extension discovery variant table (15 paired samples): See separate.xlsx file. **Table S9 in File S2**: (a) Genes in the MAPK and Cell Cycle TP53/RB pathways. (b) Nonsynonymous mutations in the MAPK pathway in 28 paired discovery cases (c) Nonsynonymous mutations in the Cell Cycle TP53/RB pathway in 28 paired discovery cases: See separate.xlsx file. **Table S10 in File S2**: Extension analysis filtered variant table (96 unpaired samples): See separate.xlsx file. **Table S11 in File S2**: Extension analysis truncation mutation table (96 unpaired samples): See separate.xlsx file. **Table S12 in File S2**: (a)Readcounts for point mutations pictured in Figure 5. (b)Readcounts for point mutations pictured in Figure 6, MEL5 Lung sample. (c) Readcounts for point mutations pictured in Figure 6,

MEL5 Pancreas sample. (d) Readcounts for point mutations pictured in Figure 6, MEL13 Lung sample. (e)Readcounts for point mutations pictured in Figure 6, MEL13 Chest Wall sample.(f) Readcounts for point mutations pictured in Figure 6, MEL5 sample.(g) Readcounts for point mutations pictured in Figure 6, MEL13 sample: See separate.xlsx file. **Table S14 in File S2**: Nonsynonymous mutations in the four metastases samples of MEL167 along with the readcounts and annotation: See separate.xlsx file.

Author Contributions

Conceived and designed the experiments: LD RCF TJL JJM RKW JSW. Performed the experiments: MK DF HS BG. Analyzed the data: LD MK NDD KLK CL MG DF HS CAM MCW OG RCF. Wrote the paper: LD JSW. Performed statistical analysis: MCW. Prepared figures and tables: NDD KLK CL MG CAM OG RCF LD JFM. Provided samples: BG LAC GPL VKS RCF JJM JSW. Supervised research: LD RKW JSW.

References

1. Siegel R, Naishadham D, Jemal A (2013) Cancer statistics, 2013. CA: a cancer journal for clinicians 63: 11–30.
2. Chapman PB, Hauschild A, Robert C, Haanen JB, Ascierto P, et al. (2011) Improved survival with vemurafenib in melanoma with BRAF V600E mutation. The New England journal of medicine 364: 2507–2516.
3. Hodi FS, O'Day SJ, McDermott DF, Weber RW, Sosman JA, et al. (2010) Improved survival with ipilimumab in patients with metastatic melanoma. The New England journal of medicine 363: 711–723.
4. Davies MA, Samuels Y (2010) Analysis of the genome to personalize therapy for melanoma. Oncogene 29: 5545–5555.
5. Curtin JA, Fridlyand J, Kageshita T, Patel HN, Busam KJ, et al. (2005) Distinct sets of genetic alterations in melanoma. The New England journal of medicine 353: 2135–2147.
6. Bauer J, Curtin JA, Pinkel D, Bastian BC (2007) Congenital melanocytic nevi frequently harbor NRAS mutations but no BRAF mutations. The Journal of investigative dermatology 127: 179–182.
7. Pollock PM, Harper UL, Hansen KS, Yudt LM, Stark M, et al. (2003) High frequency of BRAF mutations in nevi. Nature genetics 33: 19–20.
8. Michaloglou C, Vredeveld LC, Soengas MS, Denoyelle C, Kuilman T, et al. (2005) BRAFE600-associated senescence-like cell cycle arrest of human naevi. Nature 436: 720–724.
9. Wajapeyee N, Serra RW, Zhu X, Mahalingam M, Green MR (2008) Oncogenic BRAF induces senescence and apoptosis through pathways mediated by the secreted protein IGFBP7. Cell 132: 363–374.
10. Bennett DC (2003) Human melanocyte senescence and melanoma susceptibility genes. Oncogene 22: 3063–3069.
11. Pleasance ED, Cheetham RK, Stephens PJ, McBride DJ, Humphray SJ, et al. (2010) A comprehensive catalogue of somatic mutations from a human cancer genome. Nature 463: 191–196.
12. Hocker T, Tsao H (2007) Ultraviolet radiation and melanoma: a systematic review and analysis of reported sequence variants. Human mutation 28: 578–588.
13. Prickett TD, Agrawal NS, Wei X, Yates KE, Lin JC, et al. (2009) Analysis of the tyrosine kinome in melanoma reveals recurrent mutations in ERBB4. Nature genetics 41: 1127–1132.
14. Ceol CJ, Houvras Y, Jane-Valbuena J, Bilodeau S, Orlando DA, et al. (2011) The histone methyltransferase SETDB1 is recurrently amplified in melanoma and accelerates its onset. Nature 471: 513–517.
15. Stark MS, Woods SL, Gartside MG, Bonazzi VF, Dutton-Regester K, et al. (2011) Frequent somatic mutations in MAP3K5 and MAP3K9 in metastatic melanoma identified by exome sequencing. Nature genetics 44: 165–169.
16. Wei X, Walia V, Lin JC, Teer JK, Prickett TD, et al. (2011) Exome sequencing identifies GRIN2A as frequently mutated in melanoma. Nature genetics 43: 442–446.
17. Prickett TD, Wei X, Cardenas-Navia I, Teer JK, Lin JC, et al. (2011) Exon capture analysis of G protein-coupled receptors identifies activating mutations in GRM3 in melanoma. Nature genetics 43: 1119–1126.
18. Hodis E, Watson IR, Kryukov GV, Arold ST, Imielinski M, et al. (2012) A landscape of driver mutations in melanoma. Cell 150: 251–263.
19. Berger MF, Hodis E, Heffernan TP, Deribe YL, Lawrence MS, et al. (2012) Melanoma genome sequencing reveals frequent PREX2 mutations. Nature 485: 502–506.
20. Turajlic S, Furney SJ, Lambros MB, Mitsopoulos C, Kozarewa I, et al. (2012) Whole genome sequencing of matched primary and metastatic acral melanomas. Genome research 22: 196–207.
21. McKenna A, Hanna M, Banks E, Sivachenko A, Cibulskis K, et al. (2010) The Genome Analysis Toolkit: a MapReduce framework for analyzing next-generation DNA sequencing data. Genome research 20: 1297–1303.
22. Ye K, Schulz MH, Long Q, Apweiler R, Ning Z (2009) Pindel: a pattern growth approach to detect break points of large deletions and medium sized insertions from paired-end short reads. Bioinformatics 25: 2865–2871.
23. Li H, Handsaker B, Wysoker A, Fennell T, Ruan J, et al. (2009) The Sequence Alignment/Map format and SAMtools. Bioinformatics 25: 2078–2079.
24. Chen K, Wallis JW, McLellan MD, Larson DE, Kalicki JM, et al. (2009) BreakDancer: an algorithm for high-resolution mapping of genomic structural variation. Nat Methods 6: 677–681.
25. Larson DE, Harris CC, Chen K, Koboldt DC, Abbott TE, et al. (2011) SomaticSniper: Identification of Somatic Point Mutations in Whole Genome Sequencing Data. Bioinformatics.
26. Ding L, Ley TJ, Larson DE, Miller CA, Koboldt DC, et al. (2012) Clonal evolution in relapsed acute myeloid leukaemia revealed by whole-genome sequencing. Nature 481: 506–510.
27. Shah SP, Morin RD, Khattra J, Prentice L, Pugh T, et al. (2009) Mutational evolution in a lobular breast tumour profiled at single nucleotide resolution. Nature 461: 809–813.
28. Ding L, Ellis MJ, Li S, Larson DE, Chen K, et al. (2010) Genome remodelling in a basal-like breast cancer metastasis and xenograft. Nature 464: 999–1005.
29. Huang FW, Hodis E, Xu MJ, Kryukov GV, Chin L, et al. (2013) Highly recurrent TERT promoter mutations in human melanoma. Science 339: 957–959.
30. Horn S, Figl A, Rachakonda PS, Fischer C, Sucker A, et al. (2013) TERT promoter mutations in familial and sporadic melanoma. Science 339: 959–961.
31. Killela PJ, Reitman ZJ, Jiao Y, Bettegowda C, Agrawal N, et al. (2013) TERT promoter mutations occur frequently in gliomas and a subset of tumors derived from cells with low rates of self-renewal. Proceedings of the National Academy of Sciences of the United States of America 110: 6021–6026.
32. Dees ND, Zhang Q, Kandoth C, Wendl MC, Schierding W, et al. (2012) MuSiC: Identifying mutational significance in cancer genomes. Genome research 22: 1589–1598.
33. Ding L, Getz G, Wheeler DA, Mardis ER, McLellan MD, et al. (2008) Somatic mutations affect key pathways in lung adenocarcinoma. Nature 455: 1069–1075.
34. Brennan CW, Verhaak RG, McKenna A, Campos B, Noushmehr H, et al. (2013) The somatic genomic landscape of glioblastoma. Cell 155: 462–477.
35. Gray-Schopfer V, Wellbrock C, Marais R (2007) Melanoma biology and new targeted therapy. Nature 445: 851–857.
36. Smalley KS (2010) Understanding melanoma signaling networks as the basis for molecular targeted therapy. The Journal of investigative dermatology 130: 28–37.
37. Forbes SA, Bindal N, Bamford S, Cole C, Kok CY, et al. (2011) COSMIC: mining complete cancer genomes in the Catalogue of Somatic Mutations in Cancer. Nucleic acids research 39: D945–950.
38. Koboldt DC, Zhang Q, Larson DE, Shen D, McLellan MD, et al. (2012) VarScan 2: Somatic mutation and copy number alteration discovery in cancer by exome sequencing. Genome research.
39. Shi H, Moriceau G, Kong X, Lee MK, Lee H, et al. (2012) Melanoma whole-exome sequencing identifies (V600E)B-RAF amplification-mediated acquired B-RAF inhibitor resistance. Nat Commun 3: 724.
40. Grafstrom E, Egyhazi S, Ringborg U, Hansson J, Platz A (2005) Biallelic deletions in INK4 in cutaneous melanoma are common and associated with decreased survival. Clinical cancer research: an official journal of the American Association for Cancer Research 11: 2991–2997.
41. Kamb A, Shattuck-Eidens D, Eeles R, Liu Q, Gruis NA, et al. (1994) Analysis of the p16 gene (CDKN2) as a candidate for the chromosome 9p melanoma susceptibility locus. Nature genetics 8: 23–26.
42. Turajlic S, Furney SJ, Lambros MB, Mitsopoulos C, Kozarewa I, et al. (2012) Whole genome sequencing of matched primary and metastatic acral melanomas. Genome Res 22: 196–207.
43. Wei X, Walia V, Lin JC, Teer JK, Prickett TD, et al. (2011) Exome sequencing identifies GRIN2A as frequently mutated in melanoma. Nat Genet 43: 442–446.
44. Prickett TD, Wei X, Cardenas-Navia I, Teer JK, Lin JC, et al. Exon capture analysis of G protein-coupled receptors identifies activating mutations in GRM3 in melanoma. Nat Genet 43: 1119–1126.
45. Mardis ER, Ding L, Dooling DJ, Larson DE, McLellan MD, et al. (2009) Recurring mutations found by sequencing an acute myeloid leukemia genome. N Engl J Med 361: 1058–1066.

11

Novel Roles for P53 in the Genesis and Targeting of Tetraploid Cancer Cells

Batzaya Davaadelger, Hong Shen, Carl G. Maki*

Department of Anatomy and Cell Biology, Rush University Medical Center, Chicago, Illinois, United States of America

Abstract

Tetraploid (4N) cells are considered important in cancer because they can display increased tumorigenicity, resistance to conventional therapies, and are believed to be precursors to whole chromosome aneuploidy. It is therefore important to determine how tetraploid cancer cells arise, and how to target them. P53 is a tumor suppressor protein and key regulator of tetraploidy. As part of the "tetraploidy checkpoint", p53 inhibits tetraploid cell proliferation by promoting a G1-arrest in incipient tetraploid cells (referred to as a tetraploid G1 arrest). Nutlin-3a is a preclinical drug that stabilizes p53 by blocking the interaction between p53 and MDM2. In the current study, Nutlin-3a promoted a p53-dependent tetraploid G1 arrest in two diploid clones of the HCT116 colon cancer cell line. Both clones underwent endoreduplication after Nutlin removal, giving rise to stable tetraploid clones that showed increased resistance to ionizing radiation (IR) and cisplatin (CP)-induced apoptosis compared to their diploid precursors. These findings demonstrate that transient p53 activation by Nutlin can promote tetraploid cell formation from diploid precursors, and the resulting tetraploid cells are therapy (IR/CP) resistant. Importantly, the tetraploid clones selected after Nutlin treatment expressed approximately twice as much *P53* and *MDM2* mRNA as diploid precursors, expressed approximately twice as many p53-MDM2 protein complexes (by co-immunoprecipitation), and were more susceptible to p53-dependent apoptosis and growth arrest induced by Nutlin. Based on these findings, we propose that p53 plays novel roles in both the formation and targeting of tetraploid cells. Specifically, we propose that 1) transient p53 activation can promote a tetraploid-G1 arrest and, as a result, may inadvertently promote formation of therapy-resistant tetraploid cells, and 2) therapy-resistant tetraploid cells, by virtue of having higher *P53* gene copy number and expressing twice as many p53-MDM2 complexes, are more sensitive to apoptosis and/or growth arrest by anti-cancer MDM2 antagonists (e.g. Nutlin).

Editor: Andrei L. Gartel, University of Illinois at Chicago, United States of America

Funding: This work was supported by National Institutes of Health Grant 1RO1CA137598-01A1 from the National Cancer Institute (NCI) (to C.G.M.). The funders had no role in study design, data collection and analysis, decision to publish, or preparation of the manuscript.

* Email: Carl_Maki@rush.edu

Introduction

Tetraploid cells contain twice the normal amount of DNA and are rare in most normal tissues. However, tetraploid cells are relatively common in cancer and are thought to contribute to tumor development, aneuploidy, and therapy resistance [1]. Direct evidence for the tumorigenic potential of tetraploid cells was provided by Fujiwara et al. [2] who isolated binucleated, tetraploid mammary epithelial cells from p53-null mice. Remarkably, these cells were more susceptible to carcinogen-induced transformation (soft-agar growth) than diploid counterparts, and the tetraploid cells formed tumors in nude mice while diploid cells did not. Other studies have linked tetraploidy to radiation and chemotherapy resistance. For example, Castedo et al. [3,4] isolated tetraploid and diploid clones from two human cancer cell lines with wild-type p53. Importantly, tetraploid clones were resistant to radiation and multiple chemotherapy agents compared to diploid counterparts. Finally, there is mounting evidence that aneuploid cancer cells are generated from either asymmetric division or progressive chromosomal loss from tetraploid precursors. Early evidence for this came from studies in premalignant

Barrett's esophagus. In these studies, the appearance of tetraploid cells correlated with p53 loss and preceded gross aneuploidy and carcinogenesis [5,6]. In sum, tetraploid cells can have higher tumorigenic potential, be therapy and radiation-resistant, and be precursors to cancer aneuploidy. It is therefore important to identify how tetraploid cells arise and how they can be targeted for cancer treatment.

P53 is a tumor suppressor and important regulator of tetraploidy [7]. p53 is kept at low levels by MDM2, an E3-ligase that binds p53 and promotes its degradation [8,9]. DNA damage and other stresses disrupt p53-MDM2 binding, causing p53 levels to increase. Increased p53 stops proliferation by inducing expression of genes that promote G1-arrest (*P21*) or apoptosis (*PUMA, NOXA, BAX*) [10]. Evidence that p53 functions in a "tetraploidy checkpoint" comes largely from studies using microtubule inhibitors (MTIs) that block cells in metaphase. Cells arrested in metaphase by prolonged MTI exposure can eventually exit mitosis and enter a pseudo-G1 state, referred to as tetraploid G1 [11,12]. Endoreduplication refers to the case in which these tetraploid cells enter S-phase and replicate their DNA, giving rise to high ploidy cells with increasing duplications of the genome

(8N, 16N, 32N, etc). Cells lacking p53/p21 are more prone to MTI-induced endoreduplication than wild-type cells [11–14]. This supports a p53–p21 dependent tetraploidy checkpoint that blocks S-phase entry by 4N cells.

Tetraploid G1-arrest is typically characterized by depletion of G2/M proteins (e.g. Cyclins A/B, CDC2) and increased expression of G1 arrest proteins (e.g. P53, p21) in 4N cells [15–19]. DNA damaging agents e.g. ionizing radiation (IR) activate p53 and arrest cells in G1 and G2-phases. Two early reports found IR caused p53 and p21-dependent depletion of CDC2, Cyclin A, and Cyclin B mRNA and protein [20,21]. In one case, depletion of these G2/M markers coincided with endoreduplication 1–6 days after treatment [21]. These findings suggested that p53–p21 activation in IR-treated cells could promote a tetraploid G1 arrest followed subsequently by endoreduplication. Nutlin-3a (Nutlin) is a preclinical drug that stabilizes p53 by blocking p53-MDM2 binding. We reported that Nutlin could promote a tetraploid G1 arrest in multiple p53 wild-type cell lines, characterized by depletion of G2/M proteins and increased expression of p53 and p21 in 4N cells [18,19]. Upon Nutlin removal, p53 and p21 decreased and, in some cases, cells underwent endoreduplication [18]. p53 and p21 were required for both the tetraploid G1 arrest and for endoreduplication after Nutlin removal. Importantly, stable tetraploid clones could be isolated from Nutlin treated cells, and these tetraploid clones were resistant to IR and cisplatin (CP)-induced apoptosis [19]. These studies suggest p53–p21 activated by Nutlin can promote a tetraploid G1-arrest and, when p53 levels decrease (Nutlin removal) the cells may replicate their DNA (endoreduplication) and give rise to tetraploid cells that are IR and CP-resistant. However, a caveat of these previous studies is that they were done in cancer cell lines, which may include a mixture of diploid and tetraploid cells, some of which may be inherently IR/CP-resistant. It is possible that we had isolated IR/CP-resistant tetraploid clones that were already present in the population, or that the tetraploid clones we isolated were derived from diploid cells that were already IR/CP-resistant. Thus, it remained unclear whether transient p53 activation by Nutlin could convert diploid cells into tetraploid cells and, if yes, whether the resulting tetraploid clones would display increased resistance to therapy-induced apoptosis. The current study was initiated to address these questions, and to test the possibility that tetraploid cancer cells may be especially sensitive to MDM2 antagonists.

Materials and Methods

Cell Lines and Culture Conditions

p53 wild-type and p53-null HCT116 cells (described in [11] were obtained from Dr. Bert Vogelstein (John Hopkins University). D3 and D8 have been described [22] and are diploid clones that were isolated from p53 wild-type HCT116 cells by limiting dilution. The cells were grown in McCoy's 5A medium with 10% fetal bovine serum (FBS), penicillin (100 U/mL) and streptomycin (100 µg/mL). Cells were plated 24 hours before being treated with Nutlin-3 (10 µmol/L; Sigma), irradiation (10 Gy), Cisplatin (Bedford Laboratory), or as indicated.

Immunoblotting

Whole cell extracts were prepared by resuspending cell pellets in lysis buffer (150 mM NaCl, 5 mM EDTA, 0.5% Nonidet P-40, 50 mM Tris, pH 7.5), resolved by SDS-PAGE, and transferred to polyvinylidene difluoride membranes (NEN Life Science Products). Antibodies to p21 (H-164), Cyclin A (H-432), Geminin (FL-309), Cdk2 (M2) and p53 (AB-6) were from Santa Cruz Biotechnology; antibodies to Cyclin B1 (V152), CDC2 (POH1),

and Tyr-15 Phospho-CDC2 were from Cell Signaling; antibodies to MDM2 (2A10) were from Calbiochem. Antibody to Cyclin E (HE-2) was from BD Pharmingen. Primary antibodies were detected with goat anti-mouse (Pierce) or goat anti-rabbit (Life Technologies) secondary antibodies conjugated to horseradish peroxidase, using Clarity chemiluminescence (Bio-Rad).

Flow Cytometry Analysis and Live Cell Sorting

For cell cycle analysis, cells were harvested and fixed in 25% ethanol overnight. The cells were then stained with propidium iodide (25 µg/ml; Calbiochem). Flow cytometry analysis was performed on Gallios flow cytometer (Beckman Coulter) and analyzed with CellQuest (Becton Dickinson) and FlowJo 8.7 (Treestar, Inc). For each sample, 10,000 events were collected. For live cell sorting, cells were incubated with Hoechst 33342 (2 µmol/liter; Invitrogen) for 90 min at 37°C and then harvested. Cell sorting was performed based on DNA content using a MoFlo cytometer equipped with a UV excitation laser. The sorted cells were plated at low density.

MRNA analysis

Quantitative real-time PCR (qRT-PCR) was performed to measure mRNA levels for Cyclin A2, Cyclin B1, CDC2, P21, Bax, Noxa, PUMA, P53R2, Actin in cells that were untreated or treated with Nutlin, Cisplatin, or IR as indicated. The quantitative real-time PCR reaction was run in a 7300 Real Time PCR System (Applied Biosystems, Foster, CA) using SybrGreen PCR master mix, (Applied Biosystems, Foster City, CA) following manufacturer's instructions. Thermocycling was done in a final volume of 20 µL containing 2 µL of cDNA and 400 nmol/L of primers (Primers are listed in Table S1). All samples were amplified in triplicate using the following cycle scheme: 95°C for 2 minutes, 40 cycles of 95°C for 15 seconds and 55°C for 60 seconds. Fluorescence was measured in every cycle and mRNA levels were normalized using the Actin values in all samples. A single peak was obtained for targets, supporting the specificity of the reaction.

Metaphase Spreads

Cell were incubated with Colcemid (50 ng/µL) (Sigma-Aldrich) for an hour at 37°C, harvested and washed with PBS and hypotonic solution (2 parts of 0.075 M KCl and 1 part of 0.7% NaCitrate). The cells were then fixed in fixative (3 part methanol and 1 part acetic acid) and spread onto a slide. The spreads were examined under a phase contrast microscope.

FISH Analysis

Cells were harvested and washed with PBS. Then incubated with hypotonic solution (2 parts of 0.075 M KCl and 1 part of 0.7% NaCitrate) for 20mins at 37°C. The cells were then fixed in fixative (3 part methanol and 1 part acetic acid). The slides were prepared and aged at 55°C for 3 mins on a thermobrite. The slides were incubated with 0.05% Pepsin (Fisher #P53-100) for 16 mins at 37°C and then washed with PBS and dehydrated in 70%, 85% and 100% ethanol for 1 minute each at room temperature. Then the probe Vysis LSI TP53 SpectrumOrange/CEP 17 Spectrum-Green (Abbott Molecular Inc # 01N17-020) was added hybridize at 73°C for 5 min and 37 degree for 18 hours. Then the slides were washed with SSC buffer (Gibco #15557-044) with 0.5% NP-40 (Fisher #P53-100) for 2 mins at 73°C and counterstained with DAPI II (Abbott Molecular Inc # 30-804818). Slides were examined under fluorescent microscope.

Clonogenic Assay. Cells were plated 24 hours before being untreated or treated with CP, IR, or Nutlin in appropriate dilutions to form 50–100 colonies. After 2–3 weeks colonies were fixed with formaldehyde and stained with 0.05% crystal violet. The colonies were counted and normalized with the plating efficiency of untreated cells.

Results

Nutlin promotes a tetraploid G1-arrest in diploid cells

We wished to ask if transient p53 activation by Nutlin could promote tetraploid cell formation from diploid precursors and, if yes, whether the resulting tetraploid cells would have increased resistance to therapy-induced apoptosis. HCT116 is a human colon cancer cell line that expresses wild-type p53. In our previous study, HCT116 were plated at single cell density and ten individual diploid clones were isolated (clones designated D1–D10) that varied in their relative sensitivity to cisplatin (CP)-induced apoptosis [22]. We chose diploid clones D3 and D8 for the current study because these clones showed a relatively high sensitivity to CP in our previous study. D3 and D8 treated with Nutlin for 24 hrs accumulated with 2N and 4N DNA content (Fig 1A). Immunoblots showed that Nutlin treatment in both clones caused complete or near complete depletion of Cyclin A, Cyclin B1, and Tyr-15 phosphorylated CDC2 (pCDC2), and partial depletion of total CDC2 protein (Fig 1B). Cyclin A, Cyclin B1, and CDC2 mRNA were also depleted in Nutlin treated cells, suggesting the decreased level of these G2/M proteins may result from decreased transcription (Fig 1C). In contrast, p53 and p21 protein levels were markedly increased by Nutlin treatment (Fig 1B). Our previous studies showed p53 and p21 were increased in both the 2N and 4N Nutlin-arrested cells [18]. Thus, Nutlin causes depletion of G2/M proteins and increased expression of G1-arrest proteins (p53, p21) in the 4N arrested cells, indicative of a tetraploid G1 arrest. Importantly, Nutlin had no effect on Cyclin A, Cyclin B1, pCDC2, and total CDC2 levels in p53-null HCT116 cells (Fig 1B), and did not cause a 2N and 4N arrest in these cells (Fig 1A), demonstrating the tetraploid G1 arrest induced by Nutlin is p53-dependent. We also monitored levels of geminin, a DNA replication inhibitor that is absent in G1 and accumulates in S, G2, and M-phase [23]. Geminin was depleted by Nutlin in D3 and D8 cells, consistent with the cells being G1-arrested (Fig 1B). Cdk2 levels were unchanged by Nutlin, while levels of G1-phase cyclin proteins Cyclin D1 and Cyclin E were increased, also consistent with cells being arrested in G1-phase. In sum, the results demonstrate that Nutlin causes a p53-dependent tetraploid G1-arrest in diploid HCT116 clones D3 and D8.

Transient Nutlin treatment promotes formation of therapy resistant tetrapolid clones from diploid precursor cells

We previously showed 4N arrested cells can reinitiate DNA synthesis after Nutlin removal and replicate their DNA without an intervening mitotic division, a process known as endoreduplication [18,19]. To ask if D3 and D8 undergo endoreduplication after Nutlin removal, the cells were Nutlin treated for 24 hrs, followed by Nutlin removal for various time points. Flow cytometry was used to monitor cell cycle profiles and immunoblots used to monitor protein levels after Nutlin removal (Fig 2). D3 and D8 accumulated with 2N and 4N DNA content when Nutlin treated for 24 hrs, as in Fig 1. Both clones underwent endoreduplication after Nutlin removal, evidenced by the pronounced emergence of >4N cells 16 hrs after Nutlin removal and their transient accumulation in an 8N peak. The emergence of >4N cells and their accumulation in an 8N peak is indicative of 4N cells initiating DNA synthesis, replicating their DNA, and entering mitosis.

Cyclin A2, Cyclin B1, and CDC2 levels returned 16 hrs after Nutlin removal, consistent with the cells being mostly in G2/M at this time point. Cyclin E-CDK2 activity is believed to promote or be required for DNA synthesis initiation [24,25], and p21 binds to and inhibits Cyclin E-CDK2 activity [26]. Notably, the emergence of >4N, DNA replicating cells 12–16 hrs after Nutlin removal coincided with a sharp decrease in p21 protein levels. We speculate the decreased levels of p21 after Nutlin removal allowed activation of Cyclin E-CDK2 complexes that could then drive DNA synthesis.

Next, we wished to ask if endoreduplication after Nutlin removal could give rise to stable tetraploid clones and, if yes, whether the resulting tetraploid clones would be resistant to CP or ionizing radiation (IR)-induced apoptosis. To this end, D3 and D8 were Nutlin treated for 24 hrs followed by Nutlin removal for an additional 16 hrs. The cells were then labeled with the live-cell DNA stain Hoechst 33342. 2N and 8N cells were isolated by flow cytometry and replated at low density for isolation of individual clones (Fig 3A). A total of 11 D3 clones and 12 D8 clones were obtained from isolated 8N populations that arose after Nutlin removal. 7 of the 11 D3 clones (63%) and 8 of the 12 D8 clones (66%) grew as tetraploid 4N cells. This was evidenced by: 1) the G1 peak of these clones overlapping the G2/M peak of the D3 and D8 diploid cells (Fig 3B), 2) chromosome counting of metaphase spreads which supported the tetraploid clones having twice the DNA content of the diploid D3 and D8 precursors (Fig 3C), and 3) FISH analysis using P53 and chromosome 17-specific probes. This FISH analysis showed tetraploid clones have 4 copies of chromosome 17 and P53, while diploid D3 and D8 cells have 2 copies of chromosome 17 and P53 (Fig 3D). Finally, we tested whether tetraploid clones that arose after Nutlin treatment were more resistant to CP and IR-induced apoptosis than diploid counterparts. First, 5 tetraploid clones and 5 diploid clones isolated from Nutlin treated D3 or D8 cells were exposed to CP (20 μM) or IR (10 Gy), and apoptosis monitored 48 hrs later by sub-G1 DNA content. As shown in Fig 4A, the tetraploid clones as a group were significantly more resistant to CP and IR-induced apoptosis than parental cells and diploid clones isolated after Nutlin treatment. Individual tetraploid clones (T3 and TD6) were also more resistant to CP and IR-induced apoptosis compared to diploid counterparts (D3 and D81B), evidenced by a lower percent sub-G1 cells after CP and IR treatment (Fig 4B) and lower expression of cleaved PARP and cleaved caspase-3 (Fig 4D). These results are consistent with reports by us and others that showed tetraploid cells may be therapy resistant [3,19]. Previous studies have reported that p53 and p21 can contribute to CP and IR-resistance in HCT116 and other cells, most likely by inducing or enforcing a cell cycle arrest that blocks CP or IR-treated cells from proliferating and attempting to divide [27–31]. Notably, we found p53, MDM2, and p21 proteins were induced to comparable levels in CP and IR-treated diploid and tetraploid clones (Fig 4C), and that p53-responsive cell cycle arrest genes (P21), DNA repair genes (P53R2), and apoptotic genes (PUMA, Noxa, Bax), were also induced comparably in the CP and IR-treated diploid and tetraploid clones (Fig 4E). These results suggest CP and IR-resistance in the tetraploid clones is not associated with increased p53 levels or activity. In total, the results indicate transient p53 activation by Nutlin can promote a tetraploid G1 arrest and endoreduplication after Nutlin removal in diploid precursor cells, leading to the generation of therapy (CP/IR)-resistant tetraploid clones.

A.

B.

C.

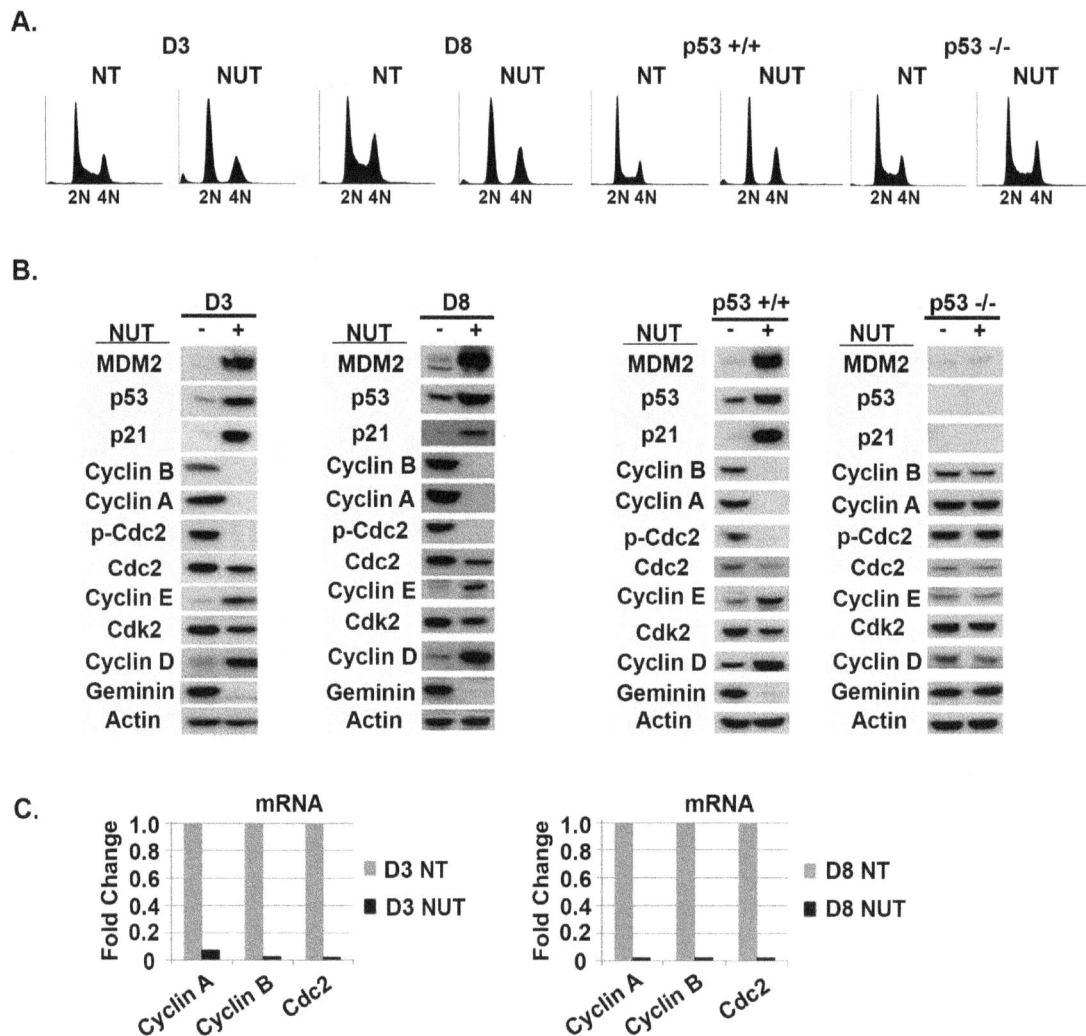

Figure 1. Nutlin-3 causes a p53-dependent tetraploid G1-arrest in diploid HCT116 clones. A) HCT116 diploid clones D3 and D8, and HCT116 that express wild-type p53 (p53+/+) or are p53-null (p53-/-) were untreated or treated with Nutlin (NUT 10 μM) for 24 hrs, and cell cycle profile determined by flow cytometry. **B)** Expression of the indicated proteins was monitored by immunoblot in cells untreated or treated with Nutlin (NUT 10 μM) for 24 hrs. **C)** mRNA levels for the indicated factors in untreated cells or cells treated with Nutlin (NUT 10 μM) for 24 hrs was determined by qRT-PCR.

Tetraploid clones are more susceptible to p53-dependent apoptosis and growth arrest induced by Nutlin

It was somewhat surprising that p53 was induced comparably by CP and IR in diploid and tetraploid clones, despite the fact that the tetraploid clones harbor twice as many copies of the *P53* gene (FISH, Fig 3D). This indicates the level to which p53 is induced by CP and IR is not dependent on *P53* copy number, but instead may be limited by other factors, perhaps the amount of damaged DNA or the activation level of stress-induced kinases that stabilize p53. Nevertheless, we speculated the level to which p53 is induced by MDM2 antagonists (e.g. Nutlin) may depend on the *P53* copy number, and that tetraploid cells may therefore express higher p53 levels in response to Nutlin and be hyper-sensitive to Nutlin-mediated growth inhibition. To test this, diploid D3 and D81B and their tetraploid counterparts T3 and TD6 were Nutlin treated for 24 hrs. P53 and MDM2 protein levels were determined by immunoblot and mRNA levels determined by qRT-PCR. The results showed that the tetraploid clones express elevated (~2X

more) p53 and MDM2 mRNA and protein than diploid clones, both basally (Fig 5A) and after 24 hr Nutlin treatment (Figs 5C). To assess p53-MDM2 complex levels, cells were treated with proteasome inhibitor MG132 to block p53 degradation, and p53-MDM2 complexes determined by co-immunoprecipitation. These studies showed tetraploid clones had more p53-MDM2 complexes after MG132 treatment than the diploid clones from which they arose (Fig 5B). Thus, tetraploid clones express higher levels of p53 and MDM2, and contain higher levels of p53-MDM2 complexes. We speculated in tetraploid clones this would translate to higher p53 levels after Nutlin treatment, and a consequent increase in p53-dependent G1 arrest and apoptosis. To test this, we compared the relative sensitivity of diploid and tetraploid clones to Nutlin-induced cell cycle arrest or apoptosis. First, diploid and tetraploid clones were treated with increasing doses of Nutlin (1.0–5.0 μM) for 24 hrs. Increased G1/S ratio, reflecting depletion of S-phase cells and an accumulation of cells in G1-phase, was used as a measure of Nutlin-induced G1 arrest. As shown in Fig 6A, the tetraploid clones as a group were more susceptible to Nutlin-induced G1 arrest than

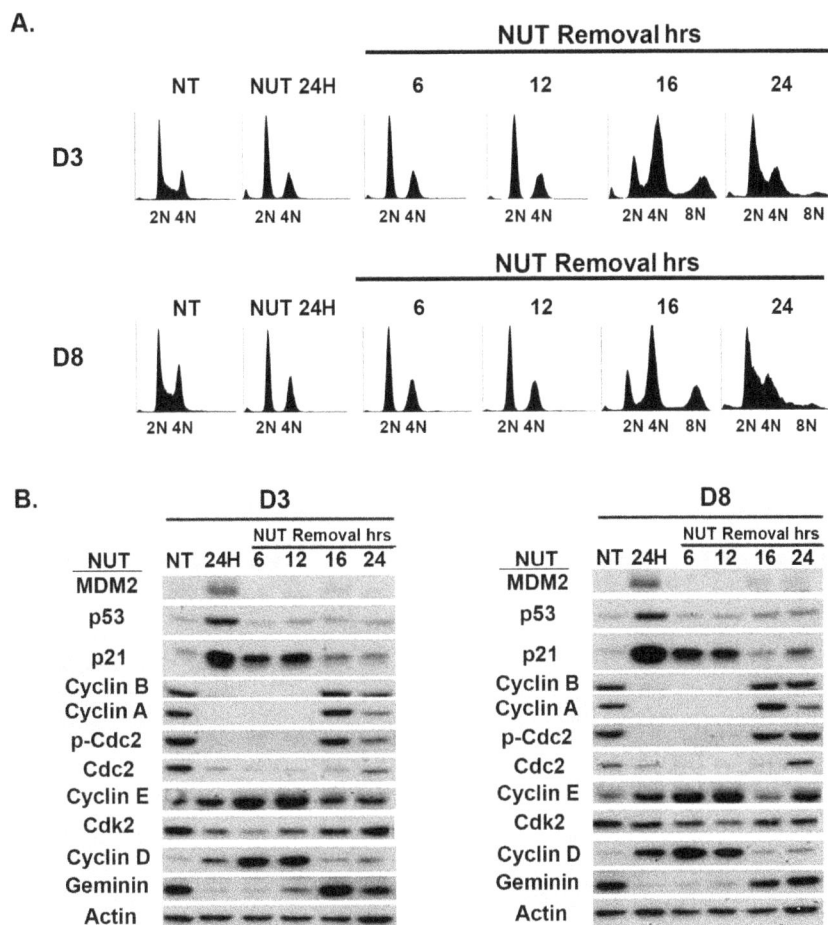

Figure 2. Transient Nutlin-3 treatment of diploid HCT116 clones induces the appearance of cells with>4N DNA content. A) Diploid HCT116 clones D3 and D8 were untreated (NT) or exposed to Nutlin (NUT 10 μM) for 24 hrs, followed by Nutlin removal. The cells were harvested at the indicated times after Nutlin removal. Fixed cells were stained with propidium iodide (25 μg/ml) and subjected to flow cytometry analysis. **B)** Cells were untreated (NT) or exposed to Nutlin (NUT 10 μM) for 24 hrs, followed by Nutlin removal. Cell lysates were collected at the indicated time points and analyzed by immunoblotting with the indicated antibodies. Actin was as a loading control. *P-Cdc2*, phosphor-Cdc2 (Tyr-15).

parental cells or diploid clones (G1/S ratio increased more with increasing Nutlin doses in tetraploid clones than parental or diploid clones). Individual tetraploid clones T3 and TD6 were also more susceptible to Nutlin-induced G1 arrest than diploid counterparts D3 and D81B (Fig 6B). As shown in Fig 6C, p53 and p21 protein were induced to higher levels and at lower Nutlin doses in the tetraploid clones compared to the diploid clones. These results indicate the tetraploid clones express higher levels of p53 and p21 after Nutlin treatment and are more susceptible than diploid clones to Nutlin-induced G1 arrest. Next, diploid and tetraploid clones were treated with a higher Nutlin dose (20 μM) for 72 hrs, and apoptosis monitored by percent cells with sub-G1 DNA content and/or expression of cleaved PARP and cleaved caspase-3. As shown in Fig 6D, tetraploid clones as a group were more susceptible to Nutlin-induced apoptosis than diploid clones. Individual tetraploid clones (T3 and TD6) were also more susceptible to Nutlin-induced apoptosis than diploid counterparts (D3 and D81B), evidenced by a higher percent sub-G1 cells after Nutlin treatment and increased expression of cleaved PARP and cleaved caspase-3 (Figs 6E and F).

Finally, we carried out clonogenic survival assays to complement the apoptosis studies. As shown in Figs 7A and B, tetraploid clones (T3, TD6) showed significantly higher colony forming

ability after CP and IR treatment than their diploid counterparts (D3, D81B). In contrast, the tetraploid clones showed lower colony forming ability than diploid counterparts when exposed continuously to 1 μM Nutlin (Fig 7C) or when exposed transiently to 20 μM Nutlin for 3 days (Fig. 7D). The results demonstrate the tetraploid clones are more resistant than diploid counterparts to CP and IR, and more sensitive than diploid counterparts to Nutlin, in both apoptosis and colony formation assays.

Discussion

Tetraploid (4N) cells contain twice the DNA content of diploid (2N) cells. Tetraploid cells are considered important in cancer because they can display increased tumorigenicity, resistance to conventional therapies, and are believed to be precursors to whole chromosome aneuploidy [1–6]. It is therefore important to determine how tetraploid cancer cells arise, and how to target them. Nutlin-3a (Nutlin) is a small molecule MDM2 antagonist and activator of p53. In our previous studies, transient (24 hr) Nutlin treatment promoted a tetraploid G1 arrest in multiple cancer cell lines that express wild-type p53 [18,19]. This tetraploid G1 arrest was characterized by depletion of G2/M proteins (e.g. Cyclins A/B, CDC2) and increased expression of G1-arrest

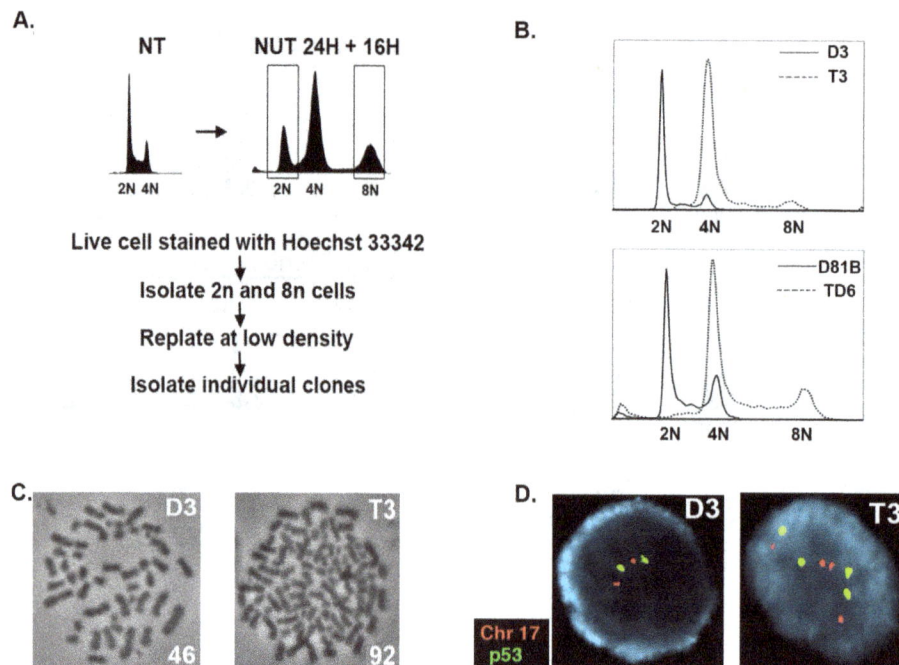

Figure 3. Stable tetraploid clones were isolated from diploid D3 and D8 clones after endoreduplication. A) Shown is the procedure to isolate diploid and tetraploid clones from cells transiently exposed to Nutlin. D3 and D8 were untreated (NT) or treated with 10 µM Nutlin (NUT) for 24 hrs, followed by Nutlin removal for an additional 16 hrs. The cells were then live-stained with Hoechst 33342 (5 µg/ml). Cell sorting was performed on a MoFlo cytometer equipped with a UV excitation wavelength laser. Sorted 2N and 8N cells were plated at low density in normal medium (minus Nutlin) and individual clones isolated. **B)** *Top*, the comparison of the DNA profiles between D3 and a representative tetraploid clone isolated from D3. *Bottom*, comparison of the DNA profiles between a representative diploid clone (D81B) isolated from D8 cells, and a representative tetraploid clone (TD6) isolated from D8 cells. **C)** Representative metaphase spread from diploid D3 cells and tetraploid T3 cells. The number in the bottom right indicates the number of chromosomes counted. **D)** FISH analysis with chromosome 17 (Chr 17) and p53-specific probes shows tetraploid (T3) cells contain 4 copies of p53 and Chr 17, while diploid (D3) cells contain 2 copies of p53 and Chr 17.

proteins (e.g. p53, p21) in 4N cells. Upon Nutlin removal, 4N cells reinitiated DNA synthesis and replicated their DNA without an intervening mitosis, a process known as endoreduplication [18,19]. Both the tetraploid G1 arrest and endoreduplication after Nutlin removal were dependent on p53 and p21. Finally, stable tetraploid clones could be isolated from Nutlin treated cells, and these cells were resistant to ionizing radiation (IR) and cisplatin (CP)-induced apoptosis [19]. These findings suggested transient p53 activation by Nutlin can promote endoreduplication and the generation of therapy resistant tetraploid cells. However, a caveat is that these previous studies were done in cancer cell lines, which could include a mixture of diploid and tetraploid cells, some of which may be inherently IR/CP-resistant. It is possible that we had isolated IR/CP-resistant tetraploid clones that were already present in the population, or that the tetraploid clones we isolated were derived from individual diploid cells that were already IR/CP-resistant. Therefore, in the current report we asked if Nutlin could promote tetraploid cell formation from diploid precursors and, if yes, whether the resulting tetraploid cells would show increased resistance to IR and/or CP. Individual diploid clones (D3, D8) isolated from the HCT116 colon cancer cell line by limiting dilution were treated with Nutlin for 24 hrs, followed by Nutlin removal. Nutlin promoted a tetraploid G1 arrest in the diploid clones that was p53-dependent. The clones underwent endoreduplication after Nutlin removal, giving rise to stable tetraploid clones that showed increased resistance to IR/CP-induced apoptosis compared their diploid counterparts. These findings demonstrate that transient p53 activation by Nutlin can promote tetraploid cell formation from diploid precursors, and the

resulting tetraploid cells are therapy (IR/CP) resistant. Importantly, tetraploid clones selected after Nutlin treatment expressed twice as much *P53* and *MDM2* mRNA as diploid cells, expressed twice as many p53-MDM2 protein complexes (by co-immunoprecipitation), and were more susceptible to p53-dependent apoptosis and growth arrest induced by Nutlin. Based on these findings, we propose that p53 plays a role in both the formation and targeting of tetraploid cells. Specifically, we propose that 1) transient p53 activation can promote a tetraploid-G1 arrest and, as a result, may inadvertently promote formation of therapy-resistant tetraploid cells, and 2) therapy-resistant tetraploid cells, by virtue of having higher *P53* gene copy number and expressing twice as many p53-MDM2 complexes, are more sensitive to apoptosis and/or growth arrest by anti-cancer MDM2 antagonists (e.g. Nutlin).

In order for endoreduplication to occur, 4N (G_2 or M phase) cells must first assume a G_1-like state from which they can enter S phase, referred to as tetraploid G_1. Typically, tetraploid G_1 is characterized by depletion/loss of G_2/M marker proteins (e.g. Cyclins A/B, CDC2) and increased expression of G_1 phase markers in 4N cells [15–19]. Diploid HCT116 clones treated with Nutlin for 24 hrs arrested with 2N and 4N DNA content, coincident with increased expression of p53 and p21 protein, and complete or near-complete depletion of Cyclin A, Cyclin B, and CDC2. We previously showed p53 and p21 are expressed in both the 2N and 4N Nutlin-arrested cell pools [18]. Thus, Nutlin causes depletion of G2/M proteins and increased expression of G1-arrest proteins (p53, p21) in 4N cells, indicative of a tetraploid G1 arrest. We previously showed p53 and p21 are required for depletion of

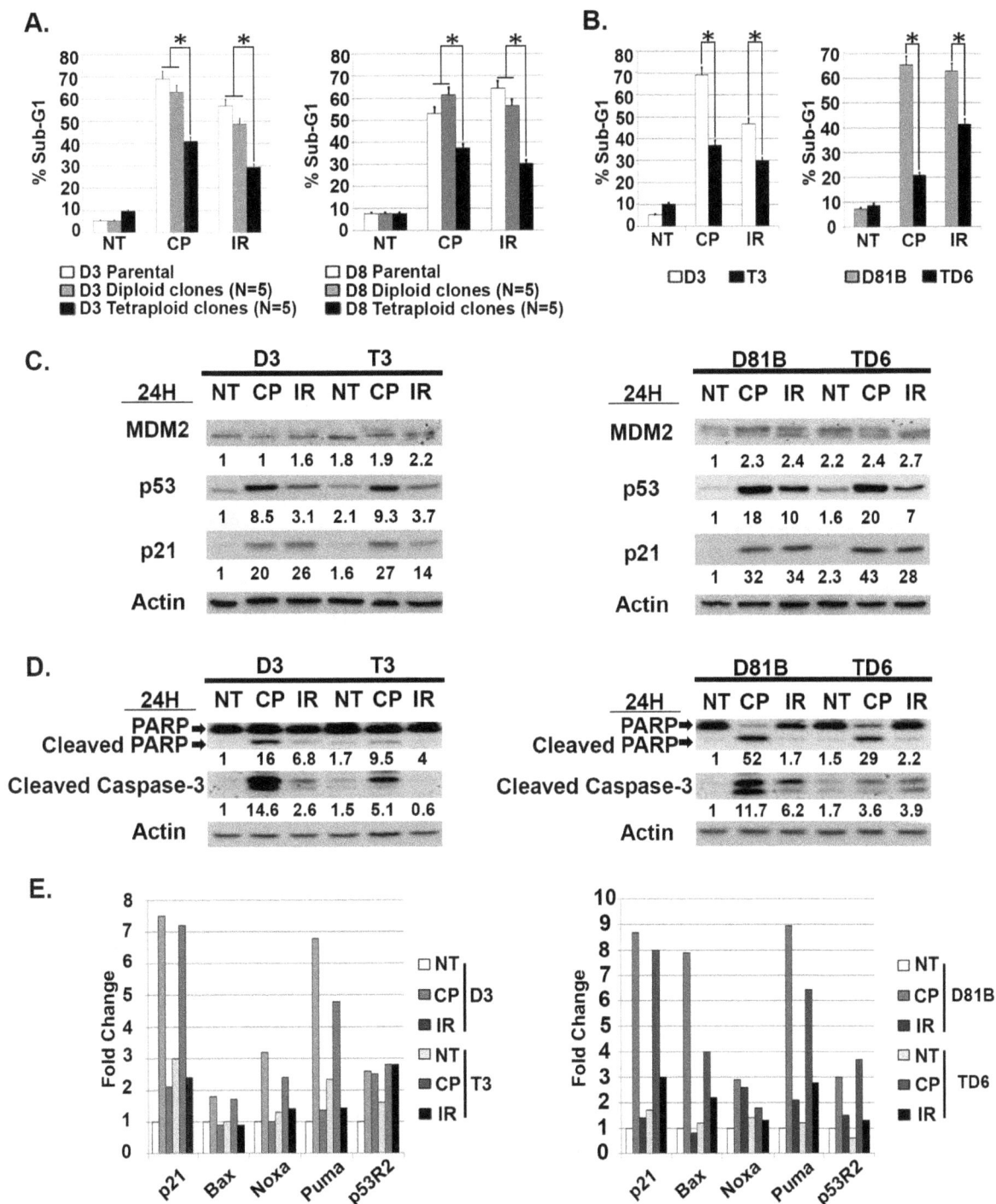

Figure 4. Tetraploid clones show resistance to cisplatin (CP) and ionizing radiation (IR)-induced apoptosis. A) Five diploid clones isolated from Nutlin treated D3 cells (D3 Diploid) and D8 cells (D8 Diploid), and five tetraploid clones isolated from Nutlin treated D3 cells (D3 Tetraploid) and D8 cells (D8 Tetraploid) were untreated (NT) or exposed to CP (20 μM) or IR (10 Gy) for 48 hrs. The percentage of cells with sub-G1 DNA was determined. Shown are the mean results from three separate experiments, *bars*, Standard error (SE). * significance value (P<0.05). **B)** Tetraploid clones (T3, TD6) and their diploid counterparts (D3, D81B) were untreated (NT) or exposed to CP (20 μM) or IR (10 Gy) for 72 hrs. The percentage of cells with sub-G1 DNA (propidium iodide staining) was determined. Shown are the mean results from three separate experiments, *bars*, Standard error (SE). * significance value (P<0.05). **C)** The indicated diploid and tetraploid clones were untreated (NT) or exposed to CP (20 μM) or IR (10 Gy) for 24 hrs. p53, p21, and MDM2 protein levels were determined by immunoblotting and quantified. Numbers indicate the relative level of each protein. Actin was used as a loading control. **D)** The indicated diploid and tetraploid clones untreated (NT) or exposed to CP (20 μM) or IR (10 Gy) for 48 hrs. Cleaved PARP and Caspase-3 protein levels were determined by immunoblotting and quantified relative to the untreated. **E)** qRT-PCR was used to determine mRNA levels for the indicated genes in diploid (D3, D81B) and tetraploid (T3, TD6) clones that were either untreated (NT) or exposed to CP (20 μM) or IR (10 Gy) for 24 hrs. The level of each mRNA transcript in untreated diploid clones (D3 NT, D81B NT) was considered "1.0", and all other values are plotted relative to it.

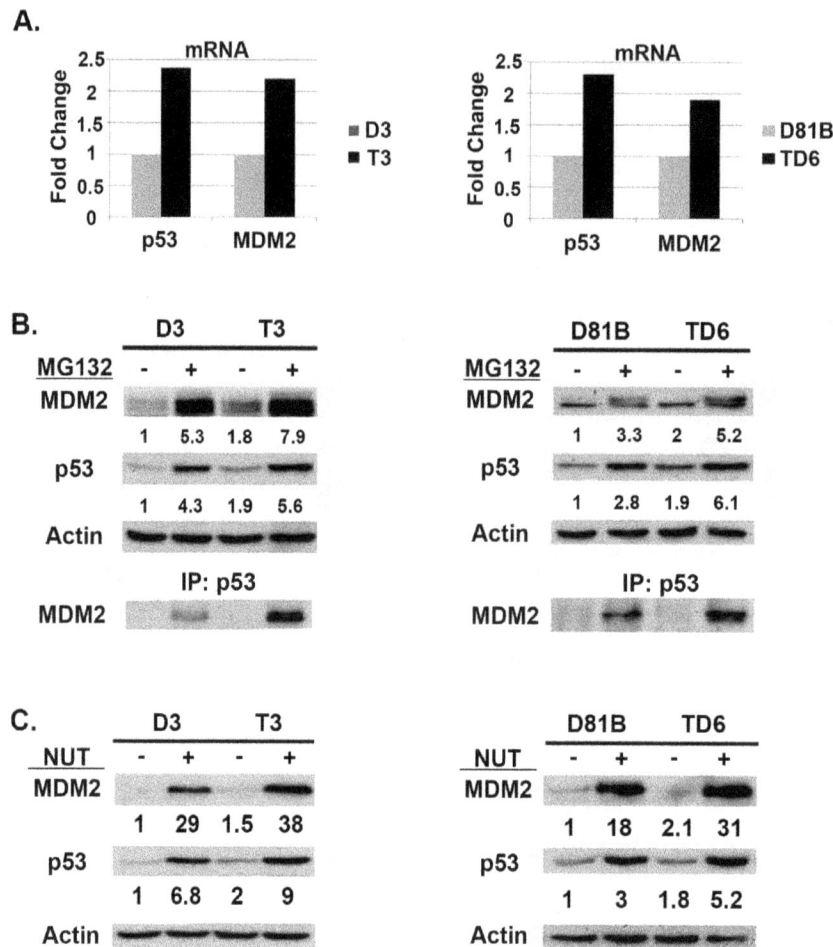

Figure 5. Tetraploid clones express more p53-MDM2 complexes and more p53 after Nutlin treatment than diploid counterparts. A) P53 and MDM2 mRNA levels were compared in untreated tetraploid clones (T3, TD6) and their diploid counterparts (D3, D81B). **B)** Diploid (D3, D81B) and tetraploid (T3, TD6) clones were untreated or treated with proteasome inhibitor MG132 (10 μM) for 8 hrs. *Upper* Levels of MDM2, p53, and actin (loading control) in untreated and MG132 treated cells are shown. *Lower* To determine levels of p53-MDM2 complexes in diploid and tetraploid cells, protein lysates were immunoprecipitated with anti-p53 antibody, followed by immunoblotting for MDM2. **C)** Diploid (D3, D81B) and tetraploid (T3, TD6) clones were untreated or treated with Nutlin (10 μM) for 24 hrs. p53 and MDM2 protein levels were detected by immunoblotting and quantified using Image-J software. The relative amount of p53 and MDM2 protein in the untreated diploid clones was given a value of "1.0". Numbers indicate the relative level of each protein. Actin was used as a loading control.

Cyclins A/B and CDC2 protein by Nutlin [18,19], and in the current study we also observed mRNA levels for Cyclin A, Cyclin B, and CDC2 are drastically reduced in Nutlin treated cells. This suggests the decrease in Cyclin A, Cyclin B, and CDC2 protein levels may result from decreased gene transcription. The Cyclin A, Cyclin B, and CDC2 gene promoters harbor E2F binding sites and are E2F-responsive [32]. pRb proteins (pRb, p107, p130) can block expression of E2F responsive genes by binding and sequestering E2F proteins away from the promoter, or by forming transcription repressor complexes with E2F on DNA gene promoters [32]. Active Cyclin-CDK complexes can phosphorylate pRb proteins, blocking their interaction with E2Fs. In contrast, p21 activates pRb proteins by binding and inhibiting the activity of Cyclin-CDKs [26]. Thus, one possibility is that p53–p21 induced by Nutlin activates pRb, p107, and/or p130, which may then bind E2F complexes and inhibit Cyclins A/B and CDC2 mRNA and protein expression, contributing to a tetraploid G1-arrest.

A second requirement for endoreduplication is that DNA replication origins are "licensed" in tetraploid G1 cells. Origin

"licensing" involves the sequential binding of the origin recognition complex (ORC), CDC6 and Cdt1, and the replicative helicase MCM2-7 [33]. Geminin binds and inhibits Cdt1, preventing assembly of pre-replication origin licensing complexes [23]. Geminin is normally expressed in S, G2, and M-phases, but not expressed in G1 when origin licensing occurs [23]. We observed a pronounced depletion of geminin in Nutlin-arrested 2N and 4N cells, consistent with the cells being arrested in a G1 state. Notably, the geminin gene promoter is also E2F-responsive [34]. Thus, geminin expression may also be reduced in Nutlin treated cells at the mRNA level, resulting from p21 activation of pRb/p107/p130 and inhibition of E2F, as described above. Given that geminin is depleted in Nutlin treated cells, we speculate that pre-replicative origin licensing complexes are formed in the 2N and 4N Nutlin-arrested cells. Subsequent origin firing/S-phase entry occurs upon recruitment of the DNA replication machinery and activation of Cyclin E-CDK2 [24,25]. In the current study, endoreduplicating (>4N) cells emerged 12–16 hrs after Nutlin removal, coincident with a sharp decrease in p21 protein levels. We speculate the most

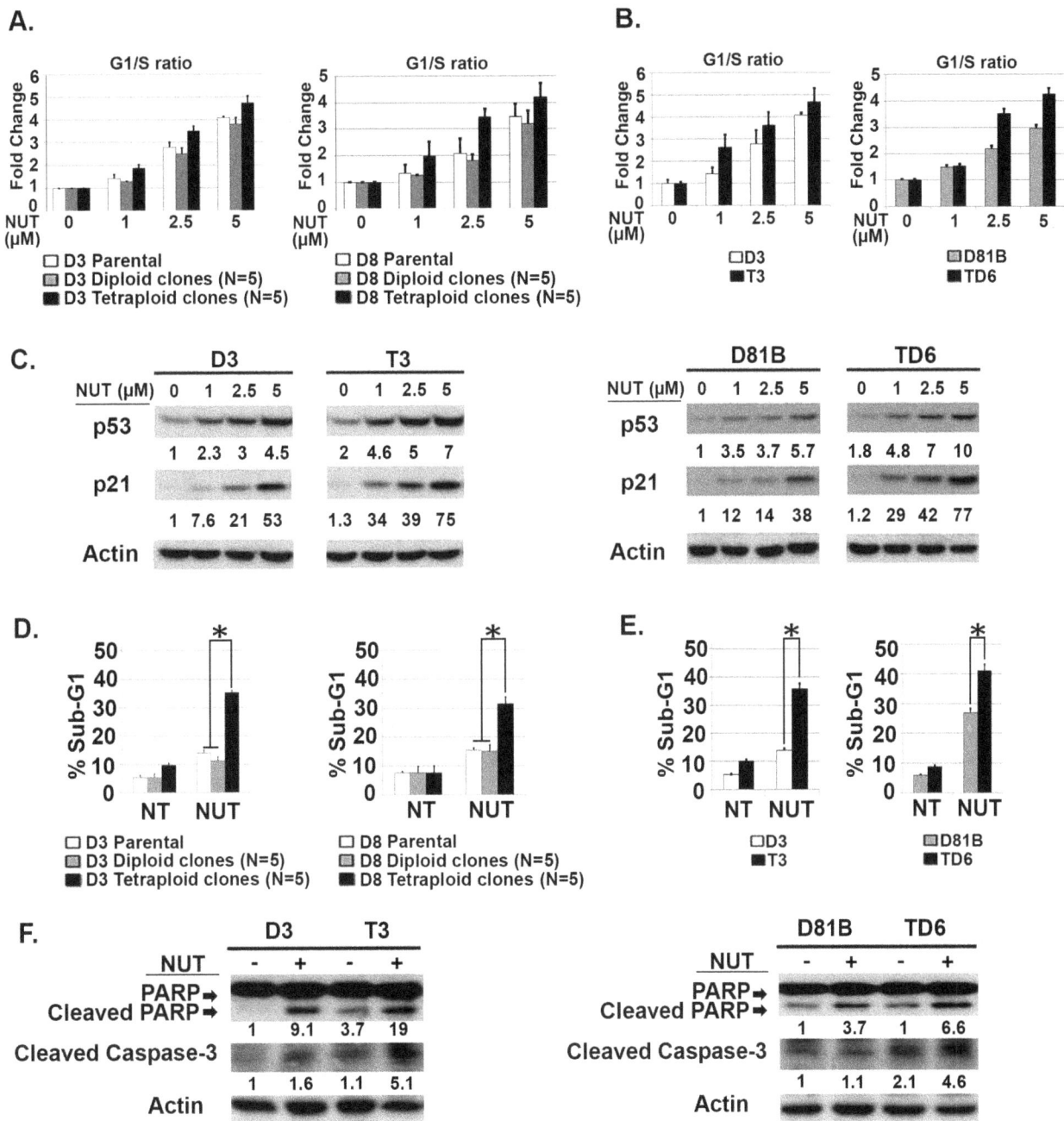

Figure 6. Tetraploid clones are more susceptible to Nutlin-induced cell cycle arrest and apoptosis. A) Five diploid clones isolated from Nutlin treated D3 cells (D3 Diploid) and D8 cells (D8 Diploid), and five tetraploid clones isolated from Nutlin treated D3 cells (D3 Tetraploid) and D8 cells (D8 Tetraploid) were untreated or treated with increasing Nutlin (1.0–5.0 μM) for 24 hrs. The percent G1 and S-phase cells was determined by flow cytometry, and the fold change of G1/S ratio is plotted. Untreated diploid clones G1/S ratio was given a value of 1.0. Shown is the average of three separate experiments, +/- SE. **B)** Tetraploid clones (T3, TD6) and diploid counterparts (D3, D81B) were untreated or treated with Nutlin (1.0–5.0 μM) for 24 hrs. The fold change of G1/S ratio is plotted. Shown is the average of three separate experiments, +/- SE. **C)** Diploid (D3, D81B) and tetraploid (T3, TD6) clones were untreated or treated with increasing Nutlin (1.0–5.0 μM) 24 hrs. P53 and p21 protein levels were quantified using Image-J software. Numbers indicate the relative levels of each protein. P53 and p21 levels in untreated diploid clones was given a value of "1.0". Actin was used as loading control. **D)** Five diploid clones isolated from Nutlin treated D3 cells (D3 Diploid) and D8 cells (D8 Diploid), and five tetraploid clones isolated from Nutlin treated D3 cells (D3 Tetraploid) and D8 cells (D8 Tetraploid) were untreated (NT) or treated with Nutlin (20 μM) 72 hrs and apoptosis determined. Shown are the mean results from three separate experiments, *bars*, Standard error (SE). *($P<0.05$). **E)** Representative diploid (D3, D81B) and tetraploid (T3, TD6) clones were untreated (NT) or treated with Nutlin (20 μM) for 72 hrs. The percentage of cells with sub-G1 DNA content (propidium iodide staining) was determined. Shown are the mean results from three separate experiments, *bars*, Standard error (SE). * significance value ($P<0.05$). **F)** The indicated diploid and tetraploid clones were untreated (NT) or exposed to Nutlin (20 μM) for 48 hrs. Cleaved PARP and Caspase-3 protein levels were determined by immunoblotting and quantified relative to the untreated.

A.

B.

C.

D.

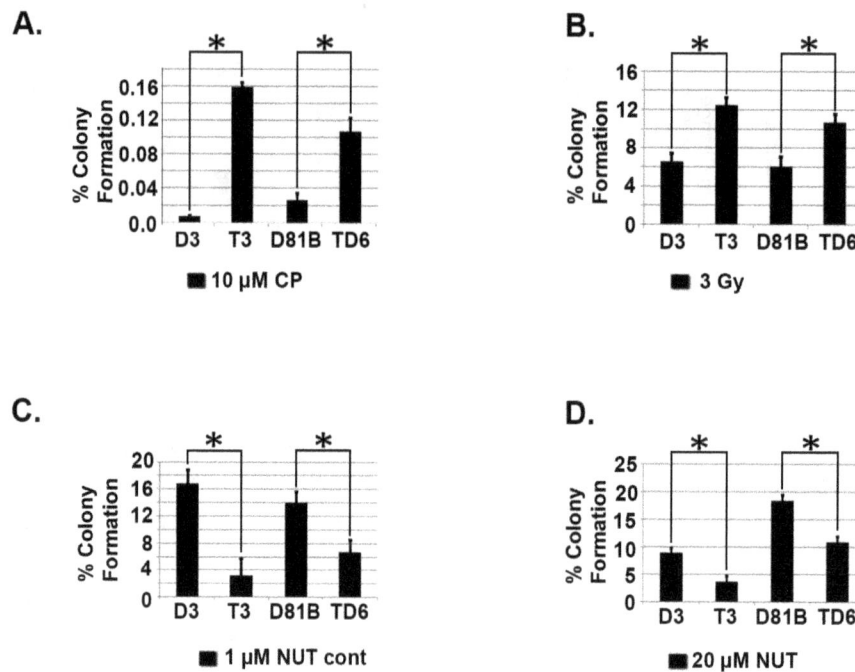

Figure 7. Tetraploid clones are more resistant than diploid clones to Cisplatin (CP) and ionizing radiation (IR) and more sensitive to Nutlin in a long term survival assay. A) Diploid (D3, D81B) and tetraploid (T3, TD6) clones were untreated or treated with CP (10 µM) for 24 hrs. Cells were rinsed and re-fed with drug free media and stained after 2–3 weeks. The colonies were counted and normalized with plating efficiency of untreated controls. Shown are the mean results from three separate experiments. **B**) Indicated clones were untreated or exposed to IR (3 Gy). Colonies were counted 2–3 weeks later and normalized with plating efficiency of untreated controls. Shown are the mean results from three separate experiments. **C**) Diploid and tetraploid clones were grown in normal medium (minus Nutlin) or grown in continuous Nutlin (1 µM) and stained after 2–3 weeks. Colony number was normalized with the plating efficiency of untreated controls. **D**) Indicated clones were untreated or treated with Nutlin (20 µM) for 72 hrs after which the cells were rinsed and re-fed with normal medium (minus Nutlin) for 2–3 weeks. The colonies were then counted and normalized with the plating efficiency of untreated controls. Shown are the mean results from three separate experiments. *bars* indicate standard error (SE). * significance value (P<0.05).

likely scenario is that the reduction in p21 levels allowed activation of Cyclin E-CDK2 complexes, which then triggered origin firing/S-phase entry by 4N cells.

Tetraploid clones that arose after Nutlin treatment were resistant to IR and CP compared to the diploid cells from which they came. Given that the Nutlin effects are p53-dependent, the results demonstrate that transient p53 activation can promote formation of therapy resistant tetraploid cells from diploid precursors. The basis for IR/CP-resistance in the tetraploid cells is unclear. Castedo et al. compared radiation and chemosensitivity of diploid RKO and HCT116 cells with tetraploid clones that arose after prolonged nocodazole treatment [3,4]. Tetraploid clones were significantly more resistant to IR, CP, and other agents. In their study, tetraploid clones expressed increased basal and therapy-induced levels of P53R2, a p53-responsive ribonucleotide reductase DNA repair enzyme [3,4]. Knockdown of P53R2 in their study sensitized tetraploid clones to apoptosis by IR and other agents. In the current study, we found P53R2 was expressed comparably at the mRNA level in the diploid and tetraploid clones either basally or after IR or CP treatment, suggesting IR and CP resistance in the tetraploid clones in our study is not associated with increased P53R2 expression. Notably, p53-null cells are sometimes more sensitive than p53 wild-type HCT116 cells to IR and CP-induced killing, probably because p53 can activate/enforce cell cycle arrests that block IR/CP treated cells from continuing to replicate with damaged DNA [27–31]. Given they have twice as many *P53* gene copies, we considered

tetraploid clones might express more p53 protein than diploid cells after IR or CP treatment and therefore be more resistant. However, we found p53 was induced to a comparable level in IR and CP-treated diploid and tetraploid cells. Thus, the increased IR and CP-resistance in tetraploid cells in our study does not appear to be associated with increased p53 levels or activity.

Targeting tetraploid cancer cells is an important goal given the potential involvement of these cells in tumor progression, therapy resistance, and aneuploidy. The tetraploid clones that arose after Nutlin treatment in the current study contained twice as many *P53* gene copies (by FISH), expressed approximately twice as much *P53* and *MDM2* mRNA as diploid cells, and expressed approximately twice as many p53-MDM2 protein complexes (by p53 and MDM2 co-immunoprecipitation). We speculated that if tetraploid cells contain more p53-MDM2 complexes than diploid cells, they might be hypersensitive to MDM2 antagonists that disrupt these complexes. Indeed, we found p53 (and p21) were induced to a higher level in Nutlin-treated tetraploid cells vs Nutlin-treated diploid cells, and the tetraploid cells were more sensitive than diploid cells to Nutlin in apoptosis, colony formation, and cell cycle arrest assays. In sum, the results suggest therapy-resistant tetraploid cells, by virtue of having higher *P53* gene copy number and expressing twice as many p53-MDM2 complexes, are more sensitive than diploid cells to anti-cancer MDM2 antagonists (e.g. Nutlin).

Conclusions

Transient p53 activation can promote a tetraploid-G1 arrest and, as a result, may inadvertently promote formation of therapy-resistant tetraploid cells. Therapy-resistant tetraploid cells, by virtue of having higher *P53* gene copy number and expressing twice as many p53-MDM2 complexes, are more sensitive to apoptosis and/or growth arrest by anti-cancer MDM2 antagonists (e.g. Nutlin).

Author Contributions

Conceived and designed the experiments: BD HS CGM. Performed the experiments: BD HS. Analyzed the data: BD HS CGM. Wrote the paper: BD CGM.

References

1. Davoli T, de Lange T (2011) The causes and consequences of polyploidy in normal development and cancer. Annu Rev Cell Dev Biol 27: 585–610.
2. Fujiwara T, Bandi M, Nitta M, Ivanova EV, Bronson RT, et al. (2005) Cytokinesis failure generating tetraploids promotes tumorigenesis in p53-null cells. Nature 437: 1043–1047.
3. Castedo M, Coquelle A, Vitale I, Vivet S, Mouhamad S, et al. (2006) Selective resistance of tetraploid cancer cells against DNA damage-induced apoptosis. Ann N Y Acad Sci 1090: 35–49.
4. Castedo M, Coquelle A, Vivet S, Vitale I, Kauffmann A, et al. (2006) Apoptosis regulation in tetraploid cancer cells. EMBO J 25: 2584–2595.
5. Galipeau PC, Cowan DS, Sanchez CA, Barrett MT, Emond MJ, et al. (1996) 17p (p53) allelic losses, 4N (G2/tetraploid) populations, and progression to aneuploidy in Barrett's esophagus. Proc Natl Acad Sci U S A 93: 7081–7084.
6. Reid BJ (2010) Early events during neoplastic progression in Barrett's esophagus. Cancer Biomark 9: 307–324.
7. Aylon Y, Oren M (2011) p53: guardian of ploidy. Mol Oncol 5: 315–323.
8. Haupt Y, Maya R, Kazaz A, Oren M (1997) Mdm2 promotes the rapid degradation of p53. Nature 387: 296–299.
9. Kubbutat MH, Jones SN, Vousden KH (1997) Regulation of p53 stability by Mdm2. Nature 387: 299–303.
10. Carvajal LA, Manfredi JJ (2013) Another fork in the road—life or death decisions by the tumour suppressor p53. EMBO Rep 14: 414–421.
11. Bunz F, Dutriaux A, Lengauer C, Waldman T, Zhou S, et al. (1998) Requirement for p53 and p21 to sustain G2 arrest after DNA damage. Science 282: 1497–1501.
12. Stewart ZA, Leach SD, Pietenpol JA (1999) p21(Waf1/Cip1) inhibition of cyclin E/Cdk2 activity prevents endoreduplication after mitotic spindle disruption. Mol Cell Biol 19: 205–215.
13. Niculescu AB 3rd, Chen X, Smeets M, Hengst L, Prives C, et al. (1998) Effects of p21(Cip1/Waf1) at both the G1/S and the G2/M cell cycle transitions: pRb is a critical determinant in blocking DNA replication and in preventing endoreduplication. Mol Cell Biol 18: 629–643.
14. Khan SH, Wahl GM (1998) p53 and pRb prevent rereplication in response to microtubule inhibitors by mediating a reversible G1 arrest. Cancer Res 58: 396–401.
15. Itzhaki JE, Gilbert CS, Porter AC (1997) Construction by gene targeting in human cells of a "conditional' CDC2 mutant that rereplicates its DNA. Nat Genet 15: 258–265.
16. Mihaylov IS, Kondo T, Jones L, Ryzhikov S, Tanaka J, et al. (2002) Control of DNA replication and chromosome ploidy by geminin and cyclin A. Mol Cell Biol 22: 1868–1880.
17. Bellanger S, de Gramont A, Sobczak-Thepot J (2007) Cyclin B2 suppresses mitotic failure and DNA re-replication in human somatic cells knocked down for both cyclins B1 and B2. Oncogene 26: 7175–7184.
18. Shen H, Maki CG (2010) Persistent p21 expression after Nutlin-3a removal is associated with senescence-like arrest in 4N cells. J Biol Chem 285: 23105–23114.
19. Shen H, Moran DM, Maki CG (2008) Transient nutlin-3a treatment promotes endoreduplication and the generation of therapy-resistant tetraploid cells. Cancer Res 68: 8260–8268.
20. de Toledo SM, Azzam EI, Keng P, Laffrenier S, Little JB (1998) Regulation by ionizing radiation of CDC2, cyclin A, cyclin B, thymidine kinase, topoisomerase IIalpha, and RAD51 expression in normal human diploid fibroblasts is dependent on p53/p21Waf1. Cell Growth Differ 9: 887–896.
21. Badie C, Itzhaki JE, Sullivan MJ, Carpenter AJ, Porter AC (2000) Repression of CDK1 and other genes with CDE and CHR promoter elements during DNA damage-induced G(2)/M arrest in human cells. Mol Cell Biol 20: 2358–2366.
22. Shen H, Perez RE, Davaadelger B, Maki CG (2013) Two 4N cell-cycle arrests contribute to cisplatin-resistance. PLoS One 8: e59848.
23. Wohlschlegel JA, Dwyer BT, Dhar SK, Cvetic C, Walter JC, et al. (2000) Inhibition of eukaryotic DNA replication by geminin binding to Cdt1. Science 290: 2309–2312.
24. Jackson PK, Chevalier S, Philippe M, Kirschner MW (1995) Early events in DNA replication require cyclin E and are blocked by p21CIP1. J Cell Biol 130: 755–769.
25. Sherr CJ (1996) Cancer cell cycles. Science 274: 1672–1677.
26. Xiong Y, Hannon GJ, Zhang H, Casso D, Kobayashi R, et al. (1993) p21 is a universal inhibitor of cyclin kinases. Nature 366: 701–704.
27. Pestell KE, Hobbs SM, Titley JC, Kelland LR, Walton MI (2000) Effect of p53 status on sensitivity to platinum complexes in a human ovarian cancer cell line. Mol Pharmacol 57: 503–511.
28. Fan S, Chang JK, Smith ML, Duba D, Fornace AJ, et al. (1997) Cells lacking CIP1/WAF1 genes exhibit preferential sensitivity to cisplatin and nitrogen mustard. Oncogene 14: 2127–2136.
29. Fan S, Smith ML, Rivet DJ 2nd, Duba D, Zhan Q, et al. (1995) Disruption of p53 function sensitizes breast cancer MCF-7 cells to cisplatin and pentoxifylline. Cancer Res 55: 1649–1654.
30. Polyak K, Waldman T, He TC, Kinzler KW, Vogelstein B (1996) Genetic determinants of p53-induced apoptosis and growth arrest. Genes Dev 10: 1945–1952.
31. Gudkov AV, Komarova EA (2010) Pathologies associated with the p53 response. Cold Spring Harb Perspect Biol 2: a001180.
32. Cobrinik D (2005) Pocket proteins and cell cycle control. Oncogene 24: 2796–2809.
33. Mechali M (2010) Eukaryotic DNA replication origins: many choices for appropriate answers. Nat Rev Mol Cell Biol 11: 728–738.
34. Yoshida K, Inoue I (2004) Regulation of Geminin and Cdt1 expression by E2F transcription factors. Oncogene 23: 3802–3812.

12

Suppression of Proteoglycan-Induced Autoimmune Arthritis by Myeloid-Derived Suppressor Cells Generated *In Vitro* from Murine Bone Marrow

Júlia Kurkó[1,2], András Vida[1], Tímea Ocskó[1], Beata Tryniszewska[1], Tibor A. Rauch[1], Tibor T. Glant[1], Zoltán Szekanecz[2◉], Katalin Mikecz[1◉]*

1 Section of Molecular Medicine, Department of Orthopedic Surgery, Rush University Medical Center, Chicago, Illinois, United States of America, **2** Department of Rheumatology, University of Debrecen, Faculty of Medicine, Debrecen, Hungary

Abstract

Background: Myeloid-derived suppressor cells (MDSCs) are innate immune cells capable of suppressing T-cell responses. We previously reported the presence of MDSCs with a granulocytic phenotype in the synovial fluid (SF) of mice with proteoglycan (PG)-induced arthritis (PGIA), a T cell-dependent autoimmune model of rheumatoid arthritis (RA). However, the limited amount of SF-MDSCs precluded investigations into their therapeutic potential. The goals of this study were to develop an in vitro method for generating MDSCs similar to those found in SF and to reveal the therapeutic effect of such cells in PGIA.

Methods: Murine bone marrow (BM) cells were cultured for 3 days in the presence of granulocyte macrophage colony-stimulating factor (GM-CSF), interleukin-6 (IL-6), and granulocyte colony-stimulating factor (G-CSF). The phenotype of cultured cells was analyzed using flow cytometry, microscopy, and biochemical methods. The suppressor activity of BM-MDSCs was tested upon co-culture with activated T cells. To investigate the therapeutic potential of BM-MDSCs, the cells were injected into SCID mice at the early stage of adoptively transferred PGIA, and their effects on the clinical course of arthritis and PG-specific immune responses were determined.

Results: BM cells cultured in the presence of GM-CSF, IL-6, and G-CSF became enriched in MDSC-like cells that showed greater phenotypic heterogeneity than MDSCs present in SF. BM-MDSCs profoundly inhibited both antigen-specific and polyclonal T-cell proliferation primarily via production of nitric oxide. Injection of BM-MDSCs into mice with PGIA ameliorated arthritis and reduced PG-specific T-cell responses and serum antibody levels.

Conclusions: Our in vitro enrichment strategy provides a SF-like, but controlled microenvironment for converting BM myeloid precursors into MDSCs that potently suppress both T-cell responses and the progression of arthritis in a mouse model of RA. Our results also suggest that enrichment of BM in MDSCs could improve the therapeutic efficacy of BM transplantation in RA.

Editor: Oliver Frey, University Hospital Jena, Germany

Funding: This study was supported by grants AR062332 and AR064206 (to KM) and AR064948 (to TAR) from the National Institute of Health http://www.nih.gov., and by the TÁMOP 4.2.4.A/2-11-1-2012-0001 National Excellence Program co-financed by the European Union and Hungary (to ZS). The funders had no role in study design, data collection and analysis, decision to publish, or preparation of the manuscript.

Competing Interests: The authors have declared that no competing interests exist.

* Email: Katalin_Mikecz@rush.edu

◉ These authors contributed equally to this work.

Introduction

Rheumatoid arthritis (RA) is a chronic autoimmune inflammatory disease that leads to painful joint destruction and disability [1,2]. Despite novel treatment strategies, not all patients respond to therapy. Although cell-based therapy such as transplantation of autologous bone marrow (BM) or hematopoietic stem cells is a promising option in both refractory RA [3] and therapy-resistant juvenile idiopathic arthritis [4], clinical remission in transplant recipients is still incomplete. Exploration of novel therapeutic options is needed in order to control immune responses that drive inflammation in these cases.

Research in recent years has uncovered a heterogeneous population of immature myeloid cells, called myeloid-derived suppressor cells (MDSCs). MDSCs with immunosuppressive capacity were initially described in tumor-bearing mice [5]. Although the vast majority of data comes from cancer research (reviewed in [6,7]), accumulating evidence supports the role of MDSCs in chronic inflammatory and autoimmune disorders. A common feature of these pathological conditions is the release of a broad array of inflammatory mediators (growth factors and cytokines) that not only exert their effects on the affected organs, but also disturb myelopoiesis in the BM. While some of these mediators promote the expansion of MDSCs through stimulation of myelopoiesis, others inhibit full differentiation of myeloid precursors, thus contributing to the accumulation of MDSCs around malignant tumors or at sites of inflammation (reviewed in [8]). As the microenvironment under different pathological conditions varies, the phenotypic and the functional properties of MDSCs can be diverse [9,10]. MDSCs do not constitute a homogenous cell population, rather, they are a mixture of "immature" forms of monocytes and granulocytes. What classifies them as an integrated system is their shared ability to suppress adaptive immune responses [8,10].

MDSCs in mice express the common myeloid markers CD11b (α chain of $\alpha_M\beta_2$ integrin, an adhesion molecule present on monocytes and granulocytes) [11] and Gr-1 [8,12]. The epitope of the widely used anti-Gr-1 monoclonal antibody (mAb, clone RB6-8C5) is present on two molecules, Ly6G and Ly6C, which are encoded by separate genes and expressed in granulocytic and monocytic cells, respectively. Based on cell surface staining with mAbs against CD11b, Gr-1, Ly6G, and Ly6C, the following subtypes of murine MDSCs have been identified: CD11b$^+$Gr-1$^+$Ly6G^{hi}Ly6C^{lo} granulocytic, and CD11b$^+$Gr-1$^+$Ly6G$^-$Ly6C^{hi} monocytic MDSCs [12,13]. These two subsets also display distinct cellular morphology and may employ different strategies to suppress immune responses in malignant, infectious, and autoimmune disease models [13–15]. The ultimate in vitro tools for identifying MDSCs are functional assays testing the ability of "MDSC-like" cells to suppress T-cell responses [8].

Although MDSCs in cancer patients inhibit anti-tumor immunity and thus promote tumor progression [6,16], the immunosuppressive ability of MDSCs could be exploited to limit further tissue damage in disorders like RA, where the pathological mechanism revolves around the excessive activation of the adaptive immune system [9]. This statement is corroborated by several recently published studies involving successful adoptive cell transfer of MDSCs in animal models of inflammatory bowel disease [17], autoimmune uveoretinitis [18], type I diabetes [19], multiple sclerosis (MS) [14] and collagen-induced arthritis (CIA) [20]. All these studies depended on one key factor: a good source of MDSCs.

In a previous study [21] we reported that synovial fluid (SF) in the joints of mice with proteoglycan (PG)-induced arthritis (PGIA) [22,23], an autoimmune animal model of RA, contained a cell population with a granulocytic phenotype and a biological activity resembling MDSCs. Upon co-culture of SF cells with T cells in the presence of antigen (Ag)-loaded dendritic cells (DCs), T-cell proliferation was profoundly inhibited, thereby confirming the suppressor activity of SF-MDSCs. Experiments employing inhibitors of MDSC products revealed that these cells exerted their suppressive effect via nitric oxide (NO) and reactive oxygen species (ROS) production [21]. SF-MDSCs also significantly inhibited the maturation of DCs through down-regulation of the major histocompatibility class II (MHC II) molecule and the co-stimulatory molecule CD86, resulting in impaired Ag presentation by the affected DCs. Phenotypic characterization revealed that the SF cell population was dominated by CD11b$^+$Gr-1$^+$ Ly6G^{hi}Ly6-C$^{int/lo}$ (granulocyte-like) MDSCs, but cells with CD11b$^+$Gr-1$^+$Ly6G$^-$Ly6C^{hi} (monocytic) phenotype were also detectable [21]. Interestingly, unlike in a recently published study [20] where CD11b$^+$Gr-1$^+$ cells isolated from the spleens of mice with CIA had suppressor activity toward T cells, splenic CD11b$^+$Gr-1$^+$ cells from

mice with PGIA did not suppress T-cell proliferation in vitro, only SF cells did [21]. This led us to the conclusion that the inflammatory microenvironment (e.g., locally produced cytokines and growth factors) within the affected joints has the utmost importance in not only promoting the recruitment of granulocytic precursors, but also keeping these cells in an immature state and endowing them with immune modulatory properties.

Our data suggested that SF-MDSCs could be exploited for therapeutic purposes to prevent the expansion of pathogenic T cells in vivo. However, the amount of SF that could be harvested from the small mouse joints was a serious limiting factor for cell transfer-based therapeutic studies, which prompted us to explore alternative sources of MDSCs. We found that murine BM was an excellent source of MDSCs and their precursors that could be expanded in culture in an appropriate cytokine milieu. We chose in vitro enrichment instead of antibody (Ab)-based positive selection of BM-MDSCs because Ab binding to either CD11b or Gr-1 on the cell surface has been shown by us and others to impair the trafficking, survival, and suppressive function of myeloid cells [11,24,25]. In the present study, we report on the development of a method for generating large amounts of MDSCs from BM (BM-MDSCs) whose characteristics are partially similar to those found in the SF of mice with PGIA, but are more powerful than SF-MDSCs in suppressing the Ag-independent, polyclonal proliferation of T cells. We have also found that BM-MDSCs are able to inhibit the progression of adoptively transferred PGIA following injection of these cells into mice with early arthritis symptoms.

Materials and Methods

Mice, immunization, and assessment of arthritis

Adult female BALB/c mice were obtained from the National Cancer Institute (Frederick, MD). Enhanced green fluorescent protein-lysozyme M transgenic (EGFP-LysM-Tg) mice [26] were back-crossed to the BALB/c background for 10 generations [21,27]. Spleens of naïve PG-specific T cell receptor transgenic (PG-TCR-Tg) BALB/c mice (recognizing a dominant epitope within the G1 domain of human PG [28]) were used as a source of PG/G1-specific T cells. BALB/c mice with the severe combined immunodeficiency (scid) mutation (SCID mice) [27,29] were purchased from the National Cancer Institute.

To induce arthritis, adult female wild type (wt) BALB/c mice were injected intraperitoneally (i.p.) with 100 μg of human PG protein [22] emulsified in dimethyl-dioctadecyl-ammonium bromide (Sigma-Aldrich, St Louis, MO) adjuvant in sterile phosphate buffered saline (PBS) 3 times 3 weeks apart [23,30]. PG was extracted from human cartilage as described previously [21–23]. Cartilage was donated by patients undergoing joint replacement surgery. Written informed consent was obtained from each patient. Collection of surgical specimens was approved by the Institutional Review Board of Rush University Medical Center (Chicago, IL). After the second injection of human PG, the limbs of immunized mice were examined for signs of arthritis. A standard visual scoring system (based on swelling and redness, ranging from 0 to 4 for each paw, 0–16 per mouse) was used for the assessment of disease severity. All experiments involving animals were conducted in accordance with the recommendations of the Guide for the Care and Use of Laboratory Animals of the National Institutes of Health. The animal studies were approved by the Institutional Animal Care and Use Committee of Rush University Medical Center (Permit Number: 11–046).

Collection of serum and cells/organs from mice, and histology

Blood for cell analysis and measurement of anti-PG Ab titers was drawn from mice under deep anesthesia induced by intramuscular injection of a Ketamine-Xylazine cocktail. Mice were then euthanized by carbon dioxide inhalation, and spleen, BM, joint-draining lymph nodes (LNs) were collected. SF was harvested in heparin containing tubes from arthritic ankles, knees, and forepaws at the peak of the disease (inflammation score: 8–16 per mouse) after puncturing and gently pressing of the joints. Red blood cells were eliminated by hypotonic lysis and single cell suspensions were prepared from the harvested tissues and fluids.

For histology, hind limbs were dissected, fixed with formalin, decalcified, and embedded in paraffin. Sagittal sections, cut from the paraffin-embedded tissue, were stained with hematoxylin and eosin and examined under a Nikon Microphot light microscope (Nikon, Melville, NY). Microphotographs of the ankle (tibio-talar) joints were prepared using a digital color CCD camera (Coolsnap; RS Photometrics, Tucson, AR) and MetaMorph software (Molecular Devices, Sunnyvale, CA).

Generation of MDSCs and DCs from BM

MDSCs were generated from BM of naïve wt or EGFP-LysM-Tg BALB/c mice. Femurs and tibias were collected under aseptic condition, and BM was flushed out with sterile PBS. After red blood cell lysis, BM cells were counted (the number of cells was usually $3-4 \times 10^7$ per mouse), and seeded in Petri dishes at a density of 5×10^5 cells per ml of Dulbecco's Modified Eagle Medium (DMEM; Sigma-Aldrich) containing 10% fetal bovine serum (FBS) (Hyclone, Logan, UT). In preliminary dose-finding experiments the BM cells were cultured for 3 to 7 days in the presence of varying doses of recombinant murine granulocyte macrophage colony stimulating factor (rmGM-CSF; Peprotech, Rocky Hill, NJ) and recombinant murine interleukin-6 (rmIL-6; Peprotech), or with a combination of rmGM-CSF, rmIL-6, and recombinant murine granulocyte colony stimulating factor (rmG-CSF; Peprotech). On the basis of phenotypic and functional characteristics, the optimal protocol for BM-MDSC generation was found to be a 3-day culture of BM cells in the presence of rmGM-CSF, rmIL-6, and rmG-CSF (10 ng/ml each).

DCs, as Ag-presenting cells (APCs), were also generated from BM of naïve wt BALB/c mice by culturing BM cells for 9 days in the presence of 40 ng/ml rmGM-CSF, as described previously [21,31].

Phenotypic analysis of cells by flow cytometry

To assess the effect of BM-MDSCs on DC maturation, DCs were cultured alone or in the presence of BM-MDSCs for 2–3 days prior to flow cytometric measurement of the levels of MHC II and CD86 expression. Similarly, T cells were co-cultured with Ag-loaded DCs in the absence or presence of BM-MDSCs (as described below for T-cell proliferation assays), and the effect of BM-MDSCs on regulatory T cell (Treg) differentiation and cytokine production was determined by flow cytometry after intracellular staining (see below).

After harvesting the cells of interest, cells were suspended in flow staining/washing buffer (PBS containing 0.05% bovine serum albumin and 0.05% sodium azide). Prior to surface staining with fluorochrome-tagged mAbs, Fc receptors were blocked with purified anti-CD16/CD32 mAb (Fc block; rat mAb, clone 2.4G2; BD Biosciences, San Diego, CA) for 10 minutes at 4°C. Immunostaining was performed using fluorochrome-conjugated mAbs (obtained from BD Biosciences, eBioscience, or BioLegend,

San Diego, CA) against the following cell surface markers: CD11b (rat mAb, clone M1/70), Gr-1 (rat mAb, clone RB6-8C5), Ly6C (rat mAb, clone HK1.4), Ly6G (rat mAb, clone 1A8), F4/80 (rat mAb, clone RM8), CD115 (rat mAb, clone AFS98), CD80 (hamster mAb, clone 16-10A1), CD11c (hamster mAb, clone N418), MHC II (I-A^d/I-E^d) (rat mAb, clone M5/114.15.2), CD86 (rat mAb, clone GL-1), CD3 (hamster mAb, clone 145-2C11), CD4 (rat mAb, clone GK1.5), and CD25 (rat mAb, clone PC61). Separate cell aliquots were stained with fluorochrome-labeled isotype-matched rat or hamster control IgGs. For detection of Tregs, cells were first stained for CD4 and CD25, permeabilized, and stained for intracellular FoxP3 using a mouse FoxP3 staining kit (Cat. No. 8-8111-40; eBioscience). A staining protocol and a fixation/permeabilization kit (Cytofix/Cytoperm kit with Golgi-Stop from BD Biosciences) were employed to detect intracellular cytokines. In brief, the cells (2×10^6/ml culture medium) were first incubated with 10 ng/ml phorbol-13-myristate acetate (PMA, Sigma), 1 μg/ml Ionomycin (Invitrogen, Grand Island, NY), and 1 μl/ml GolgiStop (2 μM Monensin) for 4 hours. After surface staining for CD4 (rat mAb, clone GK1.5) the cells were fixed, permeabilized, and stained with fluorochrome-conjugated mAb to murine interferon gamma (IFNγ) (rat mAb, clone XMG1.2; BioLegend) or IL-10 (rat mAb, clone JES5-16E3; eBioscience). Flow cytometry was performed using a BD FACS Canto II instrument, and data were analyzed with FACS Diva software (BD Flow Cytometry Systems, San Jose, CA).

Immunofluorescence imaging and cytospin preparations

Occasionally, BM-MDSCs, generated from EGFP-LysM-Tg mice (which express EGFP only in cells of myeloid origin) [31] were used for fluorescence imaging. In brief, after BM-MDSCs were immunostained with fluorochrome-labeled mAbs to Ly6G and Ly6C (specified above), a small aliquot of cell suspension was placed in a 0.5 mm-deep imaging chamber (Invitrogen). The cells were visualized using a Prairie Ultima two-photon microscope system (Prairie Technologies, Middleton, WI), and images were created with Imaris software (Bitplane, South Windsor, CT) as described previously [27].

For analysis of cell morphology, BM-MDSCs or SF cells were spun onto glass slides, air dried, and stained with Wright-Giemsa solution (Sigma-Aldrich). Cytospin preparations were viewed and photographed as described for joint histology.

Measurement of GM-CSF, IL-6, and G-CSF levels in mouse serum and SF

Concentrations of GM-CSF, IL-6, and G-CSF in serum and cell-free SF samples of arthritic mice were measured using sandwich enzyme-linked immunosorbent assay (ELISA) kits from Peprotech (Cat. No. 900-M55, 900-M50, and 900-K103, respectively). Serially diluted (1:50–1:400) serum and SF samples and the appropriate standards were incubated in plates coated with anti-GM-CSF, anti-IL-6, or anti-G-CSF Abs, and plate-bound material was detected according to the manufacturer's instructions. Absorbance at 450 nm was read by a Synergy 2 ELISA reader (BioTek Instruments, Winooski, VT).

Purification of T cells and depletion of Ly6C^{hi} monocytic MDSCs

T cells were purified from the spleens of naive PG-TCR-Tg BALB/c mice by negative selection using an EasySep Mouse T Cell Enrichment Kit (Cat. No. 19751; StemCell Technologies, Vancouver, BC, Canada). The purity of T cells, verified by flow cytometry, was greater than 95% in all cases.

Depletion of Ly6C^{hi} (monocytic) cells from the total BM-MDSC population was carried out using an EasySep Biotin Selection Kit (StemCell Technologies). Unwanted cells were targeted with a biotinylated mAb against Ly6C (rat mAb, clone HK1.4), followed by immunomagnetic depletion of the mAb-tagged cells. This resulted in the removal of essentially all Ly6C^{hi} (but not Ly6C$^{int/lo}$Ly6G^{hi}) BM-MDSCs, as confirmed by flow cytometry.

Assays for determining MDSC-mediated suppression of T-cell proliferation

For assessment of suppression of Ag-dependent T-cell proliferation, first the DCs were cultured overnight with the recombinant G1 domain of human PG (rhG1; 7.5 μg/ml) [32] as Ag in the absence or presence of BM-MDSCs, Ly6C^{hi} cell-depleted BM-MDSCs, or SF cells (as suppressors) in quadruplicate wells of 96-well plates. T cells purified from the spleens of naive PG-TCR-Tg mice were added and co-cultured for 5 days at a T cell:DC:suppressor cell ratio of 1:0.3:1. Background controls included the following: T cells and DCs co-cultured without Ag (rhG1) and each suppressor population cultured alone for the same length of time (5 days). The cells were pulsed with [^{3}H]thymidine (Perkin Elmer, Waltham, MA) at 1 μCi/well for the last 18 hours of culture, and isotope incorporation (counts per minute: cpm) was measured in a MicroBeta scincillation counter (Perkin Elmer).

To assess Ag-independent suppression of T-cell proliferation, 96-well plates were first coated with purified mAbs against CD3 (hamster mAb, clone 145-2C11) and CD28 (hamster mAb, clone 37.51) (1 μg of each per well in 100 μl sterile sodium carbonate buffer, pH 9.6). T cells were added to the coated wells alone, or with an equal number of BM-MDSCs, Ly6C^{hi} cell-depleted BM-MDSCs, or SF cells as suppressors. Background controls were T cells cultured in uncoated wells and suppressors cultured in anti-CD3/CD28-coated wells. T-cell proliferation was measured on day 4 of culture as described above.

In all cases, the results of proliferation assays were expressed as percent suppression [15] (after correction for backroud proliferation) using the following equation:

% suppression = 100– [(cpm with suppressors/cpm without suppressors) ×100].

To inhibit MDSC-mediated suppression, the following inhibitors of MDSC products [21] were added to the co-cultures of T cells, Ag-loaded DCs, and BM-MDSCs (or anti-CD3/CD28-stimulated T cells and BM-MDSCs): N$^\omega$-hydroxy-nor-arginine (nor-NOHA; 0.5 mM), an inhibitor of arginase 1; N^G-monomethyl-L-arginine acetate (L-NMMA; 0.5 mM) and 1400W (0.1mM), inhibitors of inducible nitric oxide synthase (iNOS); Z-VAD-FMK (0.1 mM), an inhibitor of caspases and caspase-mediated apoptosis (all inhibitors were purchased from Calbiochem, Gibbstown, NJ); or the ROS scavenger catalase (1,000 U/ml, Sigma-Aldrich). Cell proliferation results were expressed as % suppression in the presence and absence of each inhibitor.

Reverse transcription-polymerase chain reaction (RT-PCR)

As described in our previous study [21], the transcript for murine iNOS (Nos2) was expressed at much lower levels in spleen cells than SF cells obtained from arthritic mice. Therefore, we used spleen cells as a reference control to determine if the Nos2 gene was also upregulated in BM-MDSCs. Total RNA was isolated from BM-MDSCs and control spleen cells using TRI reagent (Sigma-Aldrich) according to the manufacturer's instructions. cDNA was synthesized employing a SuperScript First Strand kit (Invitrogen), and PCR was performed using HotStart Taq Plus enzyme (Qiagen, Carlsbad, CA) in 35 cycles (95°C for 20 sec, 57°C for 30 sec, and 72°C for 45 sec) with a final extension at 72°C for 10 min in a C1000 Thermal Cycler (Bio-Rad, Hercules, CA). A murine Nos2-specific primer pair (Nos2 forward 5′-CCCTTCCGAAGTTTCTGGCAGCAGC-3′, and Nos2 reverse 5′-GGCTGTCAGAGCCTCGTGGCTTTGG-3′) was used to detect the Nos2 transcript, and an Actb gene-specific primer pair (Actb forward 5′-TGGCTCCTAGCACCATGAAGATCA-3′ and Actb reverse 5′-ATCGTACTCCTGCTTGCTGATCCA-3′) served for detection of the housekeeping gene encoding β-actin. After amplification, samples were loaded onto 1.5% agarose gels.

Western blot

BM-MDSCs and control spleen cells were lysed in cold RIPA buffer containing a Halt protease inhibitor mixture (Pierce/Thermo Fisher, Rockford, IL), and the protein content was determined using the bichinconic acid assay (Pierce). Proteins from cell lysates (20 μg protein each) were loaded onto and resolved in 7.5% SDS-PAGE gels (Bio-Rad) under reducing conditions. The proteins were then transferred to nitrocellulose membranes. The membranes were blotted with an anti-mouse iNOS mAb (mouse mAb, Cat. No. sc-7271; Santa Cruz Biotechnology, Dallas, TX) at a 1:500 dilution. Horseradish peroxidase (HRP)-conjugated rabbit anti-mouse IgG1 (Invitrogen) was used as a secondary Ab at a 1:10,000 dilution. The protein bands were visualized using the enhanced chemiluminescence method (Amersham/GE Healthcare Life Sciences, Piscataway, NJ). The membranes were stripped and re-probed with a HRP-conjugated mAb to β-actin (mouse mAb, clone mAbcam 8226; Abcam, Cambridge, MA) at a 1:5,000 dilution to ensure equal sample loading.

Measurement of iNOS activity

To measure iNOS enzymatic activity (NO production) in the supernatants of 2-day co-cultures of murine BM-MDSCs, DCs and T cells, a nitrite/nitrate colorimetric assay was performed according to the manufacturer's protocol (Cayman Chemical, Ann Arbor, MI). Supernatants of spleen cell cultures, containing the same number of cells as the co-cultures, were used as a reference. Samples were run on a BioTek microplate reader and absorbance was measured at 540 nm. A standard curve was generated using nitrate standards serially diluted between 5 μM and 35 μM. Results were expressed as total nitrate concentration (μM).

Induction of adoptively transferred PGIA in SCID mice and BM-MDSC transfer

Adoptive cell transfer from wt BALB/c to SCID BALB/c mice is an ideal tool for investigating the in vivo distribution and effects of donor cells, as the syngeneic SCID mice exhibit complete tolerance to the wt donor cells, allowing these cells (e.g., lymphocytes) to expand rapidly in vivo [33]. SCID BALB/c mice also develop PGIA after spleen cell transfer from arthritic donors in a more uniform and synchronous manner than wt BALB/c mice following PG immunization [27,33]. To induce adoptively transferred PGIA in SCID BALB/c mice, spleen cells from arthritic wt BALB/c donors were injected intravenously into SCID recipients (~10^7 cells/mouse). At the time of spleen cell transfer, SCID mice also received 100 μg of human PG (without adjuvant) i.p. to re-activate the donor cells in vivo [27,33]. When arthritis started to develop (day 15 after the first splenocyte transfer), mice were divided into two groups (n = 10 mice/group) with mean disease scores of 2.0 and 2.05, respectively. One group received a second transfer of 10^7 splenocytes with 100 μg of human PG i.p., and the other group was co-injected i.p. with spleen cells and BM-MDSCs (~10^7 of each cell type/mouse) together with the same dose of PG. Control mice (injected with only splenocytes and PG twice) and BM-MDSC-treated mice (also receiving BM-MDSCs with the second injection of spleen cells and PG) were examined twice a week for disease severity and scored as described for the primary form of PGIA. Mice were sacrificed on day 34 after the first cell transfer for determination of joint histopathology and PG-specific immune responses. Arthritis severity data were collected from 2 independent experiments, each having 5 mice per group (10 mice total per group).

We also carried out a separate experiment on a limited number of SCID mice (n = 3) to assess the distribution of transferred BM-MDSCs in various tissues. In order to distinguish the transferred cells from the recipients' own MDSCs, BM-MDSCs were generated from EGFP-LysM-Tg BALB/c mice that express EGFP in myeloid cells [27,31]. As described above, SCID mice injected with PG and 10^7 spleen cells from arthritic wt BALB/c mice on days 0 and 15 also received 5×10^6 EGFP$^+$BM-MDSCs i.p. on day 15. This amount of BM-MDSCs (half of the therapeutic dose) only weakly inhibited arthritis progression, which enabled us to detect donor cells in the SF of recipient mice. On day 34, blood, SF, BM, spleen, and joint-draining LNs were harvested from the recipient mice. The cells were immunostained for CD11b, Ly6C, and Ly6G, and the subset composition of EGFP$^+$CD11b$^+$ donor cells was determined by flow cytometry.

Measurement of PG-specific T-cell responses and serum Abs in SCID mice

Spleens of SCID mice were harvested and splenocytes were seeded in 96-well culture plates at a density of 2×10^5 cells per well in DMEM containing 10% FBS in the presence or absence of purified human PG (25 μg/ml) as Ag in triplicate wells. Cells were cultured for 5 days, and proliferation was measured on the basis of [^{3}H]thymidine incorporation. Results were expressed as stimulation index (SI), which is a ratio of isotope incorporation (cpm) by PG-stimulated and non-stimulated cultures.

Concentrations of PG-specific Abs in the sera of SCID mice were determined by ELISA as described elsewhere [27,30,32]. Briefly, MaxiSorp ELISA plates (Nunc, Denmark) were coated with human PG (0.75 μg/well) overnight. Unbound material was washed out, and the wells were blocked with 1.5% fat-free milk in PBS. Serially diluted (1:100–1:200,000) serum samples from individual mice, and internal standard samples (pooled serum from arthritic BALB/c mice, containing known amounts of PG-specific IgG1 and IgG2a) were incubated with the immobilized PG. PG-specific IgG1 and IgG2a were detected using HRP-conjugated secondary Abs (Invitrogen), followed by HRP substrate and o-phenylene-diamine (Sigma) as chromogen. Optical densities were measured at 490 nm in an ELISA reader. Data were expressed in mg/ml serum (PG-specific IgG1) or μg/ml serum (PG-specific IgG2a).

Statistical analysis

Results are expressed as the means ± SEM unless noted otherwise. Statistical analysis was performed using GraphPad Prism 6 program (GraphPad Software, La Jolla, CA). For comparison of two groups of data, the parametric Student's t test or the non-parametric Mann-Whitney U test was employed. Multiple comparisons were performed using the Kruskal-Wallis test followed by Dunn's multiple comparisons test. Data resulting from repeat measurements over time were analyzed using two-way repeated measures analysis of variance. P values of less than 0.05 were accepted as statistically significant.

Results

Murine BM cells cultured in the presence of G-CSF, GM-CSF, and IL-6 give rise to a cell population resembling SF-MDSCs

The primary goal of this study was to establish a culture method by which BM cells can be enriched in myeloid cells resembling SF-MDSCs in both their phenotype and function. We chose BM because it is the body's largest reservoir of myeloid precursors from which large numbers of MDSCs can be generated under appropriate conditions. GM-CSF is essential for the survival and suppressor activity of MDSCs [34], and one study reported successful generation of MDSCs from human blood in 7 days with a combination of GM-CSF and IL-6 [35], factors that are also present in the SF of RA patients [36]. In preliminary experiments, we sought to determine whether BM cells cultured for 3 to 7 days in the presence GM-CSF and IL-6 acquire an SF-MDSC-like phenotype. Although the BM culture became enriched in CD11b$^+$ cells under this condition as determined by flow cytometry, unlike SF-MDSCs, only a small proportion of these myeloid cells expressed Ly6G, a marker of granulocytic MDSCs (data not shown). We added G-CSF as a booster of the granulocytic lineage to the BM culture, which resulted in the rise of cell populations expressing Ly6G alone, or co-expressing Ly6G (at high levels) with low-to-intermediate levels of the monocytic MDSC marker Ly6C (Fig. 1A, and Fig. 1C [left panel]). This overall phenotype was achieved in 3 days of culture in the presence of GM-CSF, IL-6, and G-CSF (10 ng/ml each); longer culture or higher doses of G-CSF did not result in increases in Ly6G$^+$ or double Ly6C$^+$Ly6G$^+$ cells (data not shown). In comparison with CD11b$^+$ SF cells (Fig. 1B and D), BM-MDSCs contained fewer double Ly6-C$^+$Ly6G$^+$ cells and higher proportions of subsets expressing only one of these markers (Fig. 1A and C). However, cells co-expressing Ly6C and Ly6G clearly represented a dominant population in both the BM-MDSC cultures and freshly harvested SF (Fig. 1A–D). Our choice of the combination of growth factors to generate SF-MDSC-like cells from BM was supported by the finding that SF from mice with PGIA contained high amounts of GM-CSF and G-CSF, and detectable amounts of IL-6. In each case, the SF concentrations of these factors exceeded the serum levels (Table S1).

Immunofluorescence staining of BM-MDSC-like cells generated from EGFP-LysM mice (cultured for 3 days as described above), followed by imaging with TPM demonstrated that the majority of myeloid (EGFP$^+$) cells expressed either Ly6G or Ly6C, or both markers (Fig. 1C, middle panel). Both polymorphonuclear granulocyte (neutrophil)-like cells (Fig. 1C, right panel: arrows) and large precursor-like cells (Fig. 1C, right panel: arrowheads) were seen in the cytospin preparations of such cells.

Overall, the flow cytometry profile and morphology of BM-MDSC-like cells (Fig. 1C) demonstrated greater heterogeneity than those of fresh SF cells (Fig. 1D), suggesting that in addition to the dominant population of double-positive Ly6G^{hi}Ly6C$^{int/lo}$ cells (also present in SF), BM-MDSC cultures contained a variety of immature myeloid cells with intermediate phenotypes.

Ly6G is highly expressed by both mature neutrophils and granulocytic MDSCs in mice [21,37], and no additional surface markers are available to distinguish between these two types of cells. On the other hand, among monocytic cells, classical (or "inflammatory") monocytes are characterized by high expression of Ly6C, whereas non-classical (also termed "patrolling" or "anti-inflammatory") monocytes/macrophages express Ly6C at low levels [15]. We used additional mAbs against monocyte/macrophage markers, including F4/80, CD115, and CD80 to identify

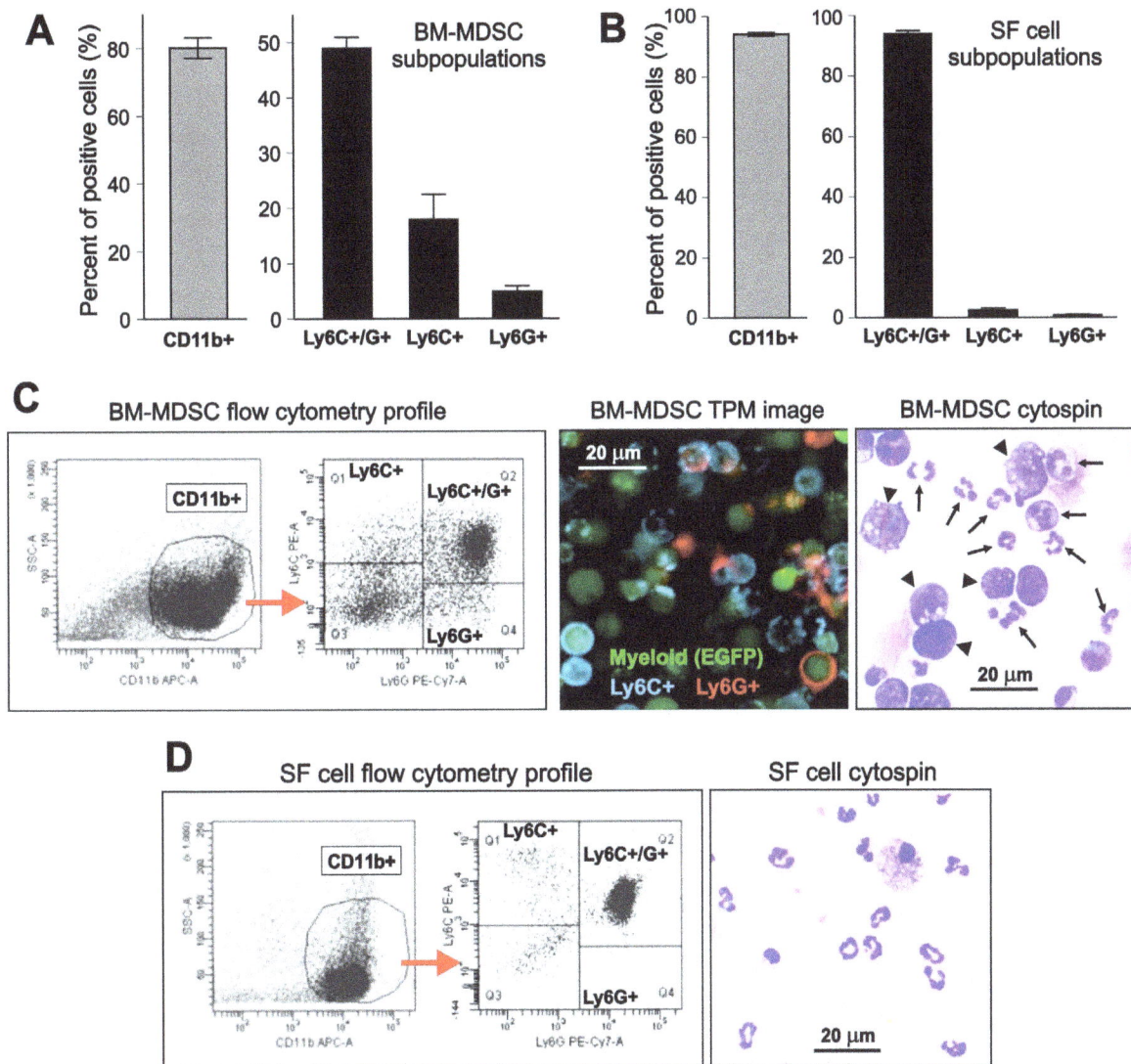

Figure 1. Phenotype and morphology of myeloid-derived suppressor cell (MDSC)-like cells generated in vitro from murine bone marrow (BM) in comparison with synovial fluid (SF) cells. (**A**) Phenotype of MDSCs arising from growth factor-cytokine treated BM cell cultures as determined by flow cytometry. BM cells were cultured in the presence of GM-CSF, IL-6, and G-CSF (10 ng/ml each). On day 3, cells were immunostained for CD11b, Ly6C, and Ly6G. Approximately 80% of the cells expressed the common myeloid marker CD11b (gray bar). Gating on CD11b$^+$ cells revealed that the majority of them co-expressed Ly6C (marker of the "monocytic" subset) and Ly6G (marker of the "granulocytic" subset), but cells expressing only one marker were also present (black bars). The results are the means ± SEM of 7 independent BM cultures. (**B**) In SF, the vast majority of the CD11b$^+$ myeloid population (gray bar) was found to be cells co-expressing Ly6G and Ly6C, and lower proportions of cells expressed Ly6C or Ly6G only (black bars) than in the BM-MDSC cell cultures. The results are the means ± SEM of 7 separate pools of SF cells freshly harvested from arthritic mouse joints. (**C**) Flow cytometry profile of BM-MDSCs (left panels) is shown as an example of subset identification after gating on CD11b$^+$ cells. Fluorescence image of EGFP$^+$ BM-MDSCs (middle panel) after surface staining with a blue fluorescent antibody to Ly6C and a red fluorescent antibody to Ly6G shows cells expressing one or both markers. BM for culture was obtained from an EGFP-LysM-Tg mouse expressing EGFP (green fluorescence) in myeloid cells. Imaging was performed using two-photon microscopy (TPM). Morphology of BM-MDSCs (right panel) was visualized by Wright-Giemsa staining of a cytospin preparation, which shows both polymorphonuclear granulocyte (neutrophil)-like cells (arrows) and large precursor-like cells (arrowheads). (**D**) Flow cytometry profile (left) and morphology (right) of SF cells harvested from the arthritic joints of mice with PGIA. While the CD11b$^+$ myeloid population is large in both the BM-MDSC culture and arthritic SF, and is dominated by Ly6C/Ly6G double positive cells in both samples (analyzed simultaneously), BM-MDSCs show greater heterogeneity in morphology than SF cells.

distinct subsets within the two monocytic cell categories. However, we found that only a few percent of cultured BM-MDSCs or freshly harvested SF cells expressed F4/80 and CD80, and less than 1% of them was CD115$^+$ (Fig. S1). The highest proportions of F4/80$^+$ and CD80$^+$ cells were detected within the Ly6C$^{lo/-}$ population among BM-MDSCs (4–5%) (Fig. S1A) and in the Ly6C$^{hi/int}$ population among SF cells (0.5–2%) (Fig. S1B).

CD115$^+$ cells represented 0.2% of both Ly6C$^{hi/int}$ and Ly6C$^{lo/-}$ BM-MDSCs (Fig. S1A), and 0.7% of Ly6C$^{lo/-}$ SF cells (Fig. S1B).

As CD11b$^{lo/-}$Ly6C^{hi}CD115$^+$ osteoclast precursors have been identified in the BM of mice with inflammatory arthritis [38], we also screened the CD11b$^{lo/-}$ populations of cultured BM-MDSCs and freshly harvested SF for the presence of such osteoclast precursor-like cells. However, we could not detect CD115$^+$ cells in either the Ly6C$^{hi/int}$ or the Ly6C$^{lo/-}$ fraction of CD11b$^{lo/-}$ BM-MDSCs (Fig. S2A) or SF cells (Fig. S2B).

BM-MDSCs have the ability to suppress both Ag/DC-dependent and -independent proliferation of T cells in vitro

To study the effect of BM-MDSC-like cells on Ag-specific T-cell proliferation, we cultured Ag (rhG1)-loaded DCs with T cells isolated from the spleens of naive PG-TCR-Tg mice in the presence or absence of BM-MDSCs as suppressors. Additional "suppressors" (as comparators) were SF cells, and BM-MDSCs depleted in Ly6C^{hi} cells. Ag-dependent T-cell proliferation was dramatically reduced in the presence of BM-MDSCs, i.e., BM-MDSC-mediated suppression reached nearly 100% (Fig. 2A, red bar). Compared with SF cells (Fig. 2A, gray bar) BM-MDSCs were equally potent in suppressing T-cell proliferation. As also reported for SF cells [21], depletion of the Ly6C^{hi} monocytic subset from the BM-MDSCs (Fig. 2A, black bar) did not reduce their suppressive capacity. BM-MDSC-mediated suppression of Ag-specific T-cell proliferation was accompanied by significant decreases in the percentage of CD4$^+$ T helper (Th) cells containing intracellular cytokines (IFNγ in Th1 and IL-10 in Th2 cells) as well as in the percentage of Tregs (CD4$^+$CD25$^+$ cells containing FoxP3) (Fig. 2B).

Since we found previously that SF-MDSCs from arthritic mice suppressed the maturation and Ag presenting capacity of DCs [21], we investigated the effect of BM-MDSCs on the expression levels of DC maturation markers MHC II and CD86. However, we could not detect significant changes in the expression level of either marker in DCs upon co-culture with BM-MDSCs (Fig. S3). Since BM-MDSCs failed to decrease the surface expression of these molecules by the DCs, this experiment also suggested that the observed suppression of the proliferation and cytokine/FoxP3 content of T cells was not due to release of cytotoxic substances from the BM-MDSCs.

To determine whether the suppressive effect of BM-MDSCs on T-cell proliferation was Ag-dependent (for which the presence of DCs was required) or Ag-independent, we stimulated the PG-TCR-Tg T cells with anti-CD3 and anti-CD28 mAbs in the presence or absence of BM-MDSCs. In this Ag/DC-independent system, BM-MDSCs also exhibited potent suppressor activity in (Fig. 2C, red bar), whereas suppression by SF cells was very weak (Fig. 2C, gray bar), consistent with our previous report [21]. As expected, depletion of Ly6C^{hi} cells did not reduce the capacity of BM-MDSCs to suppress the anti-CD3/CD28-induced proliferation of T cells (Fig. 2C, black bar).

The suppressive effects of BM-MDSCs on T cells can be reversed by iNOS inhibitors in vitro

To reveal the possible mechanism of the suppressive activity of the BM-MDSCs, we repeated the Ag-dependent and Ag-independent T-cell proliferation assays with and without various inhibitors of MDSC-related effector molecules such as arginase 1 (nor-NOHA), iNOS (L-NMMA and the more selective 1400W), and ROS (catalase). A caspase (apoptosis) inhibitor (Z-VAD-FMK) was used as a MDSC-unrelated control. Both Ag (rhG1)- and anti-CD3/CD28-induced T-cell proliferation remained suppressed in the presence of the arginase 1 inhibitor, the ROS scavenger, or the caspase inhibitor (Fig. 3A and B). However, BM-MDSCs lost much of their ability to suppress T-cell proliferation in both induction systems in the presence of the iNOS inhibitors (Fig. 3), suggesting that the main MDSC product mediating T-cell suppression was NO.

BM-MDSCs exhibit upregulated iNOS expression and elevated NO production

To corroborate the results of T-cell proliferation assays indicating a role for NO in the suppressor activity of BM-MDSCs, we performed RT-PCR to assess expression of iNOS (Nos2) mRNA in BM-MDSCs in comparison with spleen cells harvested from arthritic mice [21]. BM-MDSCs demonstrated significant up-regulation of Nos2 mRNA as compared with spleen cells (Fig. 4A), while the housekeeping gene (Actb, encoding β-actin) was expressed at equal levels. The results of Western blot were consistent with the results of RT-PCR, showing a large amount of iNOS protein ($\sim$130 kDa) in BM-MDSCs, but not in spleen cells (Fig. 4B).

The enzymatic activity of iNOS was assessed by measuring nitrite/nitrate concentrations (as indicators of NO production) in supernatants of BM-MDSCs (cultured in the presence of DCs and rhG1 with or without T cells) and spleen cell cultures. Consistent with the iNOS expression data, much higher levels of NO were detected in the supernatants of BM-MDSCs-containing cultures (Fig. 4C, orange bar) than in those of spleen cell cultures (Fig. 4C, green bar).

Injection of BM-MDSCs into SCID mice reduces Ag-specific immune responses and ameliorates adoptively transferred arthritis

To test whether BM-MDSCs could affect the development of arthritis, an adoptive transfer model of PGIA was employed. On day 0, spleen cells from arthritic wt BALB/c donor mice were injected with Ag (human PG) into SCID recipients. When the clinical signs of arthritis started to develop (15 days after the first injection), the SCID mice were divided into 2 groups with similar mean disease scores, and a second injection was administered. The first (control) group received only arthritic spleen cells and PG, while the second group received the same plus BM-MDSCs. Arthritis severity scores in the control group increased further (Fig. 5A, black line), while, in sharp contrast, the scores of SCID mice transferred with BM-MDSCs remained low until the end (day 34) of the monitoring period (Fig. 5A, red line). Histopathology revealed massive leukocyte infiltration and synovial hyperplasia as well as cartilage erosion in the ankle (tibio-talar) joints of control SCID mice transferred with spleen cells from arthritic donors (Fig. 5B, left panel). In contrast, only mild synovial hyperplasia was observed without evidence of gross inflammation or cartilage damage in the ankle joints of SCID mice co-transferred with spleen cells and BM-MDSCs (Fig. 5B, right panel).

To determine whether the BM-MDSC-mediated protection from arthritis progression was associated with reduced Ag-specific T-cell responses and Ab production, we compared the PG-specific T-cell responses and serum IgG1 and IgG2a Ab levels in control and BM-MDSC-injected SCID mice. PG-specific T-cell proliferation was significantly lower in the BM-MDSC-injected group (Fig. 5C). Serum levels of IgG1 isotype (but not of IgG2a isotype) anti-PG Abs were also significantly reduced in the BM-MDSC recipient group (Fig. 5D).

In a separate experiment, we assessed the distribution and subset composition of transferred EGFP$^+$ BM-MDSCs in various fluids and tissues (blood, SF, BM, spleens, and LNs) of SCID mice with adoptively transferred PGIA (induced as described above) 19 days after BM-MDSC injection. The donor EGFP$^+$ BM-MDSCs (injected at half of the optimal therapeutic dose) were found in considerable amounts in the blood (Fig. 6A), SF (Fig. 6B), and BM (Fig. 6C) of the recipient mice. The spleen (Fig. 6D) contained a

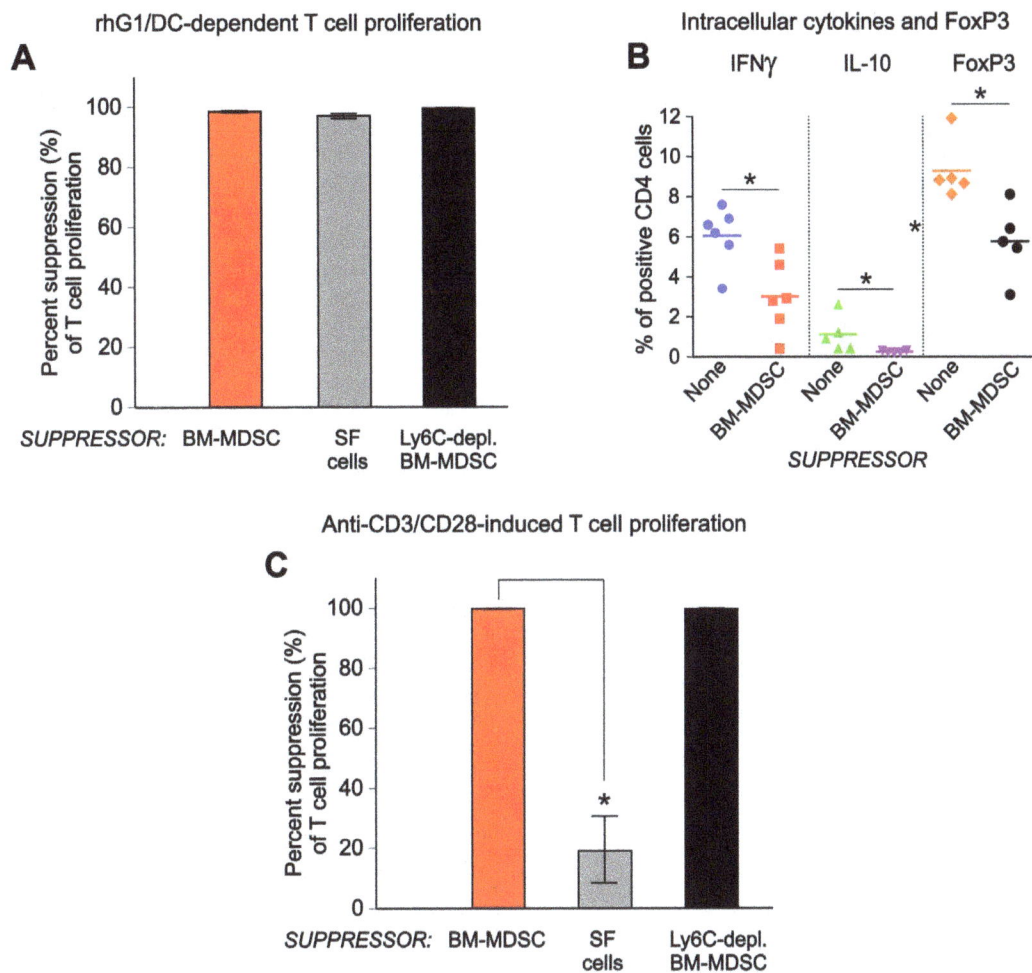

Figure 2. Suppression of antigen (Ag)-specific and non-specific T-cell responses by BM-MDSCs. (**A**) T cells, purified from the spleens of mice expressing a PG-specific T cell receptor transgene (PG-TCR-Tg) were cultured for 5 days with dendritic cells (DCs) loaded with recombinant G1 domain of human PG (rhG1) in the absence or presence of the following "suppressors": BM-MDSCs (red bar), arthritic SF cells (gray bar), or Ly6C^{hi} (monocytic) cell-depleted BM-MDSCs (black bar). The ability of suppressors to inhibit Ag (rhG1)-specific T-cell proliferation (which is also dependent on Ag presentation by DCs) was assessed on the basis of inhibition of [^{3}H]thymidine incorporation by the T cells. Percent suppression was calculated as described in the Methods. All suppressors exhibited robust inhibition of T-cell proliferation. The results shown are from 5 independent experiments. (**B**) T cells from PG-TCR-Tg mice were cultured for 2 days with rhG1-loaded DCs and BM-MDSCs as described for panel A. The percent of CD4$^+$ T cells containing IFNγ, IL-10, or FoxP3 (CD4$^+$CD25$^+$FoxP3$^+$ T regulatory cells, Tregs) was determined by flow cytometry. The results shown are the individual values (n = 5–6) and the means. On average, the percentages of IFNγ$^+$ cells, IL-10$^+$ cells, and Tregs were lower in the presence of BM-MDSCs (*p<0.001, 0.001, and 0.05, respectively; Mann-Whitney U test) than in their absence (None). (**C**) T cells from PG-TCR-Tg mice were cultured in anti-CD3/CD28-coated plates for 4 days in the absence or presence of the listed suppressors. Percent suppression was calculated and results expressed as described for panel A. Non-depleted BM-MDSCs and BM-MDSCs depleted in Ly6C^{hi} cells were equally potent in suppressing anti-CD3/CD28-induced T-cell proliferation, while arthritic SF cells exhibited much weaker inhibition (*p<0.01, n = 5; Kruskal-Wallis test followed by Dunn's multiple comparisons test) in this induction system.

much lower percentage of these cells, and the LNs (Fig. 6E) were virtually free of BM-MDSCs. In each tissue or fluid, the granulocytic subset (Ly6G^{hi}Ly6C^{int}) dominated, although small populations of monocytoid (Ly6C^{hi}Ly6G$^-$) MDSCs were also present (Fig. 6).

Discussion

MDSCs have been described as innate immune cells with a remarkable capacity to control adaptive immune responses [8]. Although first described in tumor-bearing animals and cancer patients [5,39], MDSCs have been recently identified in a variety of autoimmune conditions [14,18–20] that are characterized by excessive activation of the adaptive immune system.

We have reported previously that the SF of mice with PGIA, an animal model of RA, contains a population of cells that meets the criteria of MDSCs [21]. Our data suggested that the inflamed joint in PGIA is a supportive microenvironment in which myeloid cells survive and acquire a mainly granulocytic MDSC-like phenotype and potent suppressor activity toward DCs and Ag-specific T cells. By limiting the expansion of pathogenic T cells and the maturation of DCs locally or in lymphoid organs, MDSCs present in the SF could suppress local inflammation or prevent the spreading of arthritis to other joints. This hypothesis would be best tested by transferring SF-MDSCs to mice at the early phase of PGIA. However, the number of cells that can be collected from murine SF is limited, and SF-MDSCs do not expand in culture [21]. Thus, we sought an alternative source of cells to generate large quantities

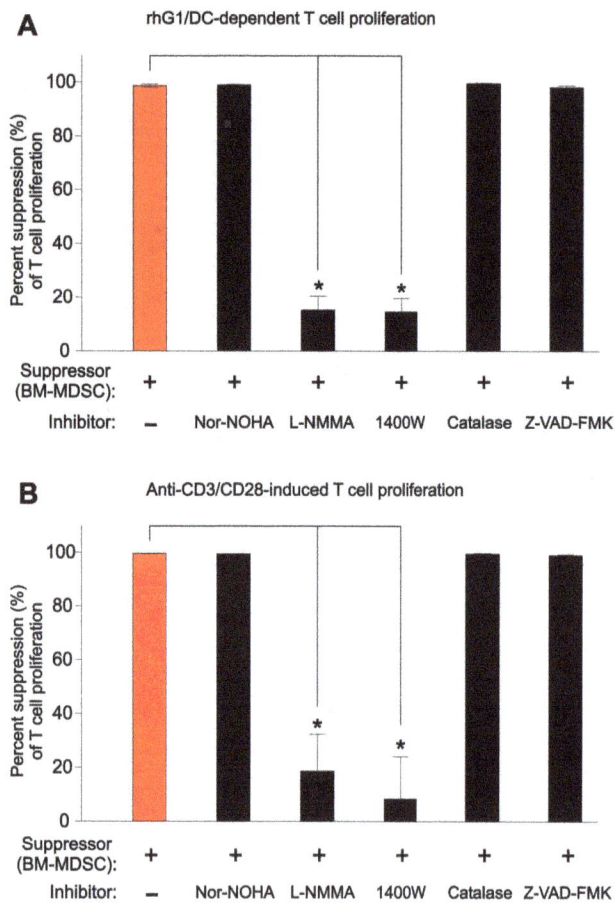

Figure 3. Reversal of the suppressive effect of BM-MDSCs on T-cell proliferation by inhibitors of inducible nitric oxide synthase (iNOS). Various inhibitors of MDSC effector molecules, including the arginase 1 inhibitor nor-NOHA, iNOS inhibitors L-NMMA and 1400W, the reactive oxygen species (ROS) scavenger catalase, and the caspase/apoptosis inhibitor Z-VAD-FMK, were used to inhibit the BM-MDSC-mediated suppression of (**A**) Ag (rhG1)-induced/DC-dependent and (**B**) anti-CD3/CD28-induced proliferation of PG-TCR-Tg T cells. The results (compiled from 2 independent series of experiments, each with 2 co-cultures) are expressed as percent suppression of T-cell proliferation in the presence (black bars) or absence (red bar) of inhibitors. While suppression of T-cell proliferation in both induction systems was significantly reversed by the iNOS inhibitors L-NMMA and 1400W (*p<0.0001 in all cases; Kruskal-Wallis test followed by Dunn's multiple comparisons test), none of the other inhibitors had a significant effect on BM-MDSC-mediated suppression of T cells.

of MDSCs for potential therapeutic intervention (via cell transfer) in PGIA. Studies by others reported accumulation of MDSCs in the spleens of mice with cancer [13] or autoimmune diseases [14,20]. However, previously we found that in mice with PGIA, the splenic myeloid population was modest in size and lacked suppressor activity [21]. Because the BM contains considerable amounts of myeloid precursors, it appeared to be a plausible source of cells with a potential to become SF-MDSC-like cells under appropriate culture conditions.

Enrichment of murine BM in immature myeloid cells in vitro in the presence of GM-CSF was described before, but these cells showed a tendency to become myeloid DCs if no other factor was added [40]. On the other hand, Lechner et al. [35] reported generation of cells with monocytic MDSC phenotype and immunosuppressive capacity by culturing normal human PBMCs

in the presence of GM-CSF and IL-6. As we have shown in this study, murine BM cells treated with a combination of GM-CSF, IL-6, and G-CSF for 3 days give rise to a dominant population of Ly6G^{hi}Ly6C$^{int/lo}$ granulocytic SF-MDSC-like cells, although such cultures also contain smaller populations of cells with intermediate phenotypes. Further phenotypic analysis of the monocytic subset using mAbs to monocyte/macrophage markers F4/80, CD115, and CD80 failed to detect a distinct population expressing these markers, although approximately 5% of BM-MDSCs in the Ly6C$^{lo/-}$ fraction could be defined as CD11b^{+}F4/80^{+}CD80^{+} macrophages. We found a small population (0.7%) of cells expressing CD115, the receptor for macrophage colony stimulating factor among CD11b^{+}Ly6C$^{lo/-}$ SF cells. A previous study reported increased number of Ly6C^{hi}CD115^{+} osteoclast precursor cells in the BM of mice with inflammatory arthritis [38]. Such precursors were identified within the CD11b$^{lo/-}$ population of BM cells and also had myeloid suppressor activity [38]. In our case, CD11b$^{lo/-}$Ly6C$^{hi/int}$ fractions of both BM-MDSCs and SF were devoid of CD115^{+} cells. Although we cannot rule out the possibility that the few CD115^{+} cells found in the CD11b^{+}Ly6C$^{lo/}$$^{-}$ SF population may differentiate into mature osteoclasts, the findings suggest that neither the BM-MDSC culture condition nor the SF milieu is conducive to the development of CD11$^{lo/}$$^{-}$Ly6C^{hi}CD115^{+} osteoclast precursors.

We have suggested previously [21] that MDSC precursors entering the joints may acquire a maturation-resistant phenotype in a milieu rich in myelopoietic growth factors and cytokines. Indeed, we found high levels of GM-CSF and G-CSF, and detectable levels of IL-6 in the cell-free SF of mice with PGIA. Similar to the findings reported by Wright et al. [36] in SF and serum samples from RA patients, we detected much higher levels of these factors in the SF than in the serum of mice with PGIA. These observations indicate that GM-CSF, G-CSF, and IL-6 (besides other pro-inflammatory mediators) are produced locally by joint-resident cells in both RA and PGIA, and likely support the survival and suppressor activity of MDSCs in the SF.

The MDSC-like cells that we generated from murine BM under the conditions described were true MDSCs, as they exerted profound suppressive effects on both the Ag-specific and non-specific (polyclonal) proliferation of T cells in vitro. While both BM-MDSCs and SF cells inhibited the expansion of Ag-stimulated T cells to a comparable degree, BM-MDSCs were much more potent than SF cells in suppressing anti-CD3/CD28-induced polyclonal T-cell proliferation. This suggests that the suppressive ability of SF cells is selective, while BM-MDSCs are capable of inhibiting T-cell responses to either Ag-specific or non-specific stimuli. With regard to BM-MDSCs, our finding is consistent with a recently published study [41] in which CD11b^{+}Gr-1^{+} "immature" myeloid cells, isolated from the BM of normal mice, have been found to suppress the Ag-independent (anti-CD3/CD28- or mitogen-induced) proliferation of T cells.

Although it has been suggested that Ly6G^{-}Ly6C^{hi} monocytic MDSCs suppress T-cell activity more strongly than the granulocytic subset [34], we found that upon depletion of Ly6C^{hi} subpopulation, the BM-MDSCs retained their suppressive ability toward T cells. This supports our previous [21] and other authors' [14,20] conclusions that Ly6G^{+} granulocytic MDSCs represent a subset with potent suppressor activity.

Our experiments elucidating the molecular mechanisms of BM-MDSC-mediated suppression revealed that inhibitors of iNOS were able to reverse both the Ag-specific and non-specific suppression of T-cell proliferation. Consistent with this observation, iNOS was upregulated in BM-MDSCs at both mRNA and at protein levels, and NO was present in high quantities in the

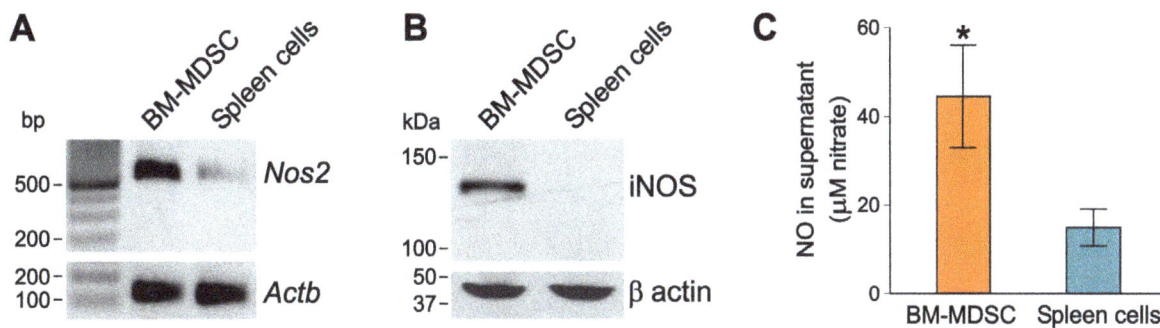

Figure 4. Expression and activity of iNOS in BM-MDSCs. (A) Comparison of murine iNOS (*Nos2*) transcript levels in BM-MDSCs and spleen cells revealed that iNOS mRNA was upregulated in BM-MDSCs. The housekeeping gene (*Actb*, encoding β-actin) was expressed at equal levels. Results of one of 2 replicate experiments (with similar results) are shown. **(B)** Western blot using an antibody against murine iNOS demonstrated the presence of iNOS protein in BM-MDSCs, but not in spleen cells. The β-actin control blot shows equal sample loading. One of 3 independent Western blots is shown. **(C)** iNOS activity was assayed on the basis of NO release into the supernatants of cultures containing BM-MDSCs (orange bar) or spleen cells (green bar), and expressed as total nitrate concentration (μM). BM-MDSC-containing cultures produced significantly higher amounts of NO than spleen cells did (*$p < 0.05$, n = 5 cultures/cell type; Mann-Whitney U test). Molecular markers: bp, base pairs; kDa, kilodalton.

supernatants of BM-MDCS cultures. NO can suppress T-cell function via multiple mechanisms including chemical alteration of the TCR and inhibition of kinases and transcription factors involved in the IL-2 receptor signaling pathway [8,42]. Granulocytic MDSCs have been shown to exert suppression on T cells via an arginase 1-dependent [20] or ROS-dependent [13] mechanism. However, MDSCs with a granulocytic phenotype that are present in the SF of arthritic mouse joints [21] or in the BM [41] as well as those generated from murine BM ex vivo (this study) are clearly capable of inhibiting T-cell responses in a NO-dependent manner. In relevance to RA, elevated concentrations of NO were found in the serum and SF of RA patients with the SF levels exceeding those in serum, suggesting NO production locally in the joint [43]. Since cells with a granulocytic phenotype constitute the major cell population in RA SF [44,45], they could be the primary source of NO, thus functioning as local granulocytic MDSCs.

Although detailed characterization of the T-cell signaling pathways altered by BM-MDSCs was beyond the scope of our investigations, intracellular levels of IFNγ and IL-10 in CD4⁺ Th cells and the induction of Tregs were assessed. Intracellular concentration of IFNγ, the pro-inflammatory cytokine produced by the Th1 subset of CD4⁺ cells, was reduced by BM-MDSCs, but so was the anti-inflammatory Th2 cell-derived cytokine IL-10. Although MDSC-mediated induction of Treg cell differentiation has been reported in vitro and in tumor-bearing mice [46], we found that the proportion of Treg cells was actually reduced in the presence of BM-MDSCs. Our observations suggest that the suppressive effect of BM-MDSCs is not selective and may extend to several T cell subsets.

Studies have reported successful intervention in various diseases by in vivo transfer of MDSCs. Highfill et al. [47] generated MDSCs in vitro from the BM of tumor-free mice in the presence of GM-CSF, G-CSF, and IL-13. Such cells inhibited responses to allogeneic cells in vitro and in graft-versus-host disease [47]. In an animal model of inflammatory bowel disease, transfer of sorted CD11b⁺Gr-1⁺ cells abrogated enterocolitis, indicating a direct immune regulatory effect via NO production [17]. In another study, it was found that MDSCs significantly delayed or prevented type I diabetes onset by suppressing autoreactive T cells and inducing the differentiation of Tregs [19]. In a mouse model of MS, transfer of spleen-derived granulocytic MDSCs delayed the onset and reduced the severity of nervous system disease through suppression of encephalitogenic Th1 and Th17 cells [14]. More

recently, Fujii et al. [20] reported accumulation of MDSCs (mainly of the granulocytic phenotype) in the spleens of mice with CIA at the peak of the disease. This finding is congruent with our previous observation that MDSCs accumulate in autoimmune arthritis, although we identified suppressive MDSCs in the SF, not in the spleens, of mice with PGIA [21]. Granulocytic MDSCs, isolated from the spleens of mice with CIA suppressed anti-CD3/CD28-induced T-cell proliferation, but their effects on Ag (type II collagen)-specific immune responses were not investigated [20]. In PGIA, SF-MDSCs exerted suppression on T cells in an Ag-specific manner and were not effective in the Ag-independent system, whereas CD11b⁺ myeloid cells isolated from the spleens at the peak of PGIA were not suppressive in either of these in vitro settings [21]. As described in the present study, MDSCs generated from the BM of naïve mice were able to suppress both Ag-specific and non-specific T-cell responses. These apparent discrepancies may be explained by the functional heterogeneity of MDSCs [10]. It is likely that distinct and overlapping modes of suppressive ability exist, depending not only on the experimental model studied, but also on the specific cytokine milieu supporting and fine-tuning the MDSCs.

Using the adoptively transferred model of PGIA, we found that a single injection of BM-MDSCs into SCID mice after the first signs of arthritis suppressed disease progression and prevented further joint damage. In order to determine whether BM-MDSCs exerted immune modulatory effects in vivo, Ag-specific T-cell proliferation and serum Abs were measured in the recipient mice. The results confirmed that both T- and B-cell responses were significantly inhibited in the BM-MDSC-injected group of mice. In vivo tracking of transferred BM-MDSCs revealed that these MDSCs preferentially accumulated in the BM and SF, sites where their survival was best supported by locally produced myelopoietic growth factors and cytokines. The presence of MDSCs in the blood 19 days after their transfer also indicated active trafficking of these cells between the BM and SF. It is likely, therefore, that BM-MDSCs suppressed arthritis progression by inhibiting the expansion of pathogenic T cells in the BM, the peripheral joints, and, to a lesser degree, in the secondary lymphoid organs of recipient mice.

It was reported earlier that transplantation of syngeneic BM restored immune homeostasis and reduced arthritis severity in mice with PGIA [48]. In this particular case, BM transfer was associated with accumulation of Treg cells in the recipient mice.

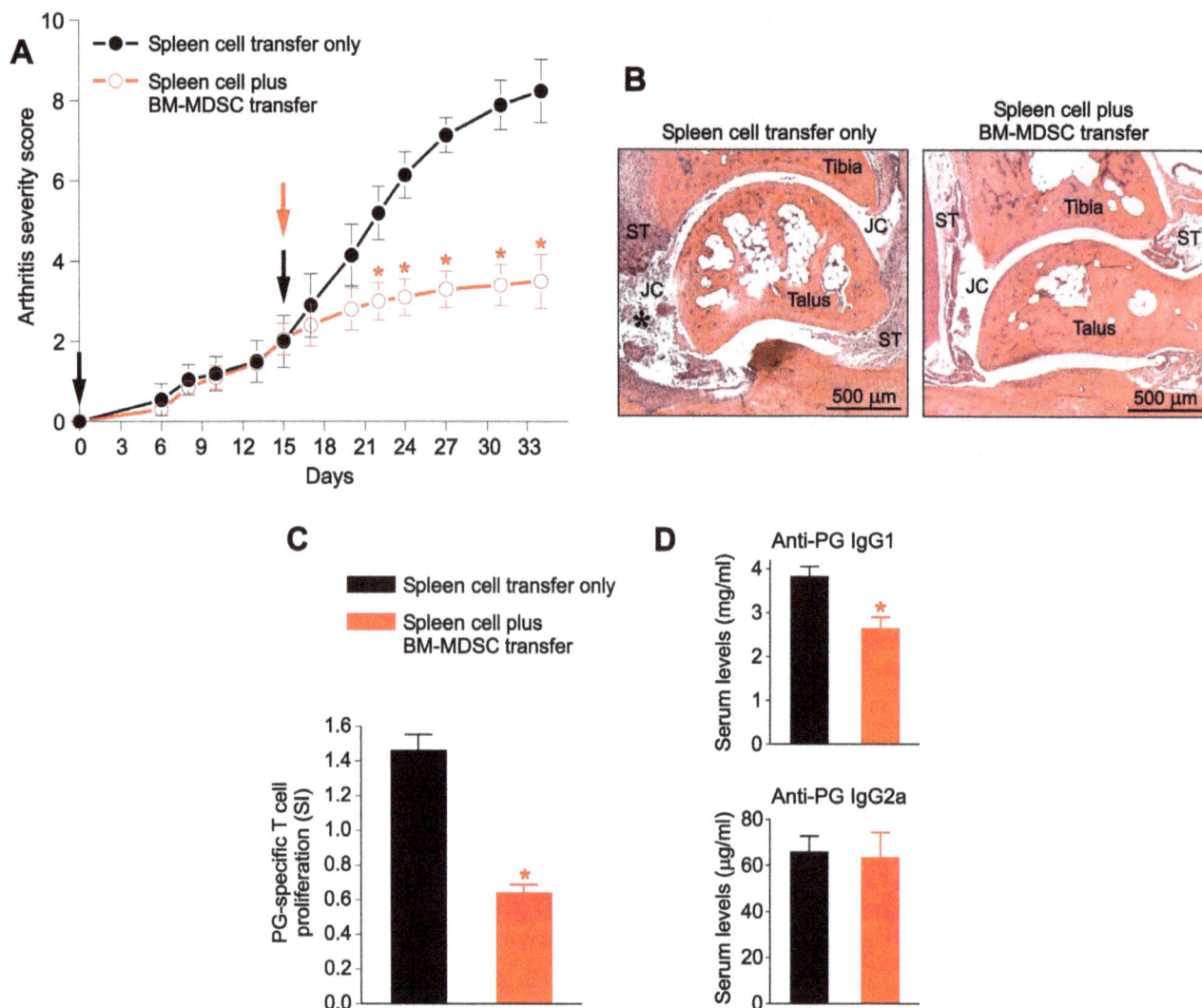

Figure 5. Effects of BM-MDSCs on arthritis severity and Ag (PG)-specific immune responses in SCID mice with PGIA. (**A**) Effect of BM-MDSC transfer on arthritis severity. Arthritis was induced in SCID mice via 2 transfers of spleen cells (black arrows) from wild type mice with PGIA as described in the Methods. At the early phase of arthritis, one group of the SCID recipients was co-injected with BM-MDSCs (red arrow). Disease severity scores were monitored until day 34. Arthritis progressed rapidly in the control group (black line), but not in the BM-MDSC-treated group (red line) (*$p < 0.05$, n = 10 mice/group; two-way repeated measures analysis of variance). (**B**) Joint histopathology of control (left panel) and BM-MDSC-treated (right panel) mice on day 34. The ankle joint of the control mouse demonstrated massive leukocyte infiltration (star) in the joint cavity (JC) and synovial tissue (ST) as well as synovial hyperplasia. The articulating surfaces appeared rough due to cartilage damage. In the ankle joint of the BM-MDSC-treated mouse only mild synovial hyperplasia was seen, suggesting the resolution of initial (previous) inflammation. Representative hematoxylin-eosin-stained tissue sections from both groups are shown. (**C**) Antigen (PG)-specific T-cell responses of control and BM-MDSC-treated mice. T-cell responses were compared between the two groups on day 34 by measuring spleen cell proliferation in the presence or absence of PG in vitro. Results are expressed as stimulation index (SI), a ratio of [³H]thymidine incorporation by PG-stimulated and non-stimulated cultures. The SI of the BM-MDSC-injected group (red bar) was significantly lower than the SI of the control group (black bar) (*$p < 0.0001$, n = 10 mice/group; Student's t test). (**D**) Serum levels of anti-PG antibodies in the control and BM-MDSC-treated groups as determined by ELISA. The levels of IgG1 anti-PG antibodies (top) were significantly lower in the sera of BM-MDSC-injected mice than in control mice (*$p < 0.01$, n = 5 samples/group; Mann-Whitney U test), while the levels of IgG2a anti-PG antibodies (bottom) were similar.

Although it was not clear if the Treg cells were of donor or recipient origin, the suggested mechanism of disease suppression was a BM-mediated induction of Treg differentiation [48]. While the BM may act as a reservoir of Treg cells [49], it also contains significant amounts of MDSCs and their precursors [41,49]. It is likely, therefore, that BM-derived MDSCs contributed to the reduction of autoimmune responses and disease severity upon BM transplantation into mice with PGIA.

A recent study reported increased frequency of MDSC-like cells in the peripheral blood of RA patients as compared with the blood of healthy control individuals, but the suppressive properties of these cells were not tested [50]. Most recently, we identified granulocytic MDSCs in the SF of RA patients; these RA SF-MDSCs moderately suppressed the anti-CD3/CD28-induced proliferation of autologous T cells, but potently suppressed alloAg-induced T-cell proliferation in vitro [51]. It is likely that SF-MDSCs inhibit the expansion of joint-homing (pathogenic) T cells in both RA and animal models of the disease. Notably, SF of the arthritic joints of both RA patients and mice has been shown to contain very low proportions of T cells [2,27,44,52,53]. In

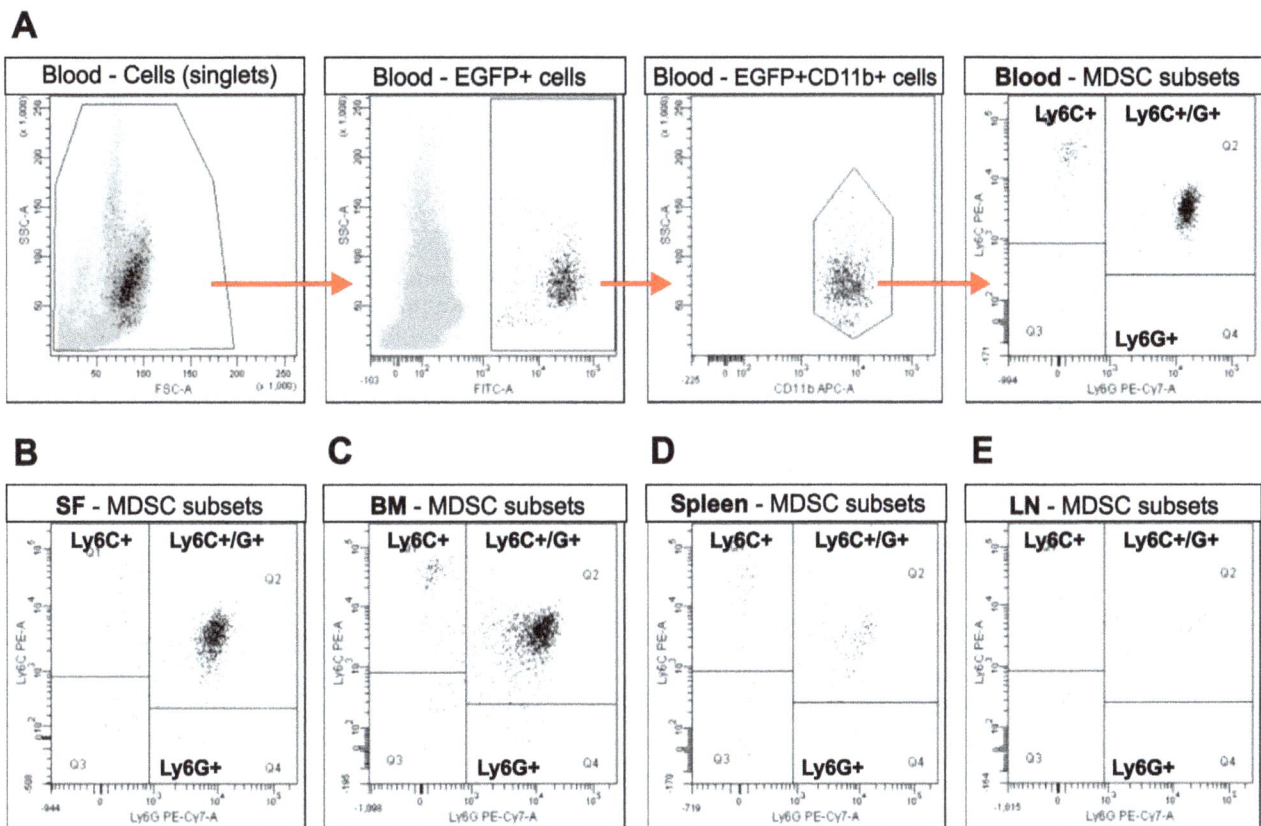

Figure 6. Tissue distribution of EGFP⁺ BM-MDSCs injected into SCID mice with adoptively transferred PGIA. BM-MDSCs, generated from EGFP-LysM-Tg mice (expressing EGFP in myeloid cells only) were co-injected with arthritic spleen cells into SCID mice at the early phase of adoptively transferred PGIA, as described in the Methods. To assess the tissue distribution of fluorescent donor cells, 19 days after the co-transfer of the cells (**A**) peripheral blood, (**B**) synovial fluid (SF), (**C**) BM, (**D**) spleen, and (**E**) joint-draining lymph nodes (LN) were harvested from the SCID recipients, immunostained, and subjected to flow cytometry. The gating strategy, as demonstrated on the blood cells (top panels), involved gating first on single cells, then on EGFP⁺ cells, followed by gating on the CD11b⁺ myeloid population (red arrows). Subset composition of EGFP⁺CD11b⁺ cells was determined on the basis of Ly6C and Ly6G expression. Peripheral blood, SF, and BM contained very well detectable populations of Ly6C^{int}Ly6G^{hi} (granulocytic) cells and much smaller populations of Ly6C^{hi}Ly6G⁻ (monocytic) cells. Cells belonging to either subset were less frequent in the spleen, and nearly undetectable in the LNs. Representative flow cytometry dot plots of cells from 1 of 3 mice (except for SF, which was pooled from all of the 3 mice) are shown.

addition, T cells isolated from the SF of RA patients exhibit hypo-responsiveness to mitogenic stimuli as compared to blood T cells of the same patients [2,54]. It is possible, therefore, that SF-MDSCs limit the expansion of T cells locally, thus contributing to the resolution of joint inflammation. Indeed, it was reported that in vivo depletion of MDSCs (using the anti-Gr-1 mAb RB6-8C5) delayed the resolution of arthritis in mice with CIA [20]. However, upon entering the joints at the early phase of arthritis, MDSCs may also cause collateral tissue damage through the release of NO and other noxious products, thereby acting as a "double-edged sword" [9]. Elucidation of the properties and function of MDSCs present in RA patients at distinct anatomical sites (e.g., peripheral blood, BM, SF, and secondary lymphoid organs) would greatly advance our understanding of the role of these cells in the regulation of autoimmunity and joint pathology in RA.

In summary, herein we describe an in vitro method for generating large quantities of MDSCs from murine BM in a controlled and reproducible manner. We show that murine BM-MDSCs, partially resembling MDSCs present in the SF of mice with PGIA, potently inhibit T-cell responses in vitro and in vivo. These results provide insights into an innate control mechanism that is involved in the regulation of immune responses and arthritis

severity in an animal model of RA and most likely in human patients as well. Although further studies are warranted, our results also suggest that in vitro enrichment of the BM in MDSCs could improve the therapeutic efficacy of autologous BM transplantation [3] in patients with severe, treatment-resistant RA.

Supporting Information

Figure S1 Analysis of monocyte/macrophage marker expression in BM-MDSC-like and SF-MDSC-like cells. The CD11b⁺ myeloid populations of (**A**) BM-MDSC-like cells and (**B**) SF cells were analyzed by flow cytometry for cells expressing the monocyte/macrophage markers F4/80, CD115, and CD80. (**A**) F4/80⁺ and CD80⁺ cells were more frequent among CD11b⁺Ly6C$^{lo/-}$ than CD11b⁺Ly6C$^{hi/int}$ BM-MDSCs, but very few CD115⁺ cells were detected in either of these populations. (**B**) SF contained much fewer F4/80⁺ and CD80⁺ cells within both the CD11b⁺Ly6C$^{hi/int}$ and CD11b⁺Ly6C$^{lo/-}$ fractions, but slightly more CD115⁺ cells within the CD11b⁺Ly6C$^{lo/-}$ population than BM-MDSCs. Initial gating on CD11b⁺ cells is indicated by red arrows. For subsequent gating, the horizontal line was set to separate the Ly6C$^{hi/int}$ and Ly6C$^{lo/-}$ populations, and the vertical lines were set at the highest levels of background staining with fluorochrome-tagged control IgGs matching the isotypes of F4/80, CD115, and CD80 mAbs. The representative samples show flow dot plots of cells from 1 of 5 independent BM-MDSC cultures, and from 1 of 3 separate pools of SF cells.

Figure S2   Screening of BM-MDSCs and SF cells for the presence of osteoclast precursor-like cells.

Flow cytometry analysis was performed on the same (**A**) BM-MDSC and (**B**) SF samples described in Figure S1, but with gating on CD11b$^{lo/-}$ cells (red arrows) containing putative Ly6C^{hi}CD115$^+$ osteoclast precursors. CD115$^+$ osteoclast precursor-like cells were not detected in either the Ly6C$^{hi/int}$ or Ly6C$^{lo/-}$ fraction of (**A**) CD11b$^{lo/-}$ BM-MDSCs (**B**) or CD11b$^{lo/-}$ SF cells. The representative samples show flow dot plots of cells from 1 of 5 independent BM-MDSC cultures, and from 1 of 3 separate pools of SF cells.

Figure S3   Effects of BM-MDSCs of the expression levels of dendritic cell (DC) maturation markers MHC II and CD86.

DCs and BM-MDSCs were generated from BM as described in the Methods. DCs were cultured for 3 days with or without BM-MDSCs. The densities of major histocompatibility complex class II (MHC II) and CD86 maturation markers on the surface of DCs (CD11c$^+$ cells) were determined by flow cytometry and the results expressed as mean fluorescence intensity (MFI). (**A**)

Expression level of MHC II on the DCs (open bar) slightly increased in the presence of BM-MDSCs (closed bar), but this increase did not reach statistical significance (ns, not significant; p = 0.059; Mann-Whitney U test). (**B**) There was no significant difference in the expression level of CD86 on the DCs either when these cells were cultured without (open bar) and with (closed bar) BM-MDSCs (ns; p = 0.667; Mann-Whitney U test). Data shown are from 5 independent experiments.

Table S1   Concentrations of GM-CSF, IL-6, and G-CSF in synovial fluid (SF) and serum collected from arthritic (PGIA) mice.

Acknowledgments

The authors would like to thank Dr. Larry Thomas for help with the cytospin preparations.

Author Contributions

Conceived and designed the experiments: JK TTG ZS KM. Performed the experiments: JK AV TO BT TAR TTG ZS KM. Analyzed the data: JK AV TO BT TAR TTG ZS KM. Contributed reagents/materials/analysis tools: AV TTG. Contributed to the writing of the manuscript: JK AV TTG ZS KM. Revised the manuscript for important intellectual content: JK AV TTG ZS KM.

References

1. Fox DA (2005) Etiology and Pathogenesis of Rheumatoid Arthritis. In: Koopman WJ, Moreland LW, editors. Arthritis and Allied Conditions: A Textbook of Rheumatology. Philadelphia: Lippincott Williams & Wilkins. 1085–1102.
2. Firestein GS (2005) Rheumatoid arthritis: Etiology and pathogeneis of rheumatoid arthritis. In: Ruddy S, Harris ED, Sledge CB, Kelley WN, editors. Kelley's Textbook of Rheumatology. Philadelphia, PA: W.B.Saunders Co. 996–1045.
3. Moore JJ, Snowden J, Pavletic S, Barr W, Burt R (2003) Hematopoietic stem cell transplantation for severe rheumatoid arthritis. Bone Marrow Transplant 32 Suppl 1: S53–S56.
4. Brinkman DM, de Kleer IM, ten Cate R, van Rossum MA, Bekkering WP, et al. (2007) Autologous stem cell transplantation in children with severe progressive systemic or polyarticular juvenile idiopathic arthritis: long-term follow-up of a prospective clinical trial. Arthritis Rheum 56: 2410–2421.
5. Subiza JL, Vinuela JE, Rodriguez R, Gil J, Figueredo MA, et al. (1989) Development of splenic natural suppressor (NS) cells in Ehrlich tumor-bearing mice. Int J Cancer 44: 307–314.
6. Nagaraj S, Gabrilovich DI (2010) Myeloid-derived suppressor cells in human cancer. Cancer J 16: 348–353.
7. Serafini P, Borrello I, Bronte V (2006) Myeloid suppressor cells in cancer: recruitment, phenotype, properties, and mechanisms of immune suppression. Semin Cancer Biol 16: 53–65.
8. Gabrilovich DI, Nagaraj S (2009) Myeloid-derived suppressor cells as regulators of the immune system. Nat Rev Immunol 9: 162–174.
9. Pastula A, Marcinkiewicz J (2011) Myeloid-derived suppressor cells: a double-edged sword? Int J Exp Pathol 92: 73–78.
10. Youn JI, Gabrilovich DI (2010) The biology of myeloid-derived suppressor cells: the blessing and the curse of morphological and functional heterogeneity. Eur J Immunol 40: 2969–2975.
11. Lowell CA, Berton G (1999) Integrin signal transduction in myeloid leukocytes. J Leukoc Biol 65: 313–320.
12. Ribechini E, Greifenberg V, Sandwick S, Lutz MB (2010) Subsets, expansion and activation of myeloid-derived suppressor cells. Med Microbiol Immunol 199: 273–281.
13. Youn JI, Nagaraj S, Collazo M, Gabrilovich DI (2008) Subsets of myeloid-derived suppressor cells in tumor-bearing mice. J Immunol 181: 5791–5802.
14. Ioannou M, Alissafi T, Lazaridis I, Deraos G, Matsoukas J, et al. (2012) Crucial role of granulocytic myeloid-derived suppressor cells in the regulation of central nervous system autoimmune disease. J Immunol 188: 1136–1146.
15. Movahedi K, Guilliams M, Van den Bossche J, Van den Bergh R, Gysemans C, et al. (2008) Identification of discrete tumor-induced myeloid-derived suppressor cell subpopulations with distinct T cell-suppressive activity. Blood 111: 4233–4244.
16. Nagaraj S, Gupta K, Pisarev V, Kinarsky L, Sherman S, et al. (2007) Altered recognition of antigen is a mechanism of CD8+ T cell tolerance in cancer. Nat Med 13: 828–835.
17. Haile LA, von Wasielewski R, Gamrekelashvili J, Kruger C, Bachmann O, et al. (2008) Myeloid-derived suppressor cells in inflammatory bowel disease: a new immunoregulatory pathway. Gastroenterology 135: 871–81, 881.
18. Kerr EC, Raveney BJ, Copland DA, Dick AD, Nicholson LB (2008) Analysis of retinal cellular infiltrate in experimental autoimmune uveoretinitis reveals multiple regulatory cell populations. J Autoimmun 31: 354–361.
19. Yin B, Ma G, Yen CY, Zhou Z, Wang GX, et al. (2010) Myeloid-derived suppressor cells prevent type 1 diabetes in murine models. J Immunol 185: 5828–5834.
20. Fujii W, Ashihara E, Hirai H, Nagahara H, Kajitani N, et al. (2013) Myeloid-derived suppressor cells play crucial roles in the regulation of mouse collagen-induced arthritis. J Immunol 191: 1073–1081.
21. Egelston C, Kurko J, Besenyei T, Tryniszewska B, Rauch TA, et al. (2012) Suppression of dendritic cell maturation and T cell proliferation by synovial fluid myeloid cells from mice with autoimmune arthritis. Arthritis Rheum 64: 3179–3188.
22. Glant TT, Mikecz K, Arzoumanian A, Poole AR (1987) Proteoglycan-induced arthritis in BALB/c mice. Clinical features and histopathology. Arthritis Rheum 30: 201–212.
23. Glant TT, Finnegan A, Mikecz K (2003) Proteoglycan-induced arthritis: immune regulation, cellular mechanisms and genetics. Crit Rev Immunol 23: 199–250.
24. Hutas G, Bajnok E, Gal I, Finnegan A, Glant TT, et al. (2008) CD44-specific antibody treatment and CD44 deficiency exert distinct effects on leukocyte recruitment in experimental arthritis. Blood 112: 4999–5006.
25. Ribechini E, Leenen PJ, Lutz MB (2009) Gr-1 antibody induces STAT signaling, macrophage marker expression and abrogation of myeloid-derived suppressor cell activity in BM cells. Eur J Immunol 39: 3538–3551.
26. Faust N, Varas F, Kelly LM, Heck S, Graf T (2000) Insertion of enhanced green fluorescent protein into the lysozyme gene creates mice with green fluorescent granulocytes and macrophages. Blood 96: 719–726.
27. Angyal A, Egelston C, Kobezda T, Olasz K, Laszlo A, et al. (2010) Development of proteoglycan-induced arthritis depends on T cell-supported autoantibody production, but does not involve significant influx of T cells into the joints. Arthritis Res Ther 12: R44.
28. Berlo SE, Guichelaar T, ten Brink CB, Van Kooten PJ, Hauet-Broere F, et al. (2006) Increased arthritis susceptibility in cartilage proteoglycan-specific T cell receptor-transgenic mice. Arthritis Rheum 54: 2423–2433.
29. Bosma MJ, Carroll AM (1991) The SCID mouse mutant: definition, characterization, and potential uses. Annu Rev Immunol 9: 323–350.
30. Hanyecz A, Berlo SE, Szanto S, Broeren CPM, Mikecz K, et al. (2004) Achievement of a synergistic adjuvant effect on arthritis induction by activation of innate immunity and forcing the immune response toward the Th1 phenotype. Arthritis Rheum 50: 1665–1676.
31. Lutz MB, Kukutsch N, Ogilvie AL, Rossner S, Koch F, et al. (1999) An advanced culture method for generating large quantities of highly pure dendritic cells from mouse bone marrow. J Immunol Methods 223: 77–92.

32. Glant TT, Radacs M, Nagyeri G, Olasz K, Laszlo A, et al. (2011) Proteoglycan-induced arthritis and recombinant human proteoglycan aggrecan G1 domain-induced arthritis in BALB/c mice resembling two subtypes of rheumatoid arthritis. Arthritis Rheum 63: 1312–1321.

33. Bardos T, Mikecz K, Finnegan A, Zhang J, Glant TT (2002) T and B cell recovery in arthritis adoptively transferred to SCID mice: Antigen-specific activation is required for restoration of autopathogenic CD4+ Th1 cells in a syngeneic system. J Immunol 168: 6013–6021.

34. Dolcetti L, Peranzoni E, Ugel S, Marigo I, Fernandez GA, et al. (2010) Hierarchy of immunosuppressive strength among myeloid-derived suppressor cell subsets is determined by GM-CSF. Eur J Immunol 40: 22–35.

35. Lechner MG, Liebertz DJ, Epstein AL (2010) Characterization of cytokine-induced myeloid-derived suppressor cells from normal human peripheral blood mononuclear cells. J Immunol 185: 2273–2284.

36. Wright HL, Bucknall RC, Moots RJ, Edwards SW (2012) Analysis of SF and plasma cytokines provides insights into the mechanisms of inflammatory arthritis and may predict response to therapy. Rheumatology (Oxford) 51: 451–459.

37. Peranzoni E, Zilio S, Marigo I, Dolcetti L, Zanovello P, et al. (2010) Myeloid-derived suppressor cell heterogeneity and subset definition. Curr Opin Immunol 22: 238–244.

38. Charles JF, Hsu LY, Niemi EC, Weiss A, Aliprantis AO, et al. (2012) Inflammatory arthritis increases mouse osteoclast precursors with myeloid suppressor function. J Clin Invest 122: 4592–4605.

39. Almand B, Clark JI, Nikitina E, van Beynen J, English NR, et al. (2001) Increased production of immature myeloid cells in cancer patients: a mechanism of immunosuppression in cancer. J Immunol 166: 678–689.

40. Rossner S, Voigtlander C, Wiethe C, Hanig J, Seifarth C, et al. (2005) Myeloid dendritic cell precursors generated from bone marrow suppress T cell responses via cell contact and nitric oxide production in vitro. Eur J Immunol 35: 3533–3544.

41. Forghani P, Harris W, Giver CR, Mirshafiey A, Galipeau J, et al. (2013) Properties of immature myeloid progenitors with nitric-oxide-dependent immunosuppressive activity isolated from bone marrow of tumor-free mice. PLOS ONE 8: e64837.

42. Bingisser RM, Tilbrook PA, Holt PG, Kees UR (1998) Macrophage-derived nitric oxide regulates T cell activation via reversible disruption of the Jak3/STAT5 signaling pathway. J Immunol 160: 5729–5734.

43. Farrell AJ, Blake DR, Palmer RM, Moncada S (1992) Increased concentrations of nitrite in synovial fluid and serum samples suggest increased nitric oxide synthesis in rheumatic diseases. Ann Rheum Dis 51: 1219–1222.

44. Bjelle A, Norberg B, Sjogren G (1982) The cytology of joint exudates in rheumatoid arthritis. Morphology and preparation techniques. Scand J Rheumatol 11: 124–128.

45. Yamamoto T, Nishiura H, Nishida H (1996) Molecular mechanisms to form leukocyte infiltration patterns distinct between synovial tissue and fluid of rheumatoid arthritis. Semin Thromb Hemost 22: 507–511.

46. Huang B, Pan PY, Li Q, Sato AI, Levy DE, et al. (2006) Gr-1+CD115+ immature myeloid suppressor cells mediate the development of tumor-induced T regulatory cells and T-cell anergy in tumor-bearing host. Cancer Res 66: 1123–1131.

47. Highfill SL, Rodriguez PC, Zhou Q, Goetz CA, Koehn BH, et al. (2010) Bone marrow myeloid-derived suppressor cells (MDSCs) inhibit graft-versus-host disease (GVHD) via an arginase-1-dependent mechanism that is up-regulated by interleukin-13. Blood 116: 5738–5747.

48. Roord ST, de Jager W, Boon L, Wulffraat N, Martens A, et al. (2008) Autologous bone marrow transplantation in autoimmune arthritis restores immune homeostasis through CD4+CD25+Foxp3+ regulatory T cells. Blood 111: 5233–5241.

49. Zhao E, Xu H, Wang L, Kryczek I, Wu K, et al. (2012) Bone marrow and the control of immunity. Cell Mol Immunol 9: 11–19.

50. Jiao Z, Hua S, Wang W, Wang H, Gao J, et al. (2013) Increased circulating myeloid-derived suppressor cells correlated negatively with Th17 cells in patients with rheumatoid arthritis. Scand J Rheumatol 42: 85–90.

51. Kurko J, Vida A, Glant TT, Scanzello CR, Katz RS, et al. (2014) Identification of myeloid-derived suppressor cells in the synovial fluid of patients with rheumatoid arthritis: a pilot study. BMC Musculoskelet Disord 15: 281.

52. Holmdahl R, Jonsson R, Larsson P, Klareskog L (1988) Early appearance of activated CD4+ T lymphocytes and class II antigen-expressing cells in joints of DBA/1 mice immunized with type II collagen. Lab Invest 58: 53–60.

53. Nguyen LT, Jacobs J, Mathis D, Benoist C (2007) Where FoxP3-dependent regulatory T cells impinge on the development of inflammatory arthritis. Arthritis Rheum 56: 509–520.

54. Cope AP (2002) Studies of T-cell activation in chronic inflammation. Arthritis Res 4 Suppl 3: S197–S211.

13

Targeted Sequencing of Large Genomic Regions with CATCH-Seq

Kenneth Day, Jun Song, Devin Absher*

HudsonAlpha Institute for Biotechnology, Huntsville, Alabama, United States of America

Abstract

Current target enrichment systems for large-scale next-generation sequencing typically require synthetic oligonucleotides used as capture reagents to isolate sequences of interest. The majority of target enrichment reagents are focused on gene coding regions or promoters en masse. Here we introduce development of a customizable targeted capture system using biotinylated RNA probe baits transcribed from sheared bacterial artificial chromosome clone templates that enables capture of large, contiguous blocks of the genome for sequencing applications. This clone adapted template capture hybridization sequencing (CATCH-Seq) procedure can be used to capture both coding and non-coding regions of a gene, and resolve the boundaries of copy number variations within a genomic target site. Furthermore, libraries constructed with methylated adapters prior to solution hybridization also enable targeted bisulfite sequencing. We applied CATCH-Seq to diverse targets ranging in size from 125 kb to 3.5 Mb. Our approach provides a simple and cost effective alternative to other capture platforms because of template-based, enzymatic probe synthesis and the lack of oligonucleotide design costs. Given its similarity in procedure, CATCH-Seq can also be performed in parallel with commercial systems.

Editor: Esteban Ballestar, Bellvitge Biomedical Research Institute (IDIBELL), Spain

Funding: The authors have no support or funding to report.

Competing Interests: The authors have declared that no competing interests exist.

* Email: dabsher@hudsonalpha.org

Introduction

Costs for next generation sequencing technology and affiliated methods continue to fall for prospective whole genome sequencing of individuals. Along with growing improvements in sequencing technologies is a rapid expansion of our knowledge of human genetic variation and the impact of this sequence variation on human traits and diseases. High throughput sequencing is providing a foundation for both individualized patient therapies and newly developing programs in personalized medicine [1,2]. However, individual whole genome sequencing currently remains expensive. The most practical alternative is to limit the sequencing per individual genome to select, meaningful loci by targeted enrichment to generate consistent high definition coverage around relevant regions [3].

Currently, solution hybridization based targeted capture for whole exome sequencing is perhaps the most consequential method for determination of sequence variation that directly affects human gene products [4], and similar capture methods can be used for targeting almost any region of the genome. Most large-scale targeted sequencing platforms rely upon the initial synthesis of tiled bait oligonucleotides of variable lengths across exons [5]. Synthesis of biotinylated RNA probe baits by in vitro transcription from oligonucleotide templates or direct use of biotinylated DNA oligonucleotide probe baits are used for solution hybridization capture of prepared sequencing libraries followed by binding to streptavidin beads and low to high stringency washes [6]. While

these systems are cost-effective for the capture of large numbers of disparate targets, such as exons, this represents less than 2% of the human genome and excludes gene regulatory regions. Some commercially available products have emerged that emphasize capture of promoter regions containing CpG islands in the bait design based on cancer and tissue-specific differentially methylated regions [7]. Sequencing targets outside of exons or promoter regions requires customized synthesis of tiled oligonucleotides for the production of probe baits. As projects like ENCODE begin to identify critical regulatory regions of the human genome, the application of targeted sequencing to non-coding sequences has the potential to identify disease-related variation outside of the exome. Furthermore, there is increasing evidence for epigenetic influences in human diseases, and these epigenetic marks are often located in non-coding regions of the genome [8]. Additionally, the identification of regional structural variants, such as large deletions is difficult with exome sequencing data unless the variants overlap multiple exons of a given gene. Clearly, there is a growing need for targeted sequencing approaches that can interrogate both coding and non-coding elements around genes of interest.

A recent study demonstrated the custom capture of a 1 Mb contiguous site that encompassed the human dystrophin gene with use of densely designed oligo bait probes that spanned the entire region [9]. While sequencing of large contiguous regions in this manner is feasible, synthesis of custom probe baits remains a substantial added expense to sequencing costs, especially for small numbers of samples. Some techniques such as circularizationbased methods using gap-fill padlock, MIPs or selector probe methods are feasible for targeted capture of specific regions, but also require design and synthesis of vast numbers of oligos, or careful restriction digest strategies [10–12]. Targeted amplification methods such as nested patch PCR carry the advantage of eliminating sequencing library construction along with high level multiplexing, but also requires oligonucleotides and is typically limited to relatively small composite targets [13].

Faster and more affordable alternatives are needed for targeted sequencing without the preliminary need of oligonucleotides to synthesize probe baits. Furthermore, many targets of interest in the genome are larger contiguous loci beyond the size of standard PCR amplicons, and may include coding and noncoding regions. Here we describe a new procedure using existing and commercially available genomic clones in clone adapted template capture hybridization sequencing (CATCH-Seq). Our method may be used all without oligonucleotide synthesis design to resolve copy number variation (CNV) boundaries, and, in combination with bisulfite treatment, to measure DNA methylation levels across large contiguous regions. Our method utilizes a simple approach for the targeted capture of any genomic region for which mapped clones exist, and is designed specifically for next-generation sequencing.

Materials and Methods

Illumina library construction

Concentrations of all input genomic DNAs were determined by Qubit high sensitivity double stranded DNA assays (Life Technologies), or by Picogreen assay in a 96-well 200 ul volume format using lambda DNA for a standard curve according to protocol (Life Technologies). Input genomic DNA quality was assessed by running 1–3 ul on a 1% agarose gel that contained 1× Sybr Green I dye (Life Technologies). Typical input DNA quantity for Illumina library construction ranged from 700 ng to 2.5 ug. Depending on the level of any DNA degradation before shearing, we often increased the quantity of input DNA. Genomic DNAs were adjusted to an 80 ul volume in water and placed into Covaris 96 microtube plates for shearing on an E210 E series focused ultrasonicator with installed intensifier at 4°C using fill level 6 and final run level 5. Program settings were 20% duty cycle, intensity 5, 200 cycles per burst, and treatment for 165 s. For repeat blocking optimization, we used K-562 cell line genomic DNA (ATCC CCL-243) obtained from the Myers Laboratory at HudsonAlpha Institute for Biotechnology for library construction.

Standard Illumina libraries were constructed using NEBNext End Repair, dA-Tailing and Quick Ligation modules (New England Biolabs). For cleanup steps following each enzymatic treatment, we used 1.8× reaction volume of SPRI Sera-Mag SpeedBeads (ThermoFisher) diluted 1:40 in binding buffer containing 2.5M NaCl and 20% PEG-8000 as reported previously [14]. After DNA binding, beads were washed twice in 70% ethanol and dried for 10 min prior to elution in water. Oligos were synthesized and annealed to generate 12 methylated or 24 non-methylated inline barcoded adapter sets (Oblique Bio). We used 80–100 pmoles methylated or non-methylated adapters in the ligation step (one adapter per sample) depending on whether we used standard sequencing or bisulfite sequencing of the target enriched library and the quantity of input DNA. For lower input DNA, we adjusted the molar ratio of adapter. Final libraries were eluted in 40 ul water and library concentrations were determined by Qubit HS assay or by Picogreen using 1:20 diluted libraries.

Biotinylated probe synthesis

To prepare probes for target enrichment, we typically selected both Cal Tech human BAC library D (CTD) and RPCI human BAC library 11 (RP11) clones that covered targets of interest in order to avoid potential gaps due to any deletions in one clone. All clones were purchased from Life Technologies. BAC or fosmid DNAs were purified from 200 mL cultures grown for at 37°C 16 h that were inoculated from a single colony isolated from LB agar plates containing 25 ug/mL chloramphenicol using PureLink HiPure Plasmid Midi or Maxiprep kits (Life Technologies) with additional buffer provided in HiPure BAC buffer kits (Life Technologies). DNA concentration was determined by Qubit HS assay. Clone identity was verified by PCR or by HINDIII restriction digest of 1 ug DNA loaded on a 1.5% agarose gel containing 1× Sybr green (Life Technologies).

BAC DNAs were sheared under the same conditions as sample genomic DNAs. If multiple BACs were used, we created pools of individual BAC DNAs before shearing by calculating the percent of the total target size covered by each BAC multiplied by the total mass (typically 4 ug) of input pooled BAC DNA. Sheared BAC fragments were processed similarly to preparation of Illumina libraries except for use of T7 promoter-containing adapters (composed of two annealed oligos: TAC TAC TAA TAC GAC TCA CTA TAG GGT and CCC TAT AGT GAG TCG TAT TAG TAG TA) in the ligation step. Following cleanup of the ligation, 100–200 pmoles of T7 F oligo is annealed with T7-BAC fragments in a reaction containing 10× PCR buffer and 1 ul 50 mM $MgCl_2$ in a 35 ul volume, heated to 95°C for 5 min, 50°C for 5 min, and cooled to 4°C. Following cleanup and elution into 36 ul water, three in vitro transcription (IVT) reactions each containing 12 ul template DNA was used for biotinylated RNA probe synthesis using a Megascript T7 kit (Ambion) according to manufacturer protocol including biotin-11-UTP (Life Technologies) for 1.5 h. Completed IVTs were DNAse treated according to protocol, 1 ul 0.5M EDTA was added, and then DNAse was heat inactivated for 10 min at 75°C. Reactions were pooled and a 2 ul aliquot was mixed with gel loading buffer (heated to 95°C for 2 min) and loaded on a 1.5% agarose gel with 10 ug/mL ethidium bromide to assess yield. Final pooled probe reactions were cleaned up using NucAway gel filtration spin columns (Ambion) according to manufacturer instructions to remove unincorporated nucleotides. Probe concentration was detemined by Qubit RNA assay (Life Technologies) and verified by Agilent Bioanalyzer.

Hybridization and capture

To ensure even pooling of input libraries, inline barcoded Illumina libraries were pooled equally according to their concentrations as determined by Agilent Bioanalyzer and KAPA real time PCR (KAPA Biosystems) depending on bisulfite (12-plex) or standard sequencing (24-plex) to a total of 1 or 2 ug of library, respectively. Hybridization reactions were assembled similarly to a previous report [5], except hybridization components were scaled up accordingly for final volumes of 52 ul or 104 ul for 12 or 24-plex hybridization reactions, respectively. Library pools were mixed with 40-fold human Cot-1 DNA (Life Technologies), and concentrated to a 15 or 30 ul volume by SpeedVac (Thermo). To determine probe quantity for select targets in individual hybridizations, the theoretical mass yield of target was calculated in picograms based on total BAC target size and male human diploid genome size. A 2500-fold probe:theoretical target yield mass ratio for individual targets typically yielded between 75–80% of aligned reads within target regions. Probe was brought to a final volume of 10 or 20 ul in nuclease-free water depending on final reaction volume, and 1–2 ul of SUPERase-In (Life Technologies) added for final working probe solution. For hybridization assembly, 26 ul or 52 ul 2× hybridization buffer (10× SSPE, 10× Denhardt's, 10 mM EDTA, and 0.2% SDS) volumes were preheated in 0.5 mL self-standing tubes with screwcaps containing O-rings (USA Scientific) within a hybridization oven. Library pools containing Cot-1 DNA were heated at 95°C for 5 min and then held at 65°C for 5 min on a thermal cycler before they were added to pre-heated hybridization buffer, followed by addition of probe solution preheated to 65°C for 2 min. All assembled hybridizations reactions were incubated at 65°C for 24–60 h within a hybridization oven.

For capture procedure per each 52 ul hybridization reaction, 35 ul of input Dynabeads MyOne Streptavidin C1 were used and washed three times according to manufacturer instructions with binding and wash buffer (Life Technologies). Beads were resuspended in 200 ul bead binding and wash buffer and hybridization reactions are added to the beads and incubated for 30 min with frequent pulse vortexing. After binding, beads were washed twice at room temperature in 0.5 mL hybridization wash buffer 1 (1×SSC, 0.1% SDS), followed by four wash steps at 65°C in preheated 0.5 mL hybridization wash buffer 2 (0.1×SSC, 0.1%SDS) using a heat block and a magnet for 1.7 mL microfuge tubes.

For standard sequencing, washed Dynabeads were resuspended in 0.1M NaOH and pulse vortexed periodically for 10 min to elute captured Illumina library. Eluted library was removed to a new tube containing 70 ul of 1M Tris-HCl pH 7.5 to neutralize the solution, and followed by a SPRI bead cleanup with two 0.5 mL 70% ethanol washes. SPRI beads were set for 10 min and library was eluted in 35 ul of nuclease free water. For bisulfite sequencing, washed Dynabeads were resuspended in 40 ul of EB buffer and transferred into 96 well PCR plates for bisulfite conversion with the Epitect Bisulfite Kit (QIAGEN) according to handbook protocol (conversion of unmethylated cytosines in small amounts of fragmented DNA). Before cleanup, beads were removed from the conversion reaction. Library was eluted twice with 20 ul each of preheated EB buffer.

Final captured libraries were amplified by PCR using 0.5 ul of each standard Illumina primer (25 uM each), 5 ul 5M Betaine (Sigma), 2.5 ul 10 mM dNTP mix (New England Biolabs), 5 ul 10× PCR buffer, 2 ul 50 mM MgCl$_2$, 1 ul Platinum Taq, library, and water up to a 50 ul volume. Cycling conditions were 98°C for 1 min, followed by 18–22 cycles (depending on target size) of 95°C for 30 s and 62°C for 3 min 30 s. PCR amplified libraries were cleaned up by SPRI beads. Library concentrations were determined by Agilent Bioanalyzer High Sensitivity DNA assay and real time PCR with a library quantification kit (KAPA Biosystems). For further multiplexing of libraries post-hybridization or post-bisulfite conversion, indexed primers were used in the PCR amplification step, and concentrations of final library pooled sets were determined by an additional real time PCR reaction. Final Illumina library sequencing was performed according to standard protocol using a variety of Illumina sequencer platforms over the course of 3 years (see Table S1).

Alignment and mapping

Pass filter sequence linked to each index and inline barcodes was demuxed from fastq files and assigned to individual libraries using software that was also used to design inline barcoded adapter sequences. Bordering adapter sequences were removed from reads using the AdapterRemoval software [15]. Standard sequencing fastq files were aligned with BWA in paired end format to the human hg19 reference genome. PCR duplicates were removed and sam files were filtered by q20 mapping quality. For

determination of CNV boundaries, read depth was extracted from wig files in non-overlapping 100 bp segments across the length of the target genomic coordinates and the fraction of total bases per segment was calculated. LogR values were determined across the target site by log$_2$ of individual cases read fraction divided by the median read fraction of all control samples. For determination of CpG methylation across a target region, human hg19 reference forward and reverse strand sequences were each bisulfite converted in silico and a reference for mapping was built with Bismark. In conjunction with Bowtie2, Bismark was modified to function with local mode for read mapping. Duplicate reads were removed, and methylation values were extracted using Bismark. CpGs with less than a 20× minimum depth coverage were filtered and percent methylation values were used for further data analysis. Alignment files demonstrating coverage of one chr11 target by CATCH-Seq in comparison to WGS from 15 merged individuals from 1000 genomes data within this same target region, in addition to our CATCH-Seq repeat blocking analysis on another chr11 target have been submitted and are available from the National Center for Biotechnology Information (NCBI) Sequence Read Archive (SRA) with BioProject accession number [SRP042633].

Results and Discussion

Hybrid selection method

We developed a simple approach for solution hybridization capture sequencing of large genomic targets without the need for oligonucleotide synthesis of target templates (Figure 1). BAC clones were selected across genomic coordinates of interest to generate templates for probe synthesis. PCR or restriction digest was used to first correctly identify selected clones, and BAC DNA was purified. For composite targets of interest greater than what was covered by a single clone, multiple BAC DNAs (contiguous or discontiguous regions) were pooled based on individual percent of composite size in basepairs of the target multiplied by the mass of input template DNA (Figure 1). Pooled BAC template DNA was randomly sheared, ligated with T7 promoter-containing adapters, and T7 forward adapter oligo was annealed to generate double stranded promoter regions with single stranded antisense templates. Following cleanup of the annealing reaction, in vitro transcription was used in the presence of biotin-UTP to synthesize the probes. A similar approach for probe synthesis was also recently described to enrich for ancient human DNA from environmental contaminants using the entire human genome as a template using T7-promoter containing adapters [18]. Solution hybridization and capture procedures were described previously for Illumina libraries, and our protocol is similar except with increased Cot-1 concentration, larger reaction volume, and additional wash steps [5]. We also typically hybridized library samples in 12-plex or 24-plex using inline barcoded adapters, and adjusted hybridization reaction volumes for scaling up concentration of pooled library and hybridization reagents appropriately. Before PCR enrichment of captured library, bisulfite conversion was also used to analyze DNA methylation in regions of interest by use of conversion-resistant, inline barcoded adapters used in library construction.

Our materials and methods described here represent our most current procedures with all quality assurance steps. Compared to previous methods used for solution-based enrichment using BAC DNA, our procedure does not require nick translation of whole BAC DNA and a 6 h pre-hybridization blocking of the BAC probe [16]. Our method also resembles exome capture methods at the step of probe synthesis and hybridization [5]. Random shearing of

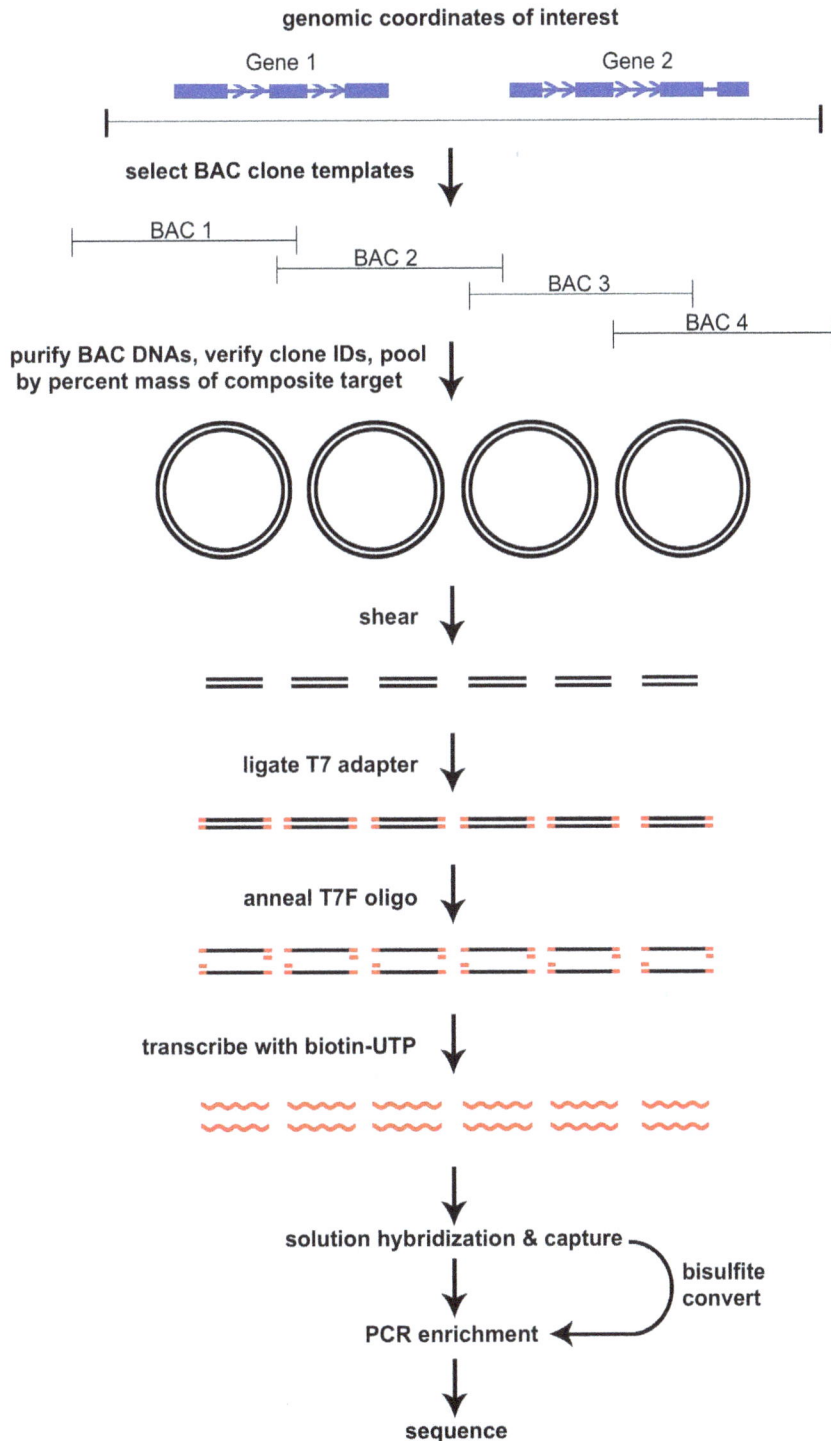

Figure 1. Overview of the clone adapted template capture hybridization sequencing procedure. BAC clone templates are selected to span genomic coordinates of interest, and pooled by percent mass of the composite target. BACs are sheared, ligated with T7 adapters to transcribe biotinylated RNA probes, and then solution hybridized with prepared libraries. Following capture, libraries are amplified by PCR, or bisulfite converted prior to amplification for analysis of DNA methylation. Target enriched libraries are pooled and sequenced.

the BAC templates and use of in vitro transcription typically yielded a 2–4 fold mass of probe relative to input BAC DNA with probe size ranging from 200–400 bp. Although we do not perform a PCR amplification step after ligation of T7 adapters, addition of this step may provide an improvement in probe yield and vastly reduce the amount of input template DNA required. This

template amplification step was used to amplify and incorporate T7-promoter into probe bait templates generated from oligos prior to in vitro transcription, and the biotinylated RNA probe yield was 10–20 fold relative to input template [5]. Overall, our method is designed specifically for next-generation sequencing without oligo synthesis.

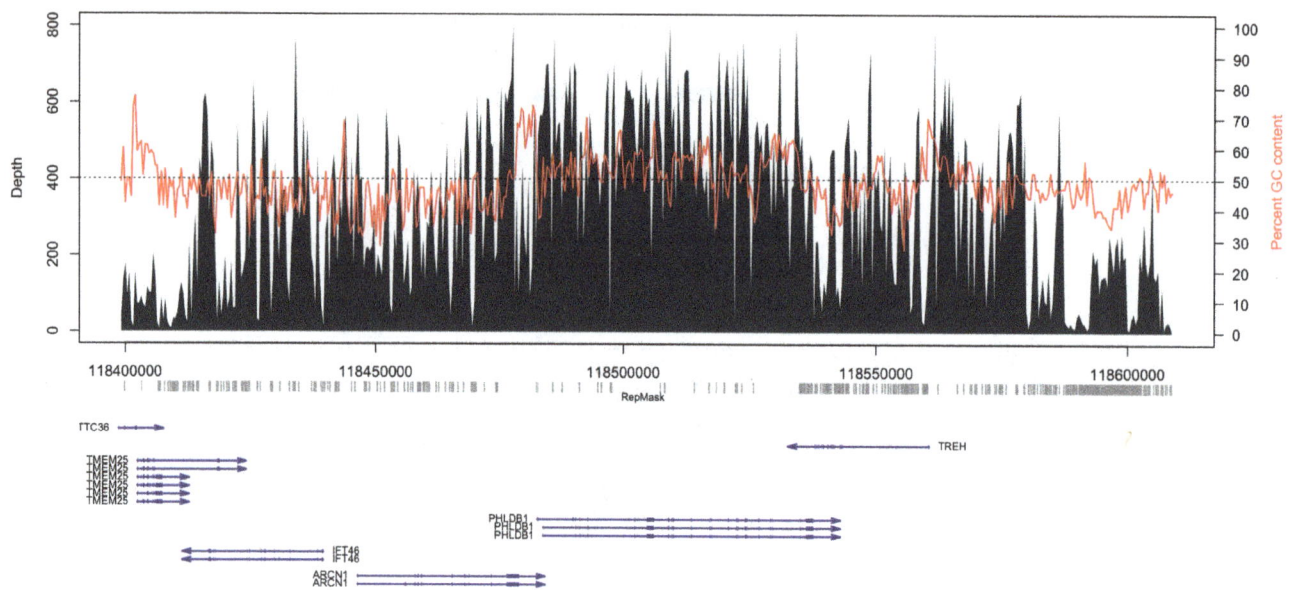

Figure 2. Read depth plot of a chromosome 11 target for a sample showing median coverage among all samples used for capture. Vertical bars indicate read depth with scale depicted on the left side of the panel. Red lines show percent GC content across non-overlapping 400 bp intervals spanning the target region with scale shown on the right side of the panel. Horizontal dotted line indicates 50% GC content. A repeat structure track (RepMask) is shown below the plot in gray derived from the UCSC genome browser for all repeats containing a Smith-Waterman score of at least 600, and larger than 200 bp in size. Genes are shown below the repeat track in dark blue and arrows depict gene orientation.

Regional capture and coverage

The percentage of uniquely mapped reads that align to captured target regions provides a metric for the specificity and efficiency of the capture method. We captured 9 independent human genomic targets of varying sizes and sample number ranges across various Illumina sequencing platforms. The mean percentage across samples of uniquely mapped reads that aligned to these target sequences ranged from ~42% to 82% when considering percentage of mapped reads with a threshold MAPQ of greater than or equal to 20 (Table S1). We found no greater than a 5% reduction in mapped reads when applying this mapping quality filter, and we consistently achieved 80–90% mapping rate of total reads depending on the target region. Our estimated PCR duplication rate ranged between 5 and 40% of sequence per sample, and was influenced by target size, the level of sample plexing within the hybridization, and whether final captures were bisulfite converted. The vast majority of samples were run in 24-plex in a single HiSeq 2000 sequencing lane. As expected, the mean target coverage varied by composite target sizes, read length, and Illumina sequencing platform. All captured targets revealed high enrichment of reads positioned directly within the boundaries of each BAC template selected for probe synthesis (Figure S1).

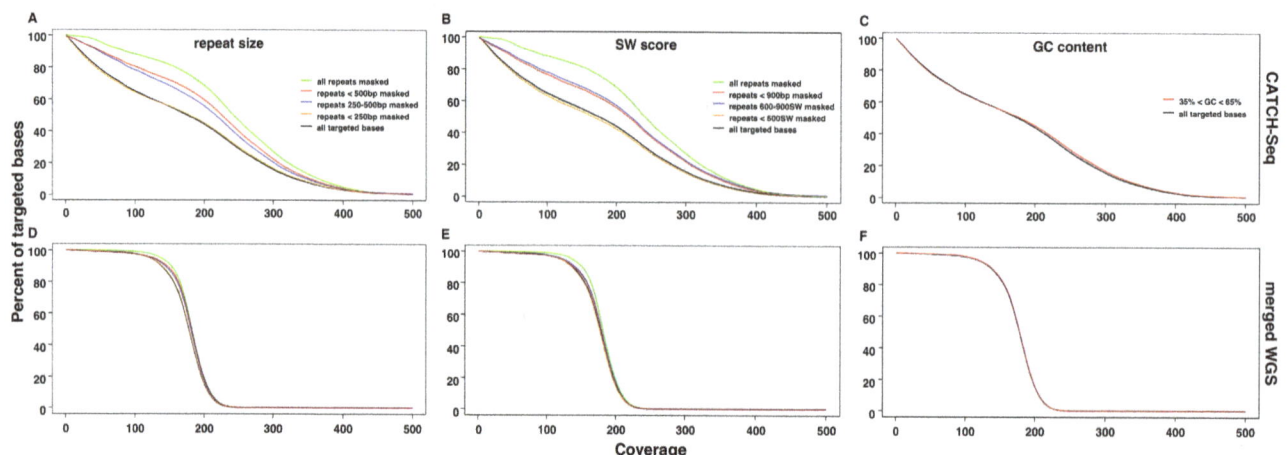

Figure 3. Capture efficiency in a sample representing the median coverage among all sequenced samples shown by the percent of total targeted bases covered at particular coverage depths in a chromosome 11 target. (A) Percent of targeted bases covered using various thresholds of repeat masking (A) by size, or (B) (SW) scores. (C) Percent of targeted bases covered based on masking of percent GC content extremes. Upper panels show coverage by CATCH-Seq within a sample that showed median coverage among all other samples used in the capture. (D–F) Lower panels show coverage within the corresponding captured region for the same number of merged reads analyzed for CATCH-Seq under the same repeat masking or percent GC content thresholds from 15 individuals sequenced for the 1000 genomes project (merged WGS).

Table 1. Repeat structure description within a chromosome 11 target.

repeat ranges	size (kb)	on target (%)[a]
total[b]	112.2	53.6
<250 bp	29.1	13.9
250 bp to 500 bp	58.8	28.1
>500 bp	24.1	11.5
<600SW	14.1	6.8
600SW to 900SW	74.9	35.8
>900SW	23.2	11.1
GC extremes[c]	18.4	8.8

[a]total captured target size is 209.2 kb, target region shown in Figure 2.
[b]all repeat hg19 coordinates, sizes, and Smith-Waterman (SW) scores obtained from RepeatMasker within the UCSC Genome Browser. For descriptions of RepeatMasker, http://www.repeatmasker.org/.
[c]total sequence within the target coordinates with 400 bp intervals containing less than 35% and greater than 65% GC percentage.

With closer inspection of individual targets, we found a variety of subregions where coverage was sparser and uneven due to large repeats, or repeats with low divergence (Figure S2).

Blocking repetitive sites is crucial for solution hybridization based capture systems, as the inclusion of repetitive sequences in a capture probe set can lead to contamination of the final sequence reads with off-target repeats [5]. For CATCH-Seq, blocking of repeats is essential because many of probes synthesized from BAC templates contain repeat regions. Based on our weaker coverage of repeats, we were interested in how both the levels of repeat divergence and repeat size influence enrichment and uniformity of coverage within target template regions. Typical commercial platforms avoid synthesis of probes within repeat regions, and

usually only consider uniformity of coverage within non-repetitive sites. We were interested in the uniformity of coverage of both repetitive and non-repetitive sequences as repeats represent a considerable proportion of the contiguous regions we targeted. We specifically analyzed a region on chromosome 11 that is one target within a composite capture of ten targets and selected the sample that represented median coverage among all of the samples we sequenced (Figure 2, Table 1). To understand the influence of repeat structures on target capture uniformity, we compared the base coverage across all targeted bases within our chromosome 11 site after repeat masking the target with increasing threshold values of repeat lengths or Smith-Waterman (SW) scores (Table 1). The variation in repeat masking thresholds gave us an indication

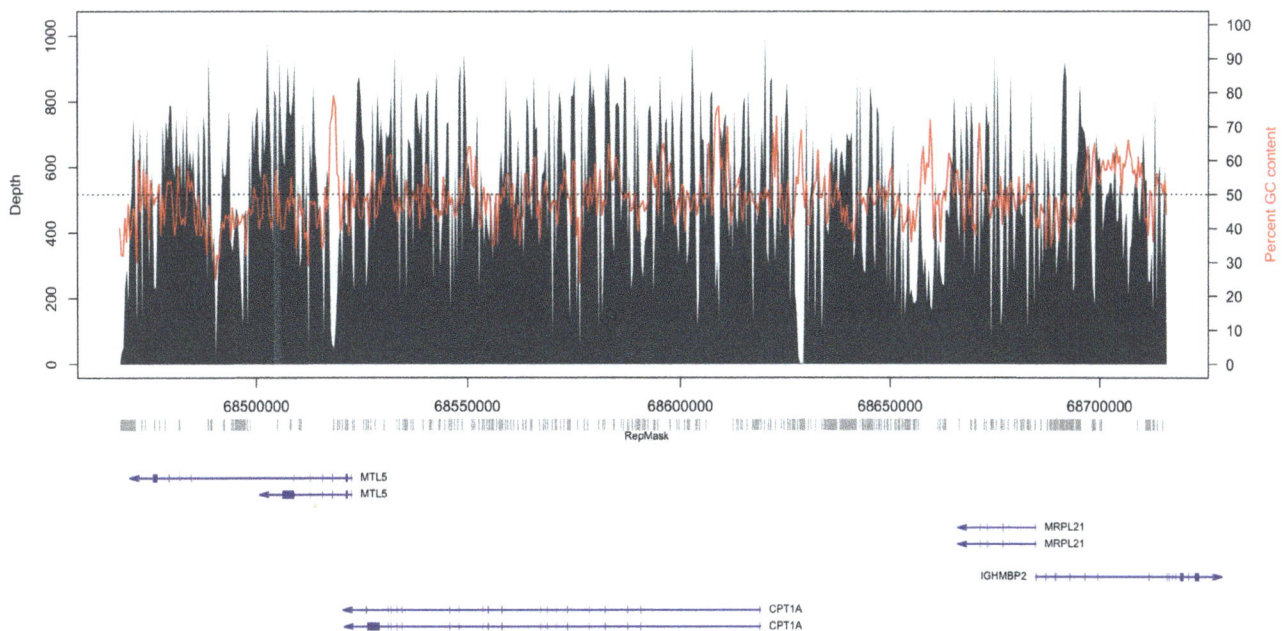

Figure 4. Read depth plot of a chromosome 11 target for a sample showing median coverage among all samples used for capture and bisulfite sequencing. Vertical bars indicate read depth with scale depicted on the left side of the panel. Red lines show percent GC content across non-overlapping 400 bp intervals spanning the target region with scale shown on the right side of the panel. Horizontal dotted line indicates 50% GC content. A repeat structure track (RepMask) is shown below the plot in gray derived from the UCSC genome browser for all repeats containing a Smith-Waterman score of at least 600, and larger than 200 bp in size. Genes are shown below the repeat track in dark blue and arrows depict gene orientation.

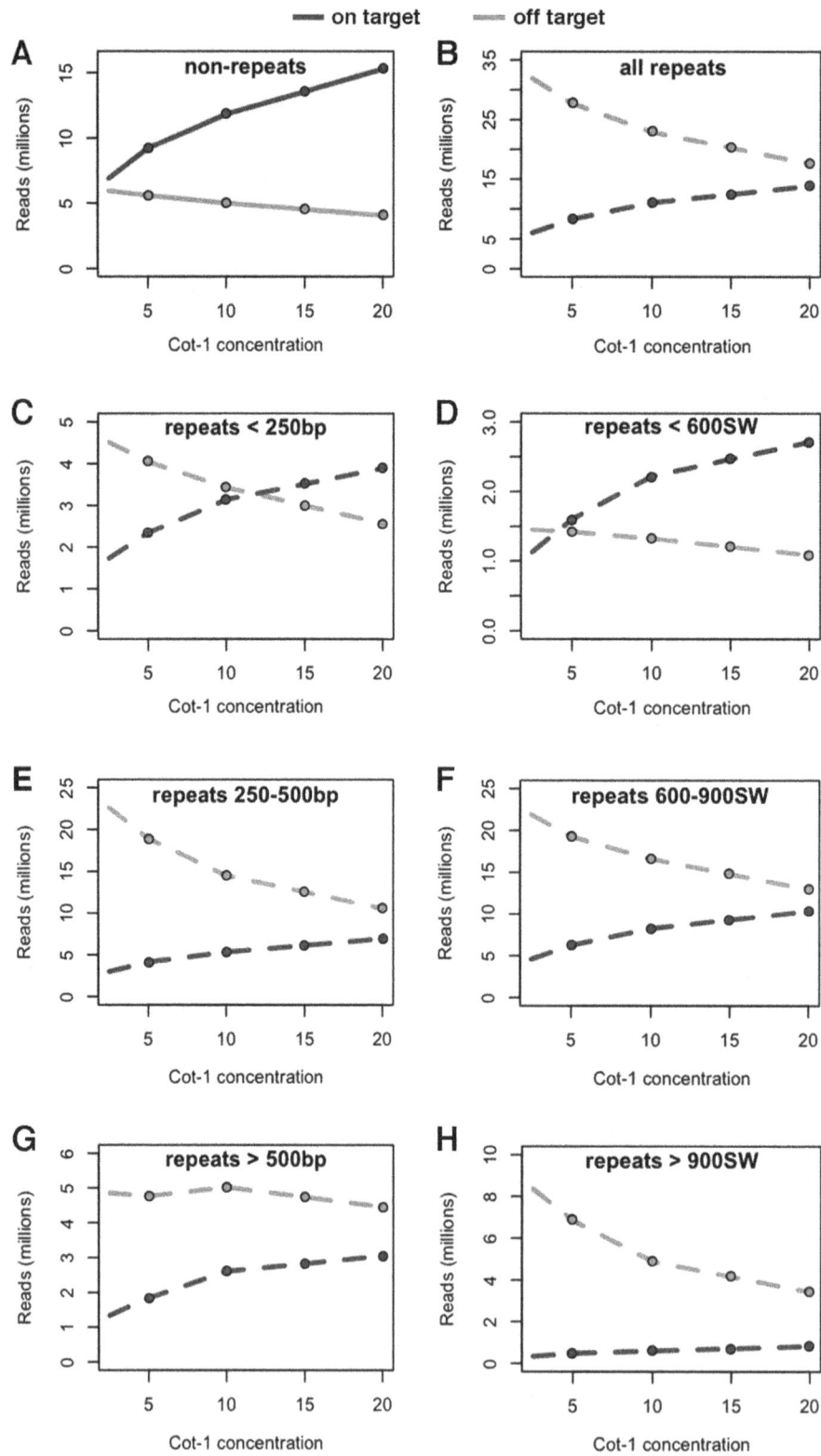

Figure 5. The effect of repeat blocking with increased concentrations of Cot-1 DNA within the CATCH-Seq hybridization step of a chromosome 11 target. Total numbers of on target and off target read yields in millions within non-repetitive sequences (A) or repetitive sequences (B). (C–H) On and off target read yields within repeat structures based on different thresholds of size (C,E,G) or divergence (D,F,H). Green and gray lines show on target and off target reads, respectively.

Table 2. Repeat structure description within a chromosome 11 target used for Cot-1 tests.

repeat ranges	size (kb)	target bases (%)[a]	on target reads (%)[b]
total[c]	129.7	52.4	47.6
<250 bp	32.0	12.9	13.4
250 bp to 500 bp	72.4	29.2	23.7
>500 bp	25.2	10.2	10.5
<600SW	19.6	7.9	9.3
600SW to 900SW	97.4	39.3	35.4
>900SW	12.6	5.1	2.8
non-repeats	118.0	47.6	52.4

[a]total captured target size is 247.6 kb; target region shown in Figure 4.
[b]out of 50 million sampled reads at 20× Cot-1 concentration.
[c]all repeat hg19 coordinates, sizes, and Smith-Waterman (SW) scores obtained from RepeatMasker within the UCSC Genome Browser. For descriptions of RepeatMasker, http://www.repeatmasker.org/.

of the proportion of our captured sequences that were uniquely mapping to repetitive versus non-repetitive regions within the target site. A mask of repeat sizes below 250 bp in length from target coverage calculations did not proportionally alter coverage rates compared to the total unmasked target, suggesting that small repeats were covered effectively. Repeat mask of sizes ranging between 250 bp and 500 bp increased our relative coverage rate, and was similar to the effect of masking all repeats less than 500 bp (Figure 3A). Masking of all repeats, regardless of size, demonstrated that non-repetitive sequences represented the majority of our capture, and indicated that our blocking approach was highly effective. Repeat masking by SW scores produced a similar trend as repeat size. Capture of repeats above a score of 600 became less efficient (Figure 3B). Extremes in GC content are also known to influence coverage of targeted bases in solution hybridization-based exon capture platforms [6]. We also masked coverage by GC content extremes in 400 bp intervals containing high (>65%) and low GC (<35%) percentages (Figure 2, Table 1). Masking stretches of extreme GC percentages did not alter the relative coverage rate (Figure 3C).

We were interested in discerning the difference between blocking of repeats in the hybridization step versus the ability to uniquely map reads within these intermediate and large repeat structures within our chromosome 11 target. We compared our target coverage with the coverage of the same region from whole genome sequencing (WGS) data. Such a comparison should reveal the effects of poor unique mapping within these target repeat regions versus capture bias produced from blocking. We used the merged WGS data of 15 individuals from the 1000 genomes project to approximate the total sequence depth of our capture experiment. We analyzed reads with a MAPQ greater than or equal to 20, and found that repeat masking WGS data made little difference in coverage rate in this same target site (Figure 3E–F). This suggests that repeat blocking within the solution hybridization step has a much greater impact on coverage of target repeat structures than difficulty in uniquely mapping captured reads within these target reference repetitive sequences. The cumulative coverage plots reveal a gradual slope of coverage rates in the target region, indicating a wider range of sequence depths compared to the more uniform coverage of WGS. However, much of this difference can be attributed to repeat blocking. Completely repeat masked coverage calculations within this target between CATCH-Seq and WGS showed very similar numbers of bases covered at 50× depth. CATCH-Seq yielded 89% of targeted bases covered at

100×, compared to 98% for WGS. Alignment files comparing CATCH-Seq and WGS can be found at NCBI SRA with BioProject accession SRP042633.

To further investigate the effect of repeat blocking on target coverage, we performed solution hybridization reactions with libraries prepared from K562 cell line genomic DNA and increasing Cot-1 DNA concentrations to test the influence of this blocking reagent on repeat coverage in another chromosome 11 target that was also captured for bisulfite sequencing of an independent sample set (Table S1, Figure 4). We typically used a 20:1 concentration of Cot-1 to library ratio and were interested in how reduction of Cot-1 DNA influenced on-target capture in both repeats and non-repeats. Approximately 50 million reads were sampled with a MAPQ of greater than or equal to 20 from the total yield of aligned reads from each hybridization with 2.5, 5, 10, and 20 fold Cot-1 to library ratio. We found that with lower input concentration of Cot-1 DNA, we compromised overall target capture efficiency in both non-repetitive sites and in repeats. Increasing concentrations of Cot-1 DNA improved both the absolute yield of reads on-target while also decreasing off-target yields (Figure 5). CATCH-Seq procedures with no Cot-1 DNA in the hybridization step yielded less than 9% of mapped reads within a target sites. In reads aligned to non-repeat sequence, we found a stronger relationship between Cot-1 concentrations and increasing on-target reads than with reduced off-target reads (Figure 5A). By comparison, increasing Cot-1 concentration produced a roughly equal exchange of reads aligned to off-target repeats as for those aligned to on-target repeats (Figure 5B). We found that this rate of exchange between off-target and on-target repeats varied depending on repeat size and SW score. Smaller repeats of less than 250 bp exhibited an equal exchange in off-target for on-target reads, while larger repeats and those with higher SW scores showed a mild increase in yield of on-target reads, while off-target yields declined (Figure 5C–H). Overall, the highest Cot-1 concentration at 20 fold produced the highest on-target read yield.

We observed that absolute on-target read yields were split almost evenly between repeats and non-repeat regions at all Cot-1 concentrations. Therefore, we expressed read yields as a percentage of total yield stratified by non-repeats, repeat size, or SW score across each hybridization experiment with increasing Cot-1 concentration (Figure S3). The largest percent increase of on-target reads was within non-repeats, and the largest percent decrease was in intermediate sized repeats or SW scores, while yields within the smallest and largest repeats changed very little

A

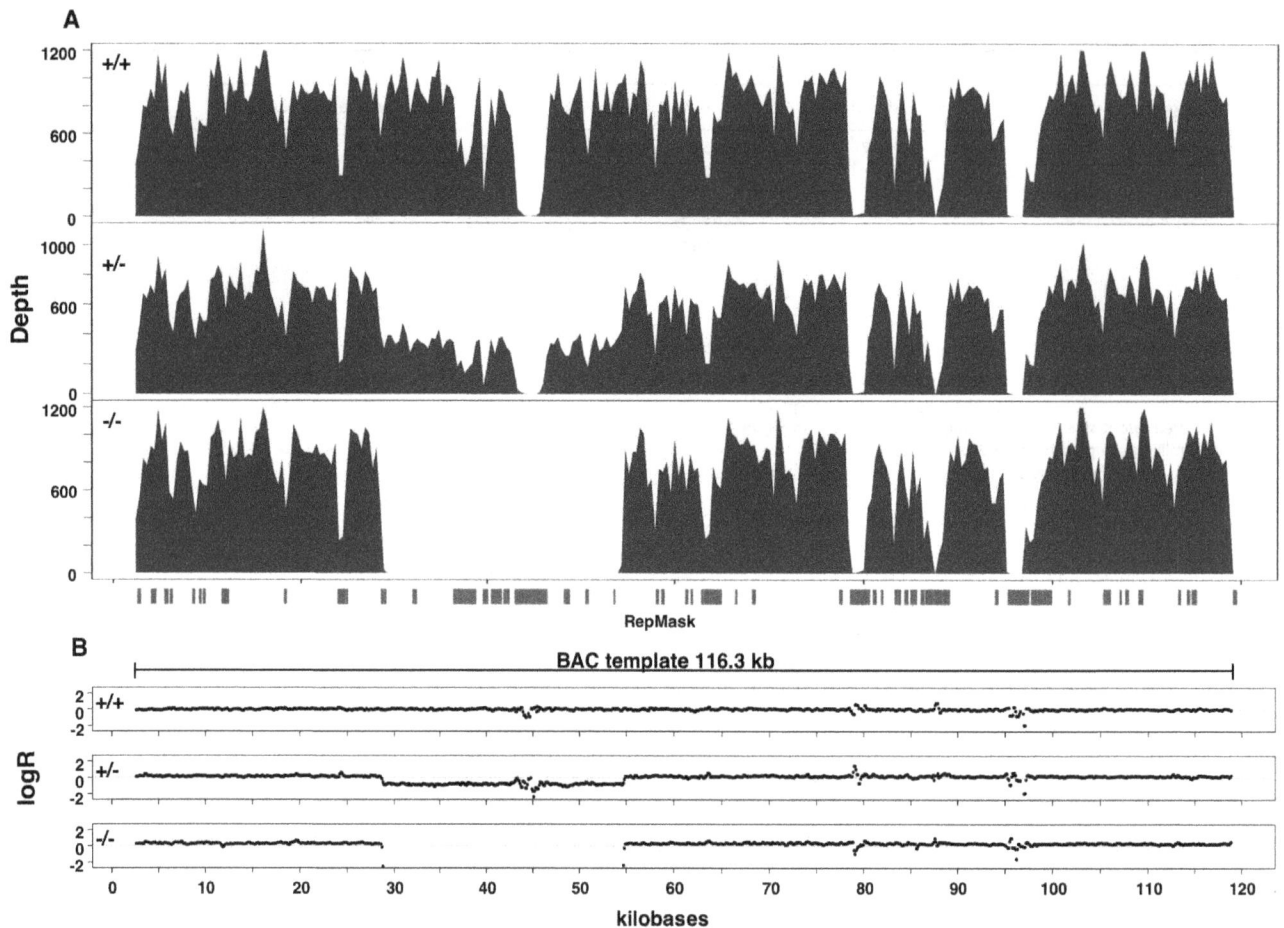

B

Figure 6. Determination of copy number variation across a CATCH-Seq target using read depth. (A) Read depths are partitioned into 100 bp segments across the length of target genomic coordinates and the fraction of total aligned bases per segment are calculated. In this target, there is a noticeable drop in read depth in two individuals shown in bottom panels compared to wild type (+/+) that indicates individuals that contain heterozygous (+/−) and homozygous (−/−) deletions in this region. (B) Log-ratio values (logR) are calculated across the target site that are normalized for read depth variance caused by capture and sequencer biases to resolve clear copy number variation boundaries. Contained within the deleted region is a repeat sequence as shown by underlying RepeatMasker track (RepMask) that is not well covered. Coverage of this repeat structure is reflected in the logR plot as a slight fluctuation from zero as indicated by the horizontal green lines. For targets containing a copy number variation that represents a large proportion of the total target sequence such as the one depicted here, often the individual base fraction normalization by the median of control samples will result in slightly elevated logR values outside the variable region that is most noticeable in the individual containing the homozygous deletion in the bottom panel. The extent of the BAC template used for CATCH-Seq is depicted just below the RepMask track.

relative to the total yield. We found that the overall percent yield proportion of on-target reads alone to one another within classified categories did not vary with Cot-1 concentrations, and these percentages of on-target reads also proportionally represented the overall repeat and non-repeat structure within this specific target site (Table 2). We did find about a 5% bias of on-target yield toward non-repeats when comparing to total percent repeats across this target site (52.4% of target bases are repeats with 47.6% of read yield, and 47.6% of target bases are non-repeats with 52.4% of read yield). These results suggest that the target sequence yield proportionally represents the diversity of the target region itself. Alignment files comparing CATCH-Seq with increasing concentrations of Cot-1 can be found at NCBI SRA with BioProject accession SRP042633.

We also found that the PCR duplication rate in our sequencing reads also increased with Cot-1 concentration. However, we have found that the duplication rate was more strongly influenced by input library quality, and this problem is not unique to our capture platform. Measurement of the true library concentration by qPCR

is critical to ensure proper library input into the hybridization for higher diversity of unique reads mapped. Our current procedure uses 40× Cot-1 DNA and with higher quality library (10–40% ligated fragments) one should expect between 5–20% duplication rate, depending on composite target size, hybridization multiplexing, and if the captured library was bisulfite converted. Based on our results, we believe the benefit of high specificity and enrichment within a target region outweighs the compromise in consequential mapped read diversity and lower coverages within intermediate to large repeat regions. Furthermore, default design conditions for commercial platforms avoid the synthesis of probes within repeat regions that would typically not be covered. We found that CATCH-Seq adequately covered all of the same sites where prospective probes would be synthesized based on commercial design within one of our targets (Figure S4). Comparison to WGS coverage from 15 lanes of sequencing yielded better uniformity across repeats, but uniformity outside of repeats was similar. These data show that CATCH-Seq has the ability to capture the same prospective target regions as

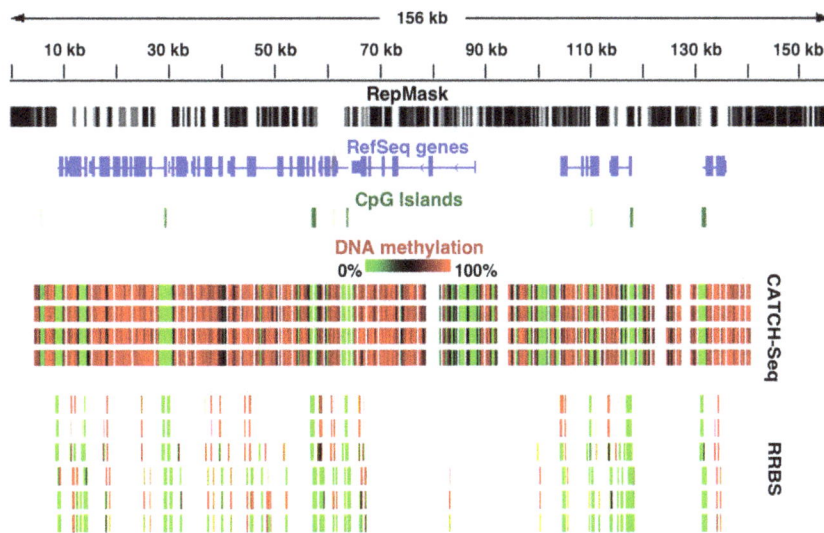

Figure 7. High density methylation data derived from bisulfite sequencing of a CATCH-Seq target. Scale of the captured region is indicated in the topmost track in kilobases (kb), followed by repeat structure in gray and black (RepMask), genes shown in blue (RefSeq), and CpG islands in green. Four CATCH-Seq tracks from the same cell type show DNA methylation levels across ~2,700 target CpGs with hypomethylation depicted in green and hypermethylation in red. Six reduced representation bisulfite sequencing (RRBS) tracks for different cell and tissue types correspond with the same captured region, and demonstrate CpGs not covered by RRBS method compared to CATCH-Seq. The four CATCH-Seq tracks are from the same cell type as the topmost RRBS track. RRBS tracks are derived from previously reported data [23]. CpGs shown within CpG islands were all typically hypomethylated across all cell and tissue types depicted.

commercial systems even with repeats included in the probe synthesis. Furthermore, while no custom capture platform will prospectively be able to cover as uniformly as WGS as a general caveat of capture sequencing, samples can be highly multiplexed and still achieve higher coverage per individual within a target compared to 15 lanes of WGS on a single sample.

Applications of CATCH-Seq

Based on our sequence enrichment procedures and our kilobase scale contiguous probe sets, we have found this technique useful for distinguishing both genetic and epigenetic variation across different target sites. Besides variant calling within target regions, we have found that the level of specificity and unique high coverage across targets allows for the resolution of large CNV boundaries (Figure 6A). By partitioning of the total reads into 100 bp segments across the length of target genomic coordinates and determining the fraction of total aligned bases per segment, we are able to calculate log-ratio values across the target site that normalize read depth variance caused by capture and sequencer biases. We calculate the log_2 of an individual's base fraction within a segment divided by the median base fraction of all control samples. With the advantage of the normalized read depth, we are able to identify both homozygous and heterozygous duplications and deletions within target sites (Figure 6B). These data show that even with varying levels of coverage across the site, we are able to effectively resolve CNVs. The majority of targets containing CNVs were pre-selected based on results from high-density genotyping arrays that implicated a CNV within the specific locus. Recent studies have shown that CNVs such as inversions, deletions, insertions, and segmental duplications may contribute another level of genetic variation that may influence human phenotypic diversity and is also associated with a variety of human diseases [19,20]. Often the most common approach of CNV determination involves SNP arrays, but often arrays cannot clearly establish

CNV boundaries [21]. Our method provides a unique validation to resolve CNV boundaries.

We have also applied CATCH-Seq to analysis of DNA methylation across large target sites containing CpG islands and multiple genes. By using methylated adapters and treating the post-capture libraries with bisulfite conversion, we can measure CpG methylation with higher coverage thresholds than what is often generated for genome-wide analysis of individual CpGs such as by reduced representation bisulfite sequencing (RRBS) (Figure 7). Furthermore, we have found that CATCH-Seq can also be used in parallel with any other practiced functional genomics approaches. A similar method to ours performed sequencing of BAC-enriched mononucleosomal fragments (known as BEM-Seq), using whole BAC labelled probes for capture of MNAse-digested fragments within a target site [17]. We have similarly used CATCH-Seq procedures with MNase-digests also from sorted mouse lymphocytes in combination with methylation analysis (unpublished). Lastly, CATCH-Seq was also used to capture gene regions associated with melanism from genomes of unsequenced Felid species using selected fosmids from *Felis catus* as templates [22] (manuscript submitted). Overall, we find that data from CATCH-Seq procedures allows for affordable, high resolution sequencing of captured genomic targets without the added cost of oligo-based probe synthesis. We have included a price per sample estimation with comparison of current commercially available custom probe synthesis platforms for two of our targets we captured (Table S2).

Supporting Information

Figure S1 Read depth plots of various BAC-template enriched sequencing reads as shown within the UCSC genome browser. Table 1 provides further details of each target depicted for which human hg19 genomic coordinates are shown above each individual target. From top to bottom black vertical lines indicate sequencing read depth, followed by genes contained within the target, and selected BACs used as templates.

Figure S2 Zoomed CATCH-Seq targets shown within the UCSC genome browser that exhibit low read depth covering repetitive sites. Black vertical lines show read depth; light gray to black tracks below indicate repeat sequences with darker shades indicating lower divergence or higher similarity to other repeats across the genome.

Figure S3 On and off target read yields expressed as a percentage of total yield stratified by non-repeats, repeat size, or SW score across each hybridization experiment that contained increased concentrations of Cot-1 DNA. On and off target read yield percentages according to repeat size thresholds (A) or by Smith-Waterman (SW) repeat scores (B). Black and shades of gray show off target reads; white and shades of green depict on target reads.

Figure S4 A zoomed in chr11 region within the UCSC genome browser (as depicted in Figure 2) showing read depth of CATCH-Seq compared to WGS that contains many SINE elements. The majority of unevenness across the capture is found within SINE repeats. Another track depicts

prospective probe baits recommended for synthesis using default parameters with NimbleDesign software for custom capture sequencing where probe is completely repeat masked. CATCH-Seq effectively covers the exact sites where probes are recommended for synthesis.

Table S1 CATCH-Seq target capture summary.

Table S2 CATCH-Seq cost per sample comparison estimation summary.

Acknowledgments

We thank Phil Dexheimer for support with inline barcoded adapter oligo design and demuxing software. We also thank Shawn Levy, Tyson DeAngelis, and Angela Jones at the Genomic Services Lab at HudsonAlpha Institute for Biotechnology for Illumina sequencing technical support.

Author Contributions

Conceived and designed the experiments: KD JS DA. Performed the experiments: KD JS. Analyzed the data: KD DA. Contributed reagents/materials/analysis tools: DA. Wrote the paper: KD DA.

References

1. Borate U, Absher D, Erba HP, Pasche B (2012) Potential of whole-genome sequencing for determining risk and personalizing therapy: focus on AML. Expert Rev Anticancer Ther 12: 1289–1297.
2. Whitcomb DC (2012) What is personalized medicine and what should it replace? Nat Rev Gastroenterol Hepatol 9: 418–424.
3. Mertes F, Elsharawy A, Sauer S, van Helvoort JM, van der Zaag PJ, et al. (2011) Targeted enrichment of genomic DNA regions for next-generation sequencing. Brief Funct Genomics 10: 374–386.
4. Goh G, Choi M (2012) Application of whole exome sequencing to identify disease-causing variants in inherited human diseases. Genomics Inform 10: 214–219.
5. Gnirke A, Melnikov A, Maguire J, Rogov P, LeProust EM, et al. (2009) Solution hybrid selection with ultra-long oligonucleotides for massively parallel targeted sequencing. Nat Biotechnol 27: 182–189.
6. Clark MJ, Chen R, Lam HY, Karczewski KJ, Euskirchen G, et al. (2011) Performance comparison of exome DNA sequencing technologies. Nat Biotechnol 29: 908–914.
7. Irizarry RA, Ladd-Acosta C, Wen B, Wu Z, Montano C, et al. (2009) The human colon cancer methylome shows similar hypo- and hypermethylation at conserved tissue-specific CpG island shores. Nat Genet 41: 178–186.
8. Heyn H, Esteller M (2012) DNA methylation profiling in the clinic: applications and challenges. Nat Rev Genet 13: 679–692.
9. Kaper F, Swamy S, Klotzle B, Munchel S, Cottrell J, et al. (2013) Whole-genome haplotyping by dilution, amplification, and sequencing. Proc Natl Acad Sci U S A 110: 5552–5557.
10. Porreca GJ, Zhang K, Li JB, Xie B, Austin D, et al. (2007) Multiplex amplification of large sets of human exons. Nat Methods 4: 931–936.
11. Johansson H, Isaksson M, Sorqvist EF, Roos F, Stenberg J, et al. (2011) Targeted resequencing of candidate genes using selector probes. Nucleic Acids Res 39: e8.
12. Diep D, Plongthongkum N, Gore A, Fung HL, Shoemaker R, et al. (2012) Library-free methylation sequencing with bisulfite padlock probes. Nat Methods 9: 270–272.
13. Varley KE, Mitra RD (2008) Nested Patch PCR enables highly multiplexed mutation discovery in candidate genes. Genome Res 18: 1844–1850.
14. DeAngelis MM, Wang DG, Hawkins TL (1995) Solid-phase reversible immobilization for the isolation of PCR products. Nucleic Acids Res 23: 4742–4743.
15. Lindgreen S (2012) AdapterRemoval: easy cleaning of next-generation sequencing reads. BMC Res Notes 5: 337.
16. Bashiardes S, Veile R, Helms C, Mardis ER, Bowcock AM, et al. (2005) Direct genomic selection. Nat Methods 2: 63–69.
17. Yigit E, Zhang Q, Xi L, Grilley D, Widom J, et al. (2013) High-resolution nucleosome mapping of targeted regions using BAC-based enrichment. Nucleic Acids Res 41: e87.
18. Carpenter ML, Buenrostro JD, Valdiosera C, Schroeder H, Allentoft ME, et al. (2013) Pulling out the 1%: Whole-Genome Capture for the Targeted Enrichment of Ancient DNA Sequencing Libraries. The American Journal of Human Genetics 93: 852–864.
19. Craddock N, Hurles ME, Cardin N, Pearson RD, Plagnol V, et al. (2010) Genome-wide association study of CNVs in 16,000 cases of eight common diseases and 3,000 shared controls. Nature 464: 713–720.
20. Henrichsen CN, Chaignat E, Reymond A (2009) Copy number variants, diseases and gene expression. Hum Mol Genet 18: R1–8.
21. Winchester L, Yau C, Ragoussis J (2009) Comparing CNV detection methods for SNP arrays. Brief Funct Genomic Proteomic 8: 353–366.
22. Schneider A, David VA, Johnson WE, O'Brien SJ, Barsh GS, et al. (2012) How the leopard hides its spots: ASIP mutations and melanism in wild cats. PLoS One 7: e50386.
23. Meissner A, Mikkelsen TS, Gu H, Wernig M, Hanna J, et al. (2008) Genome-scale DNA methylation maps of pluripotent and differentiated cells. Nature 454: 766–770.

14

The Expression of Three Opsin Genes from the Compound Eye of *Helicoverpa armigera* (Lepidoptera: Noctuidae) Is Regulated by a Circadian Clock, Light Conditions and Nutritional Status

Shuo Yan[1], Jialin Zhu[2], Weilong Zhu[1], Xinfang Zhang[1], Zhen Li[1], Xiaoxia Liu[1]*, Qingwen Zhang[1]*

1 Department of Entomology, China Agricultural University, Beijing, P.R. China, **2** Beijing Entry-Exit Inspection and Quarantine Bureau, Beijing, P.R. China

Abstract

Visual genes may become inactive in species that inhabit poor light environments, and the function and regulation of opsin components in nocturnal moths are interesting topics. In this study, we cloned the ultraviolet (UV), blue (BL) and long-wavelength-sensitive (LW) opsin genes from the compound eye of the cotton bollworm and then measured their mRNA levels using quantitative real-time PCR. The mRNA levels fluctuated over a daily cycle, which might be an adaptation of a nocturnal lifestyle, and were dependent on a circadian clock. Cycling of opsin mRNA levels was disturbed by constant light or constant darkness, and the UV opsin gene was up-regulated after light exposure. Furthermore, the opsin genes tended to be down-regulated upon starvation. Thus, this study illustrates that opsin gene expression is determined by multiple endogenous and exogenous factors and is adapted to the need for nocturnal vision, suggesting that color vision may play an important role in the sensory ecology of nocturnal moths.

Editor: Nicholas S. Foulkes, Karlsruhe Institute of Technology, Germany

Funding: This project was supported by Natural Science Foundation of China (31371943). The funders had no role in study design, data collection and analysis, decision to publish, or preparation of the manuscript.

Competing Interests: The authors have declared that no competing interests exist.

* Email: liuxiaoxia611@cau.edu.cn (XXL); zhangqingwen@263.net (QWZ)

Introduction

Vision is one of the most familiar forms of stimulus discrimination and plays numerous key roles in the performance of insect behaviors, such as searching for food and potential mates, avoiding predators and unsafe environments, and other specific behaviors [1–5]. In Lepidoptera, visual information is acquired via a highly developed visual system. Compound eyes are usually composed of several thousand ommatidia to detect and convert light into visual images, and each ommatidium contains nine photoreceptor cells [6–10]. At the molecular level, the visual pigments in photoreceptors include a vitamin A-derived chromophore, usually 11-*cis*-retinal, and a transmembrane protein, opsin [11–12]. The amino acid sequences of both the opsin and the chromophore affect their ability to absorb visual pigment [12–14]. Opsins are ancient proteins belonging to the G-protein-coupled receptor family and are characterized by a seven-transmembrane domain structure and a lysine residue in the seventh helix [15–17]. There are three types of opsins with peak absorbance at either the ultraviolet wavelengths (UV, 300–400 nm), blue wavelengths (BL, 400–500 nm) or long wavelengths (LW, 500–600 nm) [8,10,12].

Regulation of opsin mRNA levels has been studied in several species. Daily changes in opsin mRNA levels are typically observed in animals, including mice [18], fish [19], toads [19], rats [20–21], honeybees [22] and horseshoe crabs [23–25]. All

living organisms possess a circadian clock to synchronize their rhythm with the environment, and photoreceptors are clearly necessary for this synchronization [26–28]. These fluctuations in opsin mRNA levels are regulated by an endogenous circadian clock and underlie the ability of the visual system to function optimally in ambient illumination [19,21–25]. Furthermore, opsin mRNA is induced by light exposure, and organisms exposed to more light exhibit elevated opsin gene expression [14,22,29–32]. Sexual dimorphism in the expression pattern of opsin genes has also been observed in previous studies [33–34], and sexually dimorphic photoreceptors, which are adapted to differences in the behaviors of males and females, are common in butterflies [35–37]. Thus, regulation of opsin mRNA may be controlled by multiple endogenous and exogenous factors.

The role of photoreceptors in Lepidoptera is an interesting topic because of the wide variation in habits and lifestyle. Most studies on the regulation of opsin mRNA have been performed in diurnal animals, whereas almost no information exists on opsin regulation in nocturnal animals. Whether the factors influencing opsin mRNA expression are similar between diurnal and nocturnal insects remains unknown. There have been several studies on electroretinography (ERG), phototaxis, circadian rhythms and sexual behavior in the cotton bollworm [38–42], making this organism not only a worldwide pest responsible for great economic losses but also an excellent model for studying the evolution and function of opsins. Xu *et al.* [32] found no daily cycling of opsin mRNA levels in *H. armigera*. In that study, the authors collected RNA samples from whole individual moths, but we think it is more important to evaluate opsin expression patterns specifically in the compound eye. In the current study, we focused on selected genes encoding opsins in *Helicoverpa armigera* (Hübner) and investigated the determinants of opsin mRNA levels in a specific tissue (the compound eye), which was beneficial for improved understanding of the evolution and function of opsins in nocturnal moths.

Materials and Methods

The wording of our manuscript is suitable for publication. Our study was conducted in the IPM lab at China Agricultural University (40°02′N, 116°28′E). The living material sampled in our experiments consisted of cotton bollworms (*Helicoverpa armigera*). No specific permits were required for the insect collection performed for this study. Larvae of *H. armigera* collected from a cotton field in Hebei Province (China) were used in the experiments. Cotton bollworms are a common insect and are not included in the "List of Protected Animals in China".

1. Animals and experimental conditions

Larvae of *H. armigera* collected from a cotton field in Hebei Province (China) were used in the experiments. The larvae were reared on a synthetic diet [43] and maintained at $27\pm1°C$ and $75\pm10\%$ relative humidity (RH) with a 14:10 light:dark photoperiod. Zeitgeber time 0 (ZT0) was designed as lights-on, and ZT14 was designed as lights-off. Pupae were segregated by sex, placed in holding cages ($20\times25\times30$ cm) with a removable white cloth top for egg collection, and held for adult emergence. The moths were provided a 10% honey solution. Moths that emerged during the scotophase were designated as 0-day-old, 1-day-old, 2-day-old, and so forth on subsequent days.

2. Isolation and cloning of opsin-encoding cDNA

Total RNA was isolated from the compound eye of 2-day-old moths using the RNeasy Mini Kit (Qiagen, Hilden, Germany), treated with DNase I (Qiagen) to remove any residual genomic DNA, and pelleted twice by successive centrifugation through 5.7 mol/L CsCl according to Chase *et al.* [13]. A 1 μL sample of RNA was used to perform spectroscopic quantitation using a NanoDrop 2000 spectrophotometer (Thermo Fisher, USA). Then, 2 μg samples of total RNA were reverse-transcribed using M-MLV Reverse Transcriptase (Promega, USA).

Several primers (Table S1) were used to amplify fragments of the UV-sensitive opsin gene via polymerase chain reaction (PCR). Because of the conserved 5′ untranslated regions of the blue- and long wavelength-sensitive opsin genes, primers (Table S1) were designed to clone the 5′ fragments. PCR reactions were typically cycled 35 times at 94°C for 45 s, 55°C for 30 s, and 72°C for 1 min. To obtain complete sequences of the three opsin genes, rapid amplification of cDNA ends was performed using the FirstChoice RLM-RACE Kit (Ambion, Austin, TX). The amplification protocols consisted of 35 cycles of 94°C for 45 s, 58°C for 30 s, and 72°C for 90 s. TransTaq-T DNA Polymerase (TransGen Biotech, Beijing, China) was used to perform the PCR reactions. The purified fragments were cloned into the Trans1-T1 vector (TransGen Biotech) and then transformed into DH5α *Escherichia coli*; positive clones were then selected for sequencing (BGI Life Tech Co., Beijing, China).

3. Sequence analysis and construction of the phylogenetic tree

The opsin gene sequences were submitted to the NCBI website (http://www.ncbi.nlm.nih.gov), and translations of the opsin genes were performed using DNAMAN v.5.2.2 (Lynnon Biosoft, Quebec, Canada). Predictions of the transmembrane regions were made using TMpred (http://www.ch.embnet.org/software/TMPRED_form.html) [44]. The theoretical isoelectric points (pI) and molecular weights (MW) were calculated using the Compute pI/Mw tool (http://web.expasy.org/compute_pi/). The protein functional sites were predicted using PROSITE SCAN (http://npsa-pbil.ibcp.fr/cgi-bin/npsa_automat.pl?page=/NPSA/npsa_proscan.html) [45–46]. All opsin sequences, including those obtained from the present study, were retrieved from GenBank (Table S2). The phylogenetic tree was constructed using the neighbor-joining method in MEGA 4.0.

4. Validation of opsin gene expression using quantitative real-time PCR (qRT-PCR)

Tissue specificity: We extracted total RNA from various tissues (compound eye, brain, antennae, legs, thorax, abdomen, and wings) of 2-day-old moths at zeitgeber time 1 (ZT1). The entire brain was harvested. The head capsule was opened, and the tissues covering the brain were removed. Total RNA was isolated from the compound eyes of 2-day-old moths for the following tests. **Rhythmicity with photoperiod:** RNA samples were collected at 2 h intervals (ZT1, ZT3, ZT5, ZT7, ZT9, ZT11, ZT13, ZT15, ZT17, ZT19, ZT21 and ZT23, which correspond to CT1 (circadian time 1), CT3, CT5, CT7, CT9, CT11, CT13, CT15, CT17, CT19, CT21 and CT23) from the following moths. (1) Moths kept under 14L:10D, constant darkness (DD) and constant light (LL) (*H. armigera* were entrained for 7 days under 14L:10D, DD and LL). *H. armigera* reared under 14L:10D were transferred to constant DD and LL, and 2-day-old moths reared under constant DD and LL were collected according to the duration of actual clock hours. (2) Moths kept under 14L:10D and then transferred to DD (*H. armigera* were entrained for 1 day under DD). **Exposure to various wavelengths of light:** 500-lux light of three wavelengths, including UV (mainly 365 nm), blue (mainly 450 nm), and green (mainly 505 nm), using a light irradiation system made by FSL, Foshan, China were used instead of scotophase, and RNA samples were collected after 6 h of radiation at ZT20. **Starvation:** RNA was extracted from moths that were not fed after adult emergence at ZT1. **Copulation:** Twenty pairs of virgin moths were paired for mating in $20\times25\times30$ cm cages, which were checked every 15 min throughout scotophase to determine whether copulation had occurred. We then obtained RNA samples at 0 h and 3 h after moth copulation.

The circadian clock genes *cryptochrome1* and *cryptochrome2* [40] were assayed in this study. qRT-PCR was carried out on the ABI 7300 instrument (ABI, Ambion) in reactions containing 5 μl of SybrGreen (ABI, Ambion), 100 nM of forward and reverse primers, 1 μl of cDNA from the three opsin genes and 5 μl of cDNA from the circadian clock genes. The amplification protocols consisted of an initial denaturation step at 95°C for 10 min; 40 cycles of 94°C for 15 s, 55°C for 40 s, and 72°C for 35 s; and a melting curve ramp to confirm that each reaction did not produce nonspecific amplification. *EF-1α* and *RPS15* were used as the reference genes [14,22,40,47], and the amount of transcript from each gene was normalized to the abundance of *EF-1α* and *RPS15* using the $2^{-\Delta\Delta Ct}$ method described by Livak and Schmittgen [48]. Technical assays were carried out independently and in triplicate per cDNA sample, and all results were obtained from three independent RNA samples.

5. Statistical analysis

All statistical analyses were conducted using the SPSS 16.0 software (IBM, Armonk, NY). Differences in gene expression between two experimental treatments were examined using independent t-tests. Other data were analyzed using one-way ANOVA with the Tukey HSD test when the data were homoscedastic or the Games-Howell test when the data were not. In all tests, P values<0.05 were considered significant.

Results

1. Cloning and characterization of three opsin genes in cotton bollworms

We have cloned three opsin genes from the compound eyes of *H. armigera*: UV-, blue-, and long-wavelength-sensitive opsin genes, which are designated "*Ha-UV*," "*Ha-BL*," and "*Ha-LW*," respectively. Including the 5′ and 3′ UTRs, 1258 bp of *Ha-UV*, 1451 bp of *Ha-BL*, and 1574 bp of *Ha-LW* were cloned and sequenced. These three cDNAs encoded opsins of varying lengths: 379 (Ha-UV), 381 (Ha-LW) and 382 (Ha-BL) amino acids. It is possible that the full 3′ untranslated sequences of *Ha-UV*, *Ha-BL*, and *Ha-LW* were not cloned because we could not identify a polyadenylation signal with the sequence AATAAA. The molecular masses of the encoded opsins were predicted to be 41.35 kDa (Ha-UV), 43.22 kDa (Ha-BL), and 41.86 kDa (Ha-LW), and the calculated isoelectric points (pI) were 8.05 (Ha-UV), 7.02 (Ha-BL), and 7.97 (Ha-LW). The encoded amino acid sequences of *H. armigera* opsins exhibited various conserved characteristics compared with opsins in general or with insect-specific opsins, including seven membrane-spanning helical domains, visual pigment (opsin) retinal binding sites and G-protein-coupled receptors (Figure S1). *H. armigera* opsins shared putative posttranslational modification sites, including N-myristoylation sites, protein kinase C phosphorylation sites, casein kinase II phosphorylation sites, and N-glycosylation sites.

2. Sequence comparison and phylogenetic analysis

We reconstructed a molecular phylogenetic tree of insect visual opsins to clarify their evolutionary origin (Figure 1). The *H. armigera* sequences clustered in the visual opsin clades of insects: the short-wavelength (SW), middle-wavelength (MW) and long-wavelength (LW) branches. The results of the phylogenetic analysis agreed with the structure and distribution of these opsins. Moreover, the phylogenetic tree demonstrated that SW opsins exhibited a shorter genetic distance to MW opsins than to LW opsins. Based on the spectral sensitivities of compound eyes, the SW, MW and LW opsins were assigned to the UV, blue and green opsin genes, respectively. Sequence alignment revealed that *H. armigera* opsins shared significant homology with opsins identified from other insects, such as *Manduca sexta* (UV: 88% identity; blue: 90% identity; green: 90% identity), *Lycaena rubidus* (UV: 84% identity; blue: 72% identity; green: 77% identity), *Gryllus bimaculatus* (UV: 57% identity; blue: 58% identity; green: 76% identity), and *Apis mellifera* (UV: 60% identity; blue: 58% identity; green: 69% identity). The observed homology was 46% between Ha-UV and Ha-BL, 36% between Ha-UV and Ha-LW, and 35% between Ha-BL and Ha-LW.

3. Expression levels of opsin genes in various tissues

Before qRT-PCR, the expected PCR products were sequenced for confirmation, and the amplification reactions were validated using standard curve analysis as shown in Figure S2. Table S3 indicates that the expression levels of the three opsin genes were strikingly different according to the tissue in both female and male adults. The highest levels of transcription were observed in the compound eye and brain, with very low transcription levels in other tissues. Thus, compound eyes were collected in the following test to examine the visual function of the opsin genes. In compound eyes, *Ha-LW* exhibited the highest expression level, and *Ha-BL* exhibited the lowest expression level among the three opsin genes (female: $F_{2,6} = 97.541$, $P<0.001$; male: $F_{2,6} = 471.645$, $P<0.001$). The expression levels of the three opsin genes were similar between females and males.

4. Diurnal changes in opsin mRNA levels

As shown in Figure 2A, *Ha-UV* and *Ha-BL* levels were highest at ZT1 and fell off thereafter (female *Ha-UV*: $F_{11,24} = 20.522$, $P<0.001$; male *Ha-UV*: $F_{11,24} = 25.046$, $P<0.001$; female *Ha-BL*: $F_{11,24} = 8.177$, $P<0.001$; male *Ha-BL*: $F_{11,24} = 2.945$, $P = 0.013$). *Ha-LW* abundance tended to decrease during the day and then increase at night (female: $F_{11,24} = 3.246$, $P = 0.008$; male: $F_{11,24} = 6.654$, $P<0.001$). The expression patterns of *Ha-UV* and *Ha-BL* in females were similar to those in males, whereas the *Ha-LW* levels in females were significantly lower than those in males (for example, ZT1: $t = 3.036$, df$= 4$, $P = 0.039$; ZT9: $t = 3.799$, df$= 4$, $P = 0.019$; ZT21: $t = 4.579$, df$= 4$, $P = 0.010$; ZT23: $t = 3.300$, df$= 4$, $P = 0.030$). To determine the endogenous characteristics of these oscillations in the compound eye, levels of the three opsin genes were also measured under DD (*H. armigera* was entrained for 1 day under DD). Cycling persisted, exhibiting phases and amplitudes similar to those observed under 14L:10D (female *Ha-UV*: $F_{11,24} = 16.093$, $P<0.001$; male *Ha-UV*: $F_{11,24} = 10.327$, $P<0.001$; female *Ha-BL*: $F_{11,24} = 6.231$, $P<0.001$; male *Ha-BL*: $F_{11,24} = 4.146$, $P = 0.002$; female *Ha-LW*: $F_{11,24} = 3.511$, $P = 0.005$; male *Ha-LW*: $F_{11,24} = 6.799$, $P<0.001$) (Figure 2B).

As shown in Figure 3, no significant changes were observed in opsin gene levels when adults were kept under constant darkness (DD) (female *Ha-UV*: $F_{11,24} = 1.160$, $P = 0.363$; male *Ha-UV*: $F_{11,24} = 1.068$, $P = 0.425$; female *Ha-BL*: $F_{11,24} = 1.119$, $P = 0.389$; male *Ha-BL*: $F_{11,24} = 1.155$, $P = 0.366$; female *Ha-LW*: $F_{11,24} = 2.165$, $P = 0.055$; male *Ha-LW*: $F_{11,24} = 1.728$, $P = 0.127$) or constant light (LL) (female *Ha-UV*: $F_{11,24} = 1.186$, $P = 0.347$; male *Ha-UV*: $F_{11,24} = 2.083$, $P = 0.064$; female *Ha-BL*: $F_{11,24} = 2.063$, $P = 0.067$; male *Ha-BL*: $F_{11,24} = 2.034$, $P = 0.071$; female *Ha-LW*: $F_{11,24} = 1.707$, $P = 0.132$; male *Ha-LW*: $F_{11,24} = 1.154$, $P = 0.367$). Opsin gene levels tended toward upregulation in the transition from constant DD to LL, and *Ha-BL* levels were significantly up-regulated (for example, female ZT5: $t = 13.755$, df$= 4$, $P<0.001$; ZT13: $t = 4.206$, df$= 4$, $P = 0.014$; ZT17: $t = 4.456$, df$= 4$, $P = 0.011$).

To determine whether the levels of clock genes oscillated in a circadian manner, RNA was isolated from the compound eyes after *H. armigera* entrainment under standard 14L:10D or DD. Both assays gave identical results: *Ha-CRY1* and *Ha-CRY2* cycled independently, with a peak at ZT5 for *Ha-CRY1* and at ZT1 for *Ha-CRY2* (under LD: female *Ha-CRY1*: $F_{11,24} = 20.657$, $P<0.001$; male *Ha-CRY1*: $F_{11,24} = 9.466$, $P<0.001$; female *Ha-CRY2*: $F_{11,24} = 19.921$, $P<0.001$; male *Ha-CRY2*: $F_{11,24} = 13.090$, $P<0.001$; under DD: female *Ha-CRY1*: $F_{11,24} = 8.531$, $P<0.001$; male *Ha-CRY1*: $F_{11,24} = 6.279$, $P<0.001$; female *Ha-CRY2*: $F_{11,24} = 16.641$, $P<0.001$; male *Ha-CRY2*: $F_{11,24} = 15.555$, $P<0.001$) (Figure 4). However, *Ha-CRY1* and *Ha-CRY2* ceased cycling under constant DD (female *Ha-CRY1*: $F_{11,24} = 0.645$, $P = 0.774$; male *Ha-CRY1*: $F_{11,24} = 0.636$, $P = 0.781$; female *Ha-CRY2*: $F_{11,24} = 0.579$, $P = 0.827$; male *Ha-CRY2*: $F_{11,24} = 0.743$, $P = 0.689$).

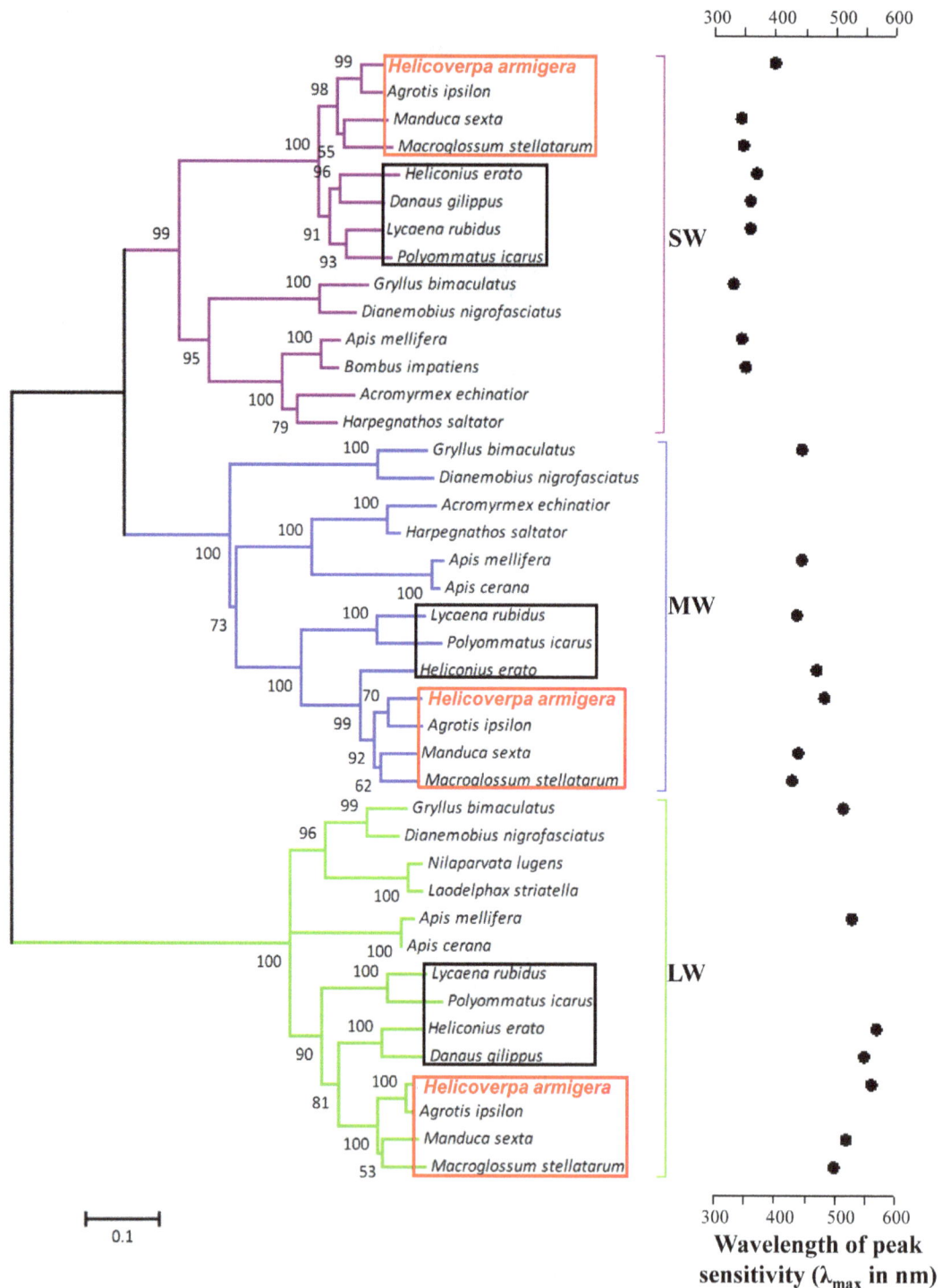

Figure 1. Evolutionary origin of opsins in *H. armigera*. The phylogenetic tree is based on aligned and full-length amino acid sequences. Numbers at the nodes indicate the bootstrap, and all nodes are supported by more than 50%. The various insect opsin lineages are colored in purple (SW = short-wavelength clade), blue (MW = middle-wavelength clade) and green (LW = long-wavelength clade). The wavelength of peak sensitivity (λ_{max}) of the respective visual pigment is given [12,39,69]. *H. armigera* is highlighted in red and bold. Moths and butterflies are highlighted in red and black boxes, respectively.

5. Effects of light exposure on opsin gene expression

As shown in Figure 5, *Ha-UV* levels increased significantly after UV or green (LW) light exposure, whereas *Ha-UV* levels were not up-regulated after 6 h of blue light exposure (female: $F_{3,8} = 9.620$, $P = 0.005$; male: $F_{3,8} = 12.912$, $P = 0.002$). In contrast, light exposure of any wavelength did not significantly up-regulate either *Ha-BL* or *Ha-LW* (*Ha-BL*: female: $F_{3,8} = 1.067$, $P = 0.416$; male: $F_{3,8} = 0.182$, $P = 0.906$; *Ha-LW*: female: $F_{3,8} = 1.042$, $P = 0.425$; male: $F_{3,8} = 1.355$, $P = 0.324$). Light exposure had

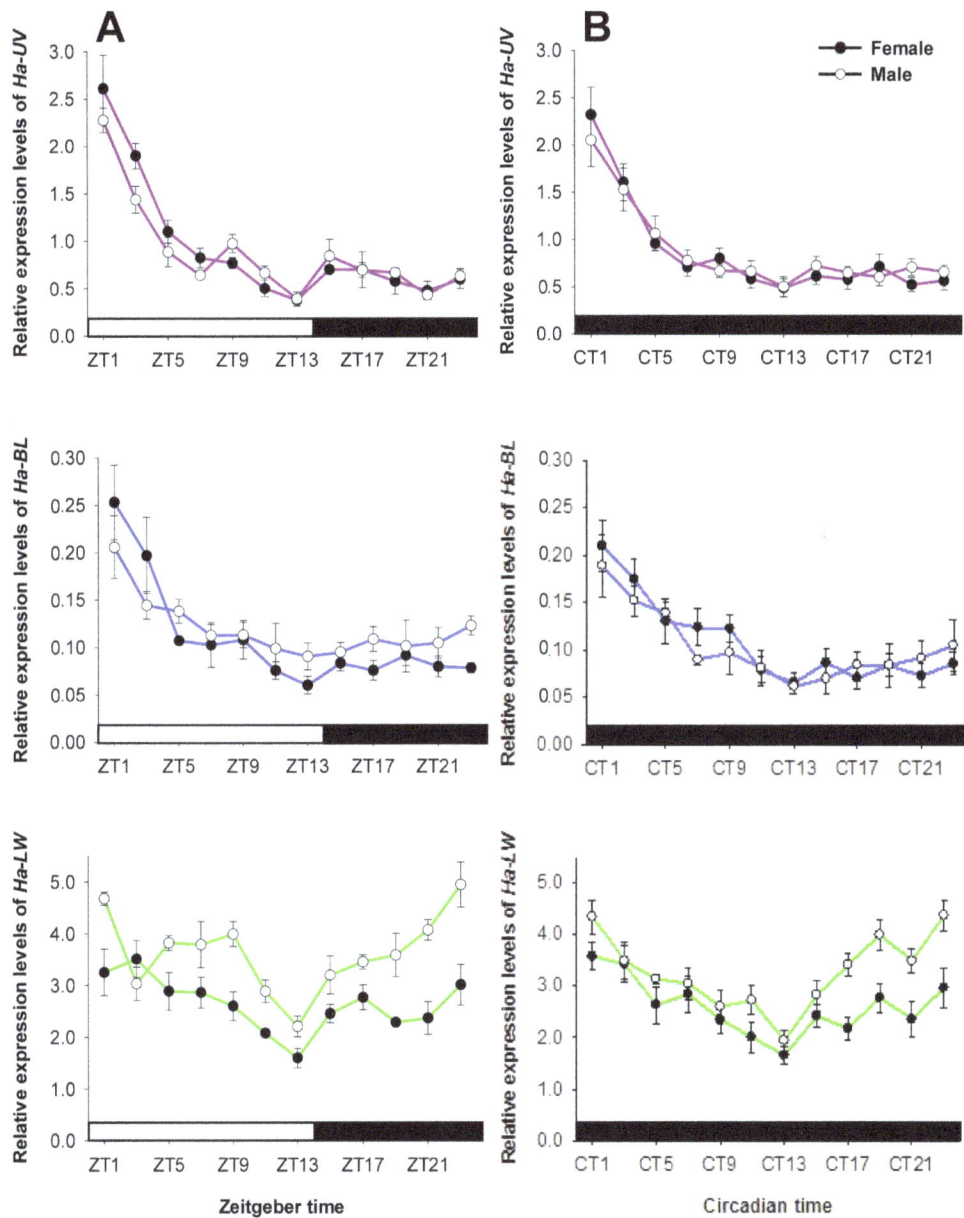

Figure 2. Diurnal changes in opsin gene levels under 14L:10D (A) and DD (B). RNA samples were collected from compound eyes at 2 h intervals. Open bar: day. Filled bar: night. Filled and open circles represent female and male moths, respectively. ZT indicates zeitgeber time; CT indicates circadian time. Each value is the mean ± SE of three collections.

similar effects on the transcription levels of the opsin genes in both female and male adults.

6. Effects of starvation on opsin gene expression

Opsin genes tended to be down-regulated following starvation (Figure 6). Ha-UV levels significantly decreased after starvation in both female and male adults (female: $t = 2.941$, df $= 4$, $P = 0.042$; male: $t = 6.146$, df $= 4$, $P = 0.004$), whereas Ha-LW levels were down-regulated only in male adults ($t = 3.172$, df $= 4$, $P = 0.034$).

7. Effects of mating on opsin gene expression

No significant differences in opsin mRNA levels were detected between mated and virgin adults at any time point in either females or males (Figure S3) (for example, Ha-UV: 0 h female: $t = 2.137$, df $= 4$, $P = 0.099$; 0 h male: $t = 0.994$, df $= 4$, $P = 0.376$;

Ha-BL: 0 h female: $t = 0.074$, df $= 4$, $P = 0.945$; 0 h male: $t = 0.147$, df $= 4$, $P = 0.890$; Ha-LW: 0 h female: $t = 0.535$, df $= 4$, $P = 0.621$; 0 h male: $t = 1.185$, df $= 4$, $P = 0.302$).

Discussion

1. Correspondence between opsin type and visual pigment in the compound eye

Compared with the methods and primers used in a previous study by Xu *et al.* (2013), we applied new approaches to clone the opsin genes from the cotton bollworm compound eye. Three opsin genes were identified: Ha-UV, Ha-BL and Ha-LW. Based on phylogenetic analysis, the opsin sequences were classified into three main opsin groups, and the alignments indicated high sequence identity among the three opsins (Figure 1). Thus, it is

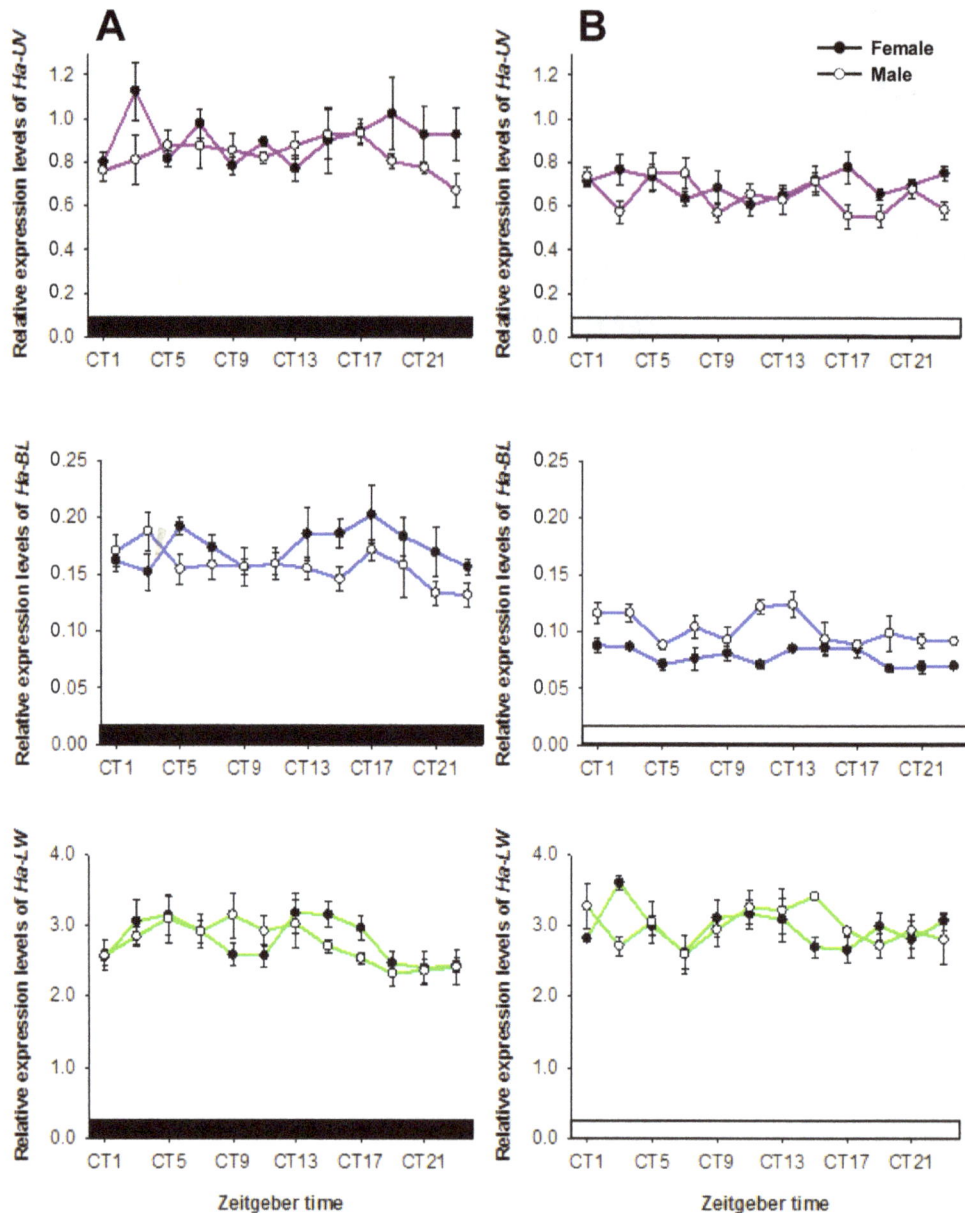

Figure 3. Cycling of opsin mRNA levels ceased under constant DD (A) and LL (B). RNA samples were collected from compound eyes at 2 h intervals. Open bar: day. Filled bar: night. Filled and open circles represent female and male moths, respectively. CT indicates circadian time. Each value is the mean ± SE of three collections.

likely that *Ha-UV*, *Ha-BL* and *Ha-LW* serve as receptors for ultraviolet, blue and green light, respectively. Our data confirmed the existence of three spectral photoreceptors in *H. armigera*, which have been identified in the compound eye of *H. armigera* from ERG recordings [39]. All three opsin genes lacked an AATAAA polyadenylation signal. Similarly, a previous study indicated that the UV- and blue-sensitive opsin genes in *Manduca sexta* lack polyadenylation signals [13]. It is likely that the primers annealed to stretches of adenines in the 3′ UTRs of opsin genes rather than to the terminal poly(A) tail.

The *H. armigera* opsins share many characteristics with other members of the opsin superfamily: (1) a Lys in the seventh transmembrane domain that interacts with the chromophore retinal [15–17], (2) two conserved Cys residues (Cys127 and Cys204 for Ha-UV, Cys127 and Cys204 for Ha-BL, and Cys131

and Cys208 for Ha-LW) that have been shown to form correct structures in bovine rhodopsin [49–50], and (3) two conserved Leu and Asn residues (Leu86 and Asn91 for Ha-UV, Leu86 and Asn91 for Ha-BL, and Leu89 and Asn94 for Ha-LW) that have been demonstrated to be crucial for rhodopsin synthesis during the nascent state [51].

2. Opsin genes were highly expressed in the compound eye and brain

The relative expression levels of the opsin genes in various tissues were analyzed to determine a suitable tissue for subsequent tests. The tissue-specific expression data revealed that opsin genes are abundant in the compound eye and brain, with the greatest abundance in the compound eye (Table S3). Opsin gene expression in the brain was previously observed in fish [52–53],

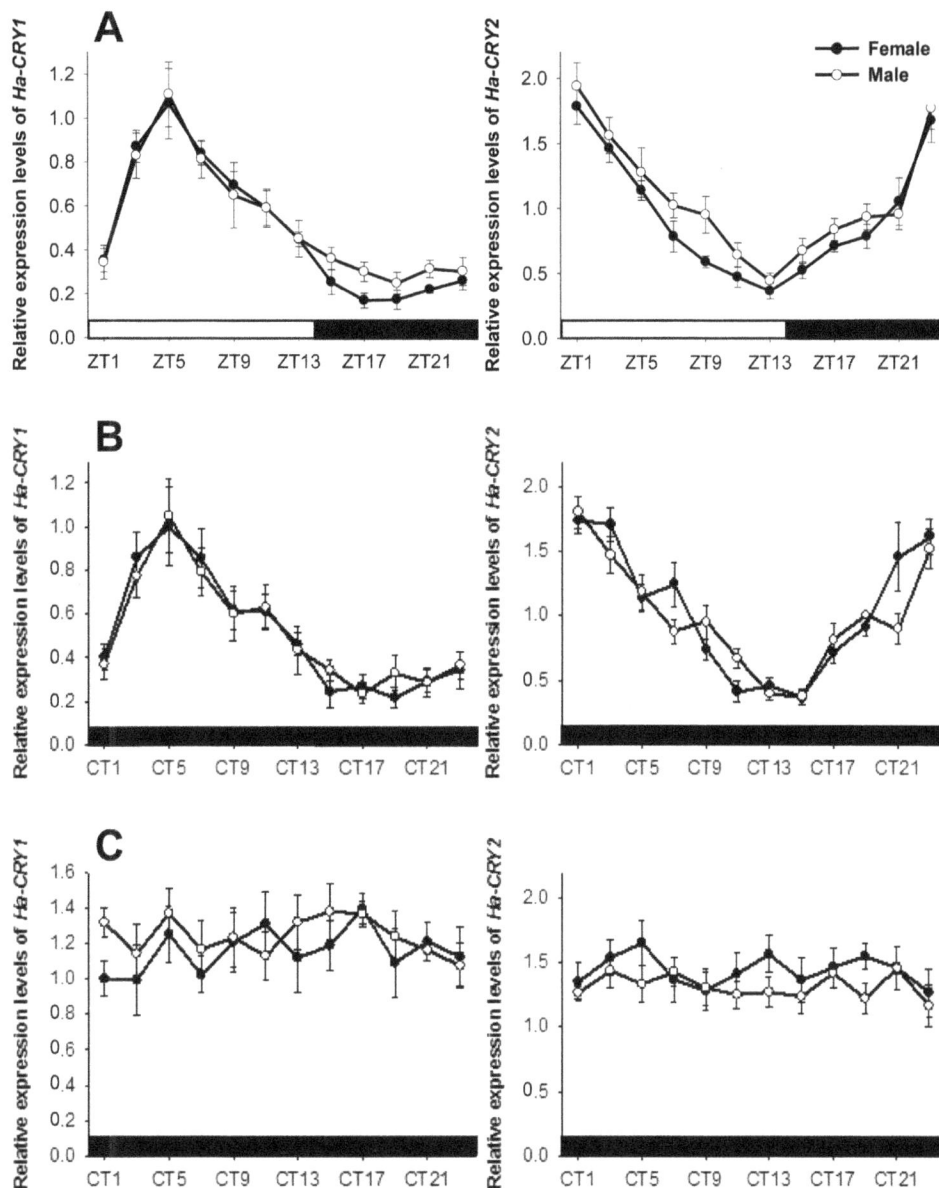

Figure 4. Temporal expression analysis of *Ha-CRY1* and *Ha-CRY2* levels in compound eyes under 14L:10D (A), DD (B), and constant DD (C). RNA samples were collected from compound eyes at 2 h intervals. Open bar: day. Filled bar: night. Filled and open circles represent female and male moths, respectively. ZT indicates zeitgeber time; CT indicates circadian time. Each value is the mean ± SE of three collections.

birds [54], spiders [55] and honeybees [56–57]. These results indicated that opsin genes might mediate not only visual function but also nonvisual function. In the current study, compound eyes were collected to examine the visual function of opsin genes. In addition, *Ha-LW* was found to be the most abundant of the three opsin genes. It is possible that *Ha-LW* is important to nocturnal moths because, of the three different wavelength lights, long-wavelength light is the strongest at night. Thus, the elevated expression of *Ha-LW* might be associated with the nocturnal activities of moths.

3. Daily changes in opsin mRNA levels were regulated by the circadian clock

Moths feed [58] and navigate [59–60] using multiple cues. Visual perception is one of the most familiar forms of stimulus

discrimination, and moths require a highly developed visual system. In contrast to butterflies, most moths are primarily nocturnal, meaning that they are active and use dim-light vision at night and rest during the day. Whether opsin mRNA levels oscillate in a circadian manner has been an interesting question for a long time.

In the current study, *Ha-UV* and *Ha-BL* levels peaked at ZT1 and then decreased, whereas *Ha-LW* levels tended to decrease during the day and increase at night (Figure 2A). However, Xu *et al.* [32] examined the expression levels of opsin genes in *H. armigera* and found that diel patterns of opsin mRNA levels did not fluctuate significantly. One possible reason for this discrepancy is that they collected RNA from whole individual moths, which may have obscured the pattern of opsin expression because it was limited to the compound eye and brain. Our result is similar to those of studies in mice [18] and toads [61], in which opsin mRNA

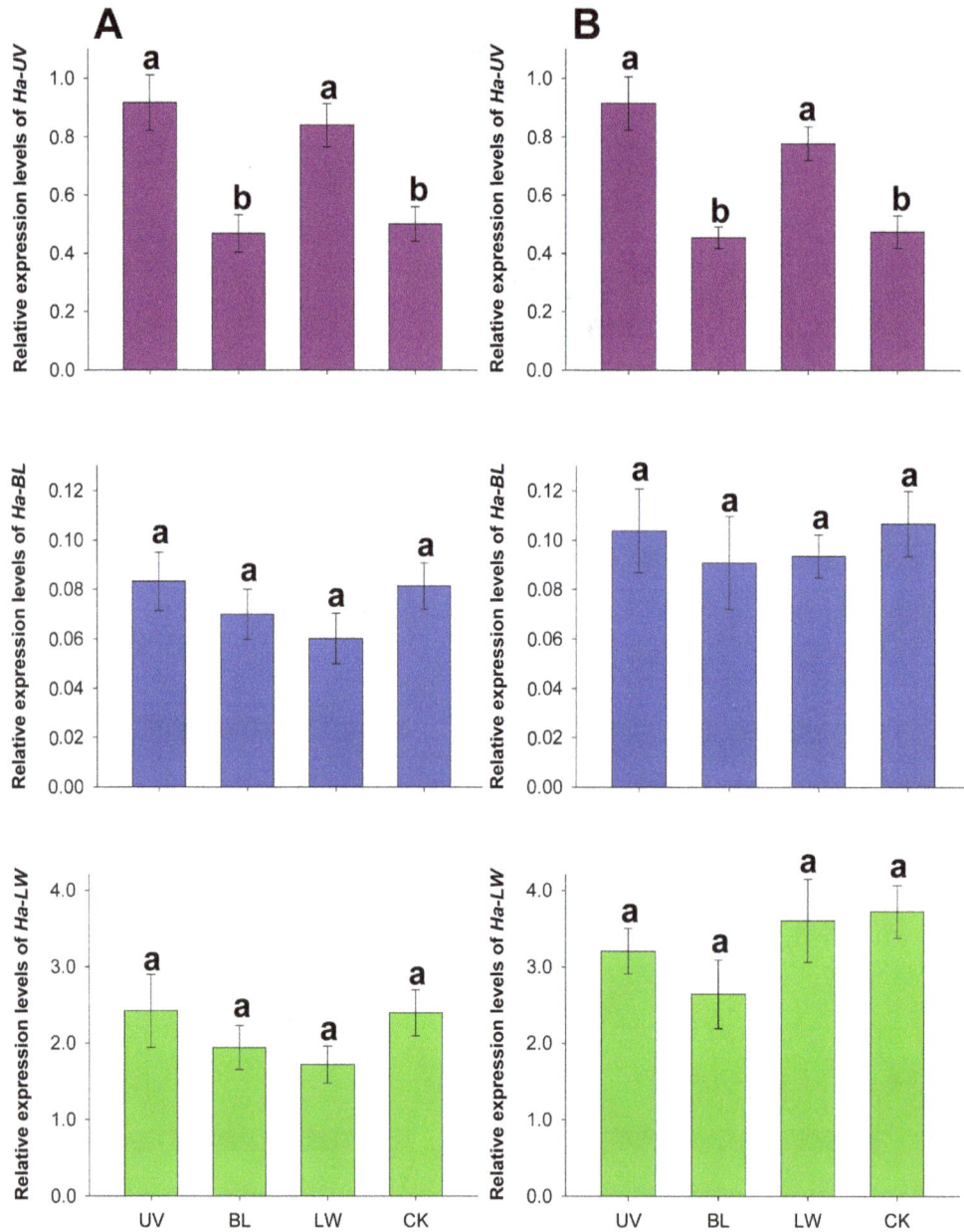

Figure 5. Effects of light exposure on opsin gene expression in female (A) and male (B) moths. RNA samples were collected from the compound eyes of 2-day-old moths after 6 h of UV, blue (BL) or green (LW) light irradiation at ZT20. RNA samples were collected at ZT20 from the compound eyes of moths reared under 14L:10D as CK. Each value is the mean ± SE of three collections. The letters above each bar indicate significant differences according to a Tukey HSD test ($P<0.05$).

levels peak after light onset. These results are in contrast to honeybee, in which long-wavelength-sensitive opsin mRNA levels peak after light onset and then subsequently decrease [22]. Increased *Ha-LW* levels at night might be associated with nocturnal behavior. The cycling of opsin gene expression in *H. armigera* under 14L:10D persisted for one day under DD conditions (Figure 2B) but did not persist further under constant DD or LL (Figure 3), demonstrating that (1) the expression profile of opsin genes depends on an endogenous circadian clock that is independent of light and that (2) opsin mRNA cycling could be disturbed by constant light or darkness. In similar previous studies, opsin mRNA cycling persisted under DD conditions in toads [61]

and honeybees [22], and photoreceptors appeared to have an endogenous pacemaker that adjusts the daily opsin mRNA levels.

Clock genes are responsible for synchronizing the endogenous rhythm to environmental conditions. The two clock genes in *H. armigera* (*Ha-CRY1* and *Ha-CRY2*) have been shown to oscillate daily in the head [40]. We determined whether *Ha-CRY1* and *Ha-CRY2* oscillate daily in the compound eye to confirm their circadian function in peripheral tissues. We found that *Ha-CRY1* levels peaked at 5 h after light onset and subsequently decreased, whereas *Ha-CRY2* exhibited an opposite expression pattern (Figure 4). The expression patterns of clock genes in the compound eye of *H. armigera* were consistent with those in the head of *H. armigera* [40] and other species [62–64]. We also

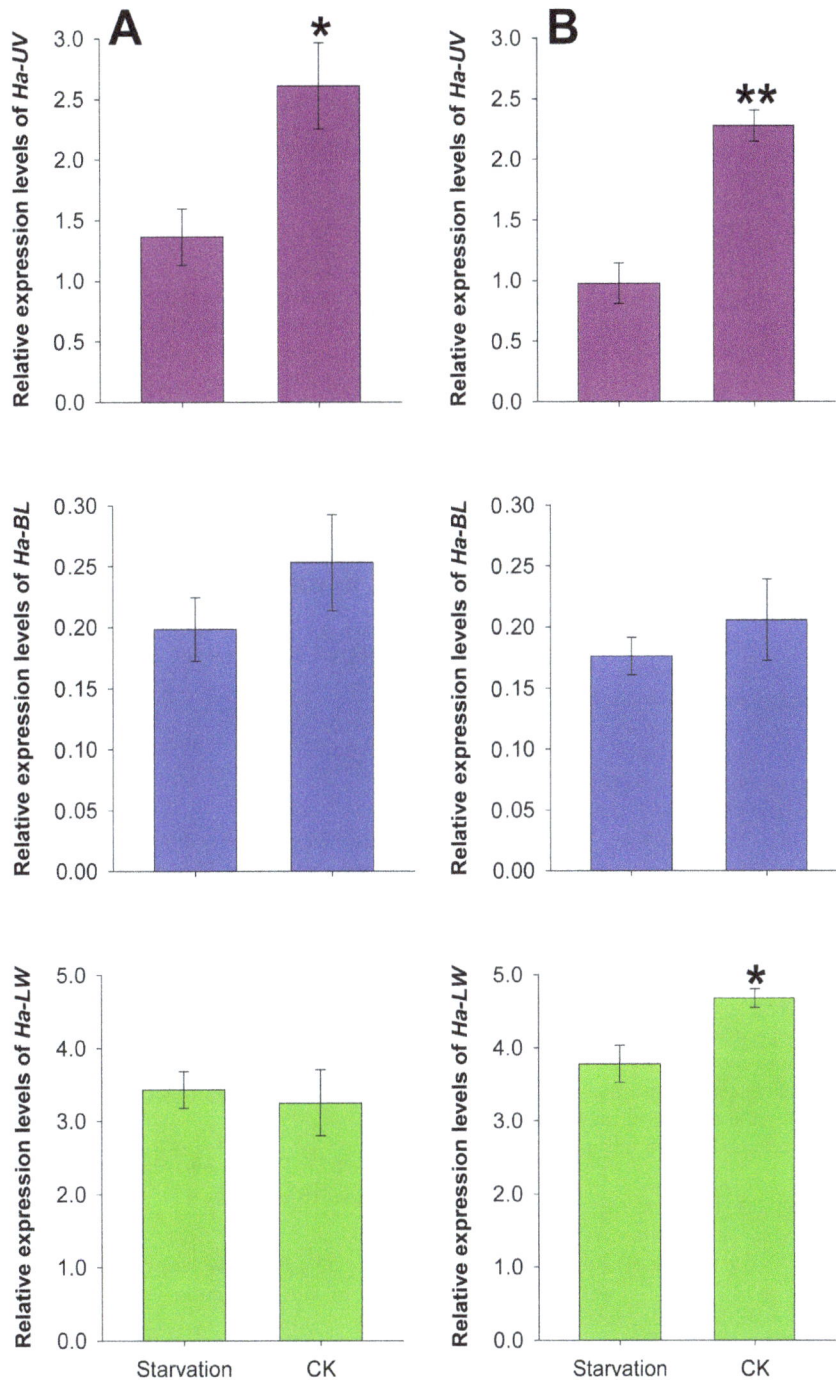

Figure 6. Effects of starvation on opsin gene expression in female (A) and male (B) moths. RNA samples were collected from the compound eyes of 2-day-old moths that were starved after adult emergence at ZT1. RNA samples were collected at ZT1 from the compound eyes of moths reared under 14L:10D as CK. Each value is the mean ± SE of three collections. The "*" and "**" indicate significant differences in opsin gene expression at $P<0.05$ and $P<0.001$, respectively, according to independent t-tests.

demonstrated that *Ha-CRY1* and *Ha-CRY2* expression cycling persisted for one day under DD conditions, illustrating that clock genes likely perform circadian functions in the compound eye. The waveform of the *Ha-CRY1* expression pattern was more rounded than that of *Ha-CRY2*, indicating that *Ha-CRY1* might be more sensitive to light. In Lepidoptera species, CRY1 is predominantly a blue-light photoreceptor that entrains the central oscillator, whereas CRY2 acts as a major transcriptional repressor but not

as a circadian photoreceptor [40]. Meanwhile, constant DD disturbed *Ha-CRY1* and *Ha-CRY2* cycling, which could be assumed to disturb opsin gene expression patterns.

What is the biological significance of daily changes in opsin mRNA levels? UV and blue light are stronger during the daytime, so increased expression of *Ha-UV* and *Ha-BL* during the day are beneficial for the recognition of short-wavelength light, which may serve to protect moths from UV damage. Moths exhibit limited diurnal behavior and are more active at night. The increase in *Ha-LW* levels at night may be useful in a dim-light environment. Under constant DD, opsin mRNA levels decrease in honeybees [22], whereas the three opsin genes in *H. armigera* increase, suggesting that (1) the mechanism controlling opsin mRNA expression in nocturnal insects differs from that in diurnal insects and that (2) opsin genes may be important for nocturnal activities in long-term darkness. However, we believe that the observed daily changes in opsin mRNA in *H. armigera* did not differ much from those in diurnal insects. In moths, these three opsin genes seem to operate under functional constraints and share conserved functions [32]. Thus, opsin genes in *H. armigera* might play an important role in the perception and discrimination of color.

4. Opsin gene levels tended to be up- and down-regulated after light exposure and starvation, respectively

Both light exposure and nutritional status influenced opsin gene expression in *H. armigera*; however, the effects of the two factors differed among the three types of opsin genes. Although opsin mRNA levels were dependent on an endogenous circadian clock, we confirmed that light exposure (UV and green light) rather than the scotophase significantly up-regulated *Ha-UV* (Figure 5). Similarly, previous studies indicated that *OPS1* (*Neurospora* rhodopsin) transcript showed enhanced expression after NUV (mainly 360 nm) irradiation in *Bipolaris oryzae* [29], and UVA (mainly 365 nm) and violet (mainly 410 nm) irradiation induced rhodopsin expression in normal human epidermal keratinocytes [30]. Organisms exposed to more light exhibited enhanced opsin gene expression, and this phenomenon has been observed in *H. armigera* [32], *Apis mellifera* [22], *Ceratosolen solmsi* [31], and *Lucania goodei* [14]. In the current study, light positively regulated the expression of *Ha-UV* in an organism-specific manner, suggesting that light-sensing in *H. armigera* might be important for synchronizing activity with the nocturnal lifestyle. Unlike the effect of light exposure on opsin gene expression, starvation down-regulated most opsin genes, especially *Ha-UV* (Figure 6). Vision plays an important role in locating food, and we initially confirmed that opsin gene levels were influenced by nutritional status. Our study revealed that *Ha-UV* was the opsin gene most affected by environmental conditions (light exposure and nutritional status).

5. Sexual dimorphism in opsin gene expression

The rhythmicity of opsin gene expression was not sexually dimorphic, except that *Ha-LW* expression was significantly lower in females than in males (Figure 2). In contrast, in *Poecilia reticulata*, the long-wavelength-sensitive opsin gene is up-regulated in females, allowing them to better discriminate male coloration and courtship displays [33]. Similar to our study, the long-wavelength-sensitive opsin gene in the butterfly *Bicyclus anynana* is up-regulated in males [34], and sexually dimorphic photoreceptors are common in other butterflies [35–37]. In numerous species of Lepidoptera, female and male moths produce sex pheromones to induce potential mating partners, and in certain species, sex pheromones play a key role in mating choice and species isolation [65–68]. Although opsin gene expression was not influenced by mating (Figure S3), increased expression of *Ha-LW* in males might be important for locating female moths and improving mating success.

Conclusion

In this paper, we presented a characterization of visual opsins in a nocturnal moth, the cotton bollworm *H. armigera*, which belongs to a markedly early branch within insect lineage. The expression of three types of opsin genes varied according to time of day and environmental conditions and was regulated by a circadian clock, light conditions and nutritional status. *Ha-UV* and *Ha-BL* were more abundant during the day than at night, which aided recognition of short-wavelength light during the day. However, *Ha-LW* levels decreased during the day and increased at night, which might be useful in a dim-light environment. *Ha-LW* levels were up-regulated in males, which might aid in locating potential sex partners and improving mating success over relatively short distances. Interestingly, daily changes in opsin mRNA levels in *H. armigera* did not differ much from those in diurnal insects, suggesting that opsin genes in *H. armigera* might play an important role in color vision, especially color perception and discrimination.

Supporting Information

Figure S1 Alignment of three opsin genes isolated as cDNA from the retina of *H. armigera*. Residues identical in two or three of the sequences are shown in red. The G-protein-coupled receptor family is boxed. The visual pigment (opsin) retinal binding site is shaded, and the arrow indicates the site of the chromophore Schiff-base linkage.

Figure S2 Standard curves for the opsin and crypto-chrome genes of *H. armigera*.

Figure S3 Effects of mating on opsin gene expression in female (A) and male (B) moths. RNA samples were collected from the compound eyes of 2-day-old moths that had completed mating 0 h (ZT15) and 3 h (ZT18) before collection. Mean ± SE.

Table S1 Primers used for gene cloning and qRT-PCR.

Table S2 The GenBank accession numbers of the genes used in this study.

Table S3 Relative expression levels of opsin genes in various adult tissues. Means ± SE in columns followed by different letters are significantly different (Tukey HSD test was used to evaluate *HeBL* in females and other comparisons were evaluated using the Games-Howell test, $P<0.05$).

Author Contributions

Conceived and designed the experiments: SY XXL QWZ. Performed the experiments: SY JLZ WLZ XFZ. Analyzed the data: SY ZL. Contributed reagents/materials/analysis tools: SY XXL QWZ. Contributed to the writing of the manuscript: SY XXL.

References

1. Kelber A, Vorobyev M, Osorio D (2003) Animal colour vision-behavioural tests and physiological concepts. Biological Reviews of the Cambridge Philosophical Society 78: 81–118.

2. Hill GE, McGraw KJ (2004) Correlated changes in male plumage coloration and female mate choice in cardueline finches. Animal Behaviour 67: 27–35.

3. Langham G (2004) Specialized avian predators repeatedly attack novel color morphs of *Heliconius* butterflies. Evolution 58: 2783–2787.

4. Endler JA, Mappes J (2005) Predator mixes and the conspicuousness of aposematic signals. The American Naturalist 163: 532–547.

5. Zhao H, Rossiter SJ, Teeling EC, Li C, Cotton JA, et al. (2009) The evolution of color vision in nocturnal mammals. Proc Natl Acad Sci USA 106: 8980–8985.

6. Gordon WC (1977) Microvillar orientation in the retina of the nymphalid butterfly. Zeitschrift Fur Naturforschung C-A Journal of Biosciences 32: 662–664.

7. Kitamoto J, Sakamoto K, Ozaki K, Mishina Y, Arikawa K (1998) Two visual pigments in a single photoreceptor cell: identification and histological localization of three mRNAs encoding visual pigment opsins in the retina of the butterfly *Papilio xuthus*. The Journal of Experimental Biology 201: 1255–1261.

8. White RH, Xu H, Münch TA, Bennett RR, Grable EA (2003) The retina of *Manduca sexta*: rhodopsin expression, the mosaic of green-, blue- and UV-sensitive photoreceptors, and regional specialization. The Journal of Experimental Biology 206: 3337–3348.

9. Frentiu FD, Bernard GD, Cuevas CI, Sison-Mangus MP, Prudic KL, et al. (2007) Adaptive evolution of color vision as seen through the eyes of butterflies. Proc Natl Acad Sci USA 104: 8634–8640.

10. Briscoe AD (2008) Reconstructing the ancestral butterfly eye: focus on the opsins. The Journal of Experimental Biology 211: 1805–1813.

11. Seki T, Vogt K (1998) Evolutionary aspects of the diversity of visual pigment chromophores in the class Insecta. Comparative Biochemistry and Physiology Part B: Biochemistry and Molecular Biology 119: 53–64.

12. Briscoe AD, Chittka L (2001) The evolution of color vision in insects. Annual Review of Entomology 46: 471–510.

13. Chase MR, Bennett RR, White RH (1997) Three opsin-encoding cDNAs from the compound eye of *Manduca sexta*. The Journal of Experimental Biology 200: 2469–2478.

14. Fuller RC, Claricoates KM (2011) Rapid light-induced shifts in opsin expression: finding new opsins, discerning mechanisms of change, and implications for visual sensitivity. Molecular Ecology 20: 3321–3335.

15. Bownds D (1967) Site of attachment of retinal in rhodopsin. Nature 216: 1178–1181.

16. Wang JK, McDowell JH, Hargrave PA (1980) Site of attachment of 11-cis-retinal in bovine rhodopsin. Biochemistry 19: 5111–5117.

17. Terakita A (2005) The opsins. Genome Biology 6: 213.

18. Bowes C, van Veen T, Farber DB (1988) Opsin, G-protein and 48-kDa protein in normal and *rd* mouse retinas: developmental expression of mRNAs and protein and light/dark cycling of mRNAs. Experimental Eye Research 47: 369–390.

19. Korenbrot JI, Fernald RD (1989) Circadian rhythm and light regulate opsin mRNA in rod photoreceptors. Nature 337: 454–457.

20. Craft CM, Whitmore DH, Donoso LA (1990) Differential expression of mRNA and protein encoding retinal and pineal S-antigen during the light/dark cycle. Journal of Neurochemistry 55: 1461–1473.

21. Bobu C, Sandu C, Laurent V, Felder-Schmittbuhl MP, Hicks D (2013) Prolonged light exposure induces widespread phase shifting in the circadian clock and visual pigment gene expression of the *Arvicanthis ansorgei* retina. Molecular Vision 19: 1060–1073.

22. Sasagawa H, Narita R, Kitagawa Y, Kadowaki T (2003) The expression of genes encoding visual components is regulated by a circadian clock, light environment and age in the honeybee (*Apis mellifera*). European Journal of Neuroscience 17: 963–970.

23. Battelle BA, Williams CD, Schremser-Berlin JL, Cacciatore C (2000) Regulation of arrestin mRNA levels in *Limulus* lateral eye: separate and combined influences of circadian efferent input and light. Visual Neuroscience 17: 217–227.

24. Battelle BA (2013) What the clock tells the eye: lessons from an ancient arthropod. Integrative and Comparative Biology 53: 144–153.

25. Battelle BA, Kempler KE, Parker AK, Gaddie CD (2013) Opsin1-2, $G_q\alpha$ and arrestin levels at *Limulus* rhabdoms are controlled by diurnal light and a circadian clock. The Journal of Experimental Biology 216: 1837–1849.

26. Cashmore AR, Jarillo JA, Wu YJ, Liu D (1999) Cryptochromes: blue light receptors for plants and animals. Science 284: 760–765.

27. Ikeno T, Numata H, Goto SG (2008) Molecular characterization of the circadian clock genes in the bean bug, *Riptortus pedestris*, and their expression patterns under long- and short-day conditions. Gene 419: 56–61.

28. Sandrelli F, Costa R, Kyriacou CP, Rosato E (2008) Comparative analysis of circadian clock genes in insects. Insect Molecular Biology 17: 447–463.

29. Kihara J, Tanaka N, Ueno M, Arase S (2009) Cloning and expression analysis of two opsin-like genes in the phytopathogenic fungus *Bipolaris oryzae*. FEMS Microbiology Letters 295: 289–294.

30. Kim HJ, Son ED, Jung JY, Choi H, Lee TR, et al. (2013) Violet light down-regulates the expression of specific differentiation markers through rhodopsin in normal human epidermal keratinocytes. PLoS ONE 8: e73678.

31. Wang B, Xiao JH, Bian SN, Niu LM, Murphy RW, et al. (2013) Evolution and expression plasticity of opsin genes in a Fig. pollinator, *Ceratosolen solmsi*. PLoS ONE 8: e53907.

32. Xu P, Lu B, Xiao H, Fu X, Murphy RW, et al. (2013) The evolution and expression of the moth visual opsin family. PLoS ONE 8: e78140.

33. Laver CRJ, Taylor JS (2011) RT-qPCR reveals opsin gene upregulation associated with age and sex in guppies (*Poecilia reticulata*)-a species with color-based sexual selection and 11 visual-opsin genes. BMC Evolutionary Biology 11: 81.

34. Everett A, Tong X, Briscoe AD, Monteiro A (2012) Phenotypic plasticity in opsin expression in a butterfly compound eye complements sex role reversal. BMC Evolutionary Biology 12: 232.

35. Arikawa K, Wakakuwa M, Qiu X, Kurasawa M, Stavenga DG (2005) Sexual dimorphism of short-wavelength photoreceptors in the small white butterfly, *Pieris rapae crucivora*. The Journal of Neuroscience 25: 5935–5942.

36. Ogawa Y, Awata H, Wakakuwa M, Kinoshita M, Stavenga DG, et al. (2012) Coexpression of three middle wavelength-absorbing visual pigments in sexually dimorphic photoreceptors of the butterfly *Colias erate*. Journal of Comparative Physiology A 198: 857–867.

37. Ogawa Y, Kinoshita M, Stavenga DG, Arikawa K (2013) Sex-specific retinal pigmentation results in sexually dimorphic long-wavelength-sensitive photoreceptors in the eastern pale clouded yellow butterfly, *Colias erate*. The Journal of Experimental Biology 216: 1916–1923.

38. Nowinszky L, Puskás J (2011) Light trapping of *Helicoverpa armigera* in India and Hungary in relation with the moon phase. Indian Journal of Agricultural Sciences 81: 152–155.

39. Wei G, Zhang Q, Zhou M, Wu W (1999) Studies on the electroretinogram of the compound eyes of *Helicoverpa armigera* (Hübner) moth. Acta Biophysica Sinica 15: 682–688.

40. Yan S, Ni H, Li H, Zhang J, Liu X, et al. (2013) Molecular cloning, characterization, and mRNA expression of two *cryptochrome* genes in *Helicoverpa armigera* (Lepidoptera: Noctuidae). Journal of Economic Entomology 106: 450–462.

41. Yan S, Li H, Zhang J, Zhu J, Zhang Q, et al. (2013) Sperm storage and sperm competition in the *Helicoverpa armigera* (Lepidoptera: Noctuidae). Journal of Economic Entomology 106: 708–715.

42. Yan S, Zhu J, Zhang J, Zhu W, Zhang Q, et al. (2012) Effects of low-dose ^{60}Co-γ radiation on the emergence, longevity, phototactic behavior and sex pheromone titer in *Helicoverpa armigera* (Lepidoptera: Noctuidae) adults. Acta Entomologica Sinica 55: 1337–1344.

43. Wu KJ, Gong PY (1997) A new and practical artificial diet for the cotton bollworm. Entomologia Sinica 4: 277–282.

44. Hofmann K, Stoffel W (1993) TMbase - a database of membrane spanning proteins segments. Biological Chemistry Hoppe-Seyler 374: 166.

45. Bairoch A, Bucher P, Hofmann K (1997) The PROSITE database, its status in 1997. Nucleic Acids Research 25: 217–221.

46. Combet C, Blanchet C, Geourjon C, Deléage G (2000) NPS@: network protein sequence analysis. Trends Biochemical Sciences 25: 147–150.

47. Yan S, Zhu JL, Zhu WL, Pan LL, Zhang QW, et al. (2013) Molecular cloning, sequence analysis and expression pattern detection of α-tubulin gene from *Helicoverpa armigera* (Hübner). Scientia Agricultura Sinica 46: 1808–1817.

48. Livak KJ, Schmittgen TD (2001) Analysis of relative gene expression data using real-time quantitative PCR and the $2^{-\Delta\Delta Ct}$ method. Methods 25: 402–408.

49. Karnik SS, Sakmar TP, Chen HB, Khorana HG (1988) Cysteine residues 110 and 187 are essential for the formation of correct structure in bovine rhodopsin. Proc Natl Acad Sci USA 85: 8459–8463.

50. Karnik SS, Khorana HG (1990) Assembly of function rhodopsin requires a disulfide bond between cysteine residues 110 and 187. The Journal of Biological Chemistry 265: 17520–17524.

51. Bentrop J, Schwab K, Pak WL, Paulsen R (1997) Site-directed mutagenesis of highly conserved amino acids in the first cytoplasmic loop of *Drosophila* Rh1 opsin blocks rhodopsin synthesis in the nascent state. The EMBO Journal 16: 1600–1609.

52. Masuda T, Iigo M, Mizusawa K, Aida K (2003) Retina-type rhodopsin gene expressed in the brain of a teleost, ayu (*Plecooglossus altivelis*). Zoological Science 20: 989–997.

53. Takeuchi Y, Bapary MAJ, Igarashi S, Imamura S, Sawada Y, et al. (2011) Molecular cloning and expression of long-wavelength-sensitive cone opsin in the brain of a tropical damselfish. Comparative Biochemistry and Physiology, Part A 160: 486–492.

54. Silver R, Witkovsky P, Horvath P, Alones V, Barnstable CJ, et al. (1988) Coexpression of opsin-and VIP-like-immunoreactivity in CSF-contacting neurons of the avian brain. Cell and Tissue Research 253: 189–198.

55. Eriksson BJ, Fredman D, Steiner G, Schmid A (2013) Characterisation and localisation of the opsin protein repertoire in the brain and retinas of a spider and an onychophoran. BMC Evolutionary Biology 13: 186.

56. Velarde RA, Sauer CD, Walden KKO, Fahrbach SE, Robertson HM (2005) Pteropsin: A vertebrate-like non-visual opsin expressed in the honey bee brain. Insect Biochemistry and Molecular Biology 35: 1367–1377.

57. Leboulle G, Niggebrügge C, Roessler R, Briscoe AD, Menzel R, et al. (2013) Characterisation of the RNA interference response against the long-wavelength receptor of the honeybee. Insect Biochemistry and Molecular Biology 43: 959–969.

58. Cutler DE, Bennett RR, Stevenson RD, White RH (1995) Feeding behavior in the nocturnal moth *Manduca sexta* is mediated mainly by violet receptors, but where are they located in the retina? The Journal of Experimental Biology 198: 1909–1917.

59. Feng HQ, Wu KM, Cheng DF, Guo YY (2004) Northward migration of *Helicoverpa armigera* (Lepidoptera: Noctuidae) and other moths in early summer observed with radar in northern China. Journal of Economic Entomology 97: 1874–1883.

60. Feng HQ, Fu XW, Wu B, Wu KM (2009) Seasonal migration of *Helicoverpa armigera* (Lepidoptera: Noctuidae) over the Bohai Sea. Journal of Economic Entomology 102: 95–104.

61. Korenbrot JI, Fernald RD (1989) Circadian rhythm and light regulate opsin mRNA in rod photoreceptors. Nature 337: 454–457.

62. Egan ES, Franklin TM, Hilderbrand-Chae MJ, McNeil GP, Roberts MA, et al. (1999) An extraretinally expressed insect cryptochrome with similarity to the blue light photoreceptors of mammals and plants. The Journal of Neuroscience 19: 3665–3673.

63. Merlin C, Lucas P, Rochat D, François MC, Maïbèche-Coisne M, et al. (2007) An antennal circadian clock and circadian rhythms in peripheral pheromone reception in the moth *Spodoptera littoralis*. Journal of Biological Rhythms 22: 502–514.

64. Nagy AD, Csernus VJ (2007) Cry1 expression in the chicken pineal gland: effects of changes in the light/dark conditions. General and Comparative Endocrinology 152: 144–147.

65. Choi MY, Tatsuki S, Boo KS (1998) Regulation of sex pheromone biosynthesis in the oriental tobacco budworm, *Helicoverpa assulta* (Lepidoptera: Noctuidae). Journal of Insect Physiology 44: 653–658.

66. Lecomte C, Thibout E, Pierre D, Auger J (1998) Transfer, perception and activity of male pheromone of *Acrolepiopsis assectella* with special reference to conspecific male sexual inhibition. Journal of Chemical Ecology 24: 655–671.

67. Hillier NK, Vickers NJ (2004) The role of Heliothine hairpencil compounds in female *Heliothis virescens* (Lepidoptera: Noctuidae) behavior and mate acceptance. Chemical Senses 29: 499–511.

68. Hillier NK, Vickers NJ (2011) Hairpencil volatiles influence interspecific courtship and mating between two related moth species. Journal of Chemical Ecology 37: 1127–1136.

69. Briscoe AD (2000) Six opsins from the butterfly *Papilio glaucus*: molecular phylogenetic evidence for paralogous origins of red-sensitive visual pigments in insects. Journal of Molecular Evolution 51: 110–121.

15

Regulation of RAB5C Is Important for the Growth Inhibitory Effects of MiR-509 in Human Precursor-B Acute Lymphoblastic Leukemia

Yee Sun Tan[1], MinJung Kim[1,3,4], Tami J. Kingsbury[1,2,3], Curt I. Civin[1,2,3,4], Wen-Chih Cheng[1,4]*

1 Center for Stem Cell Biology & Regenerative Medicine, University of Maryland School of Medicine, Baltimore, Maryland, United States of America, 2 Greenebaum Cancer Center, University of Maryland School of Medicine, Baltimore, Maryland, United States of America, 3 Department of Physiology, University of Maryland School of Medicine, Baltimore, Maryland, United States of America, 4 Department of Pediatrics, University of Maryland School of Medicine, Baltimore, Maryland, United States of America

Abstract

MicroRNAs (miRs) regulate essentially all cellular processes, but few miRs are known to inhibit growth of precursor-B acute lymphoblastic leukemias (B-ALLs). We identified miR-509 via a human genome-wide gain-of-function screen for miRs that inhibit growth of the NALM6 human B-ALL cell line. MiR-509-mediated inhibition of NALM6 growth was confirmed by 3 independent assays. Enforced miR-509 expression inhibited 2 of 2 additional B-ALL cell lines tested, but not 3 non-B-ALL leukemia cell lines. MiR-509-transduced NALM6 cells had reduced numbers of actively proliferating cells and increased numbers of cells undergoing apoptosis. Using miR target prediction algorithms and a filtering strategy, RAB5C was predicted as a potentially relevant target of miR-509. Enforced miR-509 expression in NALM6 cells reduced RAB5C mRNA and protein levels, and RAB5C was demonstrated to be a direct target of miR-509. Knockdown of RAB5C in NALM6 cells recapitulated the growth inhibitory effects of miR-509. Co-expression of the RAB5C open reading frame without its 3′ untranslated region (3′UTR) blocked the growth-inhibitory effect mediated by miR-509. These findings establish RAB5C as a target of miR-509 and an important regulator of B-ALL cell growth with potential as a therapeutic target.

Editor: Linda Bendall, University of Sydney, Australia

Funding: This work was supported by the following grants from National Foundation for Cancer Research (http://www.nfcr.org), the National Institutes of Health (http://www.nih.gov) (PO1CA70970) and Maryland Stem Cell Research Foundation/TEDCO (http://www.mscrf.org) (2007-MSCRFII-0114, 2008-MSCRF-303524 and 2010-MSCRFII-0065-00) to CIC, and Gabrielle's Angel Foundation (http://www.gabriellesangels.org) to WCC. The funders had no role in study design, data collection and analysis, decision to publish, or preparation of the manuscript.

Competing Interests: The authors have declared that no competing interests exist.

* Email: WCheng@som.umaryland.edu

Introduction

More effective and less toxic therapies are needed for precursor-B acute lymphoblastic leukemia (B-ALL), the most common childhood cancer [1–3]. To find novel therapeutic targets, deeper understanding of the mechanisms involved in leukemia cell proliferation and survival is necessary. MicroRNAs (miRs) are short non-coding RNAs which regulate expression of mRNA targets, most commonly by binding to the 3′ untranslated regions (3′UTRs) of mRNAs [4–6]. Each miR has many, often hundreds of predicted mRNA targets, and reciprocally a single mRNA may be targeted by multiple miRs. MiRs are involved in many cellular processes, and dysregulation of miRs has been linked to diseases, prominently including cancer [7]. For instance, overexpression of miR-155 has been detected in certain subtypes of acute myeloid leukemia (AML), chronic lymphoblastic leukemia, and lymphomas [8]. Transplantation of mouse bone marrow cells overexpressing miR-155 resulted in myeloproliferative disorders, and transgenic overexpression of miR-155 resulted in ALL and lymphoma in mice [9,10]. In contrast, miR-34 is a well-studied tumor suppressor miR; its expression is down-regulated in a wide range of solid and hematologic malignancies, and it targets multiple

molecules that promote cancer development and progression, including BCL2 and cyclin D1 [11,12].

Expression profiling studies, such as microarray hybridization, real-time PCR, or sequencing assays of global miR expression in leukemia cells versus normal counterpart cells, are often used to identify miRs associated with acute leukemias [13–15]. In B-ALLs, multiple miRs are known to be dysregulated [16,17], but only a few miRs, including miR-196b [18], miR-124a [19] and miR-143 [20], have been shown to inhibit B-ALL growth. Although expression profiling studies can implicate miRs as biomarkers, it is often difficult to differentiate 'passenger miRs' from 'driver miRs' [21]. As an alternative to expression profiling approaches, functional screens for miRs that drive hallmark cancer properties have successfully identified miRs involved in regulation of cellular processes including growth in melanoma [22], pancreatic cancer [23], and colon cancer [24], as well as metastasis in liver cancer [25]. We previously identified a set of miRs that regulate growth of the human lung fibroblast cell line IMR90 by a miR-high throughput functional screen (miR-HTS) [26]. In this paper, we extended our gain-of-function screening of human miRs to B-ALL cells and identified miR-509 as a novel B-ALL growth-inhibitory miR. MiR-509 inhibited growth of 2 additional B-ALL cell lines.

We went on to determine the cellular mechanism of miR-509-mediated B-ALL growth inhibition and identify RAB5C as a key B-ALL growth-promoting factor targeted by miR-509.

Material and Methods

Functional screen of miRs

Detailed description of the miR-HTS methodology was previously described [26]. Briefly, in each miR-HTS, 1.8 million NALM6 cells were infected at a multiplicity of infection (MOI) = 0.3 with the human Lenti-miR pooled virus library (System Biosciences, Mountain View, CA, USA; Cat# PMIRHPLVA-1) to achieve ~30% transduced cells. 4 μg/ml polybrene (Sigma-Aldrich, St. Louis, MO, USA) was used as the infection vehicle. On days 4, 12, 20 and 28 after infection, a fraction of the infected culture (2 million cells) was harvested and genomic DNA isolated using the DNeasy Blood & Tissue Kit (Qiagen, Valencia, CA, USA). To identify candidate growth-regulatory miRs, nested PCR, customized qPCR assays, and candidate selection were conducted as described [26]. 3 independent miR-HTS was conducted.

Cell lines

NALM6, RCH-ACV, REH, KARPAS-45 were obtained from DSMZ (Braunschweig, Germany). Jurkat and K562 cells were obtained from ATCC (Manassas, VA, USA). All cell lines were maintained according to manufacturer's protocol.

Plasmids and cloning

Overexpression of miRs was achieved by cloning each precursor miR sequence plus ~200 bp of flanking genomic sequence into the pJET1.2 plasmid (Thermo Scientific, Waltham, MA, USA) (Primers listed in Table S1). The genomic sequence of each miR was obtained from the UCSC genome browser. The miR sequences were then subcloned into our pWCC52 lentiviral vector (Empty lentiviral vector #1, EV#1) downstream of GFP driven by human EF1α promoter. MiR-509 was also subcloned into our pWCC72 lentiviral vector (empty lentiviral vector #2, EV#2) downstream of DsRed driven by human EF1α promoter. Both pWCC52 and pWCC72 were generated in our lab based on lentivectors designed to express miRs as described [27].

3 plasmids, each containing a different shRNA targeting RAB5C [shRNA#1 (TRCN0000072935), shRNA#2 (TRCN0000072933), shRNA#3 (TRCN0000072937)], were purchased from Thermo Scientific. The plasmid containing non-targeting scramble control sequence was purchased from Addgene (plasmid 1864) [28]. Next, each of the shRNA plasmids was digested with BamHI and NdeI to subclone the scramble control sequence and the shRNA containing sequences into pLKO.3G lentiviral plasmid (Addgene Plasmid 14748).

For luciferase assays, full length RAB5C 3'UTR was PCR amplified using cDNA of NALM6 as template, and cloned into pmirGLO Dual-Luciferase miRNA Target Expression vector (Promega, Madison, WI, USA). Site directed mutagenesis of RAB5C-3'UTR-luciferase deletion construct 1 (Δ1) was carried out using the QuikChange Lightning Site-Directed Mutagenesis Kit (Agilent Technologies, Santa Clara, CA, USA) according to manufacturer's protocol. For deletion of the second miR-509-3p binding site in Δ2 construct and Δ1Δ2 constructs, standard PCR was performed. Primers used to create the luciferase constructs are listed in Table S2.

A lentivector overexpressing the RAB5C was constructed by PCR amplification of the RAB5C open reading frame from NALM6 cDNA (Primers listed in Table S3). The PCR product

was then cloned into the pWCC61 plasmid (Empty lentiviral vector #3; EV#3), a dual-promoter lentiviral vector generated by our lab in which the human EF1α promoter drives RAB5C and the ubiquitin promoter drives DsRed.

Lentivirus production and transduction

Lentivector plasmids were co-transfected with purchased packaging plasmids, pMD2.G (Addgene plasmid 12259) and pCMVR8.74 (Addgene plasmid 22036), using 3 μg of polyethylenimine (Polysciences Inc., Warrington, PA, USA) per μg of DNA. Viruses were then titered in each cell line 3 days after transduction by measuring %GFP$^+$ cells using flow cytometry. Cultures transduced between 30-70% GFP$^+$ were used to calculate lentivirus titer and MOI. To increase transduction efficiency, the following amounts of polybrene was added to each cell line: 0.8 μg/ml polybrene for RCH-ACV and KARPAS-45 cells, 1.6 μg/ml polybrene for Jurkat cells, 4 μg/ml polybrene for NALM6, REH and K562 cells. Mock-transduced cells were cells treated with polybrene but no lentivirus. In all experiments with transduced cells, cells were transduced with each lentivirus to MOI = 2. All transduced cells were washed with phosphate buffered saline (PBS) at 2 days after transduction to remove the polybrene.

GFP competition assay

3 days after transduction, >80% of NALM6 cells were GFP$^+$. 7 days after transduction, the transduced cells were mixed with mock-transduced cells to obtain a cell mixture containing ~50% GFP$^+$ cells, and this time point was set as day 0 for the GFP competition assay. This cell mixture was cultured for 5 weeks, and the %GFP$^+$ cells was measured weekly by flow cytometry (Accuri C6, Becton Dickinson, New Jersey, USA), after gating on only the viable cell population based on the FSC and SSC parameters. Analysis was performed using FlowJo software (Tree Star Inc, Ashland, OR, USA).

Cell growth assays

For alamarBlue (Life Technologies, Grand Island, NY, USA) dye-based cell growth assays, cells were seeded at 5×10^3 cells/100 μl media (NALM6 and RCH-ACV cells) or at 2×10^3 cells/100 μl media (REH cells) in triplicates in 96-well plates at 3 days after transduction. At 7 days after transduction, 10 μl alamarBlue was added to each well and plates incubated (37°C, 4 h) before reading using a VictorX3 (PerkinElmer, Waltham, MA, USA; 530/580 nm excitation/emission filters). For trypan blue exclusion cell counts, 2.5×10^5 cells/ml were seeded in each well of a 96-well plate day on day 3 after transduction. 10 μl of cell suspensions were removed at each time point and counted using a hemocytometer.

RNA isolation and measurement of miR and mRNA expression levels by quantitative real-time reverse-transcription PCR (qRT-PCR)

For qRT-PCR of mature miRs, cell lysates were made using Cell Lysis Buffer (Signosis, Santa Clara, CA, USA) and reverse transcription performed using TaqMan microRNA reverse transcription kit (Life Technologies) according to manufacturer's protocol. For mRNA levels, SYBRGreen qRT-PCR assays were conducted with total RNA isolated using the miRNeasy kit (Qiagen) according to manufacturer's protocol, and reverse transcription performed using the High-capacity-RNA-to-cDNA kit (Life Technologies) according to manufacturer's protocol. Primers for qRT-PCR for genes were obtained from PrimerBank

[29] (Table S4). The TaqMan IDs are listed in Table S5 (Life Technologies). All SYBRGreen and TaqMan qRT-PCR assays were performed using the 7900 HT Real-Time PCR system (Life Technologies). All Ct values >35 were assigned a value of 35 for calculation of fold expression level change. For qRT-PCR of mature miRs, U18 was used as endogenous control. For SYBR-Green qRT-PCR of mRNA genes, GAPDH was used as endogenous control. DNA oligonucleotides (synthesized by Integrated DNA Technologies, Coralville, IA, USA) of mature miR sequences (miRBase.org) were used to create standard curves for absolute qRT-PCR miR quantitation, which was performed as described previously [30,31].

Microarray data

All microarray data has been previously deposited in NCBI Gene Expression Omnibus [32] (GEO Series accession number GSE51908; http://www.ncbi.nlm.nih.gov/geo/query/acc.cgi?acc= GSE51908). Samples used in this analysis include B-ALL cell lines ($n = 27$, replicates of 9 cell lines), primary B-ALL samples ($n = 16$), T-ALL cell lines ($n = 15$, replicates of 5 cell lines), primary T-ALL samples ($n = 8$), AML cell lines ($n = 21$, replicates of 7 cell lines), primary AML samples ($n = 15$), primary blood B lymphocytes ($n = 11$), primary mobilized blood $CD34^+$ hematopoietic stem-progenitor cells (HSPCs) ($n = 4$), primary blood granulocytes ($n = 14$), primary blood monocytes ($n = 5$) and primary blood T lymphocytes ($n = 20$).

Apoptosis and cell cycle analysis

For apoptosis assays, 10^5 NALM6 cells were stained with APC Annexin V and DNA binding dye 7-amino-actinomycin (7-AAD) (Biolegend, San Diego, CA, USA) 4 days after transduction according to manufacturer's protocol and analyzed by flow cytometry (Accuri C6, Becton Dickinson). For cell cycle analysis, at 3 days after transduction, NALM6 cells (0.5×10^6 cells/ml) were cultured for 24 h in fresh medium, then 10^6 cells were labeled with BrdU (Becton Dickinson) for 1 h. Cells were then washed twice in ice cold PBS and the pellet suspended in 500 µl PBS. Cells were fixed in 5 ml ice cold 70% ethanol overnight at $-20°C$. 2 M hydrochloric acid was then used to denature the DNA for 30 min at room temperature, and the washed pellet resuspended in 1 ml 0.1 M $Na_2B_4O_7$, pH 8.5 (Sigma-Aldrich) to neutralize the acid for 10 min. Cells were stained with 1 µl APC anti-BrdU antibody (BioLegend) in 20 µl volume for 30 min at room temperature, followed by 20 µl 7-AAD for 15 min at room temperature. APC BrdU and 7-AAD signal was then assessed by flow cytometry (Accuri C6, Becton Dickinson). FlowJo software (Tree Star Inc) was used to determine the cell cycle profile of each sample.

Caspase-3/7 assay

Transduced NALM6 cells were seeded at 500 cells/well in a 384-well plate on day 3 after transduction. On day 7 after transduction, caspase activity was measured using the Apo-ONE homogenous caspase-3/7 assay (Promega) according to manufacturer's instructions at 4 h after addition of reagent to cells, using a VictorX3 (PerkinElmer, 485/535 nm excitation/emission filters).

Luciferase assay

HEK293T cells were cultured overnight at 10^5 cells/450 µl in each well of a 24-well plate. 300 ng of plasmid was co-transfected with 50 nM of miR mimic using 2.5 µl of Lipofectamine2000 (Life Technologies) according to manufacturer's protocol. Lysates were harvested 48 h after transfection and processed using Dual luciferase reporter assay system (Promega) according to manufac-

turer's protocol. Lysates were diluted 400-fold in passive Lysis buffer Assay before plating and read using VictorX3 (PerkinElmer). Renilla luciferase values were used to normalize for transfection efficiency; the ratio of firefly/renilla luciferase is designated as relative luciferase activity.

Western blotting

Lysates of transduced cells were harvested 7 days after transduction and lysed in RIPA buffer (Sigma-Aldrich) containing 1 mM phenylmethanesulfonyl fluoride (Sigma-Aldrich) and $1\times$ complete protease inhibitor cocktail tablet (Roche Applied Science, Indianapolis, USA). Protein concentration was determined by Bio-Rad Protein assay (Bio-Rad, Hercules, CA, USA) according to manufacturer's protocol and lysates containing 30-40 µg protein loaded onto a pre-made 4-12% Bis-Tris NuPAGE gel (Life Technologies) and transferred to a PVDF membrane using an iBlot Dry Blotting system (Life Technologies). RAB5C (ab137919, Abcam, Cambridge, MA, USA) and α-tubulin (T6074, Sigma-Aldrich) antibodies were used according to manufacturer's protocol and signal detected using an ECL detection kit (Thermo Scientific) imaged by the ChemiDOC XRS+ System (Bio-Rad). Bands were analyzed and quantified using ImageLab software (Bio-Rad).

Results

Enforced miR-509 expression inhibited growth of NALM6 cells

We applied our functional miR-HTS to screen a pooled lentivirus library of 578 human miRs or miR clusters for their growth-regulatory properties in human NALM6 B-ALL cells and identified candidate miRs as previously described [26]. 4 miRs (miR-381, miR-509, miR-550a, and miR-873) and 1 miR cluster (miR-432~136) inhibited NALM6 growth in at least 2 of 3 replicate screens performed. In order to confirm the growth inhibitory effects of the candidate miRs identified from the functional screen, each of the 5 miR or miR cluster candidates was cloned into a lentiviral expression vector downstream of green fluorescent protein (GFP) (Figure 1A). We expressed the miR-432~136 cluster as a single unit rather than as 2 individual miRs, to recapitulate the way they were screened and because the 2 miRs may cooperate. The growth inhibitory potential of each candidate miR or miR cluster was then tested, by performing multiple GFP competition assays [33,34]. NALM6 cells were transduced with each of the 5 miR lentiviruses (>80% GFP⁺ cells), and each culture was then mixed with GFP⁻ cells to obtain an initial culture with ~50% GFP⁺ cells. If enforced expression of a given miR or miR cluster inhibited NALM6 growth, the %GFP⁺ cells in culture would decrease over time. For NALM6 cells transduced with the control empty vector, the %GFP⁺ cells remained stable at ~50% over the 5-week GFP competition assay (Figure 1B). Similarly, no change in %GFP⁺ cells was observed over 35 days in the GFP competition assays for miR-381, miR-550a, miR-873 and miR-432~136 (Figure S1A-S1D). In contrast, NALM6 cells transduced with miR-509 lentivirus were out-grown by the GFP⁻ cells; the %GFP⁺ cells decreased from 46% at assay day 0 to 10% 35 days later (Figure 1B).

As expected, miR-509-5p and miR-509-3p were strongly overexpressed in miR-509-transduced NALM6 cells as assayed by qRT-PCR (Figure 1C). Similarly, overexpression of miR-381, miR-550a, miR-873, and miR-432 was achieved by lentiviral transduction (Figure S1E). These results indicate that miR-381, miR-432, miR-550a, and miR-873 do not inhibit growth of NALM6. However, no expression of miR-136 was detected in

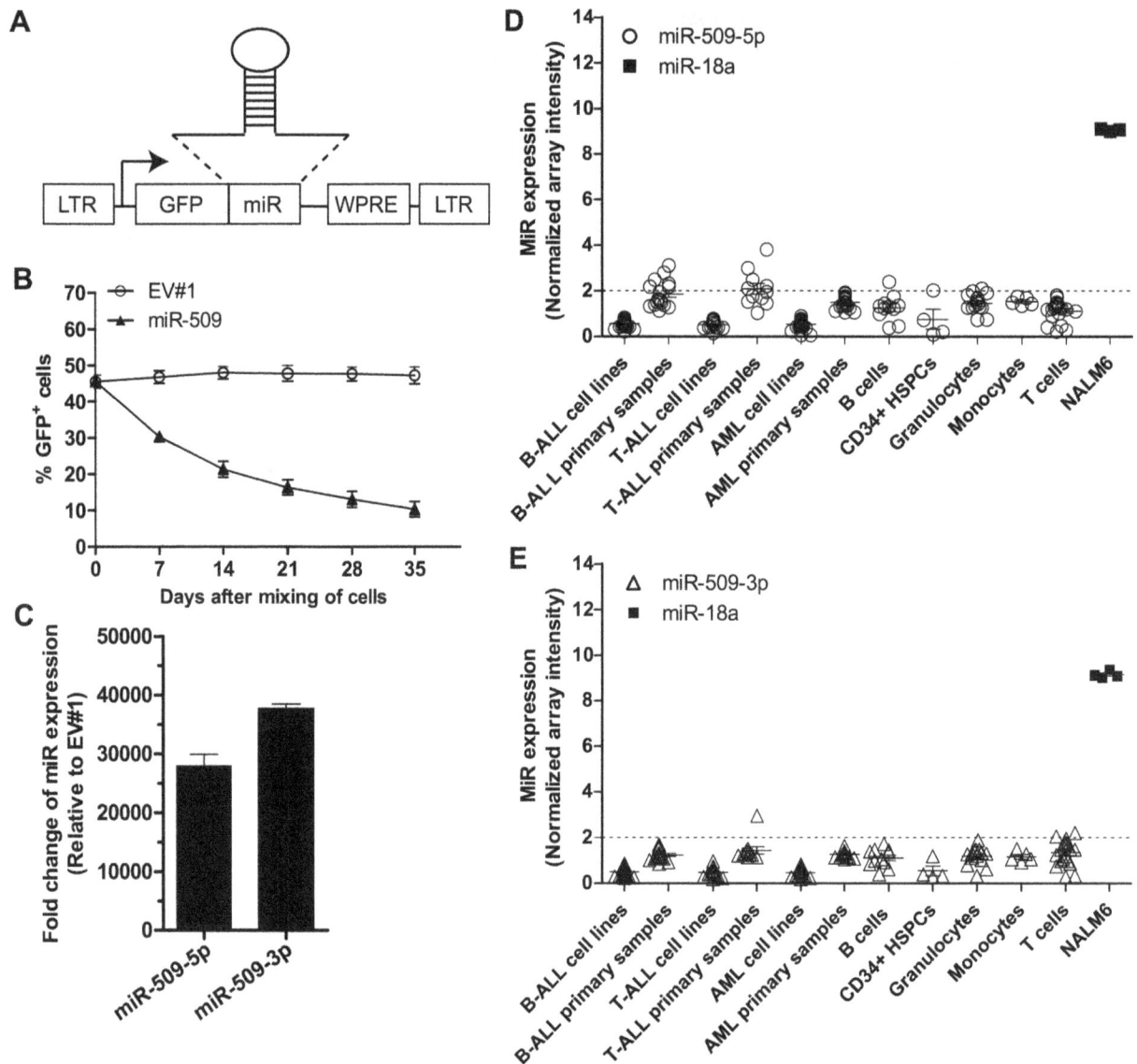

Figure 1. Enforced miR-509 expression inhibits growth of NALM6 cells. (A) Schematic of lentiviral vector used to express miRs. Arrow depicts the direction of human EF1α promoter. LTR: long terminal repeat; GFP: green fluorescent protein; WPRE: woodchuck hepatitis virus post-transcriptional regulatory element. The parental plasmid without miR is denoted as empty vector #1 (EV#1). The miR sequence consists of the native miR hairpin with ~200 bp of its flanking genomic sequences. (B) Assessment of %GFP$^+$ cells by flow cytometry in the GFP competition assay. NALM6 cells were transduced with miR-509 lentivirus or empty vector (EV#1) at MOI =2, and transduced GFP$^+$ cells were mixed with an equal number of mock-transduced cells (GFP$^-$) 7 days later to achieve an initial culture of ~50%GFP$^+$ cells; this was designated Day 0 and the %GFP$^+$ cells (pre-gated on viable cells) was assessed weekly by flow cytometry. Means ± SEMs are shown for 3 independent experiments. (C) Enforced expression of mature miR-509-5p and miR-509-3p NALM6 cells, as assayed by qRT-PCR. NALM6 cells were transduced with miR-509 lentivirus to MOI =2, and total RNA was collected at 7 days after transduction. U18 was used as the loading control, and normalized to EV#1-transduced NALM6 cells. Means ± SEMs of 3 independent experiments. (D) Expression of mature miR-509-5p was determined by miR microarray analysis in B-ALL, T-ALL and AML cell lines and primary samples, B cells, CD34$^+$ HSPCs, granulocytes, monocytes and T cells. Dotted line represents normalized microarray intensity of 2 whereby any value <2 denotes undetectable expression. Data points shown are means ± SEMs. Expression data is accessible through GEO Series accession number GSE51908 [32]. (E) Expression of mature miR-509-3p and miR-18a as determined by miR microarray analysis similar to (D). (D, E) Data shown for miR-18a is only for the NALM6 cell line.

miR-432~136 cluster-transduced NALM6 cells. This lack of miR-136 expression could be due to lack of necessary cis-regulatory elements or trans-regulatory factors required for miR-136 biogenesis; we did not investigate the possibility that an alternative approach to successfully express miR-136 in NALM6 would validate a growth inhibitory role for this miR. Instead, we decided to focus on miR-509 for further studies.

Our miR microarray expression analyses [32] (GEO Series accession number GSE51908) revealed undetectable endogenous levels of mature miR-509-5p and miR-509-3p in NALM6 and other acute leukemia cell lines (Figure 1D, 1E), as well as in primary leukemia cases and CD34$^+$ hematopoietic stem-progenitor cells (HSPCs) and blood cell types from normal human donors (Figure 1D, 1E). In absolute qRT-PCR quantifications [30,31], miR-509-transduced NALM6 cells expressed 1,814±95 copies (mean ± SEM) per cell of miR-509-5p (Table 1), comparable to levels of miR-18a, which for reference is expressed at the 70th percentile of all miRs in NALM6 cells based on our miR microarray data (Figure 1D). MiR-509-3p was expressed at $3,656 \pm 117$ copies per cell in miR-509-transduced NALM6 cells, also within the physiological range of miR copy numbers per cell (range: <10 to$>30,000$ copies per mammalian cell) [30].

MiR-509 reduced NALM6 cell growth by 2 additional independent assays

To further confirm the effect of miR-509 on NALM6 cell growth, we performed trypan blue dye exclusion cell counts and alamarBlue assays. At 8 days after transduction, cultures of miR-509-transduced NALM6 cells contained 43% fewer viable cells than empty vector-transduced cells by trypan blue counts (Figure 2A). Similarly, miR-509-transduced NALM6 cells had 48% reduced ($p<0.05$) cell growth, as compared to empty vector-transduced cells using the alamarBlue assay (Figure 2B). Since the alamarBlue dye-based assay measures the reducing environment within cells, which is linked to mitochondria metabolism [35], we examined whether miR-509 affected mitochondrial membrane potential. No difference in mitochondrial membrane potential was observed between miR-509-transduced and empty vector-transduced NALM6 cells (data not shown).

MiR-509 inhibited growth of RCH-ACV and REH B-ALL cell lines

We next examined if the growth inhibitory effects of miR-509 extended to other B-ALL (RCH-ACV and REH), T-cell ALL (T-ALL; Jurkat and KARPAS-45) or myeloid leukemia (K562) cell lines. MiR-509-transduced RCH-ACV cells had ~30% reduced growth by trypan blue on day 8 after transduction or alamarBlue assay on day 7 after transduction (Figure 2C, 2D). In addition, miR-509-transduced REH cells had 23% ($p<0.05$) reduced growth in the alamarBlue assay (Figure 2E). In contrast, no reduction in cell growth was observed in Jurkat, KARPAS-45 or K562 cells transduced with miR-509 as compared to control empty vector using alamarBlue assays (Figure S2A–S2D). This was despite documented overexpression of miR-509 in these transduced cell lines (Figure S3). Thus, miR-509 inhibited the growth of all 3 tested human B-ALL cell lines, NALM6, RCH-ACV and REH.

MiR-509-transduced NALM6 cells had a lower proportion of cells in cell cycle S-phase and increased apoptosis

To investigate the cellular mechanisms by which enforced miR-509 expression inhibits growth, we examined whether miR-509 regulates cell cycle progression by conducting BrdU/7-AAD

staining [36]. 4 days after transduction, miR-509-transduced NALM6 had fewer cells in S-phase than empty vector-transduced cells (Figure 3A), and this was statistically significant (Figure 3B, $p<0.05$). In addition, there were slightly elevated numbers of cells in the subG$_1$ and G$_2$/M phases, but these differences were not statistically significant. To investigate if miR-509 promotes cell death via apoptosis, Annexin V/7-AAD staining was performed. 4 days after transduction, miR-509-transduced NALM6 cells had 1.5-fold ($p<0.05$) higher numbers of Annexin V$^+$/7-AAD$^-$ apoptotic cells and 1.4-fold higher numbers of Annexin V$^+$ dying/dead cells ($p<0.05$), as compared to empty vector-transduced cells (Figure 3C, 3D). Consistent with these findings, we detected a 1.5-fold increase ($p<0.05$) in activated capase-3/7 activity in miR-509-transduced NALM6 cells as compared to empty vector-transduced cells (Figure 3E).

Informatics prediction of RAB5C as a target of miR-509

To identify targets of miR-509 that might mediate growth of B-ALL cells, we used a filtering strategy to prioritize the many predicted targets of miR-509 (Figure 4A). First, we downloaded the sets of predicted mRNA targets of miR-509-5p and miR-509-3p (Set 1), as well as those of the 4 miRs that we had shown not to inhibit NALM6 growth (i.e. miR-381, miR-432, miR-550a and miR-873; Set 2) from the TargetScan6.2 [37] and/or miRDB [38,39] miR target prediction databases. Since NALM6 cells transduced with miR-432~136 did not result in miR-136 overexpression, we did not include miR-136 targets in Set 2 (Figure 4A). Next, we downloaded the gene expression profile of NALM6, determined by genome-wide microarray profiling as listed in the Cancer Cell Line Encyclopedia (CCLE) [40] and focused on genes which have detectable expression in NALM6 (i.e. annotated as "marginal" or "present" in CCLE; Set 3). Then, we intersected these 3 sets of mRNAs [41] to identify the subset of genes expressed in NALM6 and predictively targeted by miR-509, but not predictively targeted by the 4 miRs that did not inhibit NALM6 growth. This resulted in a set of 395 genes (listed in Table S6). This list was subsequently reduced to 74 genes by selecting for genes known to participate in growth regulation based on annotations at NCBI's "Gene" database, DAVID bioinformatics resources [42,43], as well as our own literature searches. Of these 74 predicted targets of miR-509, 12 genes previously demonstrated in the literature to be either involved in leukemia and oncogenesis (ERLIN2, FLI1, FOXP1, MAML1, RAC1, YWHAB and YWHAG), or predicted as miR-509 targets by both TargetScan6.2 and miRDB (PGRMC1, RAB5C, RAC1, TFDP2, UHMK1, USP9X) were selected for initial qRT-PCR analysis. We used this informatic filtering strategy, as compared to performing global differential gene expression analysis such as microarray analysis, to enable us to rapidly and at low cost identify

Table 1. Absolute copy number of mature miR-509 and miR-18a RNA per NALM6 cell.

qRT-PCR assay	Copy number per NALM6 cell transduced with	
	Empty vector #1	miR-509
miR-509-5p	<10	$1,814 \pm 95$
miR-509-3p	<10	$3,656 \pm 117$
miR-18a	$1,591 \pm 105$	$1,415 \pm 53$

RNA was isolated from NALM6 cells on day 7 after transduction with either control empty vector #1 or miR-509, and absolute qRT-PCR quantification was performed for miR-509-5p, miR-509-3p or miR-18a. Copy number per cell was estimated based on standard curves of miR-509-5p, miR-509-3p or miR-18a using DNA oligonucleotides. For reverse transcription, 10 ng RNA (equivalent to 800 cells, i.e. 12.5 pg of total RNA per cell) was used in each reaction. Means ± SEMs of 3 independent experiments.

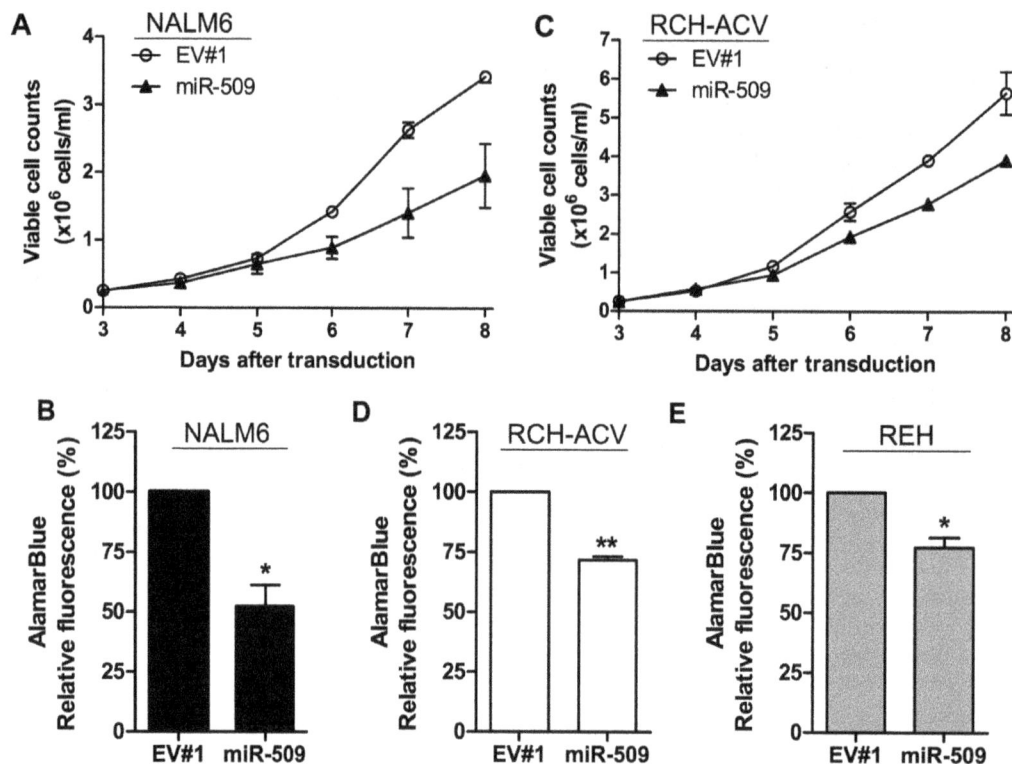

Figure 2. Enforced miR-509 resulted in inhibition of growth of 3 B-ALL cell lines, NALM6, REH and RCH-ACV. (A) Viable cell numbers measured via trypan blue dye exclusion counts of NALM6 cells transduced with either miR-509 lentivirus or empty vector (EV#1); 25,000 cells were plated for each sample starting at 3 days after transduction. (B) AlamarBlue cell growth assay on day 7 after transduction of NALM6 cells transduced with either miR-509 lentivirus or EV#1. Values for miR-509 were normalized to EV#1. (C) Viable cell counts of RCH-ACV cells based on trypan blue exclusion counts, initial plating of 25,000 cells for both samples on 3 days after transduction. Means ± SEMs are plotted, and SEMs for miR-509 were very small. (D) Cell growth of RCH-ACV transduced with either EV#1 or miR-509 overexpressing lentivirus using alamarBlue cell growth assay conducted on day 7 after transduction. Values for miR-509 were normalized to EV#1. (E) MiR-509-transduced REH cells reduced growth compared to EV#1 in an alamarBlue cell growth assay. Cells were transduced 7 days prior to addition of alamarBlue. (A to E) Means ± SEMs, 3 independent experiments done in triplicates. Statistical analysis was done by Student's t test. *$p<0.05$, **$p<0.01$.

target genes-of-interest. 3 of these 12 predicted targets (RAB5C, RAC1, and UHMK1) were down-regulated by miR-509 at the mRNA level (Figure 4B). RAB5C mRNA levels showed the greatest reduction, with a 40% lower level ($p<0.05$) in miR-509-transduced than in empty vector-transduced NALM6 cells (Figure 4B). Correspondingly, RAB5C protein was 85% ($p<0.001$) lower in miR-509-transduced cells by western blotting (Figure 4C, 4D). We also observed a ≥86% decrease in RAB5C protein levels in miR-509-transduced RCH-ACV and REH cells as compared to empty vector (Figure S4). Since RAB5 has been implicated in cell cycling [44,45] and is one of the top 3 predicted targets of miR-509-3p by both TargetScan6.2 (Total context+ score = −0.65) and miRDB (Target score = 91), we focused our subsequent studies on RAB5C.

MiR-509 directly targets RAB5C

To examine if miR-509 directly represses RAB5C, we employed RAB5C-3′UTR luciferase reporter assays. There are 2 miR-509-3p binding sequences in the 3′UTR of RAB5C (Figure 5A), as predicted by both miRDB and TargetScan6.2. Both miR-509-3p binding sequences are present in the RAB5C 3′UTR of several species including human, mouse, rat, horse and dog, suggesting that the regulation of RAB5C by miR-509 is also conserved. We cloned the full-length wild type (WT) 3′UTR of RAB5C downstream of firefly luciferase gene (*luc2*) in the pmirGLO luciferase vector and also generated 3 luciferase constructs

containing 1 (Δ1 or Δ2) or both (Δ1Δ2) deletions of miR-509-3p binding sites (Figure 5B). Co-transfection of miR-509-3p mimic and RAB5C-3′UTR WT luciferase vector resulted in 81% lower ($p<0.001$) relative luciferase activity than in cells transfected with RAB5C-3′UTR WT luciferase vector alone (Figure 5C). Co-transfection of the non-targeting miR-551b mimic plus the RAB5C-3′UTR WT luciferase vector did not repress luciferase activity. Co-transfection of either RAB5C-3′UTR-luciferase deletion construct, Δ1 or Δ2, plus miR-509-3p mimic resulted in > 50% lower ($p<0.01$) relative luciferase activity than cells transfected with only the indicated RAB5C-3′UTR deletion constructs. Co-transfection of Δ1Δ2 construct (in which both predicted miR-509-3p binding sites were deleted) with miR-509-3p mimic abolished the reduction in luciferase signal. This indicated that miR-509 directly targets the 3′UTR of RAB5C via both predicted miR-509-3p binding sites.

RAB5C mediates the growth-inhibitory effect of miR-509

We then examined if reduced RAB5C is responsible for the functional effects of miR-509. To determine if repression of RAB5C would phenocopy the growth suppressive effect of miR-509, NALM6 cells were transduced with 3 different lentiviruses, each containing a distinct shRNA against RAB5C. In alamarBlue assays, all 3 shRNAs inhibited NALM6 cell growth by ≥42% ($p<0.01$) as compared to cells transduced with the scrambled control (Figure 6A). We verified that all 3 shRNAs resulted in ≥80%

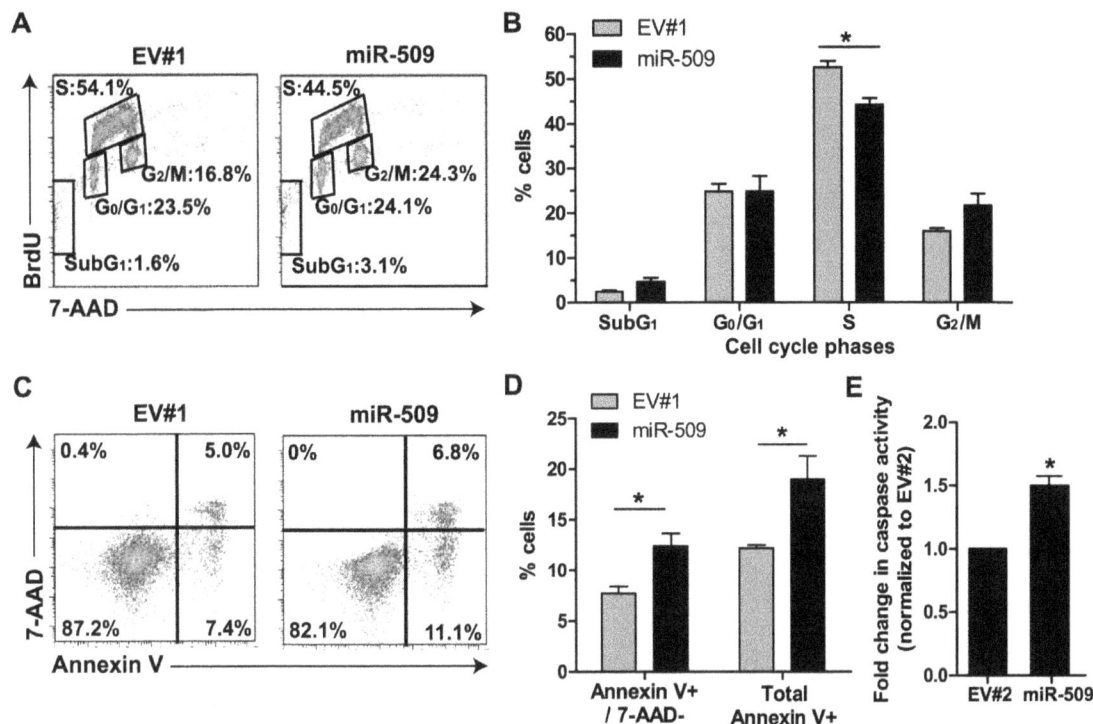

Figure 3. Enforced miR-509 expression in decreased proportion of cells in S-phase, induced apoptosis and activated caspase-3/7.
(A) Representative flow cytometric plots showing cell cycle distribution of NALM6 cells transduced with empty vector (EV#1) or miR-509 overexpressing lentivirus. On day 3 after transduction, cells were labeled with BrdU for 1 h. Cells were then fixed overnight and stained on the next day with both BrdU and 7-AAD before analysis by flow cytometry. Percent of cells at each phase of cell cycle are boxed as indicated. (B) Frequencies of cells at the different phases of cell cycle. Means $\pm$ SEMs of 3 independent experiments with statistical analysis by Student's t test. *$p<0.05$. (C) Representative flow cytometric plots of cell death distribution of NALM6 cells transduced with EV#1 or miR-509 overexpressing lentivirus. Cells were stained with both Annexin V and 7-AAD before analysis by flow cytometry on day 4 after transduction. Numbers represent the frequency in each quadrant. (D) Frequencies of apoptotic cells which are Annexin V-positive (Annexin V$^+$) and 7-AAD negative (7-AAD$^-$), as well as total Annexin V$^+$. Means $\pm$ SEMs of 3 independent experiments with statistical analysis by Student's t test. *$p<0.05$. (E) NALM6 cells were transduced with empty vector #2 (EV#2) or miR-509 overexpressing lentivirus, and cells were seeded in 384-well plate on day 3 after transduction. On day 7 after transduction, caspase-3/7 activity was measured and fold-change in caspase activity was normalized to EV#2. Means $\pm$ SEMs of 3 independent experiments done in triplicates, with statistical analysis by Student's t test. *$p<0.05$.

decreased RAB5C protein levels ($p<0.01$) in NALM6 cells (Figure 6B, 6C) using western blotting.

In order to determine whether RAB5C mediates miR-509 induced growth inhibition in NALM6 cells, we performed a rescue experiment. We cloned the RAB5C open reading frame (ORF) without its 3′UTR into a lentiviral vector. In alamarBlue assays, NALM6 cells co-transduced with miR-509 plus empty vector had 51% lower growth ($p<0.001$) than cells co-transduced with the 2 control empty lentiviral vectors (Figure 6D). In contrast, NALM6 cells co-transduced with miR-509 plus RAB5C lentiviruses had 36% greater growth than cells co-transduced with miR-509 plus the empty vector ($p<0.05$). Overexpression of RAB5C ORF in NALM6 cells co-transduced with miR-509 was confirmed by western blotting (Figure 6E, 6F). Thus, RAB5C rescued, in large part, the growth inhibitory effects of miR-509.

Discussion

In this study, we conducted a functional miR-HTS in NALM6 cells and identified miR-509 as a novel inhibitor of human B-ALL cell growth. Using NALM6 B-ALL cell line, enforced expression of miR-509 reduced NALM6 B-ALL cell growth in 3 independent growth assays. MiR-509-mediated growth inhibition was also observed in 2 additional B-ALL cell lines, REH and RCH-ACV. However, miR-509 is not a global inhibitor of cell growth, as

enforced miR-509 expression in 2 T-ALL (Jurkat and KARPAS-45) and 1 myeloid leukemia (K562) cell lines did not inhibit growth. Susceptibility to miR-509 growth inhibition is likely due to differential expression or differential dependence upon miR-509 target genes for cell proliferation and survival [46]. More extensive testing will be necessary to determine if the growth inhibitory effects of miR-509 might be specific to B-ALL cells or some molecularly-defined subset of leukemias or shared with other cancer types. MiR-509 has been reported previously to be down-regulated in renal cell carcinoma as compared to normal tissue counterparts [47,48]. Since miR-509-5p and miR-509-3p are undetectable in normal or leukemic hematopoietic cells, miR-509 does not qualify as a tumor suppressor miR for leukemias. The lack of miR-509 expression in healthy donor blood cell types and CD34$^+$ HSPCs [32] exemplifies the importance of functional screening to identify growth-suppressing miRs, as expression profiling comparing acute leukemia cases versus healthy donor samples would not have identified miR-509 as a miR capable of inhibiting leukemia cell growth.

We further observed that enforced miR-509 expression reduced the number of actively proliferating cells and increased apoptotic and dead NALM6 cells, indicating that miR-509 reduces cell proliferation and survival. Our observations in NALM6 are consistent with previous reports that both miR-509-5p [47] and

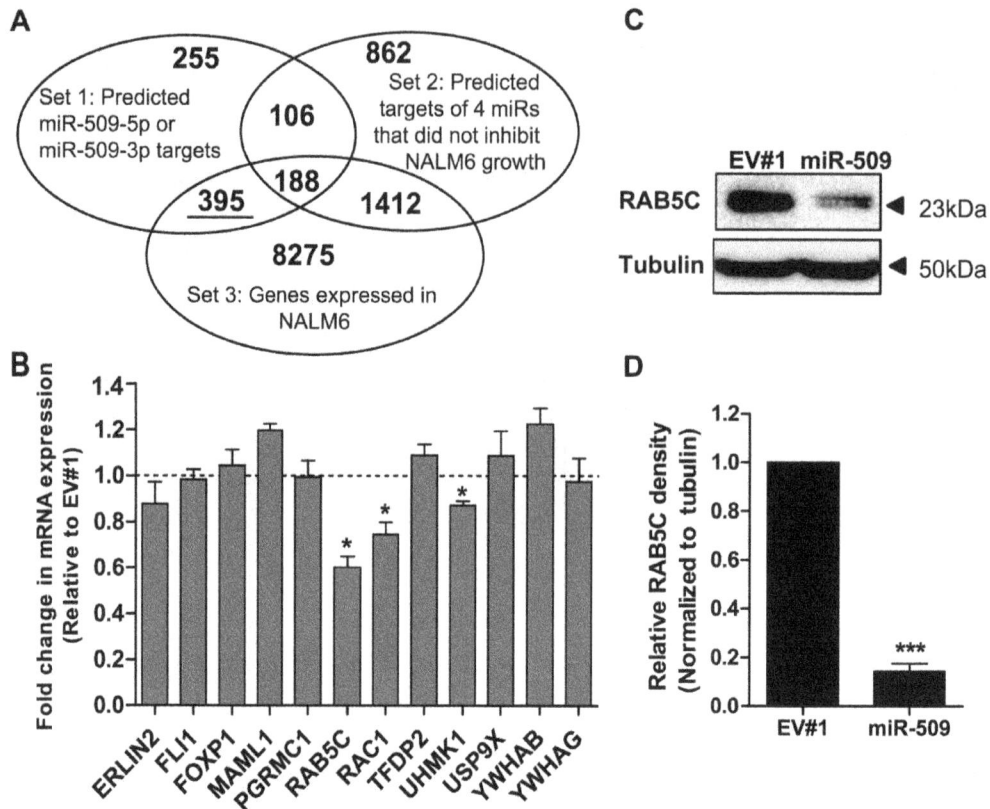

Figure 4. Identifying mRNA targets of miR-509. (A) Venn diagram showing the number of mRNAs that do not overlap, or are shared between each set in our *in silico* strategy to identify relevant targets of miR-509. Set 1 refers to the list of predicted targets of miR-509-5p or miR-509-3p from TargetScan6.2 or miRDB. Set 2 is the list of predicted targets of miRs tested to not inhibit NALM6 growth (i.e. miR-550a, miR-873, miR-381 and miR-432) from TargetScan6.2 or miRDB, while Set 3 is the list of mRNA that is expressed in NALM6, as determined by genome-wide microarray profiling downloaded from the Cancer Cell Line Encyclopedia and its expression levels are denoted in the microarray dataset as "marginal" or "present". (B) Expression levels of 12 putative targets of miR-509 as determined by qRT-PCR. RNA was isolated from NALM6 cells transduced with EV#1 or miR-509 overexpressing lentivirus at 7 days after transduction. All values were normalized to GAPDH and fold-change was calculated relative to EV#1 sample. Data represents means ± SEMs of 3 independent experiments, with statistical analysis by Student's t test. *p<0.05. (C) Representative western blots of RAB5C expression. NALM6 cells were transduced with either EV#1 or miR-509 overexpressing lentivirus, and whole cell lysates were harvested at 7 days after transduction. α-tubulin was used for loading control. (D) Densitometry analysis of RAB5C expression of western blot in (C) and 2 other independent experiments. α-tubulin was used for normalization, and relative densitometry was then calculated compared to EV#1. Data shown represent means ± SEMs, with statistical analysis by Student's t test. ***p<0.001.

miR-509-3p [48] suppressed cell growth and induced apoptosis in a human renal cancer cell line.

To identify relevant miR-509 targets, we may in the future employ biochemical or genomic techniques [49,50] to identify all of the targets of miR-509 in NALM6 cells. In the initial study herein, we instead used bioinformatics to define a subset of predicted miR-509 target genes known to be expressed in NALM6 cells but not predicted to be targeted by the miRs that failed to inhibit NALM6 cell growth. We then selected those targets known to be involved in cellular processes that regulate growth (e.g. proliferation, cell cycle, cell death, oncogenes), resulting in a set of 74 growth-related predicted miR-509 targets. Using qRT-PCR to assess levels of 12 of these 74 targets in miR-509-transduced versus empty vector-transduced NALM6 cells, 3 predicted miR-509 targets were reduced in miR-509-transduced NALM6 cells. Although the mRNAs of 9 of the 12 tested predicted miR-509 targets were not reduced in miR-509-transduced NALM6 cells, some of these may still be targets of miR-509 as they might be inhibited at the translational level [51]. However, given that reduction at the mRNA level was observed in ≥84% of miR targets with reduced protein levels [52], we decided to focus in the

study herein on predicted targets inhibited by miR-509 at the mRNA level. Moreover, RAB5C mRNA was the target most reduced in response to miR-509 expression, and therefore we focused on RAB5C for further experiments. We showed that miR-509 indeed binds to the 3'UTR of RAB5C via the 2 thermodynamically predicted sites. Upon effective knockdown of RAB5C using each of 3 shRNA constructs, NALM6 cell growth was reduced; thus, RAB5C knockdown phenocopied the growth inhibition observed by enforced miR-509 expression. Our observations that co-transduction of the RAB5C ORF lacking its 3'UTR (therefore no longer regulated by miR-509) rescues miR-509-mediated growth inhibition indicates that reduction of RAB5C is a major mechanism of miR-509-mediated NALM6 growth inhibition. Thus, while future studies may find additional targets of miR-509-5p and/or miR-509-3p that contribute to miR-509-mediated growth inhibition, our current results demonstrate that RAB5C is a novel target of miR-509 and an important driver of the growth of human B-ALL cells.

As members of the Rab family of small monomeric GTPases, RAB5 molecules are central in coordinating vesicle trafficking, particularly in the early stages of endocytosis [53–55]. In addition

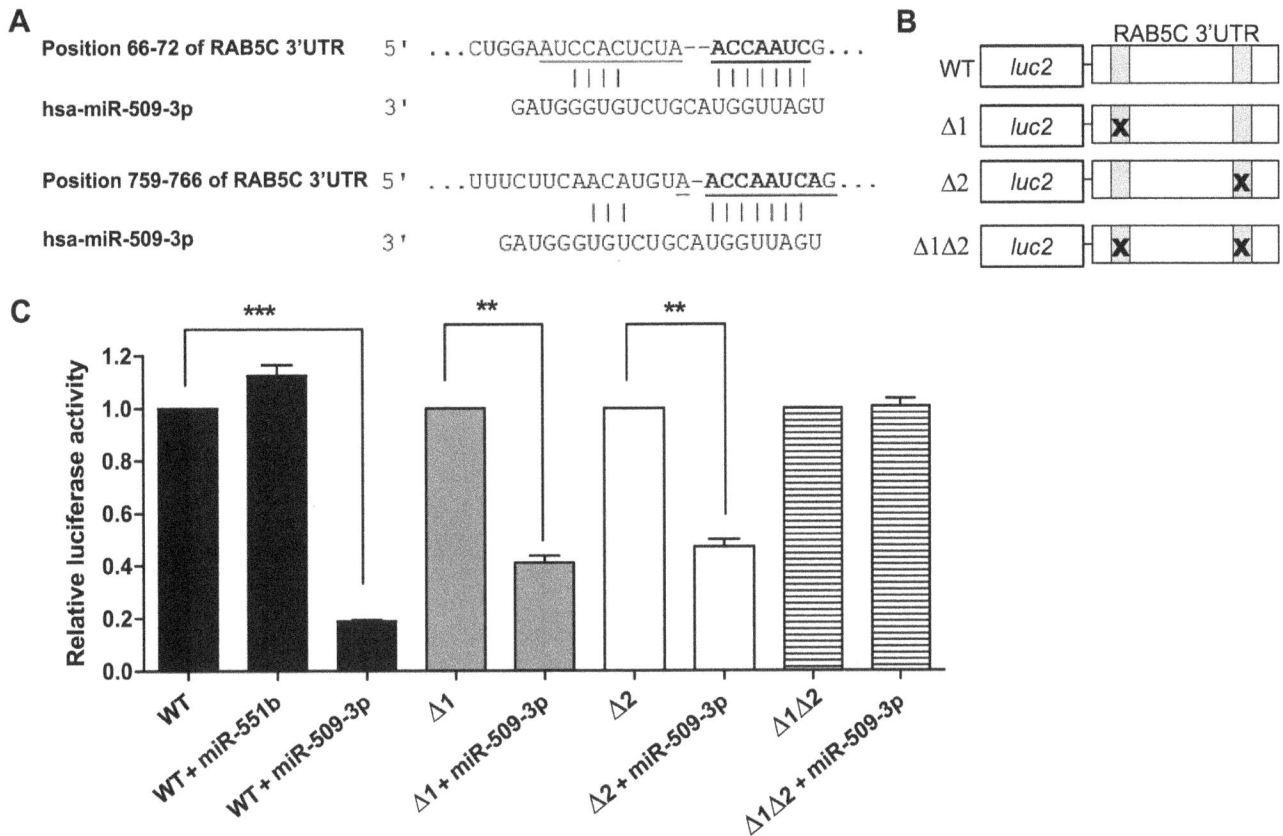

Figure 5. RAB5C is a direct target of miR-509. (A) Sequence alignment of RAB5C to miR-509-3p predicted by TargetScan6.2. The full length 3'UTR of RAB5C is 803 bases. Sequences shown in bold refer to position 66–72 and 759–766 of RAB5C 3'UTR where miR-509-3p is predicted to target. The underlined sequences were deleted in the RAB5C-3'UTR deletion constructs listed in (B) for the luciferase assay. (B) Schematic representation of luciferase vector constructs used in luciferase assay. Full length RAB5C 3'UTR was cloned downstream of the firefly luciferase gene (luc2) in the pmirGLO luciferase vector. Wild type RAB5C 3'UTR is listed as WT. Grey boxes indicate the 2 predicted miR-509-3p target sites (66–72 and 759–766), and the "X" indicates the deletion sites present in the deletion (Δ) constructs. (C) Luciferase assay demonstrates that RAB5C 3'UTR is targeted by miR-509-3p via 2 binding sites. 293T cells were transfected with the 300ng of the indicated luciferase plasmids and 50nM of miR mimics, and harvested for luciferase assay 48 h after transfection. All values were first normalized to Renilla luciferase. Relative luciferase activity was then calculated by normalizing co-transfection of miR mimics plus luciferase constructs to cells transfected with only the respective luciferase construct. MiR-551b was used as a non-targeting miR negative control. Data shown represent means $\pm$ SEMs of 3 independent experiments, with statistical analysis by Student's t test. **$p < 0.01$, ***$p < 0.001$.

to cell cycling [44,45], RAB5 has been reported to play a role in other cellular pathways including autophagy [56,57] and mTOR signaling [58,59]. In humans, the RAB5 subfamily includes 3 isoforms, which may have distinct functions [60–62]. RAB5C isoform has specifically been shown to be involved in cell migration during zebrafish gastrulation [63], cell invasion via regulation of growth factor-stimulated recycling of integrin [64], and cell motility through RAC1 [65]. Protein alignment of RAB5A and RAB5B revealed 83% and 86% sequence similarity to RAB5C protein, respectively [66]. Neither miR-509-5p nor miR-509-3p is predicted to target RAB5A, and we did not detect any change in RAB5A expression in miR-509-transduced NALM6 cells (Figure S5). TargetScan6.2 predicts a miR-509-5p binding site in the 3'UTR of RAB5B (total context+ score $= -0.04$). However, using qRT-PCR, we did not detect RAB5B expression in NALM6 cells (Figure S6).

Previously, it has been shown that knockdown of all 3 RAB5 isoforms, but not knockdown of individual isoforms, in human cells resulted in defective alignment of chromosomes, delayed progression though mitosis and defective chromosome segregation [45]. While our BrdU/7-AAD analysis did not detect significantly

elevated numbers of miR-509-transduced NALM6 cells in cell cycle phase G_2/M, this could be due to expression of the compensatory isoform RAB5A. The downstream mechanism by which RAB5C regulates B-ALL cell growth remains unclear. Given that RAB5 is a key regulator of the endosome pathway, the impaired cell growth in miR-509-transduced or RAB5C-knockdown cells might be due to aberrant recycling of surface growth receptors, such as transferrin receptor [53]. In HeLa cells, knockdown of all 3 RAB5 isoforms resulted in delayed internalization of transferrin receptor and reduced uptake of transferrin [62]. Since the transferrin receptor is important in regulation of intracellular iron concentration which in turn affects cell proliferation [67,68], efforts to examine the effects of RAB5C on transferrin receptors and/or other growth-related receptors required for B-ALL growth are ongoing.

Our data indicate that RAB5C is important for B-ALL cell growth. Therefore, we might expect RAB5C to be overexpressed in B-ALL cells as compared to normal counterpart cells. Thus, we queried RAB5C expression in leukemia using the Oncomine cancer microarray database [69], comparing specifically the 'cancer versus normal' analysis with a threshold of p-value $\leq$

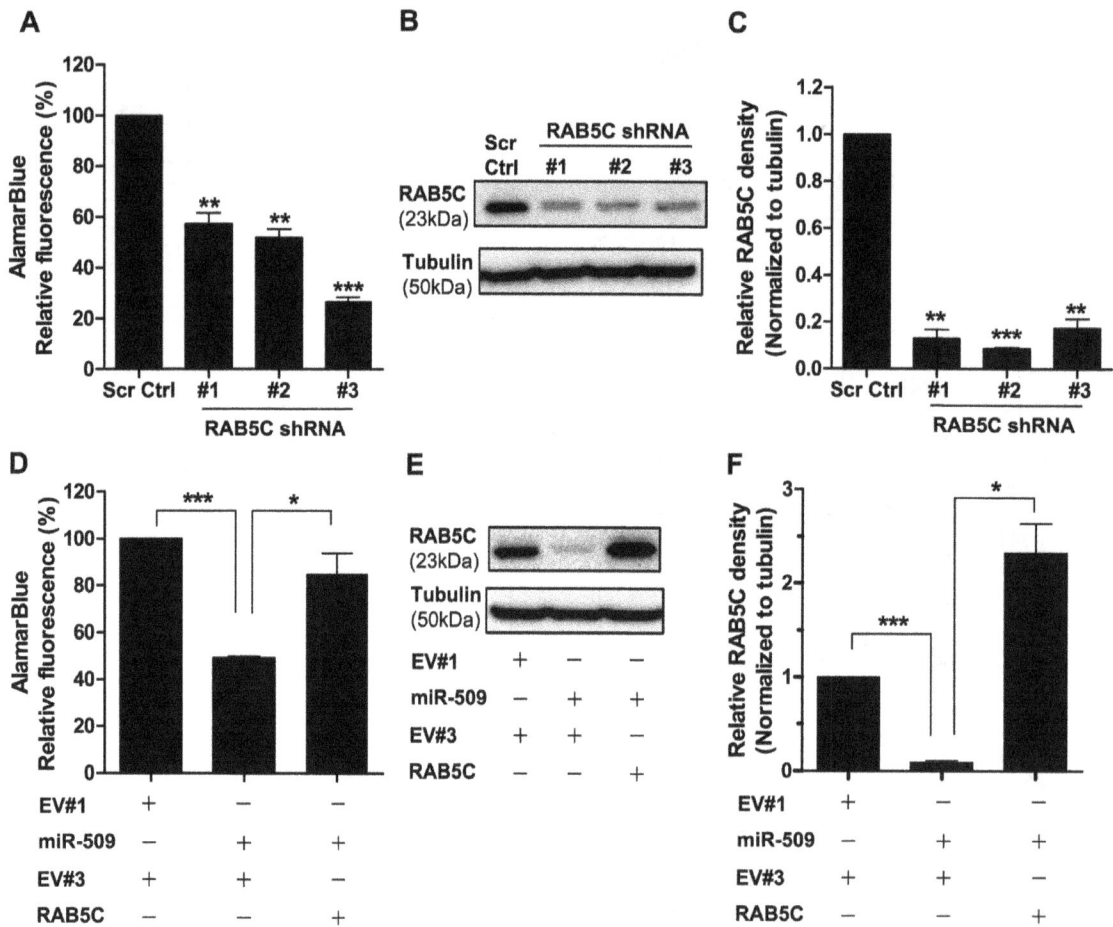

Figure 6. RAB5C mediates the growth-inhibitory effect of miR-509. (A) AlamarBlue assay of NALM6 cells transduced with either lentivirus of scrambled control (Scr Ctrl), shRNA#1, shRNA#2 or shRNA#3 for RAB5C on 7 days after transduction. Data represent means $\pm$ SEMs of 3 independent experiments with statistical analysis by Student's t test. **p<0.01 and ***p<0.001. (B) Representative western blot of NALM6 cells transduced with lentivirus of scrambled control (Scr Ctrl), shRNA#1, shRNA#2 or shRNA#3 for RAB5C. Protein lysates were harvested 7 days after transduction and α-tubulin was used as a loading control. (C) Quantitation of western blots shown in (B) and 2 other independent experiments. Relative densitometry values were calculated relative to Scr Ctrl. Results show means $\pm$ SEMs with statistical analysis by Student's t test. **p<0.01 and ***p<0.001. (D) Enforced expression of RAB5C without its 3'UTR rescues miR-509-mediated growth inhibition. NALM6 cells were co-transduced with the indicated plasmids, and alamarBlue assay was read at 7 days after transduction. The empty lentiviral vector in this experiment is EV#3, which does not have RAB5C cloned in. Results show means $\pm$ SEMs of 3 independent experiments, with statistical analysis using Student's t test. *p<0.05 and ***p<0.001. (E) Representative western blot of NALM6 cells co-transduced with lentivirus of the indicated plasmids. Protein lysates were harvested 7 days after transduction and α-tubulin was used as a loading control. (F) Quantitation of western blots shown in (E) and 2 other independent experiments. Relative densitometry values were calculated relative to EV#1 and EV#3. Results show means $\pm$ SEMs with statistical analysis by Student's t test. *p<0.05 and ***p<0.001.

0.001 and 1.5-fold over-expression. Using these criteria to assess the 14 'cancer versus normal' datasets and focusing solely on leukemia in Oncomine, RAB5C was overexpressed in the dataset of B-ALL patient samples harboring the t(12;21) chromosomal translocation (producing the TEL/AML-1 fusion protein oncogene) as compared to normal B-lymphoid precursors (pro/pre–B cells and immature B cells) from healthy donors [70]. In this TEL/AML-1 B-ALL subset, there was 1.8-fold elevated (average; Student's t test, p = 3.67^{-6}) RAB5C expression (Figure S7). The overexpression of RAB5C in this B-ALL subset, along with our findings that RAB5C supports growth of B-ALL cells, suggests that RAB5C may represent a target for treatment of the TEL/AML-1 B-ALL subset, especially if future studies reveal that growth of primary B-ALL cases harboring TEL/AML-1 is highly dependent on RAB5C. In addition, future work may include determining

whether RAB5C overexpression in hematopoietic stem cells can drive B-ALL development.

Despite using the same candidate selection criteria, we were surprised to find that the candidate validation rate (20%) of the miR-HTS conducted in NALM6 cells herein was lower than previously observed for miR-HTS conducted in the IMR90 human lung fibroblast (75% validation rate) [26]. False positives in the miR-HTS may be due to the Monte Carlo effect [71,72], where low template amounts might result in sporadic amplification at the reference time point sample but not the samples at the later time points. Consequently, such a candidate miR will not be validated. Indeed, 2 of the 4 false-positive candidate miRs in this study were detected only at the first or the first 2 time points. Neither of the other 2 false-positive candidates were detected at any of the 4 times points, including the reference time point. We designated these as candidates because we have used the same batch of lenti-miR library to conduct miR-HTS in a total of 4 cell lines (i.e. IMR90, NALM6 and 2 other cell lines) and both of these lenti-miRs were detected at the reference time point in at least one of these 4 cell lines. This suggested to us that the lenti-miR library indeed contained these 2 miR lentiviruses and should have infected the NALM6 as effectively as the other cell lines. Thus, lack of detecting cells containing these 2 lenti-miRs in NALM6 even at the reference time point might be due to a very strong growth-inhibitory effect of these 2 miR candidates. Hence, these 2 miR candidates were included in our validation analyses. However, it is possible that the lenti-miR viruses for these 2 false-positive miR candidates did not actually infect NALM6 cells.

In summary, our findings demonstrate the ability of our miR-HTS platform to identify leukemia cell growth inhibitory miRs and their molecular targets. Our observation that enforced miR-509 expression inhibits growth of B-ALL cell lines provided the clue to identifying the role of RAB5C in the growth of B-ALL cells. To our knowledge, this is the first report of RAB5C as a regulator of B-ALL cell growth. Elucidating the downstream mechanistic roles of RAB5C in growth of human B-ALL cells might suggest novel therapeutic strategies against B-ALL.

Supporting Information

Figure S1 No growth defects were observed for 4 other miR candidates using the GFP competition assay. NALM6 cells were individually transduced with lentivirus of (A) miR-381; (B) miR-550a; (C) miR-873 and (D) miR-432~136 and empty vector (EV#1) to MOI = 2. At 7 days after transduction, cells were mixed with mock-transduced cells to 50% GFP$^+$ cells and this was set as Day 0. The %GFP$^+$ cells (pre-gated on viable cells) of each culture were assessed weekly by flow cytometry for 35 days. Means ± SEMs are shown for three independent experiments. (E) Overexpression of miR candidates in NALM6 cells, as assayed by qRT-PCR. NALM6 cells were transduced with each miR lentivirus to MOI = 2, and total RNA was collected at 7 days after transduction. U18 was used as the loading control. Values shown were calculated as fold overexpression relative to EV#1-transduced NALM6 cells (EV#1). Means ± SEMs are shown for 3 independent experiments.

Figure S2 MiR-509 does not regulate the growth of Jurkat, KARPAS-45 and K562 cells. AlamarBlue cell growth assay was then performed on day 7 after transduction of (A) Jurkat, (B) KARPAS-45 and (C) K562 cells with either miR-509 lentivirus or EV#1. Each cell line was transduced with the indicated lentivirus to MOI = 2. On day 3 after transduction, cells were seeded at the indicated numbers per well/100 μl media: Jurkat (5×10^3 cells), KARPAS-45 (3×10^3 cells) and K562 (1.25×10^3 cells) in triplicates in 96-well plates. Values for miR-509 were normalized to EV#1. Means ± SEMs, ns = no statistical significance was detected by Student's t test.

Figure S3 Enforced expression of miR-509 was detected by qRT-PCR in selected T-ALL and myeloid leukemia cell lines transduced with miR-509 lentivirus. (A) Jurkat, (B) KARPAS-45 and (C) K562 cells were transduced with miR-509 lentivirus or EV#1. On day 7 after transduction, cells were collected for RNA isolation. U18 was used as the endogenous control. Values shown were calculated as fold overexpression relative to each EV#1-transduced cells. Means ± SEMs are shown for 3 independent experiments.

Figure S4 RAB5C protein levels were decreased in RCH-ACV and REH cells with enforced miR-509 expression. Representative western blot of RAB5C expression in (A) RCH-ACV and (B) REH. Cells were transduced with either EV#1 or miR-509 overexpressing lentivirus, and whole cell lysates were harvested at 7 days after transduction. α-tubulin was used for loading control. The bar graph below represents the densitometry analysis of RAB5C expression of 3 independent experiments, normalized to α-tubulin, and relative densitometry was then calculated compared to EV#1. Data shown represent means ± SEMs, with statistical analysis by Student's t test. **p<0.01, ***p<0.001.

Figure S5 RAB5A mRNA levels show no change in miR-509-transduced NALM6 cells. NALM6 cells were transduced with empty vector #1 (EV#1) to MOI = 2, and RNA was isolated at day 7 after transduction for qRT-PCR. All values were normalized to GAPDH and fold-change was calculated relative to EV#1 sample. Data represents means ± SEMs of 3 independent experiments, with statistical analysis by Student's t test, ns = no statistical significance was detected by Student's t test.

Figure S6 Expression of RAB5A and RAB5C, but not RAB5B, was detected in NALM6 cells using qRT-PCR. NALM6 cells were transduced with empty vector #1 (EV#1) to MOI = 2, and RNA was isolated at day 7 after transduction for qRT-PCR. Ratio to GAPDH (endogenous control) was calculated as 2E[-(RAB5$_{Ct}$ – GAPDH$_{Ct}$)]. Means ± SEMs, n = 3 independent experiments. Value above each bar represents the mean.

Figure S7 Scatter dot plot of RAB5C mRNA expression in B-ALL cells and pre-B lymphocytes based on Oncomine cancer microarray database. RAB5C expression in leukemia was examined using the Oncomine cancer microarray database by comparing specifically the 'cancer versus normal' analysis and setting a threshold of p-value ≤0.001 and 1.5-fold over-expression. 14 'cancer versus normal' datasets were identified and we focused solely on leukemia in Oncomine. RAB5C was overexpressed by 1.8-fold (average; Student's t test, $p = 3.67^{-6}$) in the dataset of B-ALL patient samples harboring the t(12;21) chromosomal translocation (producing the TEL/AML-1 fusion protein oncogene; n = 17) as compared to normal B-lymphoid precursors (pro/pre–B cells and immature B cells; n = 2) from healthy donors [67]. Error bars represent the mean ± SEM.

Table S1 List of primers used for cloning of miR hairpin with flanking genomic sequences. PCR products were first cloned into pJET1.2 and subcloned into empty lentiviral vector #1 (EV#1; pWCC52) downstream of GFP. MiR-509 was then subcloned from pWCC52-miR-509 into empty lentiviral vector #2 (EV#2; pWCC72) downstream of DsRed.

Table S2 Primers used for PCR of RAB5C-3′UTR and deletion of miR-509-3p binding sites. Full length RAB5C-3′UTR was cloned into pmirGLO Dual-Luciferase miRNA Target Expression vector (Promega). This plasmid was then used as a template for site-directed mutagenesis to delete the first miR-509-3p binding sites in RAB5C-3′UTR-luciferase deletion construct, Δ1 or Δ1Δ2 using primers Del56-72. For the deletion of the second miR-509-3p binding site in RAB5C-3′UTR-luciferase deletion construct, Δ2 or Δ1Δ2, standard PCR was performed using the Del758-767 primers.

Table S3   Primers used in cloning of RAB5C lacking its 3′UTR into pWCC61 lentiviral vector (Empty lentiviral vector #3, EV#3).

Table S4   List of primers used for SYBRGreen qRT-PCR. Primer sequences were obtained from PrimerBank. Fwd: Forward; Rev: Reverse.

Table S5   List of TaqMan microRNA assay ID used for qRT-PCR.

Table S6   List of the 395 predicted targets of miR-509-5p and/or miR-509-3p selected based on filtering strategy shown in Figure 4A. These targets were subjected to a filtering strategy presented in Fig. 4A and meet the following criteria: (i) They are predicted targets of miR-509-5p and/or miR-509-3p from TargetScan6.2 and/or miRDB. (ii) These targets are not

targets of miR-381, miR-550a, miR-873 and miR-432 as predicted by TargetScan6.2 and/or miRDB. (iii) These targets are expressed in NALM6 cells as determined by genome-wide microarray profiling downloaded from the Cancer Cell Line Encyclopedia and its expression levels are denoted in the microarray dataset as "marginal" or "present".

Acknowledgments

We would like to thank all members of the Civin lab and Dr. Marta Lipinski for their helpful suggestions.

Author Contributions

Conceived and designed the experiments: YST WCC. Performed the experiments: YST MK TJK WCC. Analyzed the data: YST MK CIC WCC. Wrote the paper: YST TJK CIC WCC.

References

1. Bassan R, Hoelzer D (2011) Modern therapy of acute lymphoblastic leukemia. J Clin Oncol 29: 532–543. 10.1200/JCO.2010.30.1382.
2. Hunger SP, Raetz EA, Loh ML, Mullighan CG (2011) Improving outcomes for high-risk ALL: Translating new discoveries into clinical care. Pediatr Blood Cancer 56: 984–993. 10.1002/pbc.22996.
3. Raetz EA, Bhatla T (2012) Where do we stand in the treatment of relapsed acute lymphoblastic leukemia? Hematology Am Soc Hematol Educ Program 2012: 129–136. 10.1182/asheducation-2012.1.129.
4. Bartel DP (2009) MicroRNAs: Target recognition and regulatory functions. Cell 136: 215–233. 10.1016/j.cell.2009.01.002.
5. Pasquinelli AE (2012) MicroRNAs and their targets: Recognition, regulation and an emerging reciprocal relationship. Nat Rev Genet 13: 271–282. 10.1038/nrg3162.
6. Ameres SL, Zamore PD (2013) Diversifying microRNA sequence and function. Nat Rev Mol Cell Biol 14: 475–488. 10.1038/nrm3611.
7. Croce CM (2009) Causes and consequences of microRNA dysregulation in cancer. Nat Rev Genet 10: 704–714. 10.1038/nrg2634.
8. Faraoni I, Antonetti FR, Cardone J, Bonmassar E (2009) miR-155 gene: A typical multifunctional microRNA. Biochim Biophys Acta 1792: 497–505. 10.1016/j.bbadis.2009.02.013.
9. O'Connell RM, Rao DS, Chaudhuri AA, Boldin MP, Taganov KD, et al. (2008) Sustained expression of microRNA-155 in hematopoietic stem cells causes a myeloproliferative disorder. J Exp Med 205: 585–594. 10.1084/jem.20072108.
10. Costinean S, Sandhu SK, Pedersen IM, Tili E, Trotta R, et al. (2009) Src homology 2 domain-containing inositol-5-phosphatase and CCAAT enhancer-binding protein beta are targeted by miR-155 in B cells of emicro-MiR-155 transgenic mice. Blood 114: 1374–1382. 10.1182/blood-2009-05-220814.
11. Hermeking H (2010) The miR-34 family in cancer and apoptosis. Cell Death Differ 17: 193–199. 10.1038/cdd.2009.56.
12. Lal A, Thomas MP, Altschuler G, Navarro F, O'Day E, et al. (2011) Capture of microRNA-bound mRNAs identifies the tumor suppressor miR-34a as a regulator of growth factor signaling. PLoS Genet 7: e1002363. 10.1371/journal.pgen.1002363.
13. Garzon R, Volinia S, Liu CG, Fernandez-Cymering C, Palumbo T, et al. (2008) MicroRNA signatures associated with cytogenetics and prognosis in acute myeloid leukemia. Blood 111: 3183–3189. 10.1182/blood-2007-07-098749.
14. Dixon-McIver A, East P, Mein CA, Cazier JB, Molloy G, et al. (2008) Distinctive patterns of microRNA expression associated with karyotype in acute myeloid leukaemia. PLoS One 3: e2141. 10.1371/journal.pone.0002141.
15. Schotte D, De Menezes RX, Akbari Moqadam F, Khankahdani LM, Lange-Turenhout E, et al. (2011) MicroRNA characterize genetic diversity and drug resistance in pediatric acute lymphoblastic leukemia. Haematologica 96: 703–711. 10.3324/haematol.2010.026138.
16. Ju X, Li D, Shi Q, Hou H, Sun N, et al. (2009) Differential microRNA expression in childhood B-cell precursor acute lymphoblastic leukemia. Pediatr Hematol Oncol 26: 1–10. 10.1080/08880010802378338.
17. Schotte D, Pieters R, Den Boer ML (2012) MicroRNAs in acute leukemia: From biological players to clinical contributors. Leukemia 26: 1–12. 10.1038/leu.2011.151.
18. Bhatia S, Kaul D, Varma N (2010) Potential tumor suppressive function of miR-196b in B-cell lineage acute lymphoblastic leukemia. Mol Cell Biochem 340: 97–106. 10.1007/s11010-010-0406-9.
19. Agirre X, Vilas-Zornoza A, Jimenez-Velasco A, Martin-Subero JI, Cordeu L, et al. (2009) Epigenetic silencing of the tumor suppressor microRNA hsa-miR-124a regulates CDK6 expression and confers a poor prognosis in acute lymphoblastic leukemia. Cancer Res 69: 4443–4453. 10.1158/0008-5472.CAN-08-4025.
20. Dou L, Zheng D, Li J, Li Y, Gao L, et al. (2012) Methylation-mediated repression of microRNA-143 enhances MLL-AF4 oncogene expression. Oncogene 31: 507–517. 10.1038/onc.2011.248.
21. Izumiya M, Tsuchiya N, Okamoto K, Nakagama H (2011) Systematic exploration of cancer-associated microRNA through functional screening assays. Cancer Sci 102: 1615–1621. 10.1111/j.1349-7006.2011.02007.x.
22. Poell JB, van Haastert RJ, de Gunst T, Schultz IJ, Gommans WM, et al. (2012) A functional screen identifies specific microRNAs capable of inhibiting human melanoma cell viability. PLoS One 7: e43569. 10.1371/journal.pone.0043569.
23. Izumiya M, Okamoto K, Tsuchiya N, Nakagama H (2010) Functional screening using a microRNA virus library and microarrays: A new high-throughput assay to identify tumor-suppressive microRNAs. Carcinogenesis 31: 1354–1359. 10.1093/carcin/bgq112.
24. Tsuchiya N, Izumiya M, Ogata-Kawata H, Okamoto K, Fujiwara Y, et al. (2011) Tumor suppressor miR-22 determines p53-dependent cellular fate through post-transcriptional regulation of p21. Cancer Res 71: 4628–4639. 10.1158/0008-5472.CAN-10-2475.
25. Okamoto K, Ishiguro T, Midorikawa Y, Ohata H, Izumiya M, et al. (2012) miR-493 induction during carcinogenesis blocks metastatic settlement of colon cancer cells in liver. EMBO J 31: 1752–1763. 10.1038/emboj.2012.25.
26. Cheng WC, Kingsbury TJ, Wheelan SJ, Civin CI (2013) A simple high-throughput technology enables gain-of-function screening of human micro-RNAs. BioTechniques 54: 77–86. 10.2144/000113991.
27. Amendola M, Passerini L, Pucci F, Gentner B, Bacchetta R, et al. (2009) Regulated and multiple miRNA and siRNA delivery into primary cells by a lentiviral platform. Mol Ther 17: 1039–1052. 10.1038/mt.2009.48.
28. Sarbassov DD, Guertin DA, Ali SM, Sabatini DM (2005) Phosphorylation and regulation of akt/PKB by the rictor-mTOR complex. Science 307: 1098–1101. 10.1126/science.1106148.
29. Wang X, Spandidos A, Wang H, Seed B (2012) PrimerBank: A PCR primer database for quantitative gene expression analysis, 2012 update. Nucleic Acids Res 40: D1144–9. 10.1093/nar/gkr1013.
30. Chen C, Ridzon DA, Broomer AJ, Zhou Z, Lee DH, et al. (2005) Real-time quantification of microRNAs by stem-loop RT-PCR. Nucleic Acids Res 33: e179. 10.1093/nar/gni178.
31. Heiser D, Tan YS, Kaplan I, Godsey B, Morisot S, et al. (2014) Correlated miR-mRNA expression signatures of mouse hematopoietic stem and progenitor cell subsets predict "stemness" and "myeloid" interaction networks. PLoS One 9: e94852. 10.1371/journal.pone.0094852 [doi].
32. Candia J, Cherukuri S, Doshi KA, Banavar JR, Civin CI, et al. (2014) Uncovering differential multi-microRNA signatures of acute myeloid and lymphoblastic leukemias with a machine-learning-based network approach. In Press.
33. Mavrakis KJ, Wolfe AL, Oricchio E, Palomero T, de Keersmaecker K, et al. (2010) Genome-wide RNA-mediated interference screen identifies miR-19 targets in notch-induced T-cell acute lymphoblastic leukaemia. Nat Cell Biol 12: 372–379. 10.1038/ncb2037.
34. Eekels JJ, Pasternak AO, Schut AM, Geerts D, Jeeninga RE, et al. (2012) A competitive cell growth assay for the detection of subtle effects of gene transduction on cell proliferation. Gene Ther 19: 1058–1064. 10.1038/gt.2011.191.
35. Rampersad SN (2012) Multiple applications of alamar blue as an indicator of metabolic function and cellular health in cell viability bioassays. Sensors (Basel) 12: 12347–12360. 10.3390/s120912347.
36. Kaplan IM, Morisot S, Heiser D, Cheng WC, Kim MJ, et al. (2011) Deletion of tristetraprolin caused spontaneous reactive granulopoiesis by a non-cell-

autonomous mechanism without disturbing long-term hematopoietic stem cell quiescence. J Immunol 186: 2826–2834. 10.4049/jimmunol.1002806.

37. Grimson A, Farh KK, Johnston WK, Garrett-Engele P, Lim LP, et al. (2007) MicroRNA targeting specificity in mammals: Determinants beyond seed pairing. Mol Cell 27: 91–105. 10.1016/j.molcel.2007.06.017.

38. Wang X (2008) miRDB: A microRNA target prediction and functional annotation database with a wiki interface. RNA 14: 1012–1017. 10.1261/rna.965408.

39. Wang X, El Naqa IM (2008) Prediction of both conserved and nonconserved microRNA targets in animals. Bioinformatics 24: 325–332. 10.1093/bioinformatics/btm595.

40. Barretina J, Caponigro G, Stransky N, Venkatesan K, Margolin AA, et al. (2012) The cancer cell line encyclopedia enables predictive modelling of anticancer drug sensitivity. Nature 483: 603–607. 10.1038/nature11003.

41. Hulsen T, de Vlieg J, Alkema W (2008) BioVenn - a web application for the comparison and visualization of biological lists using area-proportional venn diagrams. BMC Genomics 9: 488-2164-9-488. 10.1186/1471-2164-9-488.

42. Huang da W, Sherman BT, Lempicki RA (2009) Systematic and integrative analysis of large gene lists using DAVID bioinformatics resources. Nat Protoc 4: 44–57. 10.1038/nprot.2008.211.

43. Huang da W, Sherman BT, Lempicki RA (2009) Bioinformatics enrichment tools: Paths toward the comprehensive functional analysis of large gene lists. Nucleic Acids Res 37: 1–13. 10.1093/nar/gkn923.

44. Capalbo L, D'Avino PP, Archambault V, Glover DM (2011) Rab5 GTPase controls chromosome alignment through lamin disassembly and relocation of the NuMA-like protein mud to the poles during mitosis. Proc Natl Acad Sci U S A 108: 17343–17348. 10.1073/pnas.1103720108.

45. Serio G, Margaria V, Jensen S, Oldani A, Bartek J, et al. (2011) Small GTPase Rab5 participates in chromosome congression and regulates localization of the centromere-associated protein CENP-F to kinetochores. Proc Natl Acad Sci U S A 108: 17337–17342. 10.1073/pnas.1103516108.

46. Calin GA, Croce CM (2006) MicroRNA signatures in human cancers. Nat Rev Cancer 6: 857–866. 10.1038/nrc1997.

47. Zhang WB, Pan ZQ, Yang QS, Zheng XM (2013) Tumor suppressive miR-509-5p contributes to cell migration, proliferation and antiapoptosis in renal cell carcinoma. Ir J Med Sci 182: 621–627. 10.1007/s11845-013-0941-y.

48. Zhai Q, Zhou L, Zhao C, Wan J, Yu Z, et al. (2012) Identification of miR-508-3p and miR-509-3p that are associated with cell invasion and migration and involved in the apoptosis of renal cell carcinoma. Biochem Biophys Res Commun 419: 621–626. 10.1016/j.bbrc.2012.02.060.

49. Thomas M, Lieberman J, Lal A (2010) Desperately seeking microRNA targets. Nat Struct Mol Biol 17: 1169–1174. 10.1038/nsmb.1921.

50. Thomson DW, Bracken CP, Goodall GJ (2011) Experimental strategies for microRNA target identification. Nucleic Acids Res 39: 6845–6853. 10.1093/nar/gkr330 [doi].

51. Selbach M, Schwanhausser B, Thierfelder N, Fang Z, Khanin R, et al. (2008) Widespread changes in protein synthesis induced by microRNAs. Nature 455: 58–63. 10.1038/nature07228.

52. Guo H, Ingolia NT, Weissman JS, Bartel DP (2010) Mammalian microRNAs predominantly act to decrease target mRNA levels. Nature 466: 835–840. 10.1038/nature09267.

53. Bucci C, Parton RG, Mather IH, Stunnenberg H, Simons K, et al. (1992) The small GTPase rab5 functions as a regulatory factor in the early endocytic pathway. Cell 70: 715–728.

54. Zerial M, McBride H (2001) Rab proteins as membrane organizers. Nat Rev Mol Cell Biol 2: 107–117. 10.1038/35052055.

55. Zeigerer A, Gilleron J, Bogorad RL, Marsico G, Nonaka H, et al. (2012) Rab5 is necessary for the biogenesis of the endolysosomal system in vivo. Nature 485: 465–470. 10.1038/nature11133.

56. Ravikumar B, Imarisio S, Sarkar S, O'Kane CJ, Rubinsztein DC (2008) Rab5 modulates aggregation and toxicity of mutant huntingtin through macroautophagy in cell and fly models of huntington disease. J Cell Sci 121: 1649–1660. 10.1242/jcs.025726.

57. Dou Z, Pan JA, Dbouk HA, Ballou LM, DeLeon JL, et al. (2013) Class IA PI3K p110beta subunit promotes autophagy through Rab5 small GTPase in response to growth factor limitation. Mol Cell 50: 29–42. 10.1016/j.molcel.2013.01.022.

58. Li L, Kim E, Yuan H, Inoki K, Goraksha-Hicks P, et al. (2010) Regulation of mTORC1 by the rab and arf GTPases. J Biol Chem 285: 19705–19709. 10.1074/jbc.C110.102483.

59. Bridges D, Fisher K, Zolov SN, Xiong T, Inoki K, et al. (2012) Rab5 proteins regulate activation and localization of target of rapamycin complex 1. J Biol Chem 287: 20913–20921. 10.1074/jbc.M111.334060.

60. Bucci C, Lutcke A, Steele-Mortimer O, Olkkonen VM, Dupree P, et al. (1995) Co-operative regulation of endocytosis by three Rab5 isoforms. FEBS Lett 366: 65–71.

61. Wainszelbaum MJ, Proctor BM, Pontow SE, Stahl PD, Barbieri MA (2006) IL4/PGE2 induction of an enlarged early endosomal compartment in mouse macrophages is Rab5-dependent. Exp Cell Res 312: 2238–2251. 10.1016/j.yexcr.2006.03.025.

62. Chen PI, Kong C, Su X, Stahl PD (2009) Rab5 isoforms differentially regulate the trafficking and degradation of epidermal growth factor receptors. J Biol Chem 284: 30328–30338. 10.1074/jbc.M109.034546.

63. Ulrich F, Krieg M, Schotz EM, Link V, Castanon I, et al. (2005) Wnt11 functions in gastrulation by controlling cell cohesion through Rab5c and E-cadherin. Dev Cell 9: 555–564. 10.1016/j.devcel.2005.08.011.

64. Onodera Y, Nam JM, Hashimoto A, Norman JC, Shirato H, et al. (2012) Rab5c promotes AMAP1-PRKD2 complex formation to enhance beta1 integrin recycling in EGF-induced cancer invasion. J Cell Biol 197: 983–996. 10.1083/jcb.201201065.

65. Chen PI, Schauer K, Kong C, Harding AR, Goud B, et al. (2014) Rab5 isoforms orchestrate a "division of labor" in the endocytic network; Rab5C modulates rac-mediated cell motility. PLoS One 9: e90384. 10.1371/journal.pone.0090384; 10.1371/journal.pone.0090384.

66. Sievers F, Wilm A, Dineen D, Gibson TJ, Karplus K, et al. (2011) Fast, scalable generation of high-quality protein multiple sequence alignments using clustal omega. Mol Syst Biol 7: 539. 10.1038/msb.2011.75.

67. May WS Jr, Cuatrecasas P (1985) Transferrin receptor: Its biological significance. J Membr Biol 88: 205–215.

68. O'Donnell KA, Yu D, Zeller KI, Kim JW, Racke F, et al. (2006) Activation of transferrin receptor 1 by c-myc enhances cellular proliferation and tumorigenesis. Mol Cell Biol 26: 2373–2386. 26/6/2373 [pii].

69. Rhodes DR, Yu J, Shanker K, Deshpande N, Varambally R, et al. (2004) ONCOMINE: A cancer microarray database and integrated data-mining platform. Neoplasia 6: 1–6.

70. Maia S, Haining WN, Ansen S, Xia Z, Armstrong SA, et al. (2005) Gene expression profiling identifies BAX-delta as a novel tumor antigen in acute lymphoblastic leukemia. Cancer Res 65: 10050–10058. 10.1158/0008-5472.CAN-05-1574.

71. Karrer EE, Lincoln JE, Hogenhout S, Bennett AB, Bostock RM, et al. (1995) In situ isolation of mRNA from individual plant cells: Creation of cell-specific cDNA libraries. Proc Natl Acad Sci U S A 92: 3814–3818.

72. Bustin SA, Nolan T (2004) Pitfalls of quantitative real-time reverse-transcription polymerase chain reaction. J Biomol Tech 15: 155–166. 15/3/155 [pii].

16

Src Mutation Induces Acquired Lapatinib Resistance in *ERBB2*-Amplified Human Gastroesophageal Adenocarcinoma Models

Yong Sang Hong[1,2], Jihun Kim[1,3], Eirini Pectasides[1,4], Cameron Fox[1], Seung-Woo Hong[5], Qiuping Ma[1], Gabrielle S. Wong[1], Shouyong Peng[1,6], Matthew D. Stachler[1,7], Aaron R. Thorner[8], Paul Van Hummelen[8], Adam J. Bass[1,6,9,10]*

1 Department of Medical Oncology, Dana-Farber Cancer Institute, Boston, Massachusetts, United States of America, 2 Department of Oncology, Asan Medical Center, University of Ulsan College of Medicine, Seoul, Korea, 3 Department of Pathology, Asan Medical Center, University of Ulsan College of Medicine, Seoul, Korea, 4 Division of Hematology/Oncology, Beth Israel Deaconess Medical Center, Boston, Massachusetts, United States of America, 5 Innovative Cancer Research, Asan Institute for Life Science, Asan Medical Center, University of Ulsan College of Medicine, Seoul, Korea, 6 Cancer Program, The Broad Institute of MIT and Harvard, Cambridge, Massachusetts, United States of America, 7 Department of Pathology, Brigham and Women's Hospital, Boston, Massachusetts, United States of America, 8 Center for Cancer Genome Discovery, Dana-Farber Cancer Institute, Boston, Massachusetts, United States of America, 9 Department of Medicine, Harvard Medical School, Boston, Massachusetts, United States of America, 10 Department of Medicine, Brigham and Women's Hospital, Boston, Massachusetts, United States of America

Abstract

ERBB2-directed therapy is now a routine component of therapy for *ERBB2*-amplified metastatic gastroesophageal adenocarcinomas. However, there is little knowledge of the mechanisms by which these tumors develop acquired resistance to ERBB2 inhibition. To investigate this question we sought to characterize cell line models of *ERBB2*-amplified gastroesophageal adenocarcinoma with acquired resistance to ERBB2 inhibition. We generated lapatinib-resistant (LR) subclones from an initially lapatinib-sensitive *ERBB2*-amplified esophageal adenocarcinoma cell line, OE19. We subsequently performed genomic characterization and functional analyses of resistant subclones with acquired lapatinib resistance. We identified a novel, acquired Src^{E527K} mutation in a subset of LR OE19 subclones. Cells with this mutant allele harbour increased Src phosphorylation. Genetic and pharmacologic inhibition of Src resensitized these subclones to lapatinib. Biochemically, *Src* mutations could activate both the phosphatidylinositol 3-kinase and mitogen activated protein kinase pathways in the lapatinib-treated LR OE19 cells. Ectopic expression of Src^{E527K} mutation also was sufficient to induce lapatinib resistance in drug-naïve cells. These results indicate that pathologic activation of Src is a potential mechanism of acquired resistance to ERBB2 inhibition in *ERBB2*-amplified gastroesophageal cancer. Although *Src* mutation has not been described in primary tumor samples, we propose that the Src hyperactivation should be investigated in the settings of acquired resistance to ERBB2 inhibition in esophageal and gastric adenocarcinoma.

Editor: Guillermo Velasco, Complutense University, Spain

Funding: This work was supported by The Funderburg Research Award in Gastric Biology Related to Cancer from the American Gastroenterological Association, The Anna Fuller Fund, Phi Beta Psi, a Research Scholar Grant from the American Cancer Society and P01 CA098101 from the National Institute of Health, all to A.J.B. The funders had no role in study design, data collection and analysis, decision to publish, or preparation of the manuscript.

Competing Interests: The authors have declared that no competing interests exist.

* Email: adam_bass@dfci.harvard.edu

Introduction

Gastroesophageal (GE) adenocarcinomas are one of the leading causes of the cancer death worldwide [1]. The mainstay of systemic chemotherapy for patients with advanced or metastatic disease still consists of cytotoxic agents including fluoropyrimidines, platinum derivatives, taxanes and topoisomerase inhibitors [2–5]. However, a recent randomised trial demonstrated that trastuzumab, a humanized IgG1 monoclonal antibody targeting human epidermal growth factor receptor-2 (ERBB2 or HER2), improved overall survival by 2.7 months in patients with *ERBB2*-amplified advanced gastroesophageal cancer when combined with chemotherapy [6]. Based upon these results, trastuzumab is now a routine component of care for patients with metastatic *ERBB2*-amplified GE adenocarcinomas.

Despite the adoption of ERBB2 inhibitor therapy in clinical practice, the addition of anti-HER2 targeting strategies in patients with *ERBB2* amplified gastroesophageal cancer have been modest, attributable both to intrinsic resistance of many tumors to trastuzumab containing therapy as well as to the emergence of acquired resistance in those tumors which initially responded to treatment. The etiology of resistance to ERBB2-directed therapies has been widely investigated in breast cancer [7–15]; the accepted resistance mechanisms included constitutive activation of the PI3-K pathway [7,9], truncated p95 isoform of HER2 receptor which cannot bind to trastuzumab [15], and constitutive Src activation as a common node downstream of multiple pathways [10]. In GE cancer, the variant of dopamine and cyclic AMP-regulated phosphoprotein (t-DARPP) has been suggested as a resistance mechanism to ERBB2 inhibitors [16], and exogenous HGF administration to cell line cultures has been shown to induce *in vitro* resistance in GE adenocarcinoma [17]. In one report, NCI-N87 *ERBB2*-amplified gastric adenocarcinoma cells were found to acquire enhanced activity of Src activity following prolonged in vitro exposure to trastuzumab [18]. Beyond these reports, there is little understanding the etiology of acquired or *de novo* resistance to ERBB2 inhibition in GE cancer.

To address this problem, various clinical efforts are evaluating the empiric potential of distinct second-line agents to improve survival and clinical responses [19–22]. To guide the development of such treatment strategies, we sought to investigate the potential mechanisms of acquired resistance to ERBB2 inhibition in *ERBB2*-amplified GE adenocarcinoma cell line models. Indeed, in other tumor types, study of means of resistance in cell line models has identified resistance mechanisms subsequently validated in primary cancers [9,10,14]. Although trastuzumab is utilized in clinical practice, trastuzumab has limited efficacy in *in vitro* culture compared to direct kinase inhibitors such as lapatinib [23,24]. Therefore, we have chosen lapatinib as our tool compound to identify mechanisms by which *ERBB2*-amplified GE adenocarcinomas can bypass effective ERBB2 inhibition.

From the originally lapatinib-sensitive *ERBB2*-amplified esophageal adenocarcinoma cell, OE19, we generated several resistant subclones by prolonged exposure to lapatinib. Through genomic and functional analyses of this lapatinib-resistant model, we found that an activating mutation of *Src* was responsible for the acquired lapatinib resistance in OE19 cells. In addition, we further demonstrated that genetic or pharmacologic blockade of Src could restore ERBB2 inhibitor sensitivity in lapatinib-resistant cells. These data establish the role of oncogene *Src* as a pharmacologically tractable candidate mediator of acquired lapatinib resistance in ERBB2-positive GE adenocarcinomas.

Materials and Methods

Cell lines and Reagents

OE19 cells were obtained from the European Collection of Cell Cultures (ECACC), and OE33 cells were purchased from the Sigma (St. Louis, MO). OE19 is *ERBB2*-amplified gastroesophageal cancer cell line and sensitive to lapatinib, hence OE33 is *ERBB2-* and *MET*-amplified gastroesophageal cancer cell line and has intrinsic resistance to lapatinib. OE19 and OE33 were cultured in a humidified, 5% CO_2 atmosphere at 37°C in Roswell Park Memorial Institute (RPMI-1640; GIBCO BRL, Grand Island, NY) medium supplemented with 10% fetal bovine serum. Tyrosine kinase inhibitors (lapatinib, saracatinib, and crizotinib) were purchased from the LC laboratories and were dissolved in dimethylsulfoxide (DMSO). Trastuzumab was purchased from the Department of Pharmacy at the Dana-Farber Cancer Institute.

Generation of Lapatinib-Resistant (LR) Subclones

Working with lapatinib-sensitive *ERBB2*-amplified esophageal adenocarcinoma cell line, OE19, whose IC_{50} to lapatinib is 200 nM, we initiated the development of lapatinib resistance by culturing the cell line in the presence of progressively increasing doses of lapatinib during three months; the final concentration of lapatinib was 3 μM and some clones survived at a low density with small colonies. Following six months of culture with drug, we were able to obtain OE19 derivatives that were capable of proliferation in the presence of 3 μM of lapatinib. We could observe that

several colonies had distinct cellular morphologies at this time, and the pathologist (JK) picked some colonies with distinct morphologies and named them according to the selecting orders. We subsequently expanded distinct clonal subcultures from the resistant OE19 cells and subsequently extracted DNA from seven distinct clonal populations for genomic analysis.

Genomic DNA extraction and Targeted Exome sequencing

Genomic DNA was extracted from the parental OE19 and isolated LR subclones using DNeasy Blood & Tissue Kit (Qiagen) per manufacturer's protocol. DNA quality was evaluated by quantification using Quant-iT Pico Green dsDNAassay Kit (Invitrogen) per manufacturer's protocol. DNA from these resistance cell lines (and the parental OE19) were subjected to focused exon sequencing using the Oncopanel_v2 cancer gene panel at the Center for Cancer Genome Discovery at the Dana-Farber Cancer Institute. OncoPanel_v2 represents a targeted sequencing strategy to simultaneously detect mutations, translocations and copy-number variations in archived clinical tumor specimens. Targeted sequencing was achieved by designing RNA baits to capture the exons of 504 genes with relevance to cancer.

Sequencing libraries were prepared, as previously described [25], starting from 100 ng of genomic DNA. Libraries were quantified by QPCR (Kapa Biosystems, Inc., Woburn, MA) and pooled in equimolar concentrations to 500 ng total and enriched for the Oncopanel_v2 baitset using the Agilent SureSelect hybrid capture kit. The enriched targeted exon libraries were again quantified by QPCR (Kapa Biosystems, Inc., Woburn, MA) subsequently sequenced in one lane of a Hiseq2000 sequencer (Illumina Inc., San Diego, CA) in a 2× 100 bp pair-end mode. Sequence alignment, demultiplexing and variant calling, including SNV and Indels, was performed using PICARD, GATK tools, Mutect and IndeLocator as previously described [25]. Sequence results from the resistance subclones were compared to the genomic results from the parental OE19 cell line in order to identify putative somatic mutations and copy-number aberrations that are the potential etiology of resistance. Only candidate somatic alterations with mutant allele fractions >5% were considered.

Direct DNA Sequencing Analysis

The *Src* [E527K] mutation (g1579a) was additionally validated by direct sequencing as follows; *Src* was PCR-amplified from genomic DNA using a 2720 thermal cycler (Applied Biosystems) using OneTaq Quick-Load 2X mix (BioLabs) with each primer (forward: 5'- GGGATGGTGAACCGCGAGGT-3', reverse: 5'-TTCTCCCCGGGCTGGT-3'). DNA electrophoresis was performed and the pure amplified PCR product with 203 bp size was isolated using QIAquick Gel Extraction Kit (Qiagen) per manufacturer's protocol. We performed TA cloning with purified PCR products using TOPO TA Cloning Kit, bacterial transformation and propagation with competent *E.coli*, and extracted plasmid which contained sequence target using QIAprep Miniprep Kit (Qiagen). Direct sequencing was using the M13R sequencing primer at Genewiz, Inc.

DNA Restriction Analysis

We performed DNA restriction analysis to confirm that the *Src* [E527K] mutation is an acquired event owing to the prolonged lapatinib exposure, and it is not already present in the parental OE19 cells. Genomic DNA were extracted from the parental OE19 and two LR subclones harbouring *Src* mutant, and PCR-amplified as described above. Genomic DNA from the Het1A cells, a non-tumor esophageal cell line, was also tested as a negative control.

Aliquots from the PCR amplicons were digested separately with *Ban* II restriction enzyme (New England Biolabs Inc., Beverly, USA) at 37°C for 4 hours. Digestions were carried out per manufacturer's protocol.

DNA restriction fragments were electrophoresed on 2.0% agarose gels with ethidium bromide (0.5 μg/mL). DNA fragment sizes were estimated through comparison with EZ Load 100 bp Molecular Ruler (Bio-Rad).

Vectors and Lentiviral Infection

Lentiviruses were produced by transfecting 293T cells with FuGENE6 transfection reagent (Promega) with 300 ng of VSVG, 2.7 μg of delta-8.9 or PSPAX2 and 3 μg of each construct. Target cells were infected with each virus in the presence of polybrene (8 μg/ml) for 6 hours. Forty-two hours later, the infected target cells were selected by using a predefined concentration of puromycin (1.5 μg/ml for OE33, and 2.0 μg/ml for OE19) at least for 7 days before biological experiments.

pLKO-shSrc vectors were obtained from The RNAi Consortium, and the shRNA sequences targeting *Src* were as following; for shSrc1 (NM_198291.1, clone ID: TRCN0000038149), 5'-CCGGGACAGACCTGTCCTTCAAGAACTCGAGTTCTT-GAAGGACAGGTCTGTCTTTTTG-3'; for shSrc2 (NM_198291.1, clone ID: TRCN0000038151), 5'-CCGGGTCATGAAGAAGCTGAGGCATCTCAGATGCCT-CAGCTTCTTCATGACTTTTTG-3'. pLKO-shLacZ1650 served as control in the RNAi silencing experiments.

Wild-type pDONR223-Src (#23934) was purchased from Addgene Inc. (Cambridge, Massachusetts). pDONR223-Src^{E527K} was made by site-directed mutagenesis using QuikChange II XL Site-Directed Mutagenesis Kit from Agilent Technologies, Inc. (Santa Clara, California). Primer sequences for site-directed mutagenesis (g1579a) were following; sense 5'-ACTTCACGTC-CACCAAGCCCCAGTACCAG-3' and antisense 5'-CTGGTATGGGGCTTGGTGGACGTGAAGT-3'. pLX301-Src$^{wild-type}$ and PLX301-Src^{E527K} lentiviral vectors were made from pDONR223-Src$^{wild-type}$ and pDONR223–Src^{E527K}, respectively, by performing LR clonase reaction with pLX301 destination vector (Invitrogen). pLX301-GFP served as a control in the ectopic expression experiments.

In vitro cell proliferation assay

Cells (4,000/well) were seeded in quadruplicate in 96-well plate, were treated with either vehicle or variable doses of small molecule inhibitors after 24 hours, and then were allowed to grow for 72 hours. For cells grown in trastuzumab, cells were allowed to proliferate for five days prior to assays of cell proliferation. We used Cell-titer Glo assay (Promega, Madison, Wisconsin) to measure cell viability. Percentage of inhibition of cell proliferation was calculated as [1-(treated cells/untreated cells) × 100]. The results from the cell viability assay were compared between cell lines using repeated measures analysis of variance (ANOVA) and also were tested using student *t*-test at the specific concentration. A p value <0.05 was considered statistically significant.

Immunoblotting

Cells were lysed with RIPA lysis buffer (50 mM Tris-HCl pH 7.5, 150 mM NaCl, 0.1% SDS, 1% NP-40, 0.5% sodium deoxycholate) supplemented by protease inhibitor cocktail (Roche) and phosphatase inhibitor cocktails (Calbiochem). Lysates were separated on 7.5% or 8% Tris-Glycine SDS-polyacrylamide gel

and were transferred to PVDF membranes (Millipore). The membranes were blocked with 5% skim milk (Bio-Rad) dissolved in TBST buffer (50 mM Tris-HCl, 150 mM NaCl, 0.05% Tween-20). Then, the membranes were incubated with primary antibodies overnight at 4°C. Anti-EGFR antibody (#A300-388A) was purchased from Bethyl Laboratories. Anti-β-actin antibody (#A5441) and anti-γ-tubulin antibody (#A9044) were purchased from Sigma-Aldrich. All other antibodies including anti-phospho EGFR Y1068 (#3777), anti-phospho ERBB2 Y1221/1222 (#2243), anti-ERBB2 (#2165), anti-phospho SRC Y416 (#6943), anti-SRC (#2109), anti-phospho-ERK 1/2 T202/Y204 (#4370), anti-ERK 1/2 (#4695), anti-phospho AKT S473 (#4060), and anti-AKT (#9272) were purchased from Cell Signaling Technologies. Horseradish peroxidase-conjugated secondary antibodies (anti-rabbit: #31460, anti-mouse: #31430, Pierce) and SuperSignal West Pico Chemiluminescent Substrate (Pierce) were used to detect signals.

Results

The Novel *Src*^{E527K} mutation was found in the two lapatinib-resistant (LR) OE19 subclones

For generating lapatinib-resistant subclones, OE19 cells were treated with lapatinib, of which dose was progressively increased from 200 nM to 3 μM during 3 months. We expanded the surviving clones in the presence of 3 μM of lapatinib till 6 months and colonies with distinct cellular morphologies were selected. Each selected colony was subcultured in the different plates in the presence of 3 μM of lapatinib, and 7 subclones were selected for genomic analysis (Figure 1A).

Using DNA from these seven subclones as well as from the parental OE19 cell population, we attempted to identify acquired genomic alterations that could have induced drug resistance. DNA from these distinct populations were submitted for a focused next-generation sequencing panel wherein the coding exons from 504 distinct genes were isolated via solution hybrid capture and then sequenced with an Illumina sequencer with an average depth of 252.5× and 97.1% of targets achieving a minimum coverage of 30× (Table S1). These DNA samples were analysed for the presence of somatic mutations and copy-number alterations unique to the resistance subclones compared to the parental cell line. Across these LR subclones, we identified distinct somatic mutations affecting genes *Src*, *KEAP1* and *PHOX2B* (Figure 1A). We did not find any evidence for the mutations found in the derived mutants described as above in the parental OE19 cell line of which sequencing coverage for the *Src* 527 codon of 30×.

Within these data we initially focused upon two distinct clones, both harbouring the same acquired *Src*^{E527K} mutation (Figure 1B) which was present in ~30% of sequenced alleles in each of these two subclones. In the setting of arm-level gain of 20q in OE19 cells, a 30% allele fraction of this mutation is consistent with one of the three copies of *Src* in the cell line being mutated clonally in this population. Notably, this specific base change, an E to K substitution at codon 527, had been utilized as a means to artificially activate Src in previous biochemical studies of this kinase [10]. Comparing the two clones with the *Src* mutation, we noted that one of the clones harboured a unique *KEAP1* mutation suggesting that these two *Src*-mutant clones may not be identical. Review of the copy-number spectrum between these two *Src*-mutant subclones, however, revealed a similar spectrum of copy-number aberrations (Figure S1) suggesting that the two clones may have diverged from a common ancestor prior to the *KEAP1* event in one subclone. Given the likely shared origin, we termed these

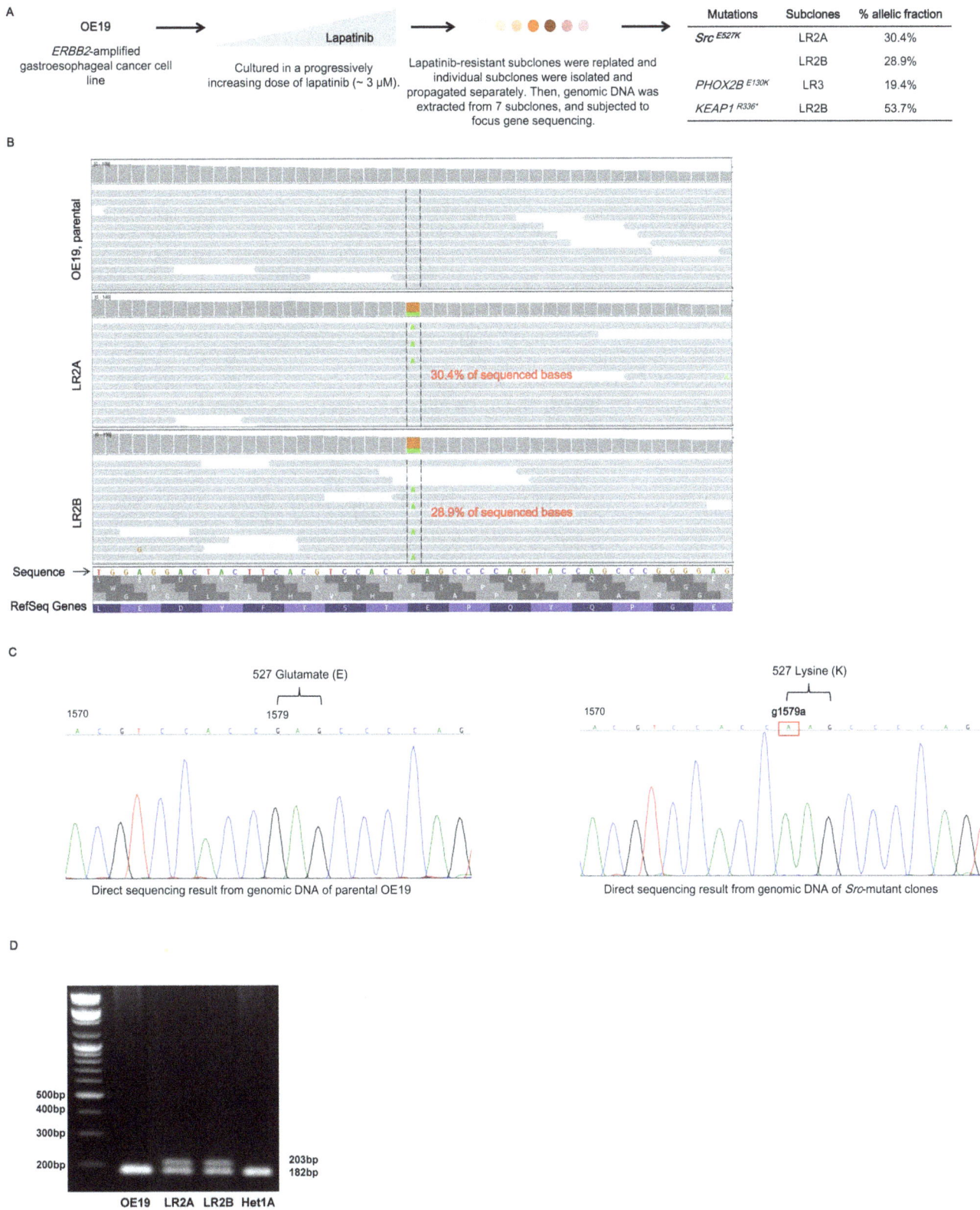

Figure 1. Generation of lapatinib-resistant (LR) OE19 subclones, and the identification of a novel, acquired Src^{E527K} mutation. A, Schematic view of the development of lapatinib resistance OE19 cells followed by subcloning of distinct clones of resistant cells for genomic characterization. At right is the listing of somatic alterations identified in the distinct subclones, compared to the genome of the parental OE19 cells. The allelic fraction of each mutation, percent of sequenced reads with the mutant allele, is listed for each candidate mutation. **B,** IGV (Integrated Genomic Viewer) snapshot of Src mutations in two LR subclones compared to the sequencing seen from this locus in the parental, lapatinib-sensitive cell line. **C,** Direct sequencing results from genomic DNA from both parental OE19 cells and Src^{E527K} mutant LR2A and LR2B subclones identifies mutation detected from next-generation sequencing. **D,** DNA restriction analysis results using Ban II enzyme, from the genomic DNA from the

parental OE19, Src E527K mutant LR2A and LR2B subclones, and Het1A (normal esophageal cell line). Ban II enzyme cuts and yields new amplicons of 182 bp in the parental OE19 and Het1A, however, the bands of 203 bp, which contained Src mutants, are still visualized in the two Src mutant subclones, LR2A and LR2B.

clones lapatinib-resistant (LR) clones LR2A and LR2B. Src^{E527K} mutation was confirmed by direct sequencing (Figure 1C).

Although we did not identify this Src mutation in the parental cell line from our sequencing, we performed additional focused analysis of this locus to detect if low-frequency Src mutant cells were present in the parent cell population. To identify possible mutations at low frequency, we performed restriction enzyme digestion of PCR-amplified DNA from the parental OE19, LR2A, LR2B, and Het1A using Ban II restriction enzyme. The target sequence of Ban II restriction enzyme is G(A/G)GC(T/C)C, which will cut wild-type Src sequence. The PCR amplicons of 203 bp from two LR subclones harboring Src mutant were not digested, in contrast those from the parental OE19 and Het1A were totally digested and yielded 182 bp products (Figure 1D). From these results, we could identify no evidence of the Src^{E527K} mutant in the parental cell population, suggesting it is an acquired event during the prolonged lapatinib exposure.

LR2A and LR2B subclones showed stable resistance to lapatinib

With these two Src-mutant subclones, we re-evaluated their lapatinib sensitivity (Figure 2A). Each of these two subclones had an IC_{50} value for lapatinib greater than 1,000 nM, far exceeding that of the parental cell line. We also evaluated these two Src-mutant subclones for their sensitivity to trastuzumab, and they showed reduced growth inhibition to trastuzumab compared to the parental OE19 cells (Figure S2).

Acquired Src^{E527K} mutation is an activating mutation

To investigate the function of mutant Src within the LR2A and LR2B clones, we evaluated first Src phosphorylation at tyrosine 416, a marker of the kinase's represent an active status [26] with immunoblotting. Both the LR2A and 2B subclones showed higher phosphorylation of Src compared to the parental OE19 cells (Figure 2B) consistent with what we would predict in the setting of an activating mutation. Additionally, in both the LR2A and LR2B clones, the expression and phosphorylation of ERBB2 and EGFR was slightly decreased relative to the parental cell line in the absence of lapatinib.

RNAi-mediated silencing of Src sensitizes Src mutant OE19 cells to lapatinib

Given the clear association between Src mutation and lapatinib resistance, we asked whether shRNA-mediated silencing of Src might restore lapatinib sensitivity in LR2A and LR2B clones. Indeed, silencing of Src by two independent small hairpin constructs sensitized both LR2A and LR2B clones to lapatinib treatment to the extent that the sensitivity profile of those two LR subclones became similar to that of parental OE19 (Figure 3A). Mock or control hairpin transduction did not impact the lapatinib sensitivity of all cell lines (Figure 3A–B).

Consistent with lapatinib resistance profile, we observed sustained p-ERK 1/2 phosphorylation even in the presence of 1 µM lapatinib in mock- or control hairpin-transduced LR2A and LR2B subclones. Notably, the p-ERK 1/2 phosphorylation was successfully blocked by lapatinib treatment in shSrc transduced

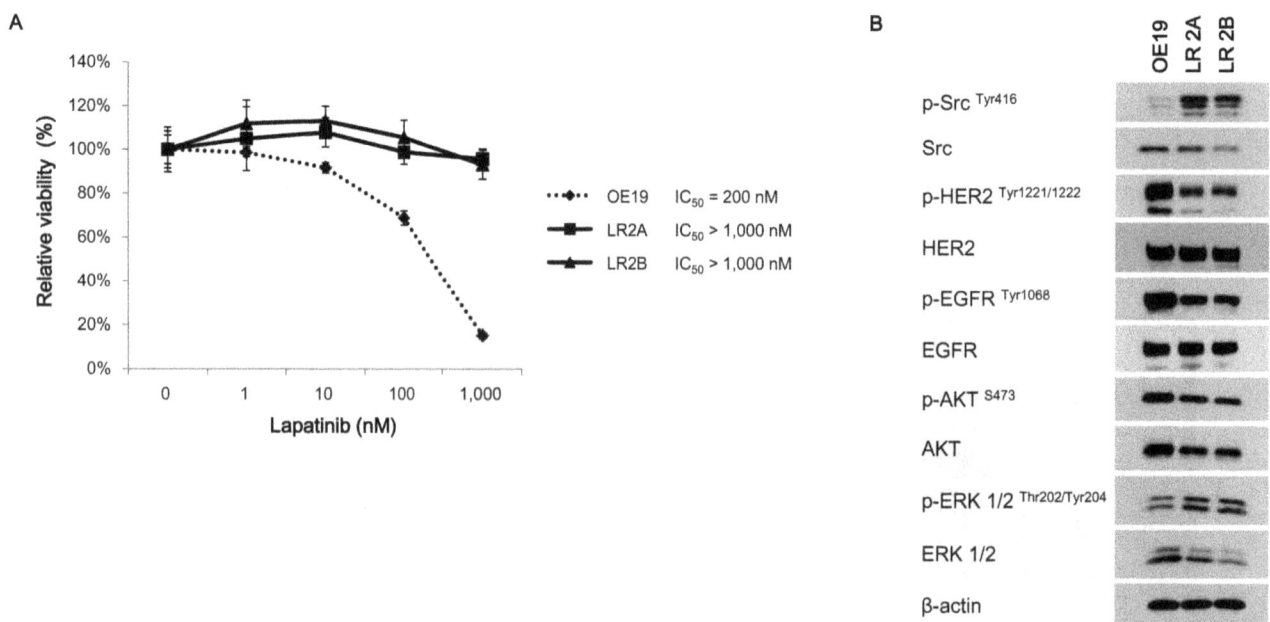

Figure 2. Comparisons of lapatinib sensitivities and the baseline signalling proteins activities between parental and lapatinib-resistant subclones. A, Lapatinib sensitivity curves in the parental OE19 and two lapatinib-resistant (LR) subclones. The calculated values of IC_{50} of lapatinib were 200 nM in parental cells and >1,000 nM in two LR subclones. Values were presented as relative cellular viability relative to vehicle-treated controls with the mean ± S.E. of quadruplicate from a representative experiment. The p values were <0.0001 for OE19 vs LR2A and OE19 vs LR2B, and was 0.129 for LR2A vs LR2B. The p values were calculated by two-way ANOVA. **B,** Immunoblots showing the phosphorylations of distinct signalling molecules in parental OE19 cells and the Src-mutant lapatinib-resistant derivatives.

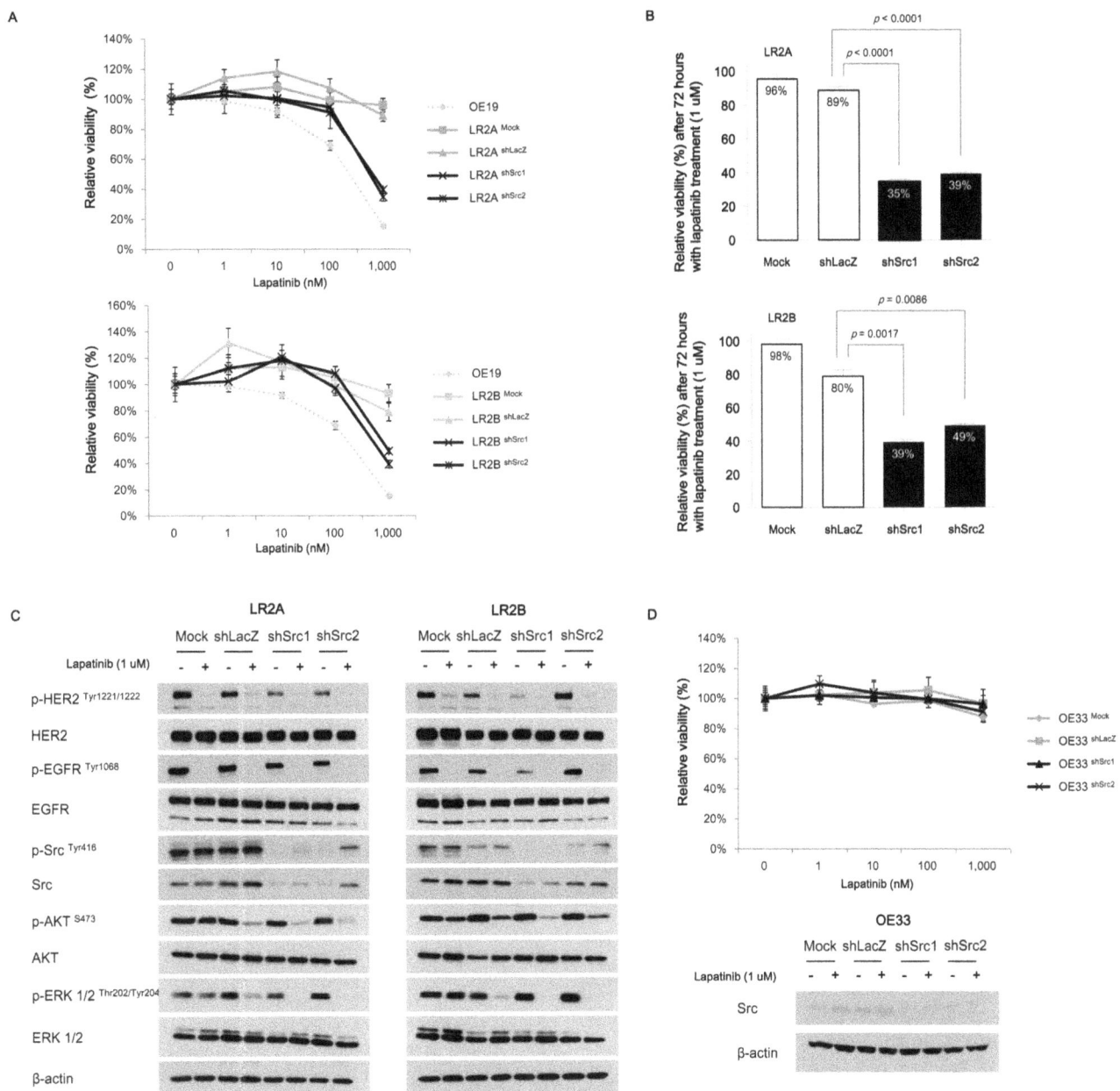

Figure 3. Effects of RNAi-mediated silencing of *Src* on lapatinib sensitivities and responses of signalling proteins. A, Lapatinib sensitivity curves for LR2A and LR2B subclones transduced with shSrc or control (shLacZ). The lapatinib sensitivities were restored after RNAi silencing of Src in the two LR subclones. The calculated values of IC$_{50}$ were following; 200 nM for parental OE19; >1,000 nM for LR2A ^{Mock}, LR2A ^{shLacZ}, LR2B ^{Mock}, and LR2B ^{shLacZ}; 432.8 nM and 403.7 nM for LR2A ^{shSrc1} and LR2A ^{shSrc2}, respectively; 276.6 nM and 444.2 nM for LR2B ^{shSrc1} and LR2B ^{shSrc2}, respectively. Values were presented as relative cellular viability relative to vehicle-treated controls with the mean ± S.E. of quadruplicate from a representative experiment. In the lapatinib sensitivity curve for LR2A and its subclones transduced with lentiviral vectors, the *p* value was 0.121 for mock *vs* control (shLacZ), <0.0001 for control *vs* shSrc1, and <0.0001 for control *vs* shSrc2. In the lapatinib sensitivity curve for LR2B and its subclones transduced with lentiviral vectors, the *p* value was 0.764 for mock *vs* control, <0.0001 for control *vs* shScr1, and 0.012 for control *vs* shSrc2. The *p* values were calculated by two-way ANOVA. **B,** Relative cell viability after 1 μM concentration of lapatinib treatment in two LR subclones with or without RNAi-mediated silencing of *Src*. The *p* values were calculated by two-tailed *t*-test. LR subclones transduced with shSRC restored lapatinib sensitivities. Values were presented as relative cellular viability relative to vehicle-treated controls with the mean ± S.E. of quadruplicate from a representative experiment. **C,** Immunoblots showing changes of various signalling proteins after treatment with 1 μM concentration of lapatinib in two LR subclones with or without RNAi-mediated silencing of *Src*. Proteins were harvested 4 hours after lapatinib or vehicle treatment. **D,** Lapatinib sensitivity in OE33 cells following regarding RNAi transduction targeting *Src*. Values were presented as relative cellular viability relative to vehicle-treated controls with the mean ± S.E. of quadruplicate from a representative experiment. Proteins were harvested 4 hours after lapatinib or vehicle treatment. There was no statistical significance between cell lines in terms of lapatinib sensitivity (0.070 for mock *vs* control [shLacZ], 0.520 for control *vs* shSrc1, and 0.753 for control *vs* shSrc2, respectively). The *p* values were calculated by two-way ANOVA.

LR subclones (Figure 3C). These data indicate that, in these *Src*-mutant LR subclones, blockade of both Src and ERBB2 is

required to completely block pathologic mitogenic signalling. The phosphorylations of AKT and ERK were slightly downregulated in the shLacZ transduced LR subclones, which were not observed dominantly in the mock-treated LR subclones, possibly attributable to the effects from lentiviral infections and puromycin selection. As an additional control to ensure that these *Src*-directed shRNA vectors did not impact lapatinib sensitivity due to non-specific off target effects, we also evaluated the impact of these constructs on the lapatinib sensitivity of OE33 cell line which is not sensitive to lapatinib through unrelated mechanism, *MET* co-amplification. In this model, introduction of the sh*Src* did not sensitize OE33 cells to lapatinib (Figure 3D).

Lapatinib resistance could be overcome by combination treatment with saracatinib

We then tested whether pharmacologic inhibition of Src by saracatinib in LR2A and LR2B subclones could restore lapatinib sensitivity. Saracatinib is a dual tyrosine kinase inhibitor targeting c-Src/Abl kinase. Saracatinib showed preclinical activity in various cancer cell lines including gastroesophageal cancers [27–34]; however, only low to modest antitumor activity were demonstrated in several phase I or II clinical trials when tested as monotherapy [35–38].

While the two *Src* ^{E527K} mutant subclones were not sensitive to either lapatinib or saracatinib alone, their growth was effectively inhibited when the two drugs were combined (Figure 4A & 4B). Consistent with this drug sensitivity profile, AKT and ERK phosphorylation was sustained when the LR subclones were treated with either lapatinib or saracatinib alone, and their sustained phosphorylation was blocked upon the combined treatment of both drugs (Figure 4C). HER2 and EGFR phosphorylations were blocked upon lapatinib treatment regardless of the presence of saracatinib in both the parental OE19 cells and the two LR subclones suggesting that the mutant *Src*-mediated resistance is independent of HER2 or EGFR signalling (Figure 4C). These data suggest that pathologic Src activation enhances survival of lapatinib resistant clones upon ERBB2 inhibition through activation of both PI3-K and MAPK pathways. Saracatinib alone could not inhibit cellular proliferation either in the parental or in the two LR subclones (Figure S3). In *MET* co-amplified OE33 cells, saracatinib treatment did not show any synergy with lapatinib treatment. By contrast, the addition of MET inhibitor crizotinib synergistically inhibited cell proliferation (Figure 4D).

Ectopic expression of *Src* ^{E527K} induces lapatinib resistance in the parental OE19 cells

Last we evaluated the ability of exogenous expression of the *Src*-mutants to impact the lapatinib sensitivity of OE19 cells. We generated the *Src* ^{E527K} plasmid using site-directed mutagenesis, and transduced either wild-type *Src*, mutant *Src* or GFP control into parental OE19 cells. Indeed, *Src* ^{E527K} transduced OE19 developed lapatinib resistance (IC$_{50}$ to the lapatinib, 1179.0 nM), while GFP or *Src* $^{wild-type}$ transduced OE19 remained sensitive to lapatinib (IC$_{50}$ to the lapatinib, 256.6 nM and 313.8 nM, respectively, Figure 5A). Furthermore, the lapatinib resistance conferred by exogenous expression of *Src* ^{E527K} was overcome by the combined treatment of saracatinib (Figure 5B). Saracatinib also reversed the mild lapatinib resistance induced by the wild-type *Src* transduction.

As in the LR subclones harbouring spontaneous *Src* mutation, expression of p-Src was increased in *Src* transduced OE19 either with wild-type or E527K mutant compared to the parental or GFP transduced cells (Figure 5C). *Src* ^{E527K} transduced OE19 showed particularly high expression of p-SRC. Although the phosphory-

lation of Src in the OE19 cells transduced with wild-type Src was not effectively inhibited by saracatinib, Src phosphorylation in *Src* ^{E527K} transduced OE19 was significantly inhibited by saracatinib regardless of the presence of lapatinib. Additionally, the levels of p-AKT and p-ERK 1/2 were not suppressed by lapatinib alone in the *Src* ^{E527K} transduced OE19, consistent with our hypothesis that constitutive Src activation might sustain both downstream signalling pathways. However, phosphorylation of both AKT and ERK 1/2 was inhibited with combination of saracatinib and lapatinib, paralleling the results from cells with spontaneously acquired *Src* mutant. Taken together, pathologic Src activation could induce lapatinib resistance in *ERBB2*-amplified GE adenocarcinoma and the resistance could be reversed by the additional Src inhibition.

Discussion

As an effort to model acquired resistance of *ERBB2*-amplified GE adenocarcinomas to ERBB2 inhibition, we generated lapatinib-resistant subclones from an initially lapatinib-sensitive *ERBB2*-amplified esophageal adenocarcinoma cell line by prolonged exposure to the inhibitor. Through genomic and functional analysis of LR subclones, we found that an activating mutation of *Src* was responsible for the acquired lapatinib resistance in two of seven isolated subclones in this *in vitro* model system. In addition, we further demonstrated that genetic or pharmacologic blockade of Src could restore ERBB2 inhibitor sensitivity in LR subclones with hyperactive Src. Although our data remain to be validated in patient samples, these data establish the role of oncogene *Src* as a pharmacologically tractable candidate mediator of acquired lapatinib resistance in ERBB2-expressed GE adenocarcinomas.

Recently, increased Src kinase activity was suggested as one of the resistance mechanisms to both trastuzumab and lapatinib in breast cancer cell lines [10,39]. c-Src is a membrane-associated tyrosine kinase and cellular homologue of the oncogenic v-Src encoded by the chicken Rous sarcoma virus [40]. Src acts as a common signalling node by interacting with multiple receptor tyrosine kinases [10,29,41]. In breast cancer, Zhang et al demonstrated that increased Src kinase activity was responsible for both the *de novo* and acquired trastuzumab resistances. Separate investigators demonstrated that lapatinib-resistant SKBR3 breast cancer cells showed increased activity of Src kinases and persistent levels of activation of ERK 1/2 and AKT and that treatment with saracatinib reduced AKT and ERK 1/2 activity and restored lapatinib sensitivity [39]. A recent study by Han et al. similarly reported that increased Src activity was observed in the trastuzumab-resistant *ERBB2*-amplified GE cancer cell line, NCI-N87, and showed that trastuzumab and saracatinib synergistically inhibited the *in vitro* growth of both parental and trastuzumab-resistant NCI-N87 [18]. We also have generated lapatinib-resistant subclones from the NCI-N87 cell line, and performed similar sequencing in those subclones; however, we could not identify any similar *Src* mutations (data not shown). Ongoing efforts are trying to identify alternative etiology of resistance mechanisms in these lapatinib-resistant NCI-N87 subclones. Although the role of increased Src activity in the resistance of *ERBB2*-amplified breast cancer and GE cancer to ERBB2 inhibition has been documented, we reported here, for the first time, the activation of a spontaneous *Src* mutation after prolonged exposure to HER2 inhibitor could induce lapatinib resistance in *ERBB2*-amplified GE adenocarcinoma and present the first evidence of acquired mutation of *Src* as an etiology of resistance. While these data clearly demonstrate the capacity of activated Src to serve as a mediator of acquired resistance to

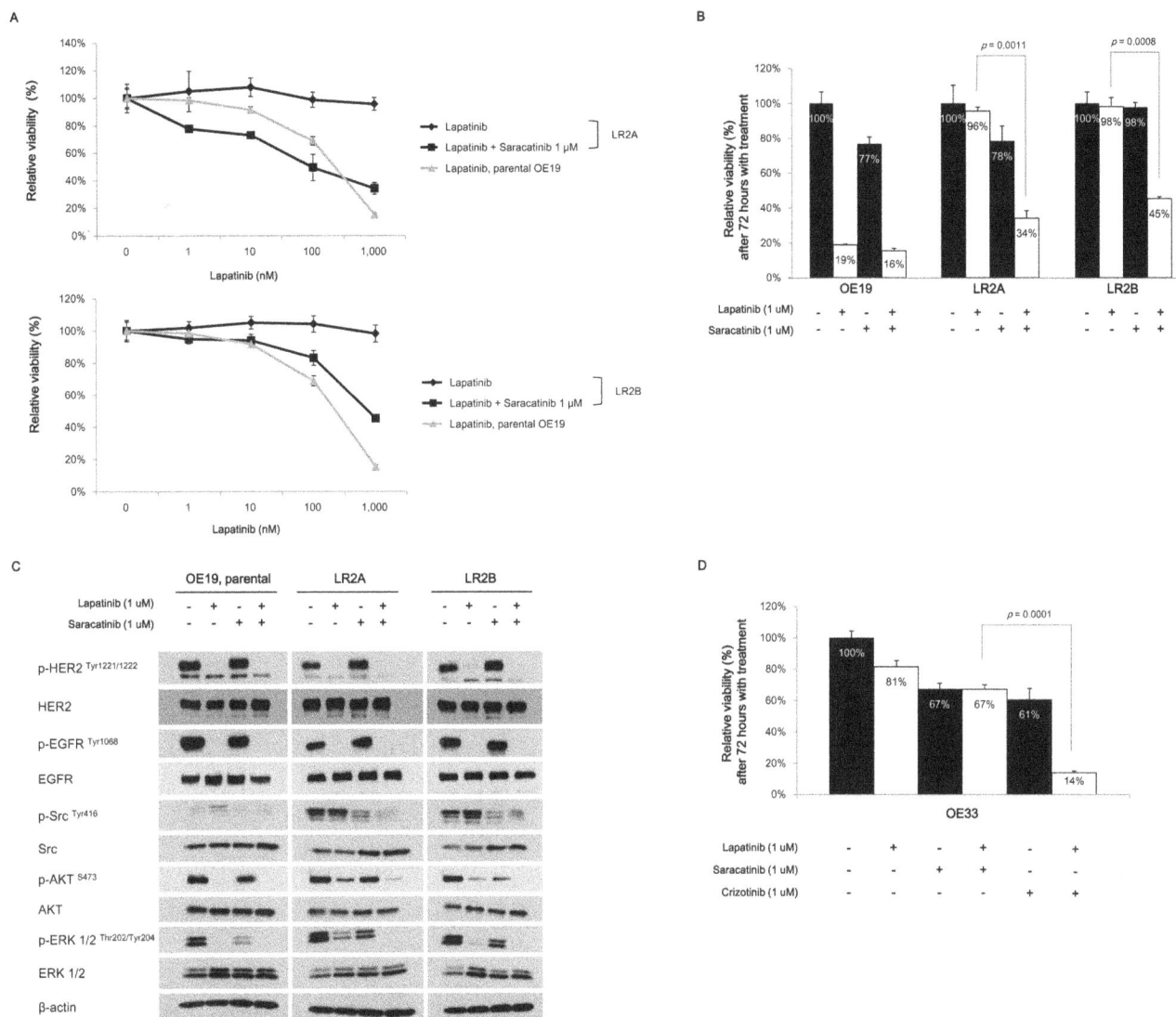

Figure 4. Effects of pharmacologic inhibition of *Src* using saracatinib in combination with lapatinib in LR subclones. A, Dose-response curves from escalated dose of lapatinib in the presence of 1 µM concentration of saracatinib. The calculated values of IC_{50} for lapatinib were following; 207.3 nM and 145 nM, respectively, after lapatinib alone and in the presence of saracatinib for parental OE19; >1,000 nM after lapatinib alone for LR2A and LR2B; 91.66 nM and 224.0 nM in the presence of saracatinib in LR2A and LR2B, respectively. Values were presented as relative cellular viability relative to vehicle-treated controls with the mean ± S.E. of quadruplicate from a representative experiment. The *p* values calculated by two-way ANOVA were <0.0001 comparing lapatinib alone with lapatinib plus saracatinib both in the LR2A and LR2B. **B,** Relative cell viability in the parental OE19 and two LR subclones after 1 µM concentration of saracatinib or lapatinib, either alone or combination. The *p* values were calculated by two-tailed *t*-test. Values were presented as relative cellular viability relative to vehicle-treated controls with the mean ± S.E. of quadruplicate from a representative experiment. **C,** Immunoblots showing changes of in intracellular signalling proteins after treatment with 1 µM concentration of lapatinib or saracatinib, either alone or combination, in LR subclones. Proteins were harvested 4 hours after each treatment. **D,** Relative cell viability in the OE33, *ERBB2* and *MET* co-amplified gastroesophageal cancer cells, after treatment with various drugs either alone or combination. The effect of combination treatment with lapatinib and saracatinib did not differ either from lapatinib alone or saracatinib alone; the combination treatment with lapatinib and crizotinib, which is a potent MET inhibitor, showed synergistic effects. The *p* values were calculated by two-tailed *t*-test. Values were presented as relative cellular viability relative to vehicle-treated controls with the mean ± S.E. of quadruplicate from a representative experiment.

ERBB2 inhibitor therapy in GE cancers, future studies will be required to query the presence of *Src* mutation or enhanced activity of this kinase in patient samples upon emergence of resistance.

Additionally, future studies will need to address potential differences between mechanisms of acquired resistance to small molecule compared to antibody ERBB2 inhibitors in GE adenocarcinomas. However, mechanisms for acquired resistance to trastuzumab have been reported to be similar to those to lapatinib in the breast cancer field. Mechanisms for *de novo* or

acquired trastuzumab resistance included constitutive activation PI3-K pathway owing to *PTEN* deficiency or *PIK3CA* gene mutation [9,42], the expression of truncated HER2 receptors [15], and overexpressions of other receptor tyrosine kinases which include EGFR, insulin-like growth factor-1 receptor, and hepatocyte growth factor receptor [43–45]. Mechanisms for lapatinib resistance were generally similar to those with the trastuzumab resistance [46–49].

Saracatinib, the c-Src/Abl dual targeting inhibitor, has shown only mild to moderate antitumor activity in phase I/II clinical

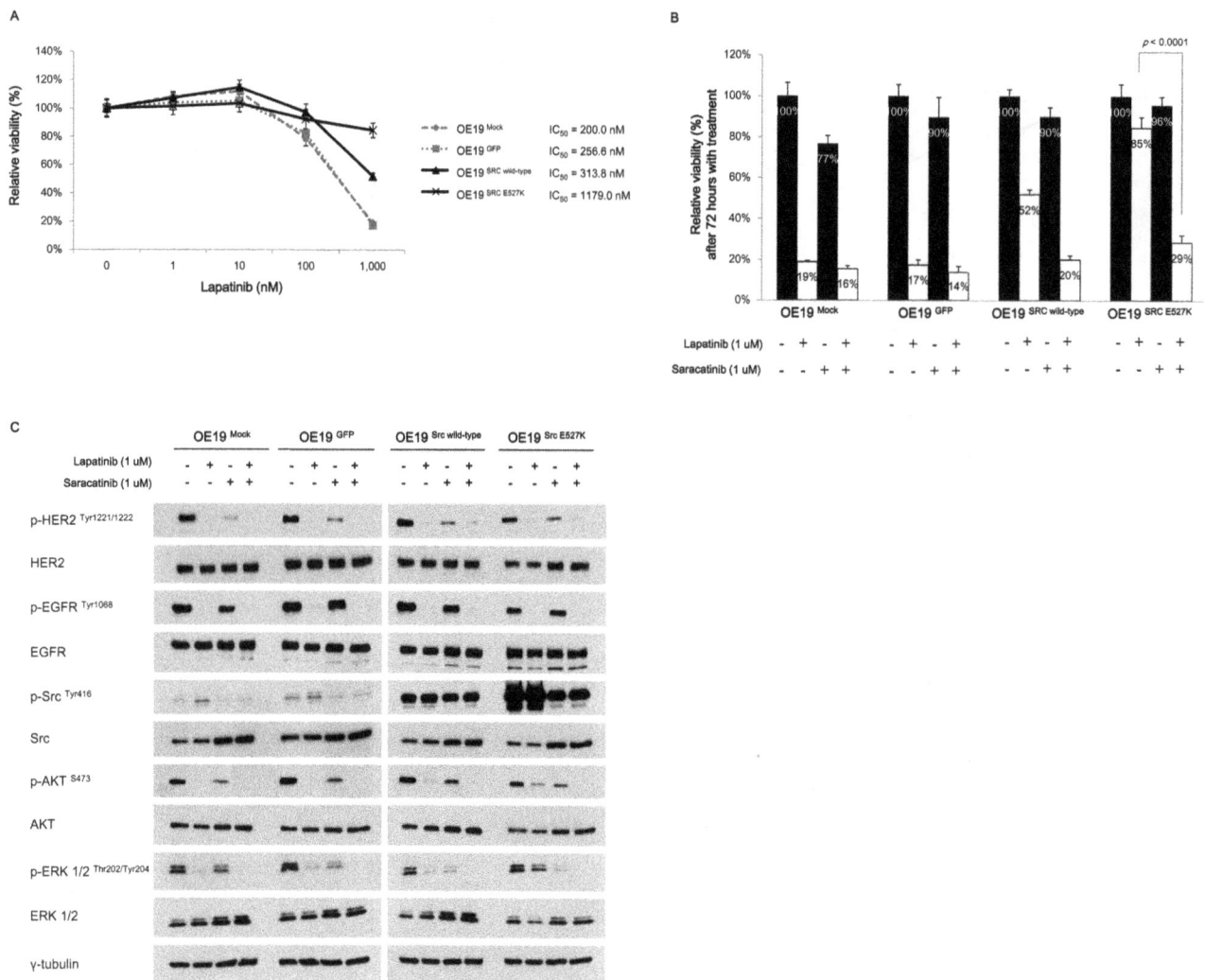

Figure 5. Impact of ectopic expression of Src^{E527K} in parental OE19 on lapatinib sensitivity and cell signalling. A, Dose-response curves for lapatinib in the non-transduced parental OE19 (OE19 ^{Mock}), transduced with GFP (OE19 ^{GFP}), wild-type Src (OE19 $^{Src\ wild-type}$) or Src^{E527K} mutation (OE19 $^{Src\ E527K}$). The calculated values of IC_{50} for lapatinib was >1,000 nM in OE19 $^{Src\ E527K}$ cells. Values were presented as relative cellular viability relative to vehicle-treated controls with the mean ± S.E. of quadruplicate from a representative experiment. The p value calculated by two-way ANOVA was 0.376 for mock vs control (GFP), <0.0001 for control vs wild-type Src, and <0.0001 for control vs Src^{E527K}. **B,** Relative cell viability lapatinib in the non-transduced parental OE19 (OE19 ^{Mock}), transduced with GFP (OE19 ^{GFP}), wild-type Src (OE19 $^{Src\ wild-type}$) or Src^{E527K} mutation (OE19 $^{Src\ E527K}$). The p values were calculated by two-tailed t-test. Values were presented as relative cellular viability relative to vehicle-treated controls with the mean ± S.E. of quadruplicate from a representative experiment. **C,** Immunoblots showing changes of various signalling proteins after treatment with 1 μM concentration of lapatinib or saracatinib, either alone or combination, in the OE19 cells with or without Src ^{E527K} transduction.

trials [35–38]. However, these trials did not utilize genomic or biochemical biomarkers to guide enrolment nor did it investigate the situation of ERBB2 inhibitor insensitivity. Based upon the results presented in this study, further focused evaluation of Src activation, either due to mutation or other means of activation, should be considered on patient tumors following the acquisition of resistance to ERBB2-directed therapy for evidence. Such testing should evaluate for both activation due to mutation or from other mechanisms. Should such Src activation or Src mutation be identified in such tumor samples, the results from this report support subsequent efforts to perform clinical trials of a combination of ERBB2 inhibition and Src inhibition. Such therapy may be able to lead to meaningful improvements in outcomes for patients whose tumors utilize Src activation as a means of bypassing ERBB2 inhibition.

Supporting Information

Figure S1 Inferred copy-number plots for LR subclones. Comparison of the copy-number profiles of the two Src-mutant LR subclones following normalization against the parental, lapatinib-sensitive OE19 cell line (x-axis: arbitrary chromosomal co-ordinates, color codes represent each chromosome in increasing order, y-axis: inferred copy-number).

Figure S2 Growth inhibition curves in the Two Src-mutant LR subclones after increasing dose of trastuzumab. Values were presented as relative cellular viability relative to vehicle-treated controls with the mean ± S.E. of quadruplicate from a representative experiment. The p values calculated by two-way ANOVA were <0.0001 both in the comparisons of viabilities of OE19 vs LR2A and OE19 vs LR2B.

Figure S3 Growth inhibition curves in the parental OE19 and in the two *Src*-mutant LR subclones after increasing dose of saracatinib. Values were presented as relative cellular viability relative to vehicle-treated controls with the mean ± S.E. of quadruplicate from a representative experiment. There was no statistical significance in terms of cell viability between cell lines.

Table S1 Next-generation sequencing panel in 8 cell lines including parental OE19 and 7 lapatinib-resistant subclones.

References

1. Siegel R, Naishadham D, Jemal A (2012) Cancer statistics, 2012. CA Cancer J Clin 62: 10–29.
2. Al-Batran SE, Hartmann JT, Probst S, Schmalenberg H, Hollerbach S, et al. (2008) Phase III trial in metastatic gastroesophageal adenocarcinoma with fluorouracil, leucovorin plus either oxaliplatin or cisplatin: a study of the Arbeitsgemeinschaft Internistische Onkologie. J Clin Oncol 26: 1435–1442.
3. Dank M, Zaluski J, Barone C, Valverre V, Yalcin S, et al. (2008) Randomized phase III study comparing irinotecan combined with 5-fluorouracil and folinic acid to cisplatin combined with 5-fluorouracil in chemotherapy naive patients with advanced adenocarcinoma of the stomach or esophagogastric junction. Ann Oncol 19: 1450–1457.
4. Kang YK, Kang WK, Shin DB, Chen J, Xiong J, et al. (2009) Capecitabine/cisplatin versus 5-fluorouracil/cisplatin as first-line therapy in patients with advanced gastric cancer: a randomised phase III noninferiority trial. Ann Oncol 20: 666–673.
5. Van Cutsem E, Moiseyenko VM, Tjulandin S, Majlis A, Constenla M, et al. (2006) Phase III study of docetaxel and cisplatin plus fluorouracil compared with cisplatin and fluorouracil as first-line therapy for advanced gastric cancer: a report of the V325 Study Group. J Clin Oncol 24: 4991–4997.
6. Bang YJ, Van Cutsem E, Feyereislova A, Chung HC, Shen L, et al. (2010) Trastuzumab in combination with chemotherapy versus chemotherapy alone for treatment of HER2-positive advanced gastric or gastro-oesophageal junction cancer (ToGA): a phase 3, open-label, randomised controlled trial. Lancet 376: 687–697.
7. Razis E, Bobos M, Kotoula V, Eleftheraki AG, Kalofonos HP, et al. (2011) Evaluation of the association of PIK3CA mutations and PTEN loss with efficacy of trastuzumab therapy in metastatic breast cancer. Breast Cancer Res Treat 128: 447–456.
8. Esteva FJ, Guo H, Zhang S, Santa-Maria C, Stone S, et al. (2010) PTEN, PIK3CA, p-AKT, and p-p70S6K status: association with trastuzumab response and survival in patients with HER2-positive metastatic breast cancer. Am J Pathol 177: 1647–1656.
9. Nagata Y, Lan KH, Zhou X, Tan M, Esteva FJ, et al. (2004) PTEN activation contributes to tumor inhibition by trastuzumab, and loss of PTEN predicts trastuzumab resistance in patients. Cancer Cell 6: 117–127.
10. Zhang S, Huang WC, Li P, Guo H, Poh SB, et al. (2011) Combating trastuzumab resistance by targeting SRC, a common node downstream of multiple resistance pathways. Nat Med 17: 461–469.
11. Dua R, Zhang J, Nhonthachit P, Penuel E, Petropoulos C, et al. (2010) EGFR over-expression and activation in high HER2, ER negative breast cancer cell line induces trastuzumab resistance. Breast Cancer Res Treat 122: 685–697.
12. Gallardo A, Lerma E, Escuin D, Tibau A, Munoz J, et al. (2012) Increased signalling of EGFR and IGF1R, and deregulation of PTEN/PI3K/Akt pathway are related with trastuzumab resistance in HER2 breast carcinomas. Br J Cancer 106: 1367–1373.
13. Pastuskovas CV, Mundo EE, Williams SP, Nayak TK, Ho J, et al. (2012) Effects of anti-VEGF on pharmacokinetics, biodistribution, and tumor penetration of trastuzumab in a preclinical breast cancer model. Mol Cancer Ther 11: 752–762.
14. Scaltriti M, Eichhorn PJ, Cortes J, Prudkin L, Aura C, et al. (2011) Cyclin E amplification/overexpression is a mechanism of trastuzumab resistance in HER2+ breast cancer patients. Proc Natl Acad Sci U S A 108: 3761–3766.
15. Scaltriti M, Rojo F, Ocana A, Anido J, Guzman M, et al. (2007) Expression of p95HER2, a truncated form of the HER2 receptor, and response to anti-HER2 therapies in breast cancer. J Natl Cancer Inst 99: 628–638.
16. Hong J, Katsha A, Lu P, Shyr Y, Belkhiri A, et al. (2012) Regulation of ERBB2 receptor by t-DARPP mediates trastuzumab resistance in human esophageal adenocarcinoma. Cancer Res 72: 4504–4514.
17. Chen CT, Kim H, Liska D, Gao S, Christensen JG, et al. (2012) MET activation mediates resistance to lapatinib inhibition of HER2-amplified gastric cancer cells. Mol Cancer Ther 11: 660–669.
18. Han S, Meng Y, Tong Q, Li G, Zhang X, et al. (2014) The ErbB2-targeting antibody trastuzumab and the small-molecule SRC inhibitor saracatinib synergistically inhibit ErbB2-overexpressing gastric cancer. MAbs 6: 403–408.
19. Fuchs CS, Tomasek J, Yong CJ, Dumitru F, Passalacqua R, et al. (2014) Ramucirumab monotherapy for previously treated advanced gastric or gastro-oesophageal junction adenocarcinoma (REGARD): an international, randomised, multicentre, placebo-controlled, phase 3 trial. Lancet 383: 31–39.
20. Kim HS, Kim HJ, Kim SY, Kim TY, Lee KW, et al. (2013) Second-line chemotherapy versus supportive cancer treatment in advanced gastric cancer: a meta-analysis. Ann Oncol 24: 2850–2854.
21. Kang YK, Muro K, Ryu MH, Yasui H, Nishina T, et al. (2013) A phase II trial of a selective c-Met inhibitor tivantinib (ARQ 197) monotherapy as a second- or third-line therapy in the patients with metastatic gastric cancer. Invest New Drugs.
22. Bang YJ, Kang YK, Kang WK, Boku N, Chung HC, et al. (2011) Phase II study of sunitinib as second-line treatment for advanced gastric cancer. Invest New Drugs 29: 1449–1458.
23. Wainberg ZA, Anghel A, Desai AJ, Ayala R, Luo T, et al. (2010) Lapatinib, a dual EGFR and HER2 kinase inhibitor, selectively inhibits HER2-amplified human gastric cancer cells and is synergistic with trastuzumab in vitro and in vivo. Clin Cancer Res 16: 1509–1519.
24. Janjigian YY, Viola-Villegas N, Holland JP, Divilov V, Carlin SD, et al. (2013) Monitoring afatinib treatment in HER2-positive gastric cancer with 18F-FDG and 89Zr-trastuzumab PET. J Nucl Med 54: 936–943.
25. Hettmer S, Teot LA, van Hummelen P, MacConaill L, Bronson RT, et al. (2013) Mutations in Hedgehog pathway genes in fetal rhabdomyomas. J Pathol 231: 44–52.
26. Roskoski R, Jr. (2005) Src kinase regulation by phosphorylation and dephosphorylation. Biochem Biophys Res Commun 331: 1–14.
27. Arcaroli JJ, Quackenbush KS, Powell RW, Pitts TM, Spreafico A, et al. (2012) Common PIK3CA mutants and a novel 3′ UTR mutation are associated with increased sensitivity to saracatinib. Clin Cancer Res 18: 2704–2714.
28. Arcaroli JJ, Touban BM, Tan AC, Varella-Garcia M, Powell RW, et al. (2010) Gene array and fluorescence in situ hybridization biomarkers of activity of saracatinib (AZD0530), a Src inhibitor, in a preclinical model of colorectal cancer. Clin Cancer Res 16: 4165–4177.
29. Bertotti A, Bracco C, Girolami F, Torti D, Gastaldi S, et al. (2010) Inhibition of Src impairs the growth of met-addicted gastric tumors. Clin Cancer Res 16: 3933–3943.
30. Cavalloni G, Peraldo-Neia C, Sarotto I, Gammaitoni L, Migliardi G, et al. (2012) Antitumor activity of Src inhibitor saracatinib (AZD-0530) in preclinical models of biliary tract carcinomas. Mol Cancer Ther 11: 1528–1538.
31. Ferguson J, Arozarena I, Ehrhardt M, Wellbrock C (2013) Combination of MEK and SRC inhibition suppresses melanoma cell growth and invasion. Oncogene 32: 86–96.
32. Morrow CJ, Ghattas M, Smith C, Bonisch H, Bryce RA, et al. (2010) Src family kinase inhibitor Saracatinib (AZD0530) impairs oxaliplatin uptake in colorectal cancer cells and blocks organic cation transporters. Cancer Res 70: 5931–5941.
33. Nam HJ, Im SA, Oh DY, Elvin P, Kim HP, et al. (2013) Antitumor activity of saracatinib (AZD0530), a c-Src/Abl kinase inhibitor, alone or in combination with chemotherapeutic agents in gastric cancer. Mol Cancer Ther 12: 16–26.
34. Rajeshkumar NV, Tan AC, De Oliveira E, Womack C, Wombwell H, et al. (2009) Antitumor effects and biomarkers of activity of AZD0530, a Src inhibitor, in pancreatic cancer. Clin Cancer Res 15: 4138–4146.
35. Baselga J, Cervantes A, Martinelli E, Chirivella I, Hoekman K, et al. (2010) Phase I safety, pharmacokinetics, and inhibition of SRC activity study of saracatinib in patients with solid tumors. Clin Cancer Res 16: 4876–4883.
36. Fujisaka Y, Onozawa Y, Kurata T, Yasui H, Goto I, et al. (2013) First report of the safety, tolerability, and pharmacokinetics of the Src kinase inhibitor saracatinib (AZD0530) in Japanese patients with advanced solid tumours. Invest New Drugs 31: 108–114.
37. Gucalp A, Sparano JA, Caravelli J, Santamauro J, Patil S, et al. (2011) Phase II trial of saracatinib (AZD0530), an oral SRC-inhibitor for the treatment of patients with hormone receptor-negative metastatic breast cancer. Clin Breast Cancer 11: 306–311.
38. Mackay HJ, Au HJ, McWhirter E, Alcindor T, Jarvi A, et al. (2012) A phase II trial of the Src kinase inhibitor saracatinib (AZD0530) in patients with metastatic

Acknowledgments

We thank Peter Hammerman for helpful discussions.

Author Contributions

Conceived and designed the experiments: YSH JHK AB. Performed the experiments: YSH JHK CF SWH QM AT PH AB. Analyzed the data: YSH JHK CF SWH AT PH AB. Contributed reagents/materials/analysis tools: YSH JHK CF SWH QM GW EP SP MS AT PH AB. Contributed to the writing of the manuscript: YSH JHK CF SWH QM GW EP SP MS AT PH AB.

or locally advanced gastric or gastro esophageal junction (GEJ) adenocarcinoma: a trial of the PMH phase II consortium. Invest New Drugs 30: 1158–1163.

39. De Luca A, D'Alessio A, Gallo M, Maiello MR, Bode AM, et al. (2013) Src and CXCR4 are involved in the invasiveness of breast cancer cells with acquired resistance to lapatinib. Cell Cycle 13.

40. Schwartzberg PL (1998) The many faces of Src: multiple functions of a prototypical tyrosine kinase. Oncogene 17: 1463–1468.

41. Olayioye MA, Badache A, Daly JM, Hynes NE (2001) An essential role for Src kinase in ErbB receptor signaling through the MAPK pathway. Exp Cell Res 267: 81–87.

42. Berns K, Horlings HM, Hennessy BT, Madiredjo M, Hijmans EM, et al. (2007) A functional genetic approach identifies the PI3K pathway as a major determinant of trastuzumab resistance in breast cancer. Cancer Cell 12: 395–402.

43. Lu Y, Zi X, Zhao Y, Mascarenhas D, Pollak M (2001) Insulin-like growth factor-I receptor signaling and resistance to trastuzumab (Herceptin). J Natl Cancer Inst 93: 1852–1857.

44. Ritter CA, Perez-Torres M, Rinehart C, Guix M, Dugger T, et al. (2007) Human breast cancer cells selected for resistance to trastuzumab in vivo overexpress epidermal growth factor receptor and ErbB ligands and remain dependent on the ErbB receptor network. Clin Cancer Res 13: 4909–4919.

45. Shattuck DL, Miller JK, Carraway KL, 3rd, Sweeney C (2008) Met receptor contributes to trastuzumab resistance of Her2-overexpressing breast cancer cells. Cancer Res 68: 1471–1477.

46. Wang XL, Chen XM, Fang JP, Yang CQ (2012) Lentivirus-mediated RNA silencing of c-Met markedly suppresses peritoneal dissemination of gastric cancer in vitro and in vivo. Acta Pharmacol Sin 33: 513–522.

47. Eichhorn PJ, Gili M, Scaltriti M, Serra V, Guzman M, et al. (2008) Phosphatidylinositol 3-kinase hyperactivation results in lapatinib resistance that is reversed by the mTOR/phosphatidylinositol 3-kinase inhibitor NVP-BEZ235. Cancer Res 68: 9221–9230.

48. Liu L, Greger J, Shi H, Liu Y, Greshock J, et al. (2009) Novel mechanism of lapatinib resistance in HER2-positive breast tumor cells: activation of AXL. Cancer Res 69: 6871–6878.

49. Wetterskog D, Shiu KK, Chong I, Meijer T, Mackay A, et al. (2013) Identification of novel determinants of resistance to lapatinib in ERBB2-amplified cancers. Oncogene.

17

Coincidental Loss of Bacterial Virulence in Multi-Enemy Microbial Communities

Ji Zhang[1,2]*, Tarmo Ketola[1], Anni-Maria Örmälä-Odegrip[2], Johanna Mappes[1], Jouni Laakso[1,2]

1 Centre of Excellence in Biological Interactions, Department of Biological and Environmental Science, University of Jyväskylä, Jyväskylä, Finland, **2** Department of Biological and Environmental Science, University of Helsinki, Helsinki, Finland

Abstract

The coincidental virulence evolution hypothesis suggests that outside-host selection, such as predation, parasitism and resource competition can indirectly affect the virulence of environmentally-growing bacterial pathogens. While there are some examples of coincidental environmental selection for virulence, it is also possible that the resource acquisition and enemy defence is selecting against it. To test these ideas we conducted an evolutionary experiment by exposing the opportunistic pathogen bacterium *Serratia marcescens* to the particle-feeding ciliate *Tetrahymena thermophila*, the surface-feeding amoeba *Acanthamoeba castellanii*, and the lytic bacteriophage Semad11, in all possible combinations in a simulated pond water environment. After 8 weeks the virulence of the 384 evolved clones were quantified with fruit fly *Drosophila melanogaster* oral infection model, and several other life-history traits were measured. We found that in comparison to ancestor bacteria, evolutionary treatments reduced the virulence in most of the treatments, but this reduction was not clearly related to any changes in other life-history traits. This suggests that virulence traits do not evolve in close relation with these life-history traits, or that different traits might link to virulence in different selective environments, for example via resource allocation trade-offs.

Editor: Boris Alexander Vinatzer, Virginia Tech, United States of America

Funding: This work was supported by the Finnish Academy to JL, 1130724 and 1255572 (URL: www.aka.fi), which had a role in study design, data collection and analysis, decision to publish and preparation of the manuscript. This work was also supported by the CoE in Biological Interactions to JM, 252411 (URL: https://www.jyu.fi/bioenv/en/divisions/coe-interactions), which had a role in study design, data collection and analysis, decision to publish and preparation of the manuscript. This work was also supported by the Finnish Cultural Foundation to JZ (URL: www.skr.fi), which had a role in data collection and analysis, decision to publish and preparation of the manuscript. This work was also supported by the Ellen and Artturi Nyyssö nen Foundation to JZ (URL: www.eans.fi), which had a role in study design, data collection and analysis.

Competing Interests: The authors have declared that no competing interests exist.

* Email: Ji.Zhang@Helsinki.Fi

Introduction

Compared to the vast knowledge on the prevention and treatment of bacterial infectious disease, relatively little is known about how the virulence of bacteria has evolved. Virulence evolution is often exemplified as a tug of war between the multicellular host and the pathogen, where the virulence (the degree of host damage or mortality caused by the pathogen) [1] evolves solely through host-pathogen interaction [2–4]. Contrary to this idea, the "coincidental evolution of virulence hypothesis" suggests that virulence evolves indirectly due to selection forces that are not related to the host-pathogen interaction *per se*, but because of selection that occurs outside host environments [2,3,5,6]. This is a plausible expectation when considering opportunistic, environmentally growing bacterial pathogens because they typically live in a complex web of interactions with biotic and abiotic selection pressures that might not be directly connected to their potential hosts [7].

In the natural environment, top-down regulation by bacteriophages and protozoans are two major biotic causes of bacterial mortality [8,9]. In order to survive, bacteria have evolved wide arrays of defence mechanisms against their natural enemies [10,11]. These adaptations have also been suggested to alter the virulence of the bacteria [11–13]. For example, a biofilm-forming ability can effectively lower predation pressure by ciliate predators that prey in the open water. However, the biofilm-forming ability of many bacteria can also be directly linked to the virulence of bacteria as it can prevent macrophage phagocytosis inside the multicellular host [11,14–16].

In addition to the means that prevent predator ingestion in the first place, bacteria have evolved ways to survive the ingestion process and even benefit from it [11]. Survival and reproduction inside protozoan predators, especially in amoebae, may have even contributed to the evolution of several bacterial pathogens [11]. Therefore, virulence could have evolved via adaptations to survive inside protozoan food vacuoles, which could then promote survival within phagocytes in the immune system [17,18]. Perhaps the most typical example of this type of evolution is *Legionella pneumophila* causing Legionnaires' disease. This species is sometimes found as a parasite of free-living amoeba [19–21]. However, infection of the human body is an evolutionary dead end for *L. pneumophila* because human-to-human transmission is unlikely [22,23]. This suggests that the virulence traits of *L. pneumophila* are not evolved from human-bacteria interaction, but rather "coincidentally" via amoeba-bacteria interaction

[2,24]. In fact this linkage is assumed strong enough that the virulence of bacterial clones are frequently assayed indirectly via amoebae resistance tests [25–31].

Bacteriophages can also have a profound impact on the evolution of bacterial virulence. Bacteriophages are known to carry important virulence genes [32–34]. For example, they have been found to contain genes encoding exotoxins and other virulence factors that can be horizontally transferred into the bacterial genome [35,36]. Moreover, bacteria can alter their cell surface antigens to evade phage adsorption [10], whilst host immune systems rely on bacterial surface antigens to identify bacterial invaders [37,38]. Thus bacteriophage selected bacterial surface antigens could indirectly affect host entry, either positively or negatively.

Although protozoan predators and bacteriophages could potentially contribute to elevated bacterial virulence, outside-host defensive adaptations can also be costly and traded off with virulence related traits [39,40]. For example, when bacteria experience protozoan predation, the motility of bacteria that is sometimes positively linked to virulence [41,42] can trade off with anti-predator traits resulting in lowered virulence [39]. It has also been shown that elevated outside-host temperature can select for higher virulence in Serratia marcescens, while coevolution with phage can counteract this effect [43]. Moreover high virulence in Salmonella typhimurium can be costly in terms of reduced growth in the outside host environment because of the expression of virulence factor (type III secretion system) in a non-host environment [44]. The nutritional conditions of the bacterial growth environment can also significantly affect bacterial metabolism and the expression of virulence factors [45–47]. For example, it has been found that the virulence of the pathogenic fungi was negatively correlated to the carbon-to-nitrogen (C:N) ratio of the culturing medium [48–50]. Therefore, if a similar correlation occurs for bacteria, then the costly virulence traits might be selected against during a prolonged period in a non-host environment. In conclusion, the environmental lifestyle can attenuate or strengthen the virulence depending on the selection forces in the system [7].

Although predators are supposed to play an important role in the evolution of virulence, experiments testing this theory are rare and the studies that do exist only consider a single predator system [40,51,52]. However, in a natural environment it is more conceivable that several predators are present simultaneously, potentially complicating the picture considerably. To test how virulence and other life-history traits evolve in complex enemy communities, we cultured the facultative pathogen S. marcescens either alone or with three types of common bacterial predators (amoeba, ciliate and bacteriophage in all seven possible combinations) in a simulated pond water environment for 8 weeks. S. marcescens is a gram-negative opportunistic pathogen infecting a broad spectrum of hosts, including plants, corals, nematodes, insects, fish and mammals [57,58]. They can also be found free-living in soil, freshwater, and marine ecosystems [59,60] making it likely that S. marcescens frequently encounters parasitic and predatory organisms. Notably, S. marcescens is also capable of re-entering the environment after decomposing the host. This creates the possibility that the pathogen virulence is selected in nature by both environmental and host-pathogen interactions. During the experiment we followed the population dynamics of the prey bacterium. Due to the presumed importance of predators on the evolution of the bacterial virulence [28,53,54], the amoeba densities were also followed throughout the experiment. After the evolution experiment, a library containing 384 differentially evolved clones was built to detect changes in virulence, growth

ability, biofilm-forming ability and amoeba resistance. The virulence of the ancestor and the evolved bacteria, S. marcescens Db11 was quantified in the fruit fly (Drosophila melanogaster) hosts via an oral infection model [55]. Since phagocytes play a vital role in the clearance of the Db11 from the hemolymph in this animal model [55], we believed that choosing the bacterial strain and infection model was relevant to our study. We hypothesized that if S. marcescens Db11 gained amoeba-resistance in the presence of amoeba predation, this resistance could be used to fight against phagocytes in the hemolymph, and thus gain higher virulence. With data from a multi-predator experiment we can test if the bacterial virulence is selected for, a result that is expected in the presence of bacterial enemies (phage, ciliate and amoebae), especially amoebae. However, it is also plausible that selection pressures by bacterial enemies could select against virulence [39,40].

Methods

Study species

Serratia marcescens Db11 [56,57] was initially isolated from a dead fruit fly and was kindly provided by Prof. Hinrich Schulenburg. The predatory particle feeding ciliate, Tetrahymena thermophila (strain ATCC 30008) has a short generation time of ca. 2 h [61] and was obtained from American Type Culture Collection. It is routinely maintained in PPY (Proteose Peptone Yeast Medium) at 25°C [51,62]. The free-living amoeba, Acanthamoeba castellanii (strain CCAP 1501/10) has a generation time ca. 7 h [62] and was obtained from Culture Collection of Algae and Protozoa (Freshwater Biological Association, The Ferry House, Ambleside, United Kingdom) and routinely maintained in PPG (Proteose Peptone Glucose Medium) [63] at 25°C. Obligatory lytic bacteriophage Semad11, capable of infecting S. marcescens Db11, was isolated from a sewage treatment plant in Jyväskylä, Finland in 2009. No specific permission was required for collection or location of the bacteriophage. Semad11 is a T7-like bacteriophage belonging to Podoviridae (A.-M. Örmälä-Odegrip, unpublished data).

The evolution experiment was performed in New Cereal Leaf-Page's Amoeba Saline Solution (NAS) medium which was prepared as follows: 1 g of cereal grass powder (Aldon Corp., Avon, NY) was boiled in 1 liter of dH_2O for 5 minutes, and then filtered through a glass fiber filter (GF/C, Whatman). After cooling, 5 ml of both PAS stock solutions I and II were added before being made up to a final volume of 1 litre with deionized water [64,65].

Before the experiment started, the organisms were cultured separately and prepared as follows: bacterial culture, a single colony of S. marcescens was seeded to 80 ml of NAS medium in a polycarbonate Erlenmeyer flask capped with a membrane filter (Corning). The flask was incubated at 25°C on a rotating shaker (120 rpm) for 48 hours. The amoeba and ciliate cells were harvested and washed twice in 40 ml of PAS (Page's Amoeba Saline) with centrifugation at $1200 \times g$ for 15 min to pellet the cells. After the centrifugation, cells were suspended in PAS and adjusted to a final concentration of ca. 10 cells μl^{-1}. To prepare the bacteriophage stock, LB-Soft agar (0.7%) from semi-confluent plates was collected and mixed with LB (4 ml per plate), and incubated for 3.5 h at 37°C. Debris was removed by centrifugation for 20 min at $9682 \times g$ at 5°C. Stock was filtered with 0.2 μm Acrodisc Syringe Filters (Pall). The bacteriophage stock was diluted 1:100,000 in NAS medium, giving approximately 10^6 plaque-forming unit (PFU) ml^{-1}.

Evolution experiment

The bacterium *S. marcescens* was either cultured alone or in a co-culture with the ciliates, amoebae and bacteriophages enemies 8 combinations (B, BA, BC, BP, BAC, BAP, BCP and BACP; B: bacteria; A: amoebae; C: ciliate; P: phage; Figure 1) for 8 weeks. Each treatment was replicated in 8 flasks. The experiment was initiated in 25 cm^2 polystyrene flasks with 0.2 μm hydrophobic filter membrane caps (Sarstedt). Each flask was inoculated with 1 ml of the appropriate microorganism suspension and then the total volume was adjusted to 15 ml with NAS medium. The static liquid cultures were incubated at 25°C and 50% of the medium were replaced weekly with fresh NAS medium, making the system a pulsed resource type [66,67]. Static liquid culture would create a spatial structuration that was similar to the pond water environment. All samples were taken just before the weekly medium renewal (Figure 1).

Measurements during the evolution experiment

Bacterial biomass dynamics. Bacterial biomass in the free water phase was measured from 5 separate 200 μl samples from each flask on 100-well Honeycomb plates (Oy Growth Curves Ab Ltd). The amount of biomass was measured as optical density (OD) at 460–580 nm wavelength using Bioscreen C spectrophotometer (Oy Growth Curves Ab Ltd). The measurements were repeated 10 times at 5 min intervals. The mean of the measurements was used in the data analysis. To measure the amount of *S. marcescens* biofilm attached to the flask walls after 8 weeks had expired, 15 ml of 1% crystal violet solution (Sigma-Aldrich) was injected to the flasks. After 10 minutes, the flasks were rinsed 3 times with distilled water, and then 15 ml of 96% ethanol was added to flasks to dissolve crystal violet from the walls for 24 hours [68]. The amount of biofilm was quantified with the OD of the crystal violet-ethanol solution at 460–580 nm with Bioscreen C spectrophotometer [66].

Amoeba population dynamics. To follow the population dynamics of the amoeba, we measured the density of amoeba cells attached on the flask well. This measurement largely reflects the amoeba population dynamics in the flasks since the proportion of floating cells and cysts would be minimal after 7 days culture in the static cultures. In brief the flasks were carefully flipped and images (total area 5.23 mm^2) of the flask wall were digitized with an Olympus SZX microscope (32× magnification). The amoeba cells

attached to the flask wall were counted with a script developed in our lab for the Image Pro Plus software (v. 7.0) (Material S1). To determine the ciliate density by the end of the experiment, 250 μl of open water sample was mixed with 10 μl Lugol solution and injected into a glass cuvette rack (depth 2.34 mm). For each sample, 8 randomly placed images (total area 41.84 mm^2) were digitized with an Olympus SZX microscope (32× magnification). The cell numbers in each image were counted with an Image Pro Plus script [69].

Detecting phage presence. To detect if the bacteriophages were present in the microcosms throughout the experiment and to detect possible contamination, we took 3 independent 500 μl samples from all flasks at the end of evolution experiment. The samples were treated with chloroform and centrifuged to remove bacteria, amoebas and ciliates. 10 μl supernatant drops were then added to 1.5% agar plates. The upper layer of the each plate was covered with 0.7% LB-agar that was mixed with 200 μl of overnight grown *S. marcescens* Db11 ancestor cells. The plates were incubated overnight in 25°C and the presence of phage plaques were checked.

Measurements after the evolution experiment

Amoeba plaque test. After the evolution experiment was finished, half of the flasks from each treatment were randomly sampled to test for any resistance of the bacteria to amoeba predation. The test was adapted from a Wildschutte *et al.* [70] briefly, the flasks from the evolution experiments were shaken vigorously before 1 ml of the culture was transferred to a new tube containing 7 ml of dH$_2$O. The tubes were mixed thoroughly and then centrifuged at 250 g for 10 min to bring down the suspended protozoan cells. 1 ml of the supernatant was spread evenly on to LN agar plates (PAS with 0.2% peptone, 0.2% glucose and 1% agar). A total of 10^5 predatory amoeba cells (washed twice in PAS) suspended in 15 μl PAS solution were added to a sterile paper disk, and then placed in the middle of the plate. All the plates were incubated at 25°C for 8 days and then photographed. The images of the plates were used to measure plaque sizes with Image Pro Plus software (v. 7.0). A large plaque size indicates a small amoeba predation resistance.

Growth and biofilm forming ability of the individual clones. After the evolution experiment was complete, liquid samples from each replicate population of all treatments were

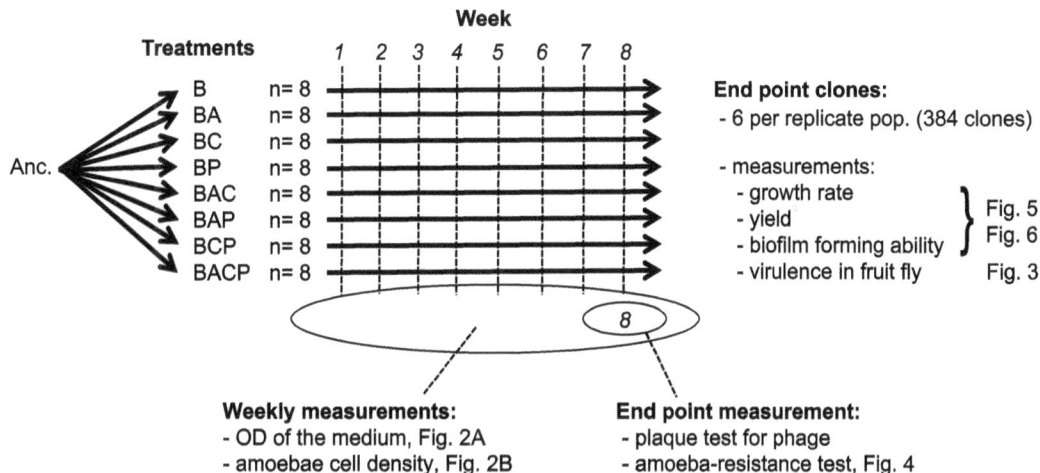

Figure 1. Schematic overview of our experimental evolution study, number of replicate populations, legends for treatments (Anc.: ancestral bacterial strain DB 11; B: bacteria; A: amoebae; C: ciliate; P: phage), and descriptions of different measurements.

streaked on three Luria–Bertani (LB) agar plates. They were incubated for 48 hours at 25°C, two bacterial colonies were randomly picked from each plate and inoculated to 5 ml LB liquid medium. The clones were grown at 25°C overnight on a shaker (120 rpm). To make stock cultures of the clone library, 200 μl of the liquid culture of each strain was mixed with 200 μl of 80% glycerol on 100-well Honeycomb plates in a randomized order and stored at −80°C. Prior to clonal growth measurements stock cultures from the clone library were inoculated to 100-well Honeycomb plates, directly from the freezer, with a plate replicator (EnzyScreen). Each well of the 100-well Honeycomb plates contained 400 μl fresh LB liquid medium. The OD of each well was measured continuously without shaking for 30 hours in 5 min intervals to estimate the maximum growth rate and maximum population size of the clones. After 30 hours, 100 μl of 1% crystal violet solution (Sigma-Aldrich) was added to each well to quantify amount of biofilm that was produced. After 10 minutes incubation in 25°C, the plates were rinsed with distilled water 3 times and then 400 μl of 96% ethanol was added to each well and left for 24 hours to dissolve the crystal violet from the walls [68]. The amount of biofilm was quantified by measuring the OD of the crystal violet-ethanol solution at 460–580 nm with Bioscreen C spectrophotometer [66].

Identical measurements were also recorded with NAS medium either with or without the protozoan predators (amoeba or ciliate). The abovementioned clones were first grown in 400 μl LB liquid medium in 100-well Honeycomb plates. After incubation at 25°C for 24 h in static cultures, 10 μl of the bacterial culture was transferred to 100-well Honeycomb plates. Each well contained 390 μl NAS mixed with or without NAS washed amoeba (5–10 cells/μl) or ciliate cells (0.5–1 cells/μl). Subsequent measurements for growth and biofilm assay were performed as described above.

Estimation of growth rate and yield was based on the Matlab script written by TK that fits linear regression to 25 time-points along a sliding data window with background correction, using ln-transformed OD data. The maximal growth rate is determined by finding the largest slope of linear regression within all fitted regressions for the particular clone. The yield was determined as the highest average OD over the 25 data point window.

Virulence of the evolved clones. Stock cultures of the clone library were inoculated to 100-well Honeycomb plates, filled with 400 μl fresh LB liquid medium with plate replicator (EnzyScreen). For a positive control the ancestor *S. marcescens* Db11 was added to two separate wells in each plate and used in the subsequent infection experiment. After 24 h incubation at 25°C without shaking, 800 μl of the bacterial culture was mixed with the 800 μl of 100 mM sucrose solution. The mixture was absorbed to cotton dental roll (Top Dent, Lifco Dental, Enköping, Sweden) folded on the bottom of a standard 75×23 mm fly vial (Sarstedt, Nümbrecht, Germany). 1600 μl of 100 mM sucrose solution was used as a negative control. Ten *D. melanogaster* adults (2–3 days old) from a large laboratory colony (Oregon R, kindly provided by Christina Nokkala from the University of Turku) were transferred to each vial and plugged with cotton. This was done for all the bacterial clones. Deaths of flies were monitored over next 4 days at 3–6 h intervals.

Statistical analysis

Changes in bacterial density and amoeba density were compared using repeated measurements ANOVA. The effects of the evolutionary treatments on bacterial virulence were quantified with Cox regression by fitting evolutionary treatment and identity of population as categorical covariates. The amoeba plaque test and the amount of biofilm at the end of the evolution experiment

were compared using ANOVA. All the analyses were done with SPSS v. 19 (IBM).

Life history and defensive traits of evolved clones were tested with ANOVA including treatment (all possible combinations of predators) as a fixed factor and population identity as a random factor. From the data we tested effects of treatments on growth rate, yield and biofilm, as well as growth rate and yield under the influence of ciliate or amoebae presence. Coevolution of traits was studied with MANOVA and subsequent eigenanalysis to reveal if changes in certain traits would lead to corresponding changes in other traits across the treatments. Thus, this analysis allows pinpointing strongly interconnected traits [71]. MANOVA and eigenanalysis was performed with MATLAB function manova1 (R2012a, Mathworks; Statistics toolbox) for population averaged trait values. Values used for MANOVA for virulence were hazard function coefficients averaged over the populations within each treatment.

Two replicated populations (one in the treatment BC and one in BAC) were found contaminated by Semad11 phages. Moreover, in two replicates of the treatment BACP phages were not detected by plaque assay. All the other samples from phage containing treatments formed phage plaques on the ancestor Db11 bacterial lawn. This confirms that phages did not go extinct during the experiment. The aforementioned 4 flasks were excluded from the data analysis.

Results

The population dynamics of bacterial prey and amoeba predators during the experiment

The presence of predators generally reduced the bacterial biomass in free water phase (OD of the medium: $F_{7, 52} = 674.620$, p<0.001; Figure 2A). The ciliates reduced the biomass most dramatically: on average by 27% during the weeks 1–8 when compared to the control (B). Biomass reduction by ciliate and phage (BCP), and amoeba and ciliate (BAC) communities was 25%. Amoeba (BA) and amoeba and phage (BAP) communities reduced bacteria biomasses by 20%. The bacteriophage (BP) reduced the biomass only by 2%. The pairwise-comparisons of the rest of the treatments were significant after Bonferroni correction, except BCP vs. BC, and BACP vs. BAC.

The amount of biofilm produced in each treatment was different at week 8 when measured directly from the microcosm walls (ANOVA, $F_{7, 52} = 39.101$, p<0.001). The highest amount of biofilm was found in the presence of ciliates (BC) and the lowest amount of biofilm was found in the treatments BAC and BACP. Detailed pairwise comparison can be found in Table S1.

The amoeba population sizes declined in all treatments after the initial increase during the first week ($F_{3, 28} = 280.257$, p<0.001; Figure 2B). The amoeba population sizes were higher in amoeba (BA) and amoeba and phage (BAP) treatments (on average 30 cells μl^{-1}) throughout the 8-week evolution experiment. Adding phage to the amoeba treatment did not change the population dynamics of the amoeba (Fisher's LSD: BA vs. BAP, p = 0.485). However, ciliates reduced the amoeba population sizes: on average only 5 cells μl^{-1} were found throughout the experiment in treatment BAC, and on average 8 cells/μl in treatment BACP.

Virulence

In order to explore if past selection with predators had influenced virulence we utilized the *Drosophila* oral infection assay. The treatment group that had evolved with ciliates and phages (BCP) had clearly lower virulence than the rest of the evolved treatment groups (p<0.01 in all pairwise comparisons;

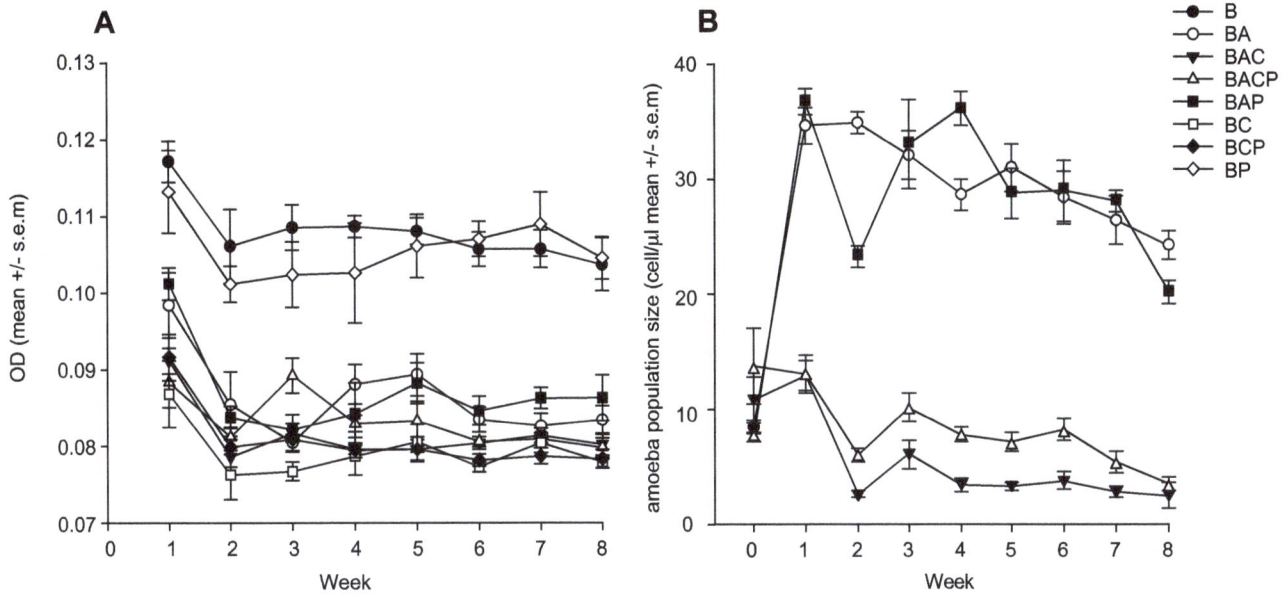

Figure 2. Bacterial biomass dynamics (A) and amoebae population dynamics (B) during the eight-week evolution experiment. The bacteria were reared alone or in several combinations of bacterial enemies (Anc.: ancestral bacterial strain DB 11; B: bacteria; A: amoebae; C: ciliate; P: phage). See Table S1 for pairwise comparisons.

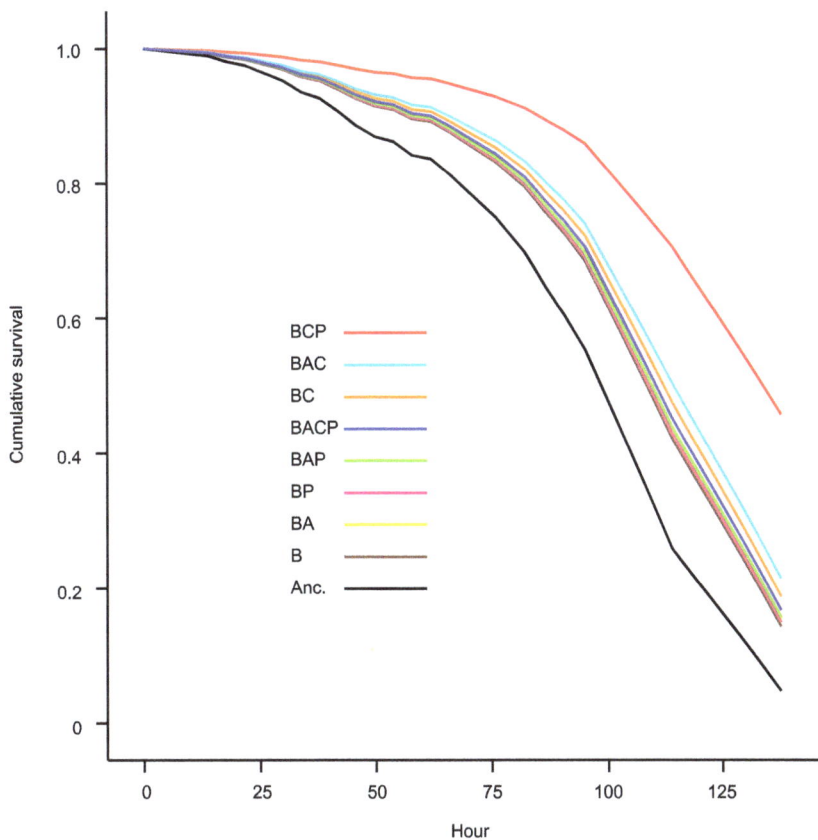

Figure 3. Cumulative survival curves of the fruit flies that were infected with evolved and ancestral bacterial clones. Anc.: ancestral bacterial strain DB 11; B: bacteria; A: amoebae; C: ciliate; P: phage. The survival curves represented the pooled survival data of the 480 fly individuals for each treatment (10 flies per vial, 6 clones per population and 8 replicates per treatment). The treatment codes are in the order of the increasing virulence. See Table S2 for pairwise comparisons.

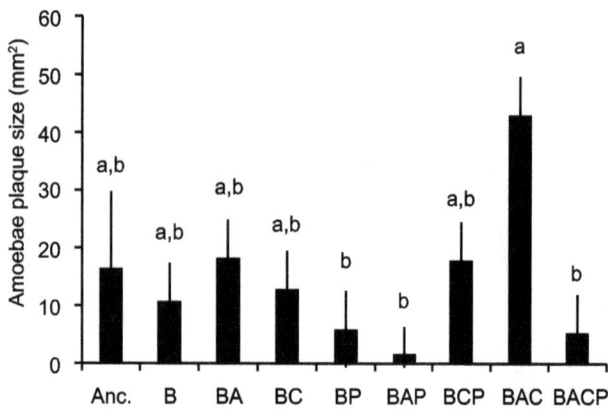

Figure 4. Sensitivity of evolved bacteria on amoebae predation measured using amoeba plaque test. Anc.: ancestral bacterial strain DB 11; B: bacteria; A: amoebae; C: ciliate; P: phage. Sensitivity is measured as a plaque size (mm²) in bacterial lawn caused by the introduced amoeba in semi-solid agar plate. Letters indicate if treatment means are statistically similar (p>0.05), after Bonferroni correction for multiple comparisons. Tests are based on the post hoc comparisons of estimated marginal means for treatments ANOVA. All bars correspond to 4 randomly picked samples from 8 replicate populations.

Figure 3; Table S2). In this group the between population variation was also very high and statistically significant. In the majority of the other groups, the between population variation was clearly non-significant (p<0.135, but in BCP, p<0.001; BAC p = 0.003). Moreover, all treatments had lower virulence than the ancestor (p<0.01 in all pairwise comparisons; Figure 3; Table S2). There was no statistical support for the difference in virulence between any other treatment groups.

Sensitivity to amoebae predation

The amoeba plaque test revealed that the sensitivity to amoeba predation (measured with the area of visible plaque formed on bacterial lawn) was highest in bacteria co-cultured with amoeba and ciliates (BAC: p<0.05 in all pairwise comparisons; Figure 4;

Table S1). The lowest amoeba sensitivity was found in the treatment where the amoeba were reared with phages (BAP; Figure 4) or with ciliates and phages (BACP; BAP vs. BACP: p = 0.66, Figure 4). The detailed result of pairwise comparisons can be found in the Table S1. Although the treatment group that had evolved with ciliates and phages (BCP) had the lower virulence than the other evolved treatment groups, its sensitivity to amoeba predation did not differ from the others in pairwise comparisons (Table S1), which suggests no clear link between amoebae predation and virulence.

Life-history traits

Treatments did not influence the maximal growth rate strongly (Table 1), however low growth rates were found in clones that had evolved with ciliates (BC; Figure 5A; Table S2). The ancestor clones had the largest yield. Evolutionary treatments did not differ greatly from each other but the clones that had evolved with phages had the lowest yield (BP; Figure 5B; Table S2). Evolutionary changes occurred most dramatically in the biofilm forming ability. The highest biofilm forming abilities were found from the clones that had evolved alone (B) or with amoeba and ciliate (BAC). Intermediate biofilm production was found in ancestral clones (Anc.) or if clones had evolved with amoebae (BA) or with all enemies (BACP). The lowest biofilm production was found if clones that had evolved with ciliate (BC), phage (BP), amoebae and phage (BAP) or with ciliate and phage (BCP) (Figure 5C; Table S1).

From the defensive traits the strongest changes were observed in growth rate and yield when the bacteria were co-cultured with amoebae. In both of the traits ancestor bacteria deviated from the clones that had undergone evolutionary treatments; ancestors had higher growth rate co-cultured with amoebae but lower yield than evolved clones (Figure 6; Table S2). Similarly, from the evolved groups the highest growth rate was found from the group that had evolved with phage and amoebae (BAP), whereas its yield with amoebae was lowest (Figure 6; Table S2). Growth rate measurements did not indicate that treatments affected the resistance of clones against ciliate predators. However, yield with ciliates was lowest if bacteria had evolved alone (B) or with amoebae and

Table 1. Estimated (ANOVA) evolutionary effects of different combinations of enemies (treatment), and population identity on bacterial virulence against *Drosophila melanogaster*, and on bacterial life-history traits, measured alone or with amoebae or ciliate.

	Treatment			Population		
	Wald	df	p	Wald	df	p
Virulence	48.6	8	<0.001	169.58	52	<0.001
	F	df1,2	p	F	df	p
Growth rate	2.465	8,32.007	0.033	0.799	52,306	0.836
Yield	2.582	8,39.627	0.023	1.349	52,306	0.066
Biofilm	12.987	8,37.001	<0.001	1.097	52,306	0.311
	F	df1,2	p	F	df	p
Growth with amoebae	9.565	8,40.485	<0.001	1.456	52,306	0.029
Yield with amoebae	3.595	8,46.032	0.003	2.885	52,306	<0.001
Growth with ciliate	1.296	8,39.223	0.274	1.304	52,306	0.091
Yield with ciliate	4.757	8,40.432	<0.001	1.449	52,306	0.031

Wald denotes Wald's test statistics, and F corresponds to F-test statistics, df denote degrees of freedom and p indicate statistical significance.

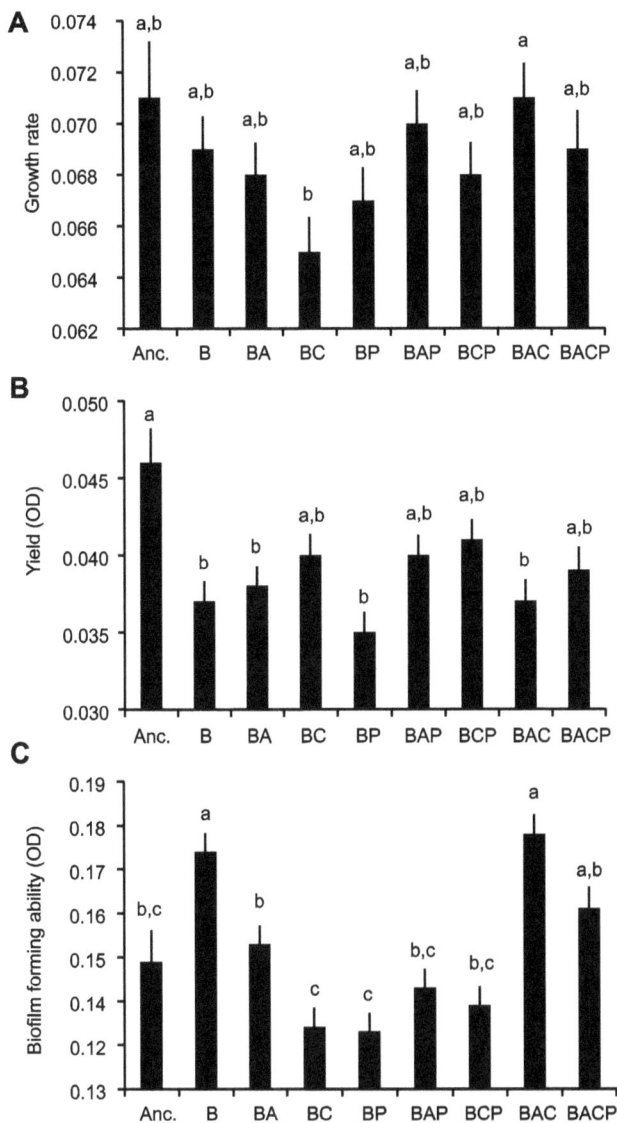

Figure 5. Growth rate (panel A), yield (panel B) and biofilm forming ability (panel C) differences between ancestral clones and clones that have evolved alone (B) or in different combinations of bacterial enemies. Anc.: ancestral bacterial strain DB 11; B: bacteria; A: amoebae; C: ciliate; P: phage. Letters indicate if treatment means are considered statistically similar (p>0.05) after Bonferroni correction for multiple comparisons. Tests are based on the post hoc comparisons of estimated marginal means for treatments of ANOVA testing the effects of treatment and population identity on these traits. Bars correspond to measurements of 6 clones from 8 replicate populations, in ancestor n = 16).

ciliate (BAC), and highest if clones had evolved with ciliate (BC) and ciliate and phage (BCP) (Figure 6C; Table 1; Table S2).

Coevolved traits

Based on the individual traits in different treatment combinations, it is difficult to get an idea of how co-ordinate traits evolved. Therefore we analysed all traits in multivariate ANOVA, followed by eigen-analysis. In this analysis we found that two dimensions dictated multivariate evolution amongst evolutionary treatments (support for two dimensions p = 0.028, for 1 dimension p<0.001). The first dimension was characterized by variation in biofilm and

ciliate defence. Those treatments that exerted strong positive selection on biofilm production had a lower yield with ciliate in free water. The second dimension of trait evolution was formed by those treatments that had increased growth rate without predators and also had a lower growth rate with ciliate. Neither of the two "major" eigen-functions linked virulence to other traits (Table 2).

When similar analysis was performed with information from the amoeba plaque test, a similar result was found, again supporting what was found for two multivariate dimensions (support for two dimensions p = 0.023, for 1 dimension p<0.001). The amoeba plaque test contributed moderately to a second eigenvector. This eigenvector had a slightly different composition than the analysis without the amoeba plaque test as the ltter contained all microcosm replicates. If anything, a higher resistance against amoebae was associated with higher biofilm forming ability and higher growth rate with ciliate. However, the amoeba plaque test clearly did not predict virulence. Since the amoeba plaque test was performed for a subset of the populations, their inclusion in more detailed measurements resulted in a smaller dataset. We base our discussion on the analysis of the larger and thus more reliable eigen-analysis without the amoeba plaque test (Table 2).

Discussion

Predators such as ciliates and amebae, and parasitic phages are expected to be the main determinants of bacterial mortality in the natural environment [8,9]. In addition to selection exerted by these predators and parasites on defensive traits, selections by bacterial natural enemies have often been suggested to lead to increased bacterial virulence [2,5,72] to such a degree that amoebae-resistance has been used as a direct proxy of strains' virulence [25,27–31,73]. However, we did not find evidence to suggest such a relationship as most of the experimental treatments had attenuated virulence. Moreover, there was no indication of coevolution of other life-history traits with virulence in evolved strains.

Contradictory to the theory that protozoan predators and phage parasites, amoebae in particular, play a strong role in the evolution of high virulence [11–13,53,72,74], we found that virulence attenuated in all of the evolved populations regardless of the presence of the enemies. This is in accordance with the previous study with *S. marcescens* that suggest that ciliates can select for attenuated virulence [39,40]. However, here we show that this is also the case with amoeba and phages [43], and under selection by multiple enemies at the same time. By far the strongest decrease in virulence was found in clones that had evolved with phages and ciliates. However, the between population variation in this group was very high and statistically significant. This was shown in separate analyses designed to test the amount of between population variance within treatments. In most of the other groups the between population variation was clearly non-significant (p<0.135, but in BCP, p<0.001; BAC p=0.003). The between population variation is often seen as a signature of drift, mutation accumulation and lack of directional selection on traits [40,75–78]. Thus, we propose that the strong decline of virulence in ciliate-phage treatment was primarily caused by the decay of unused traits through random mutation accumulation [40,76].

To find out if evolutionary changes in virulence could be linked to changes in traits that are important for fitness outside the host, we measured growth parameters and resistance against amoeba and ciliates. However, none of the traits seemed to be determining the level of virulence amongst evolved strains. Thus, it is clear that resistance to protozoan predators or changes in other measured

A

B

C

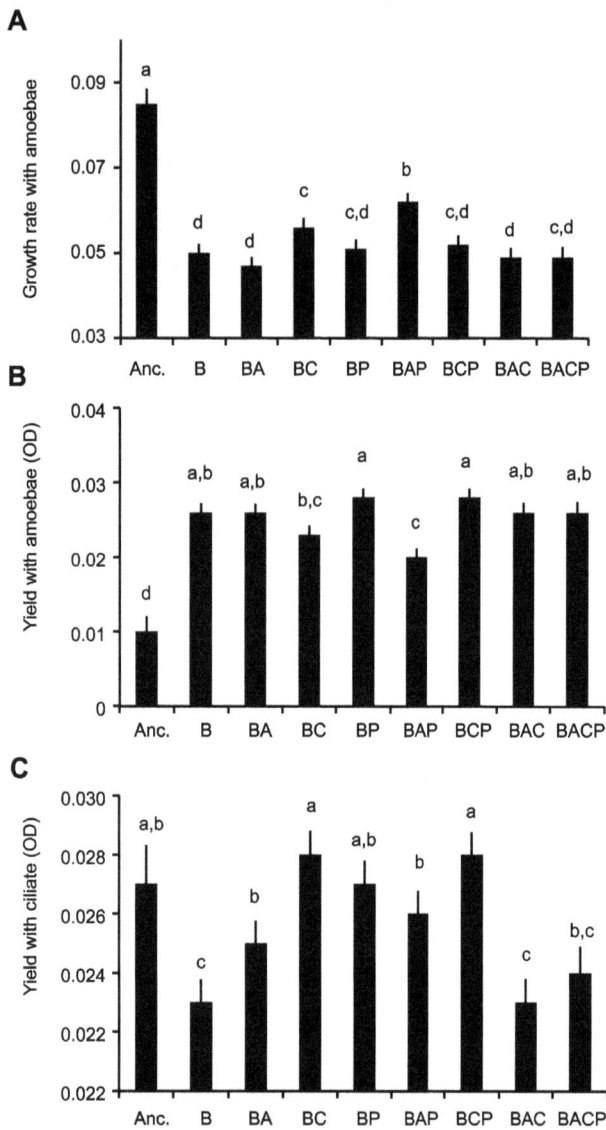

Figure 6. Growth rate with amoebae (panel A), yield with amoebae (panel B) and yield with ciliate (panel C) differences between ancestral clones and clones that have evolved alone or in different combinations of bacterial enemies. Anc.: ancestral bacterial strain DB 11; B: bacteria; A: amoebae; C: ciliate; P: phage. Letters indicate if treatment means are considered statistically similar after Bonferroni correction for multiple comparisons. Tests are based on the post hoc comparisons of estimated marginal means for treatments of ANOVA testing the effects of treatment and population identity on these traits. Bars correspond to measurements of 6 clones from 8 replicate populations, in ancestor n = 16).

life-history traits are poor indicators for virulence in *S. marcescens*. However, in other bacterial species, virulence correlated positively with bacterial defences against predators [5,12,17,26,72] (but see [39,40,77]). In addition, several lines of research have emphasized the role of growth rate with virulence [79,80] (but see [44,77]). However, it could be that virulence might be traded off with whatever trait that is under selection in the given environment. This could have led to the clear lack of a connection between virulence and life history traits in an experiment where selection pressures are different between each treatment. This is a plausible outcome if virulence is traded off with life-history traits via finite

Table 2. Eigenvectors describing dimensions of multivariate evolution under different kind of enemies (loadings, i.e. correlations, of original variables to new composite variable, based on MANOVA.

	Eig. 1.	Eig. 2.	Eig. 3.	Eig. 4.	Eig. 5.	Eig. 6.	Eig. 7.	Eig. 8.	Eig. 9.
Biofilm	**0.753**	0.326	-0.205	0.215	0.229	0.376	-0.343	-0.024	0.183
Yield	0.131	0.257	-0.268	0.671	-0.023	-0.155	0.150	-0.199	-0.149
Yield with amoeba	0.081	-0.223	-0.276	-0.200	0.330	0.094	0.452	-0.459	0.678
Yield with ciliate	**-0.579**	-0.253	-0.032	0.472	-0.203	0.111	-0.665	0.115	-0.026
Growth rate	0.018	**-0.518**	0.498	0.241	-0.664	-0.375	-0.106	0.245	0.022
Growth with amoeba	-0.150	0.228	0.489	-0.143	0.321	0.501	0.335	-0.357	0.511
Growth with ciliate	0.222	**0.610**	-0.483	-0.356	0.075	-0.209	0.184	0.319	0.460
Virulence	0.037	-0.149	0.080	0.171	0.176	0.236	0.202	0.666	0.063
Eigenvalues	4.692	0.9220	0.5689	0.1967	0.1039	0.0394	0.0173	<0.001	<0.001
% explained	71.74	14.10	8.70	3.01	1.59	0.60	0.27	<0.001	<0.001

Below the eigenvectors are the eigenvalues i.e. amount of variation explained by eigenvectors and the percentage of the total variation explained. First eigen function (Eig1) describes contribution of individual traits to the largest difference between the treatments. In eigenvectors that are considered significant (Eig 1 and 2, see results) the largest contributors to the evolutionary differences are highlighted with bold.

resources. Then, any energetically costly trait under strong selection in outside-host environments could lead to virulence attenuation. Although we did not find a strong connection with virulence traits and life-history traits, we found that regardless of the evolutionary treatments, high biofilm-forming ability was closely linked with a low yield in a condition of co-culturing with ciliates in free water, and high growth rate was linked with low growth rate with ciliates. The first eigenvector (biofilm vs. yield with ciliates) could indicate that growth in biofilms does not lead to good protection against predation in free water, or that exposure to predation leads to biofilm formation only when predators are present and thus reduces cells in a free water environment. However, these findings from the eigen analysis effectively mean that there are evolutionary constraints between different life-history traits that remain unchangeable regardless of the evolutionary treatments. Moreover, since the predators (amoeba and ciliates) were effectively reducing the population densities in the long run (Figure 2A) the lack of selection on life history traits is not a plausible explanation for the obtained results.

When we compared ancestor's life history traits to evolved strains, it seemed that ancestors grew better with amoebae and were also the most virulent clone (Figure 3). However, another amoebae-resistance measurement, yield with amoebae, was actually lower for the ancestor strain than evolved strains. Moreover, in the amoebae-plaque test the ancestor strain did not excel in comparison to evolved strains. These contrasting results from measurements that should indicate the ability to resist amoebae suggest that there is a weak indication that amoebae-resistance evolved simultaneously with virulence. Therefore we suggest that the culture conditions could attenuate virulence, without any clear changes in life history traits. Interestingly, several experimental evolution studies have previously found that bacterial virulence decreases due to the exposure to the outside-host environment [39,40,81].

Alternatively, it has been suggested that if traits are not needed under particular conditions then their alleles become harmful and accumulate which leads to unused trait decay [76]. *S. marcescens* strain Db11 was isolated from a dead *Drosophila* fly over thirty years ago and has been routinely grown in highly protein-rich LB medium. Yet, the virulence of the strain has been maintained from lab to lab [55,57,82]. It is possible that the protein-enriched culture condition like LB medium (LB containing 10 g/l tryptone and 5 g/l yeast extract) was somehow needed for *S. marcescens* Db11 to maintain its virulence. However, a more likely scenario could be that our experimental conditions (low concentration of high C:N ratio plant detritus) selected against virulence in the 8 week evolution experiment. In addition, there might be other unknown factors that could of affected the virulence. Although predators in general effectively lowered population sizes of the bacteria (Figure 2A), spatial heterogeneity and potentially other niches created by the static cultures might lower the strength of selection on defensive traits. For example, the biofilm could act as a protection against protozoan predation, and some bacteria might have not been under selection at all.

To summarize, we found no support for the idea that enemies outside the host could select for higher virulence, as all experimental treatments led towards lower bacterial virulence. Among evolved strains virulence was not linked to other life-history characters, suggesting that selective pressures from protozoan predators (ciliates and amoebae) and parasitic phages did not dictate virulence evolution. In conclusion, our dataset offered a case against coincidental evolution of the virulence hypothesis that expects outside-host selections, especially amoebae predation, would lead to higher bacterial virulence [2,3,5,72].

Supporting Information

Table S1 Pairwise comparisons of experimental treatment differences on amoeba population sizes, biomass in the free water phase and attached biofilm at the end of the experiment. Significant pairwise comparisons after Bonferroni correction are high-lighted with bold (critical α: $0.00178 = 0.05 \div 28$, but amoebae population size critical α: 0.008 (0.05/6)). (B = bacteria alone, BC = with ciliate; BA with amoebae; BP with phage etc.. Anc. stands for ancestor Db11 strain).

Table S2 Pairwise comparisons of experimental treatment differences on measured virulence, growth and defensive traits. Significant pairwise comparisons after Bonferroni correction are highlighted with bold (critical α: 0.00138 (0.05/36)) (B = bacteria alone, BC = with ciliate; BA with amoebae; BP with phage etc.. Anc. stands for ancestor Db11 strain).

Material S1 The macro for the Image-Pro Plus program to automatically count protozoan cells in microscopic images.

Acknowledgments

We thank Kalevi Viipale for intensive discussions on the conceptual issues. We also thank Angus Buckling, Anna-Liisa Laine, Waldron Samuel and Lauri Mikonranta for commenting on the manuscript.

Author Contributions

Conceived and designed the experiments: JZ TK AMÖ JM JL. Performed the experiments: JZ AMÖ. Analyzed the data: JZ TK AMÖ JM JL. Contributed reagents/materials/analysis tools: JZ TK AMÖ JM JL. Wrote the paper: JZ TK AMÖ JM JL.

References

1. Casadevall A, Pirofski LA (2003) The damage-response framework of microbial pathogenesis. Nat Rev Microbiol 1: 17–24.
2. Levin BR (1996) The evolution and maintenance of virulence in microparasites. Emerg Infect Dis 2: 93–102.
3. Levin BR, Svanborg Eden C (1990) Selection and evolution of virulence in bacteria: an ecumenical excursion and modest suggestion. Parasitology 100 Suppl: S103–115.
4. May RM, Anderson RM (1983) Epidemiology and Genetics in the Coevolution of Parasites and Hosts. P Roy Soc Lond B Bio 219: 281–313.
5. Adiba S, Nizak C, van Baalen M, Denamur E, Depaulis F (2010) From grazing resistance to pathogenesis: the coincidental evolution of virulence factors. PloS one 5: e11882.
6. Coombes BK, Gilmour MW, Goodman CD (2011) The evolution of virulence in non-o157 shiga toxin-producing *Escherichia coli*. Front Microbiol 2: 90.
7. Brown SP, Cornforth DM, Mideo N (2012) Evolution of virulence in opportunistic pathogens: generalism, plasticity, and control. Trends Microbiol 20: 336–342.
8. Jürgens K, Matz C (2002) Predation as a shaping force for the phenotypic and genotypic composition of planktonic bacteria. Antonie Van Leeuwenhoek 81: 413–434.
9. Suttle CA (2005) Viruses in the sea. Nature 437: 356–361.
10. Labrie SJ, Samson JE, Moineau S (2010) Bacteriophage resistance mechanisms. Nat Rev Microbiol 8: 317–327.
11. Matz C, Kjelleberg S (2005) Off the hook - how bacteria survive protozoan grazing. Trends Microbiol 13: 302–307.
12. Brüssow H (2007) Bacteria between protists and phages: from antipredation strategies to the evolution of pathogenicity. Mol Microbiol 65: 583–589.

13. Greub G, Raoult D (2004) Microorganisms resistant to free-living amoebae. Clin Microbiol Rev 17: 413–433.

14. Hall-Stoodley L, Costerton JW, Stoodley P (2004) Bacterial biofilms: from the natural environment to infectious diseases. Nat Rev Microbiol 2: 95–108.

15. Jousset A (2012) Ecological and evolutive implications of bacterial defences against predators. Environ Microbiol 14: 1830–1843.

16. Thurlow LR, Hanke ML, Fritz T, Angle A, Aldrich A, et al. (2011) Staphylococcus aureus biofilms prevent macrophage phagocytosis and attenuate inflammation in vivo. J Immunol 186: 6585–6596.

17. Al-Quadan T, Price CT, Abu Kwaik Y (2012) Exploitation of evolutionarily conserved amoeba and mammalian processes by Legionella. Trends Microbiol 20: 299–306.

18. Gao LY, Harb OS, AbuKwaik Y (1997) Utilization of similar mechanisms by Legionella pneumophila to parasitize two evolutionarily distant host cells, mammalian macrophages and protozoa. Infect Immun 65: 4738–4746.

19. Abu Kwaik Y, Gag LY, Stone BJ, Venkataraman C, Harb OS (1998) Invasion of protozoa by Legionella pneumophila and its role in bacterial ecology and pathogenesis. Appl Environ Microb 64: 3127–3133.

20. Ohno A, Kato N, Sakamoto R, Kimura S, Yamaguchi K (2008) Temperature-dependent parasitic relationship between Legionella pneumophila and a free-living amoeba (Acanthamoeba castellanii). Appl Environ Microb 74: 4585–4588.

21. Rowbotham TJ (1980) Preliminary report on the pathogenicity of Legionella pneumophila for freshwater and soil amoebae. J Clin Pathol 33: 1179–1183.

22. Fields BS, Benson RF, Besser RE (2002) Legionella and Legionnaires' disease: 25 years of investigation. Clin Microbiol Rev 15: 506–526.

23. Muder RR, Yu VL, Woo AH (1986) Mode of transmission of Legionella pneumophila. A critical review. Arch Intern Med 146: 1607–1612.

24. Ensminger AW, Yassin Y, Miron A, Isberg RR (2012) Experimental Evolution of Legionella pneumophila in Mouse Macrophages Leads to Strains with Altered Determinants of Environmental Survival. Plos Pathog 8.

25. Bonifait L, Charette SJ, Filion G, Gottschalk M, Grenier D (2011) Amoeba Host Model for Evaluation of Streptococcus suis Virulence. Appl Environ Microb 77: 6271–6273.

26. Cosson P, Soldati T (2008) Eat, kill or die: when amoeba meets bacteria. Curr Opin Microbiol 11: 271–276.

27. Froquet R, Lelong E, Marchetti A, Cosson P (2009) Dictyostelium discoideum: a model host to measure bacterial virulence. Nat Protoc 4: 25–30.

28. Greub G, La Scola B, Raoult D (2004) Amoebae-resisting bacteria isolated from human nasal swabs by amoebal coculture. Emerg Infect Dis 10: 470–477.

29. Hasselbring BM, Patel MK, Schell MA (2011) Dictyostelium discoideum as a Model System for Identification of Burkholderia pseudomallei Virulence Factors. Infect Immun 79: 2079–2088.

30. Lelong E, Marchetti A, Simon M, Burns JL, van Delden C, et al. (2011) Evolution of Pseudomonas aeruginosa virulence in infected patients revealed in a Dictyostelium discoideum host model. Clin Microbiol Infec 17: 1415–1420.

31. Smith MG, Gianoulis TA, Pukatzki S, Mekalanos JJ, Ornston LN, et al. (2007) New insights into Acinetobacter baumannii pathogenesis revealed by high-density pyrosequencing and transposon mutagenesis. Genes & development 21: 601–614.

32. Brüssow H, Canchaya C, Hardt WD (2004) Phages and the evolution of bacterial pathogens: from genomic rearrangements to lysogenic conversion. Microbiol Mol Biol Rev: 68: 560–602, table of contents.

33. Hacker J, Hentschel U, Dobrindt U (2003) Prokaryotic chromosomes and disease. Science 301: 790–793.

34. Hacker J, Kaper JB (2000) Pathogenicity islands and the evolution of microbes. Annu Rev Microbiol 54: 641–679.

35. Boyd EF (2012) Bacteriophage-encoded bacterial virulence factors and phage-pathogenicity island interactions. Adv Virus Res 82: 91–118.

36. Casas V, Maloy S (2011) Role of bacteriophage-encoded exotoxins in the evolution of bacterial pathogens. Future Microbiol 6: 1461–1473.

37. Bell JK, Mullen GE, Leifer CA, Mazzoni A, Davies DR, et al. (2003) Leucine-rich repeats and pathogen recognition in Toll-like receptors. Trends Immunol 24: 528–533.

38. Sahly H, Keisari Y, Crouch E, Sharon N, Ofek I (2008) Recognition of bacterial surface polysaccharides by lectins of the innate immune system and its contribution to defense against infection: the case of pulmonary pathogens. Infect Immun 76: 1322–1332.

39. Friman VP, Lindstedt C, Hiltunen T, Laakso J, Mappes J (2009) Predation on multiple trophic levels shapes the evolution of pathogen virulence. PloS one 4: e6761.

40. Mikonranta L, Friman V-P, Laakso J (2012) Life History Trade-Offs and Relaxed Selection Can Decrease Bacterial Virulence in Environmental Reservoirs. PloS one 7: e43801.

41. Josenhans C, Suerbaum S (2002) The role of motility as a virulence factor in bacteria. Int J Med Microbiol 291: 605–614.

42. Lertsethtakarn P, Ottemann KM, Hendrixson DR (2011) Motility and Chemotaxis in Campylobacter and Helicobacter. Nat Rev Microbiol, Vol 65 65: 389–410.

43. Friman VP, Hiltunen T, Jalasvuori M, Lindstedt C, Laanto E, et al. (2011) High temperature and bacteriophages can indirectly select for bacterial pathogenicity in environmental reservoirs. PloS one 6: e17651.

44. Sturm A, Heinemann M, Arnoldini M, Benecke A, Ackermann M, et al. (2011) The cost of virulence: retarded growth of Salmonella typhimurium cells expressing type III secretion system 1. Plos Pathog 7: e1002143.

45. Friedman ME, Kautter DA (1962) Effect of nutrition on the respiratory virulence of Listeria monocytogenes. J Bacteriol 83: 456–462.

46. Heckly RJ, Blank H (1980) Virulence and viability of Yersinia pestis 25 years after lyophilization. Appl Environ Microbiol 39: 541–543.

47. Midelet-Bourdin G, Leleu G, Copin S, Roche SM, Velge P, et al. (2006) Modification of a virulence-associated phenotype after growth of Listeria monocytogenes on food. J Appl Microbiol 101: 300–308.

48. Ali S, Huang Z, Ren SX (2009) Media composition influences on growth, enzyme activity, and virulence of the entomopathogen hyphomycete Isaria fumosoroseus. Entomol Exp Appl 131: 30–38.

49. Safavi SA, Shah FA, Pakdel AK, Reza Rasoulian G, Bandani AR, et al. (2007) Effect of nutrition on growth and virulence of the entomopathogenic fungus Beauveria bassiana. FEMS Microbiol Lett 270: 116–123.

50. Wu JH, Ali S, Huang Z, Ren SX, Cai SJ (2010) Media Composition Influences Growth, Enzyme Activity and Virulence of the Entomopathogen Metarhizium anisopliae (Hypocreales: Clavicipitaceae). Pak J Zool 42: 451–459.

51. Friman VP, Hiltunen T, Laakso J, Kaitala V (2008) Availability of prey resources drives evolution of predator-prey interaction. Proc Biol Sci 275: 1625–1633.

52. Hosseinidoust Z, van de Ven TG, Tufenkji N (2013) Evolution of Pseudomonas aeruginosa virulence as a result of phage predation. Appl Environ Microbiol 79: 6110–6116.

53. Molmeret M, Horn M, Wagner M, Santic M, Abu Kwaik Y (2005) Amoebae as training grounds for intracellular bacterial pathogens. Appl Environ Microbiol 71: 20–28.

54. Steinert M, Heuner K (2005) Dictyostelium as host model for pathogenesis. Cell Microbiol 7: 307–314.

55. Nehme NT, Liegeois S, Kele B, Giammarinaro P, Pradel E, et al. (2007) A model of bacterial intestinal infections in Drosophila melanogaster. Plos Pathog 3: e173.

56. Kurz CL, Chauvet S, Andres E, Aurouze M, Vallet I, et al. (2003) Virulence factors of the human opportunistic pathogen Serratia marcescens identified by in vivo screening. EMBO J 22: 1451–1460.

57. Flyg C, Kenne K, Boman HG (1980) Insect pathogenic properties of Serratia marcescens: phage-resistant mutants with a decreased resistance to Cecropia immunity and a decreased virulence to Drosophila. J Gen Microbiol 120: 173–181.

58. Grimont PA, Grimont F (1978) The genus Serratia. Annu Rev Microbiol 32: 221–248.

59. Mahlen SD (2011) Serratia infections: from military experiments to current practice. Clin Microbiol Rev 24: 755–791.

60. Sutherland KP, Porter JW, Turner JW, Thomas BJ, Looney EE, et al. (2010) Human sewage identified as likely source of white pox disease of the threatened Caribbean elkhorn coral, Acropora palmata. Environ Microbiol 12: 1122–1131.

61. Kiy T, Tiedtke A (1992) Mass Cultivation of Tetrahymena thermophila Yielding High Cell Densities and Short Generation Times. Appl Microbiol Biot 37: 576–579.

62. Kennedy GM, Morisaki JH, Champion PA (2012) Conserved mechanisms of Mycobacterium marinum pathogenesis within the environmental amoeba Acanthamoeba castellanii. Appl Environ Microbiol 78: 2049–2052.

63. Page FC (1976) An Illustrated Key to Freshwater and Soil Amoebae: With Notes on Cultivation and Ecology. Freshwater Biological Association.

64. La Scola B, Mezi L, Weiller PJ, Raoult D (2001) Isolation of Legionella anisa using an amoebic coculture procedure. J Clin Microbiol 39: 365–366.

65. Page FC (1988) A New Key to Freshwater and Soil Gymnamoebae with instructions for culture. Freshwater Biological Association.

66. Friman VP, Laakso J (2011) Pulsed-resource dynamics constrain the evolution of predator-prey interactions. Am Nat 177: 334–345.

67. Friman VP, Laakso J, Koivu-Orava M, Hiltunen T (2011) Pulsed-resource dynamics increase the asymmetry of antagonistic coevolution between a predatory protist and a prey bacterium. J Evol Biol 24: 2563–2573.

68. O'Toole GA, Kolter R (1998) Initiation of biofilm formation in Pseudomonas fluorescens WCS365 proceeds via multiple, convergent signalling pathways: a genetic analysis. Mol Microbiol 28: 449–461.

69. Laakso J, Loytynoja K, Kaitala V (2003) Environmental noise and population dynamics of the ciliated protozoa Tetrahymena thermophila in aquatic microcosms. Oikos 102: 663–671.

70. Wildschutte H, Wolfe DM, Tamewitz A, Lawrence JG (2004) Protozoan predation, diversifying selection, and the evolution of antigenic diversity in Salmonella. Proc Natl Acad Sci U S A 101: 10644–10649.

71. Potvin C (2001) ANOVA Experimental Layout and Analysis. In: Scheiner SM and Gurevitch J, editors. Design and Analysis of Ecological Experiments. New York: Oxford University Press. pp. 69–75.

72. Steinberg KM, Levin BR (2007) Grazing protozoa and the evolution of the Escherichia coli O157: H7 Shiga toxin-encoding prophage. Proc Biol Sci 274: 1921–1929.

73. Cosson P, Zulianello L, Join-Lambert O, Faurisson F, Gebbie L, et al. (2002) Pseudomonas aeruginosa virulence analyzed in a Dictyostelium discoideum host system. J Bacteriol 184: 3027–3033.

74. Casadevall A (2008) Evolution of intracellular pathogens. Nat Rev Microbiol 62: 19–33.

75. Cooper TF, Lenski RE (2010) Experimental tevolution with E. coli in diverse resource environments. I. Fluctuating environments promote divergence of replicate populations. BMC Evol Biol 10: 11.

76. Hall AR, Colegrave N (2008) Decay of unused characters by selection and drift. J Evol Bio 21: 610–617.

77. Ketola T, Mikonranta L, Zhang J, Saarinen K, Ormala AM, et al. (2013) Fluctuating temperature leads to evolution of thermal generalism and preadaptation to novel environments. Evolution 67: 2936–2944.

78. Travisano M, Mongold JA, Bennett AF, Lenski RE (1995) Experimental tests of the roles of adaptation, chance, and history in evolution. Science 267: 87–90.

79. Chesbro WR, Wamola I, Bartley CH (1969) Correlation of virulence with growth rate in *Staphylococcus aureus*. Can J Microbiol 15: 723–729.

80. West SA, Buckling A (2003) Cooperation, virulence and siderophore production in bacterial parasites. Proc Biol Sci 270: 37–44.

81. Gomez P, Buckling A (2011) Bacteria-phage antagonistic coevolution in soil. Science 332: 106–109.

82. Zhang J, Friman VP, Laakso J, Mappes J (2012) Interactive effects between diet and genotypes of host and pathogen define the severity of infection. Ecology and Evolution 2: 2347–2356.

18

Survival of Skin Graft between Transgenic Cloned Dogs and Non-Transgenic Cloned Dogs

Geon A Kim[1], Hyun Ju Oh[1], Min Jung Kim[1], Young Kwang Jo[1], Jin Choi[1], Jung Eun Park[1], Eun Jung Park[1], Sang Hyun Lim[2], Byung Il Yoon[3], Sung Keun Kang[2], Goo Jang[1], Byeong Chun Lee[1]*

[1] Department of Theriogenology & Biotechnology, College of Veterinary Medicine, Seoul National University, Seoul, Republic of Korea, [2] Central Research Institutes, K-stem cell, Seoul, Republic of Korea, [3] Laboratory of Histology and Molecular Pathogenesis, College of Veterinary Medicine, Kangwon National University, Chuncheon, Gangwon-do, Republic of Korea

Abstract

Whereas it has been assumed that genetically modified tissues or cells derived from somatic cell nuclear transfer (SCNT) should be accepted by a host of the same species, their immune compatibility has not been extensively explored. To identify acceptance of SCNT-derived cells or tissues, skin grafts were performed between cloned dogs that were identical except for their mitochondrial DNA (mtDNA) haplotypes and foreign gene. We showed here that differences in mtDNA haplotypes and genetic modification did not elicit immune responses in these dogs: 1) skin tissues from genetically-modified cloned dogs were successfully transplanted into genetically-modified cloned dogs with different mtDNA haplotype under three successive grafts over 63 days; and 2) non-transgenic cloned tissues were accepted into transgenic cloned syngeneic recipients with different mtDNA haplotypes and vice versa under two successive grafts over 63 days. In addition, expression of the inserted gene was maintained, being functional without eliciting graft rejection. In conclusion, these results show that transplanting genetically-modified tissues into normal, syngeneic or genetically-modified recipient dogs with different mtDNA haplotypes do not elicit skin graft rejection or affect expression of the inserted gene. Therefore, therapeutically valuable tissue derived from SCNT with genetic modification might be used safely in clinical applications for patients with diseased tissues.

Editor: Pascale Chavatte-Palmer, INRA, France

Funding: This study was supported by Rural Development Administration (#PJ008975022014), Korea Institute of Planning and Evaluation for Technology (#311062-04-3SB010), NATURE CELL (#2014-0082), Research Institute for Veterinary Science, Nestle Purina PetCare, Natural Balance Korea, and the BK21 plus program. The funders had no role in study design, data collection and analysis, decision to publish, or preparation of the manuscript.

Competing Interests: The authors received funding from NATURE CELL CO., LTD, Nestle Purina PetCare, and Natural Balance Korea. There are no further patents, products in development or marketed products to declare.

* Email: bclee@snu.ac.kr

Introduction

Somatic cell nuclear transfer (SCNT) produces genetically identical cloned animals [1]. Moreover, canine SCNT combined with transgenic technologies can make genetically identical cloned dogs with functional genetic modifications that could be used for gene therapy [2]. For example, transgenic cloned dogs could be used in replacement of diseased (malfunctioning/worn out) organs. However, tissues derived from transgenic cloned dogs, reprogrammed from somatic cells with enucleated oocytes, had not yet investigated whether they are immunologically identical tissues or cell sources of transplantation. Especially, effects of red fluorescent protein (RFP) expression using genetically identical animal models derived from SCNT have not been described and this is a critical subject since RFP has been used as a potential marker for clinical trials of gene therapy [3–5].

In addition, SCNT uses oocytes from animals unrelated to the prospective transplant recipient, oocyte-derived mitochondrial DNA (mtDNA) derived antigen could lead to rejection problems in kidney transplant [6] or not in skin transplant [7,8]. Although tissues derived from SCNT, using the recipient's somatic cells as nuclear donors, provide identical genetics, the absence of immune rejection has not yet been confirmed in cloned dogs.

To our knowledge, no previous report has mentioned *in vivo* skin immune responses against tissue expressing foreign gene or the capable effects of mitochondrial derived minor antigen in cloned animals. Here, we firstly evaluated the anti-foreign gene or minor antigen derived immune responses in cloned dogs with the following design: (1) for investigation of mtDNA derived antigen compatibility, skin graft was performed between transgenic cloned dogs with different mtDNA haplotypes; (2) furthermore, skin graft was also performed between transgenic cloned dogs and non-transgenic cloned dogs for examination of immunogenicity of foreign gene.

Materials and Methods

1. Animals

Two genetically identical cloned female beagles (C1, C2) were generated by SCNT using a beagle fetal fibroblast cell line (BF3) described in a previously study [9]. Transgenic cloned female beagles (R1, R2, R3 and R5) were also produced by SCNT using BF3 transfected with RFP [2].

Non-related controls (Co1, Co2) were healthy age-matched normal female beagles purchased from commercial kennels (Marshall Beijing Biotech Ltd., Beijing, China). All animals used in this study were cared for in accordance with recommendations described in "The Guide for the Care and Use of Laboratory Animals" published by the Institutional Animal Care and Use Committee (IACUC) of Seoul National University (approval number; SNU-110915-2). Dog housing facilities and the procedures performed met or exceeded the standards established by the Committee for Accreditation of Laboratory Animal Care. All surgery was performed under isoflurane anesthesia, and all efforts were made to minimize suffering.

2. DNA extractions and PCR reaction

Blood was collected from two control beagles and six female cloned beagles 4 years of age for DNA extractions, blood typing and blood cross-matching. Approximately 10 ml of blood were collected from the jugular vein into tubes containing EDTA as anticoagulant and used for peripheral blood mononuclear cell isolation and DNA extraction, and 3 ml of blood in plain tubes were collected to provide serum samples for antibody levels. Blood samples were kept at 38°C to maintain cell viability.

Freshly retrieved non-coagulated blood samples were mixed with RBC lysis buffer (Invitrogen, Carlsbad, CA, USA) at room temperature for 15 min. Genomic DNA was isolated according to the manufacturer's protocol. Extracted DNA samples were stored at −30°C. DLA class I (MHC class I) and II (MHC class II) typing analysis was performed by means of PCR and sequencing. The polymorphic exon 2 and exon 3 of the DLA-88 gene was amplified using PCR primers [10]. The polymorphic exon 2 of the *DRB1*, *DQA* and *DQB* genes was also amplified using PCR primers [11]. For PCR, Maxime PCR PreMix kit (iNtRON Biotechnology, Inc., Gyeongi, Korea) was used. In each PCR tube, 1 μl of genomic DNA, 1 μl (10 pM/μl) of forward primer, 1 μl (10 pM/μl) of reverse primer and 17 μl of sterilized distilled water were added according to the manufacturer's instructions. These components were then mixed and centrifuged briefly. PCR was done using a PCR machine (Biometra, Goettingen, Germany). PCR amplification was carried out for 1 cycle with denaturing at 94°C for 5 min, and subsequently for 30 cycles with denaturing at 94°C for 40 sec, annealing at 63°C (*DLA-DRB1*), 55°C (*DLA-DQA1*) and 66°C (*DLA-DQB1*) for 40 sec, extension at 72°C for 40 sec, and a final extension at 72°C for 5 min. Amplified PCR product was run on the gel by gel electrophoresis (Mupid-exu, Submarine electrophoresis system, Advance, Japan) at 100 V for 20 min. A 2% agarose gel was prepared using agarose (Invitrogen) and 1X TAE buffer. The stain (RedSafe, iNtRON Biotechnology Inc.) was used at a concentration of 2.5 μl per 50 ml of gel. After running gels, images were made under ultraviolet light. PCR product was sequenced directly using the Big Dye Terminator kit (Applied Biosystems, Foster City, CA, USA). Sequencing was performed on an automated DNA sequencer model 377 or capillary model 3110 (Applied Biosystems).

3. Sequencing of Mitochondrial DNA haplotype

For mitochondrial DNA analysis, the oligonucleotide primers were synthesized over the hypervariable regions (forward, 5′-CCTAAGACTTCAAGGAAGAAGC-3′; reverse, 5′-TTGACTGAATAGCACCTTGA-3′) of the complete nucleotide sequence of canine mtDNA (GenBank accession no. U96639). Isolated genomic DNA sample were dissolved in 50 ul TE buffer and used for PCR amplifications. It were performed in a 50 μl volume containing 5 μl of 10× reaction buffer containing 1.5 mM MgCl2, 0.2 mM dNTPs, 0.2 μM each primer, 1.5 U Taq DNA polymerase (Intron, Kyunggi, Korea). Starting denaturing for 1 cycle at 95°C for 3 minutes, subsequently denaturation at 94°C for 30 seconds, annealing at 57°C for 30 seconds, extension at 72°C for

Table 1. Mitochondrial DNA sequences of non-transgenic cloned dog (C2) and four transgenic cloned dogs (R1, R2, R3 and R5).

Sample	Nucleotide positions																					
	15435	15483	15508	15526	15595	15611	15612	15620	15627	15632	15639	15643	15650	15652	15781	15800	15814	15815	15912	15955	16025	16083
Reference[1]	G	C	C	C	C	T	T	T	A	C	T	A	T	G	C	T	C	T	C	T	T	A
C2	G	C	C	T	T	T	C	T	A	C	G	G	T	A	C	T	C	T	C	T	T	A
R1	G	C	C	C	C	T	T	T	A	C	T	A	T	G	C	T	T	T	C	C	T	A
R2	G	C	C	C	C	T	T	T	A	C	A	A	T	G	C	T	T	T	C	C	C	A
R3	G	C	C	C	C	T	C	C	G	C	A	A	T	G	C	T	T	T	C	T	T	A
R5	G	T	C	C	C	T	T	T	G	C	A	A	T	G	C	T	T	T	C	C	T	A

GenBank accession number :U96639 (Kim et al., 1998).

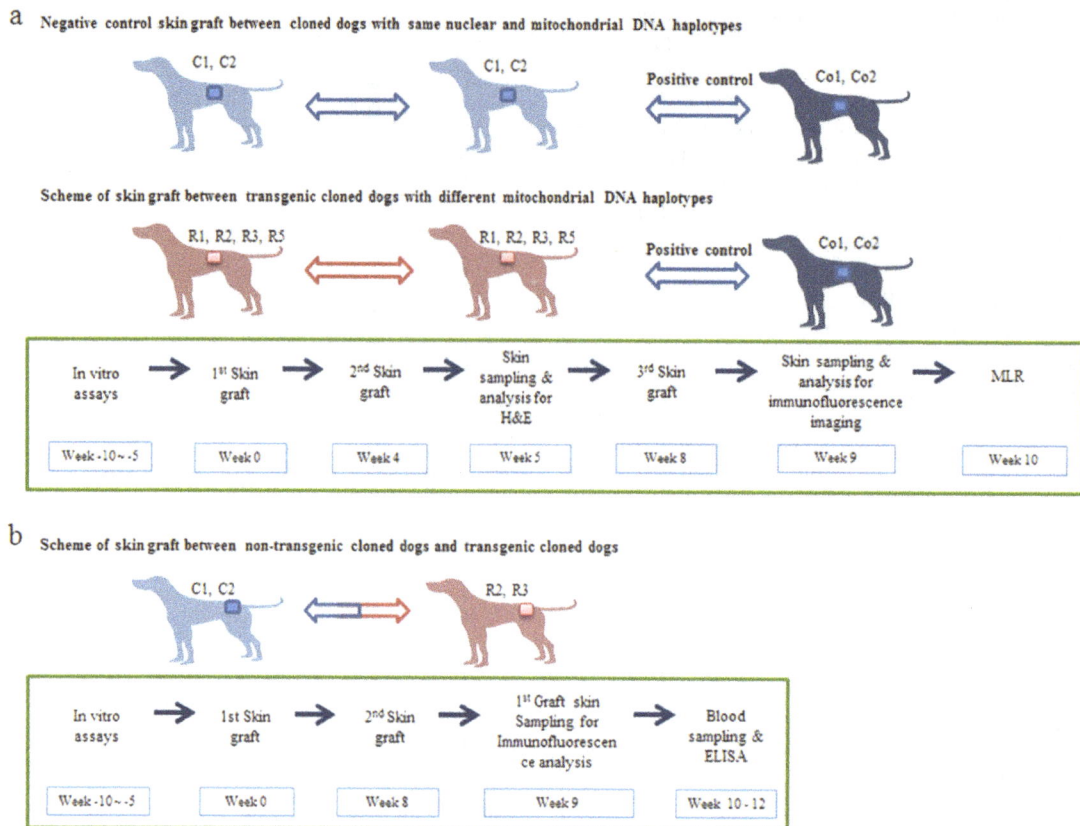

Figure 1. Experimental design and image analysis result between cloned dogs. (a) Experimental design and timeline of skin graft between cloned dogs with different mitochondrial haplotypes. As negative control, auto grafts as well as cloned dogs with same mtDNA haplotype (C1, C2) were used. Before skin graft, all *in vitro* assays were performed. For H&E staining, immunofluorescence imaging, 1st skin graft fragments were analyzed. (b) Experimental design and timeline between transgenic cloned dogs and non-transgenic cloned dogs. Before skin graft, all *in vitro* assays were performed. All dogs were tested twice for each skin graft, then skin samplings were performed. For immunofluorescence imaging, 1st skin graft fragments were analyzed. RFP expression were monitored until 63 days after skin graft.

30 seconds of 35 cycles, and a final extension at 72°C for 3 minutes were carried out. After purification of PCR products using a Gel Extraction Kit (Qiagen, Hilden, Germany), they were sequenced with an ABI3100 instrument (Applied Biosystems). Their identities with mtDNA were confirmed by BLAST search (http://blast.ncbi.nlm.nih.gov/).

4. Blood crossmatching and blood typing

Blood collection was performed from the jugular vein of all cloned dogs (R1, R2, R3, R5, C1 and C2) into an evacuated tube containing EDTA as anticoagulant. Collected samples were submitted to a commercial laboratory kit (Antech Diagnostics, Phoenix, AZ, USA). Blood type was confirmed using the tube agglutination method with antiserum; consisting of 6 types of monoclonal antibodies for canine blood typing [12].

The blood crossmatching test was done on EDTA-treated blood using the tube agglutination method. Isolated RBCs of all dogs were washed 3 times with 0.9% saline, and a 4% RBC suspension was made from the washed cells. RBC suspensions from cloned beagles (C1) were combined with equal volumes of another cloned beagle's serum (C2) and the reverse reaction was also performed. All mixtures were incubated at 37°C for 20 min, centrifuged and then assessed for hemolysis or agglutination. Agglutination was evaluated by comparing the color of supernatant in the test tube with those of the control sample. Each sample was shaken until all red blood cells in the "button" at the bottom of the tube had

become suspended. Again, the degree of RBC clumping of the test sample was compared with that of the auto-mixture of RBC and plasma. When the plasma was clear, no clumping of RBCs was detected at 400× magnification, these results were considered as negative. A positive result showed agglutination resembling stacked coins. Images were obtained using a microscope, the ProgRes Capture camera system, and the ProgRes Capture 2.6 software (JENOPTIK, Jena, Germany).

5. Peripheral blood mononuclear cell isolation and mixed lymphocyte reactions

Blood was collected from two control dogs and six female cloned dogs before and 10 weeks after skin graft. EDTA-treated whole blood was transferred to 50 ml conical centrifuge tubes. An equal volume of phosphate buffered solution (PBS, Gibco, Carlsbad, CA, USA) was mixed with the sample prior to the isolation process. Peripheral blood mononuclear cells (PBMC) were isolated from EDTA-treated blood using lymphocyte separation medium on a Ficoll-paque gradient (Ficoll-Paque Plus, GE Healthcare, Pittsburgh, PA, USA). Mixed lymphocyte reactions were modified from the previous reports [13–15]. Washed cells were diluted in culture medium (RPMI1640, Gibco) supplemented with 10% FBS to 2×10^6 cells/ml. To stimulate proliferation of lymphocytes, PBMCs were preincubated with 2 ug/ml of phytohemagglutin for 24 h before mix reaction. Then 50 ul of this cell suspension was added into each well of a 96-well

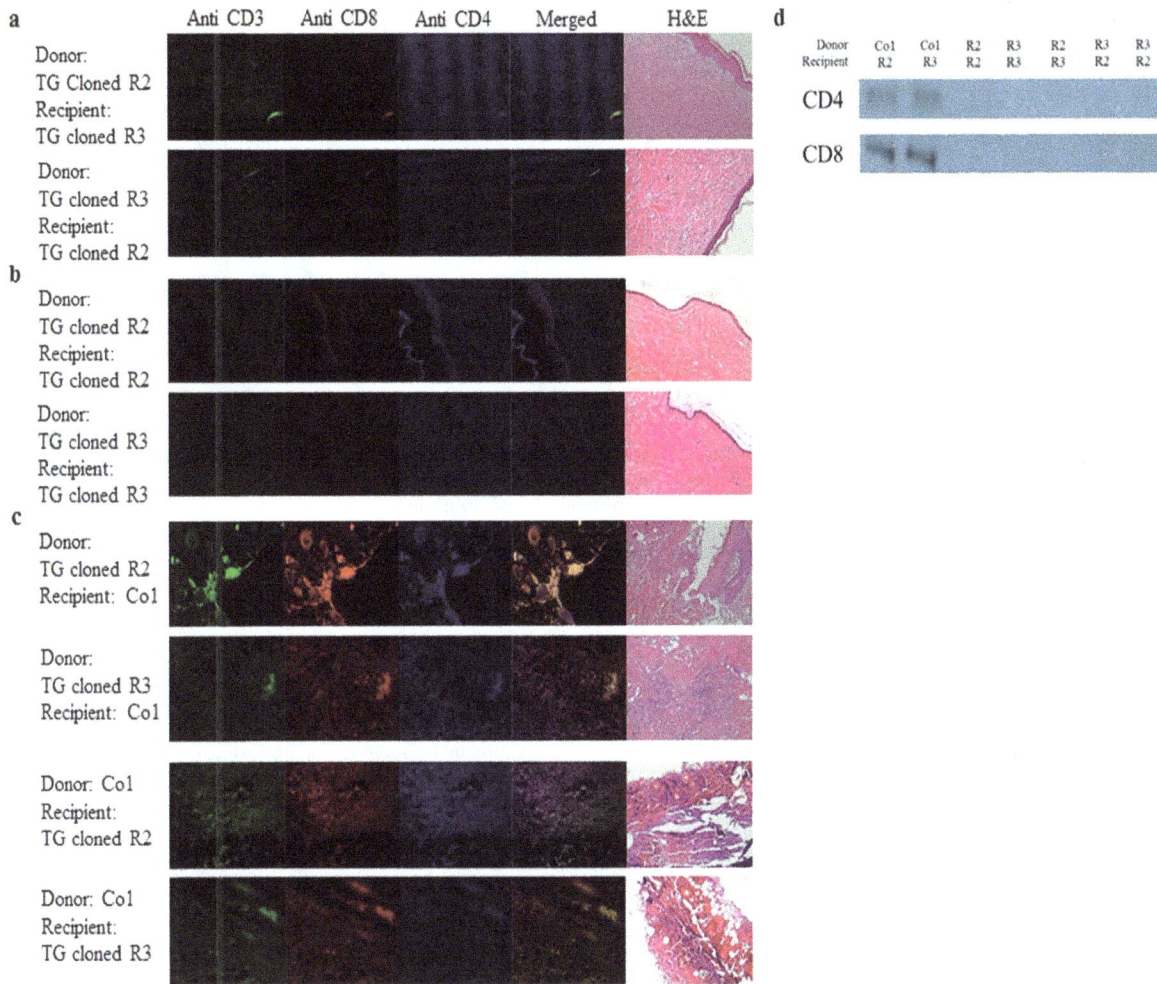

Figure 2. Absence of *in vivo* **immunogenicity in skin grafts between cloned dogs with different mitochondrial DNA sequences.** No evidence with infiltration of T cell was detected in the skin segments transplanted into the recipient dogs with different mitochondrial haplotypes (a). Sections from skin segments of autografts were used as negative controls (b). Sections from skin segments of cloned dogs transplanted into control dogs were used as positive controls (c). Western blot analysis confirms high protein levels of CD4 and CD8 in the positive controls, whereas CD4 and CD8 expression intensities were significantly lower in allograft of cloned dogs with different mitochondrial haplotypes (d). Upper lane indicates the donor dog and lower lane means recipient dog.

microplate except for the wells required for the blank and cultured at 37.5°C in a water-saturated atmosphere containing 5% CO_2. Each cell combination was tested in quadruplicate in a flat-bottomed micro plate containing 0.1 ml of culture medium per well. The mixture was cultured for 5 days and then pyrimidine analogue, bromodeoxyuridine labeling reagent (Cell proliferation ELISA, Roche Applied Science, Indianapolis, IN, USA) was added and re-incubated for 24 h. After removing the labeling medium, results are expressed as absorbance units at 450 nm wavelength read by a micro plate reader, Sunrise (Tecan Sunrise, Hayward, CA, USA). Time-course kinetics was studied by harvesting on day 7 of culture.

6. DNA walking

For confirmation of the transgene (RFP) location, PCR was performed with a DNA Walking SpeedUP Kit (Seegene Inc., Seoul, Korea) and products were gel purified (QIAquick PCR purification kit; QIAGEN, Valencia, CA, USA), and DNA strands were directly sequenced (Macrogen, Seoul, Korea; http://www.macrogen.com) using a custom-synthesized primer (5'-TCACA-GAAGTATGCCAAGCGA-3'). The sequences, except for known sequences, including primers of each product were aligned by sequence homology analysis using the Basic Local Alignment Search Tool (BLAST) at the National Center for Biotechnology Information (NCBI) GenBank (http://blast.ncbi.nlm.nih.gov/).

7. Skin graft

For skin graft procedures, experimental dogs were anesthetized with ketamine hydrochloride (6 mg/kg) after pretreatment with xylazine (0.05 mg/kg), and were maintained with 2% isoflurane in oxygen. A flank skin segment 1.5 cm×1.5 cm was excised from each donor dog. Simultaneously, the same sized skin piece was excised from recipient dogs, and the excised skin was grafted by suturing into the graft bed of the same region of an anesthetized recipient dog. Bandages were changed every day after surgery and the grafts were observed weekly.

For examination of effects mtDNA haplotypes differences among cloned dogs, skin grafts of three times were performed

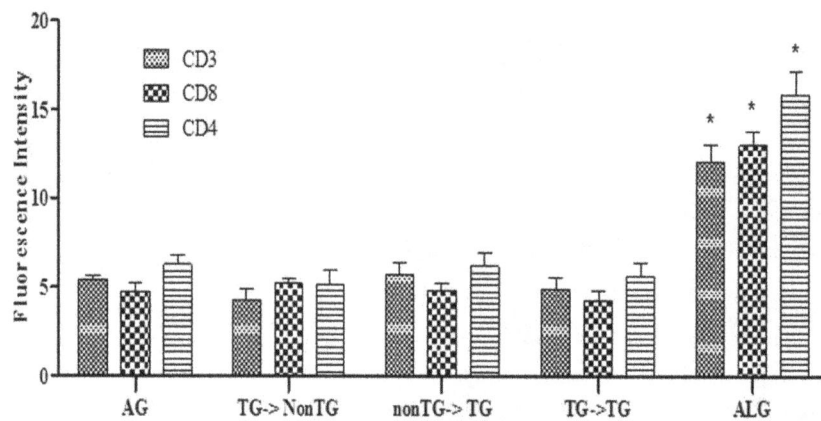

Figure 3. Expression levels of CD3, CD4 and CD8 of skin grafts between cloned dogs using fluorescence image analysis. Immunological response level of CD3, CD4 and CD8 were similar in AG (autograft), TG-> NonTG (donor: transgenic dog, recipient: non-transgenic cloned dogs), NonTG->TG (donor: non-transgenic dog, recipient: transgenic cloned dogs) and TG cloned dogs (donor: transgenic dog, recipient: transgenic cloned dogs). However, Both of ALG (allograft) between TG dogs and non-related control dogs and allograft between non-TG dogs and non-related control dogs shows significantly higher intensity of immunological response (p<0.05). Results are presented as mean ± SEM. Replication number is at least 8 times.

every 4 weeks between non-transgenic cloned dogs with same mtDNA haplotype and between transgenic cloned dogs with disparate mtDNA haplotypes. Accepted tissues were maintained until 9 weeks after skin graft. Biopsies of skin were performed after 63 days after first skin graft. A flank skin segment of 1st graft with size of 0.5 cm×1.5 cm including donor and recipient tissue were excised for H&E staining at 5 weeks of skin graft and remnant tissue were excised for immunofluorescence imaging and western blot at later.

8. Histological and immunofluorescence analysis

Immuno-staining of canine skin immune cells was carried out on formaldehyde-fixed sections using a rabbit monoclonal antibody to CD3 (1:100, ab94756, Abcam, Cambridge, MA, USA), visualized with an anti-rabbit polyclonal DyLight 488 (1:200, ab96895, Abcam) antibody. In these sections, CD4 and CD8 cells were counterstained with a CD4 (1:100, LS c122857, Lifespan Bioscience Inc., Seattle, WA, USA) and CD8 (1:200, ab22505, Abcam) specific antibody detected with a DyLight 405 (1:200, 3069-1, Abcam) and DyLight 649 (1:200, ab98389, Abcam) coupled secondary antibody. Skin sections were also processed for assessing expression of RFP using rabbit polyclonal RFP antibody (1:200, ab62341, Abcam) and visualized with an anti-rabbit polyclonal DyLight 488 (1:200, ab96895, Abcam) antibody. Sections were counterstained with 4', 6'-diamidino-2-phenylindole (DAPI).

Histology was done by fixing skin fragment in 4% neutral formalin and embedding in paraffin; sections were stained with standard hematoxylin and eosin (H&E) procedures. Fluorescent and bright field images were obtained with a Leica DMI 6000B microscope using a DFC350 camera and LAS software (Leica Microsystems Pty Ltd., North Ryde, Australia) and analyzed by a computer-assisted image analysis system (Metamorph version 6.3r2; Molecular Devices Corporation, PA, USA). To maintain a constant threshold for each image and to compensate for subtle variability of the immune-fluorescent imaging, we only counted cells that were at least 70% lighter than the average level of each positive control image after background subtraction. All image analytical procedures described above were performed blind without knowledge of the experimental scheme.

9. Western blot

Skin fragments of graft was excised and homogenized in PRO-PREP protein extraction solution (iNtRON Biotechnology, Inc.) using a tissue homogenizer. After measuring protein concentration using Nanodrop 2000 (Thermo fischer scientific, Seoul, Korea), equal amounts of proteins were loaded on 10% SDS-PAGE. Proteins were electrophoresed and blotted onto polyvinylidene fluoride membranes. The membranes were blocked with 5% skim milk in TBS with 0.1% Tween-20 and incubated with primary antibodies for 2 hours at room temperature. Monoclonal CD4 and CD8 antibodies were used as markers for immune rejection. Subsequently, membranes were incubated with goat anti-mouse IgG, anti-rat IgG (Pierce, Rockford, IL, USA) with horse radish peroxidase conjugation for 1 h at room temperature. Then, WEST-one[TM] Western blot detection system (iNtRON Biotechnology, Inc.) was added and visualized after exposing the membrane to X-ray film.

10. Statistical Analysis

The data of mixed lymphocyte reaction, image analysis of immunocytochemistry and western blot were analyzed using one-way ANOVA and a protected least significant different (LSD) test using general linear models to determine differences among experimental groups. Data were analyzed using GraphPad Prism software (GraphPad Software Inc., San Diego, CA, USA). Absorbance mean values were considered significantly different when the P-value was less than 0.05. The observations of mixed lymphocyte reaction among experimental groups were replicated at least 8 times.

Results and Discussions

It has been reported that immune rejection can occur when tissues of genetically identical SCNT cloned animals were transplanted to each other, due to the tissues having different maternally-derived antigens [6,16,17]. Antigens derived from mtDNA in accelerated skin rejection in syngeneic rodent recipients [18,19]. It has also been generally assumed that genetically-engineered tissues with insertion of a foreign gene could invoke immune-rejection by the recipient even in inbred mice [20]. Using embryonic stem cells derived from SCNT, the complete rescues of

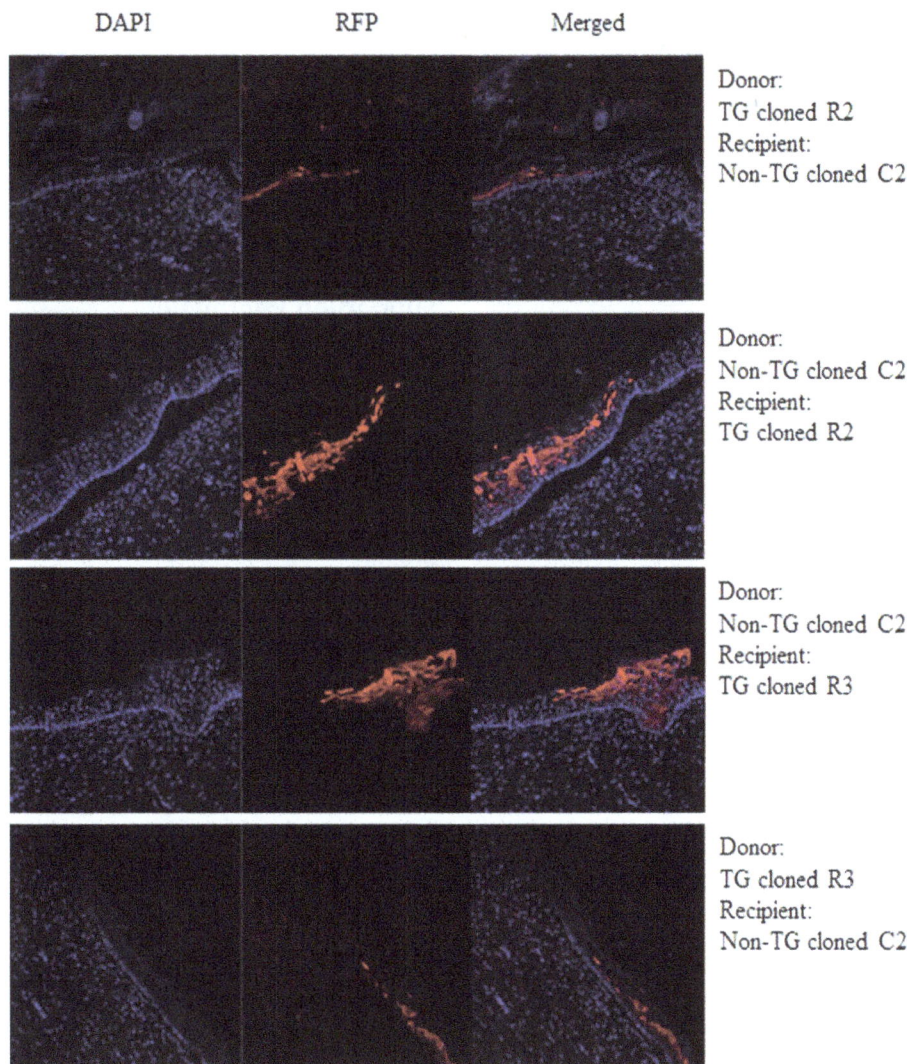

Figure 4. Maintenance of foreign gene expression between transgenic cloned beagle with foreign genes and non-transgenic cloned dogs. No expression of foreign gene in non-transgenic dog (C2) recipient was maintained in skin graft of transgenic cloned dog (R2). The limit between donor and recipient were not changed until 63 days after skin graft.

genetic defect with genetically-engineered cell therapy were not observed [21]. Engraftment of hematopoietic precursor cells differentiated from SCNT or induced pluripotent stem cells (iPSCs) was only successful in the absence of natural killer cells and immunogenicity of iPSCs was reported [21–23].

In the present study, cloned dogs produced by SCNT had different mtDNA haplotypes (Table 1), because canine SCNT used oocytes obtained from several oocyte donor dogs and the oocyte mtDNA was still present after the SCNT procedure. To examine the immunogenicity of skin tissue derived from syngeneic grafts exhibiting different mtDNA haplotypes, we initially performed *in vitro* molecular typing of dog leukocyte antigen (DLA), mixed lymphocyte reaction (MLR) and blood cross-matching using cells derived from cloned dogs with different mtDNA haplotypes (Fig. S1). Despite the different mtDNA haplotypes, they had no effects on *in vitro* immunological compatibility.

To gain insights into the therapeutic applicability of canine skin tissues with different mtDNA haplotypes, skin grafting between cloned dogs was performed to determine immunological compatibility *in vivo* (Fig. 1). Whereas allogeneic Co 1 (non-related control dogs) skin fragments were rapidly rejected in R2 (transgenic cloned dog) and R3 (transgenic cloned dog) recipients with massive infiltration of CD4+, CD8+ T cells, infiltration, edema and perivascular inflammation 7 days after 2nd skin graft, skin tissues of R2 and R3 were accepted in R3 and R2 recipients as well as autografts, without any evidence of immune rejection (Fig. 2). Likewise, skin segments from cloned dogs with different mtDNA sequences did not induce immune rejection in the recipient cloned dogs (Fig. S2). In MLR of 10 weeks after 3rd skin graft, we couldn't detect any sign of mtDNA derived minor antigen immunogenicity with no significant differences compared to those of MLR before skin graft (data not shown).

In mice, mtDNA encoded proteins could elicit rejection by innate immunity in a setting where the genomic DNA matched [24,25]. Furthermore, kidneys transplanted between cloned pigs differing in some mtDNA genes rejected those grafts [6]. Therefore, different antigenicity of grafts from different tissues could be also considered. In our experiment, despite a high level of diversity of mtDNA haplotypes heteroplasmy among domestic dogs [26], skin grafts were successfully accepted in at least 20 donor-recipient combinations. In cattle and pigs, it was shown that SCNT-derived tissues were not rejected by the immune system of the nucleus donor after SCNT in skin graft [7,27,28]. Our findings suggest that differences of canine mtDNA haplotypes could not elicit skin graft rejection among cloned dogs, as previously observed in cattle and pigs.

We also showed genetic identity between tissues of non-transgenic cloned dogs (C1, C2) derived from beagle fibroblasts (BF3) [9] and tissues of transgenic cloned dogs (R1, R2, R3 and R5) derived from BF3 transfected with RFP (Table S1.) [2]. Immunological compatibility between these dogs was completely established through *in vitro* tests such as DLA typing and MLR (Fig. S1). Skin tissues of non-transgenic cloned dogs were transplanted into transgenic cloned dogs and *vice-versa*. Skin tissues derived from cloned dogs were transplanted with no immune rejection, as determined by T cell infiltration of peri-graft skin sections after 7 days of 2nd skin graft. Despite insertion of the foreign gene RFP in transgenic cloned dogs, skin tissue from RFP transgenic cloned dogs was completely accepted in non-transgenic cloned dog recipients (Fig. 3, Fig. S3). These finding indicate that foreign gene insertion in cloned dogs did not induce a T cell-dependent skin graft rejection response in syngeneic recipients. It has been suggested that the nuclear reprogramming process in SCNT could result in surface expression of proteins and molecules unknown to the immune system of the graft recipients. In this regard, in inbred mice, enhanced GFP (eGFP) skin transplantation causes an acute reaction [29]. It was proved that eGFP also induce immune responses that interfere with its applicability in gene insertion of mouse [30]. However, our results suggest that inserted foreign gene, RFP has no immunological effects on the antigens of transgenic cloned dogs against to the non-transgenic cloned dogs. It also suggested that non-transgenic cloned dogs produced by SCNT using transfected cells have no immune regulatory effect on the host immune system and that the canine SCNT process did not result in surface expression of immunogenic molecules. Nonetheless, the possibility of immune rejection of other foreign genes, for example, pathogenically relevant transgene in clinical science remains to be confirmed.

Finally we examined whether functional expression of RFP was maintained in skin tissue grafts. During the course of this experiment, the expression level of RFP positive skin tissues were maintained for at least 63 days after surgery and RFP positive cells were detected in the epidermis, hair follicles and sebaceous glands (Fig. 4 and Table S1). It has been suggested that the co-expression of selection markers can limit or abrogate the persistence of expression of therapeutic genes [31,32]. The potential success of gene therapy or production of transgenic cloned dogs may depend on long-term transgene expression to cure or slow down the progression of disease. In addition, there were no host immune responses to the skin grafts among transgenic dogs and non-transgenic cloned dogs, and it appears that the level and duration of RFP transgene expression was not affected. This also indicates possible successful of therapeutic transplantation of tissues or cells derived from transgenic cloned dogs. In addition, the insertion site of the RFP gene into genomic DNA is not the same in all experimental dogs (Table S2). If the RFP gene insertion site can affect the immune response, it should affect the results of syngeneic skin grafting. However, no immune rejection was apparent in skin grafts with different transgene insertion sites. Our findings indicate that SCNT-derived somatic cells with or without foreign genes can be accepted in syngeneic recipients.

Our study established that tissues derived from canine SCNT can be accepted in syngeneic recipients despite different mtDNA haplotypes. We also provide evidence that skin segments containing a foreign gene are sufficiently acceptable to syngeneic recipients with or without the foreign gene. Taken together, these data indicate that SCNT using transgenic technology can support immunological compatibility between genetically engineered tissues and patients and thereby help to accelerate clinical therapeutic research and its applications.

Supporting Information

Figure S1 Immunological feature of transgenic dogs and non-transgenic dogs. (a) Molecular typing of dog leukocyte antigen, DLA-88 (MHC class I), DRB, DQA1, DQB1(MHC class II) polymorphic region in all cloned dogs (C1, C2, R1, R2, R3, and R5). (b) *In vitro* immunogenicity test using mixed lymphocyte reaction between all experimental dogs before skin graft. (c) Blood typing. (d) Analysis of blood crossmatching in all cloned dogs and control dogs.

Figure S2 Fluorescence image analysis of skin grafts between cloned dogs with different mtDNA haplotypes

Figure S3 Absence of *in vivo* immune rejection between non-transgenic dogs and transgenic dogs. (a) Positive control of skin graft, as donor skin segments were derived from non-related control dogs (Co1, Co2), they were completely rejected in the graft bed in transgenic cloned dogs (R2, R3). (b) However, skin grafts between a transgenic cloned dog, R2 and a non-transgenic cloned dog, C1 showed no apparent immune rejection. Similarly, as shown in (c) R2 - C2, (d) R3-C1, (e) R3-C2, there was no immune rejection in these grafts as well. (f) Western blot analysis of the skin graft between cloned dogs confirmed the expression of CD4 and CD8 protein only in the graft between cloned dogs and non-related control dogs.

Figure S4 Foreign gene expression between skin graft of two transgenic dogs (R2, R3). Red fluorescent protein expression in skin graft was maintained after 63 days skin graft in syngenic graft beds.

Table S1 Genetic background for microsatellite analysis of two non-transgenic cloned dogs and four transgenic cloned dogs.

Table S2 Insertion site of foreign gene, RFP in transgenic cloned dogs.

Acknowledgments

We thank Won Woo Lee for critical reading of the manuscript. We would also like to thank Dr, Barry D. Bavister for his valuable editing of the manuscript.

Author Contributions

Conceived and designed the experiments: GAK HJO MJK SKK GJ BCL. Performed the experiments: GAK YKJ JC JEP EJP SHL. Analyzed the data: GAK HJO BIY BCL. Contributed reagents/materials/analysis tools: JEP BIY BCL. Wrote the paper: GAK HJO SKK BCL.

References

1. Lee BC, Kim MK, Jang G, Oh HJ, Yuda F, et al. (2005) Dogs cloned from adult somatic cells. Nature 436: 641.

2. Hong SG, Kim MK, Jang G, Oh HJ, Park JE, et al. (2009) Generation of red fluorescent protein transgenic dogs. Genesis 47: 314–322.

3. Chang RS, Suh MS, Kim S, Shim G, Lee S, et al. (2011) Cationic drug-derived nanoparticles for multifunctional delivery of anticancer siRNA. Biomaterials 32: 9785–9795.

4. Lee CY, Li JF, Liou JS, Charng YC, Huang YW, et al. (2011) A gene delivery system for human cells mediated by both a cell-penetrating peptide and a piggyBac transposase. Biomaterials 32: 6264–6276.

5. Kinoshita Y, Kamitani H, Mamun MH, Wasita B, Kazuki Y, et al. (2010) A gene delivery system with a human artificial chromosome vector based on migration of mesenchymal stem cells towards human glioblastoma HTB14 cells. Neurol Res 32: 429–437.

6. Kwak HH, Park KM, Teotia PK, Lee GS, Lee ES, et al. (2013) Acute rejection after swine leukocyte antigen-matched kidney allo-transplantation in cloned miniature pigs with different mitochondrial DNA-encoded minor histocompatibility antigen. Transplant Proc 45: 1754–1760.

7. Martin MJ, Yin D, Adams C, Houtz J, Shen J, et al. (2003) Skin graft survival in genetically identical cloned pigs. Cloning Stem Cells 5: 117–121.

8. Theoret CL, Dore M, Mulon PY, Desrochers A, Viramontes F, et al. (2006) Short- and long-term skin graft survival in cattle clones with different mitochondrial haplotypes. Theriogenology 65: 1465–1479.

9. Hong SG, Jang G, Kim MK, Oh HJ, Park JE, et al. (2009) Dogs cloned from fetal fibroblasts by nuclear transfer. Anim Reprod Sci 115: 334–339.

10. Burnett RC, DeRose SA, Wagner JL, Storb R (1997) Molecular analysis of six dog leukocyte antigen class I sequences including three complete genes, two truncated genes and one full-length processed gene. Tissue Antigens 49: 484–495.

11. Kennedy IJ (2007) 14th International HLA and Immunogenetics Workshop: report on joint study on canine DLA diversity. Tissue Antigens 69 Suppl 1: 269–271.

12. Ogawa H, Galili U (2006) Profiling terminal N-acetyllactosamines of glycans on mammalian cells by an immuno-enzymatic assay. Glycoconj J 23: 663–674.

13. Gluckman JC (1980) [Modification of mixed lymphocyte reactivity between DLA-identical dog sibs, after in vivo sensitization]. C R Seances Acad Sci D 290: 105–108.

14. Kolb HJ, Rieder I, Grosse-Wilde H, Scholz S, Kolb H, et al. (1975) Canine marrow grafts in donor-recipient combinations with one-way nonstimulation in mixed lymphocyte culture. Transplant Proc 7: 461–464.

15. Widmer MB, Bach FH (1972) Allogeneic and xenogeneic response in mixed leukocyte cultures. J Exp Med 135: 1204–1208.

16. Do M, Jang WG, Hwang JH, Jang H, Kim EJ, et al. (2012) Inheritance of mitochondrial DNA in serially recloned pigs by somatic cell nuclear transfer (SCNT). Biochem Biophys Res Commun 424: 765–770.

17. Hiendleder S (2007) Mitochondrial DNA inheritance after SCNT. Adv Exp Med Biol 591: 103–116.

18. Chan T, Fischer Lindahl K (1985) Skin graft rejection caused by the maternally transmitted antigen Mta. Transplantation 39: 477–480.

19. Lindahl KF, Burki K (1982) Mta, a maternally inherited cell surface antigen of the mouse, is transmitted in the egg. Proc Natl Acad Sci U S A 79: 5362–5366.

20. Andersson G, Illigens BM, Johnson KW, Calderhead D, LeGuern C, et al. (2003) Nonmyeloablative conditioning is sufficient to allow engraftment of EGFP-expressing bone marrow and subsequent acceptance of EGFP-transgenic skin grafts in mice. Blood 101: 4305–4312.

21. Rideout WM 3rd, Hochedlinger K, Kyba M, Daley GQ, Jaenisch R (2002) Correction of a genetic defect by nuclear transplantation and combined cell and gene therapy. Cell 109: 17–27.

22. Hanna J, Wernig M, Markoulaki S, Sun CW, Meissner A, et al. (2007) Treatment of sickle cell anemia mouse model with iPS cells generated from autologous skin. Science 318: 1920–1923.

23. Zhao T, Zhang ZN, Rong Z, Xu Y (2011) Immunogenicity of induced pluripotent stem cells. Nature 474: 212–215.

24. Ishikawa K, Toyama-Sorimachi N, Nakada K, Morimoto M, Imanishi H, et al. (2010) The innate immune system in host mice targets cells with allogenic mitochondrial DNA. J Exp Med 207: 2297–2305.

25. Loveland B, Wang CR, Yonekawa H, Hermel E, Lindahl KF (1990) Maternally transmitted histocompatibility antigen of mice: a hydrophobic peptide of a mitochondrially encoded protein. Cell 60: 971–980.

26. Webb KM, Allard MW (2009) Mitochondrial genome DNA analysis of the domestic dog: identifying informative SNPs outside of the control region. J Forensic Sci 54: 275–288.

27. Lanza RP, Chung HY, Yoo JJ, Wettstein PJ, Blackwell C, et al. (2002) Generation of histocompatible tissues using nuclear transplantation. Nat Biotechnol 20: 689–696.

28. Oiso N, Fukai K, Kawada A, Suzuki T (2013) Piebaldism. J Dermatol 40: 330–335.

29. Lu F, Gao JH, Mizuro H, Ogawa R, Hyakusoku H (2007) [Experimental study of adipose tissue differentiation using adipose-derived stem cells harvested from GFP transgenic mice]. Zhonghua Zheng Xing Wai Ke Za Zhi 23: 412–416.

30. Stripecke R, Carmen Villacres M, Skelton D, Satake N, Halene S, et al. (1999) Immune response to green fluorescent protein: implications for gene therapy. Gene Ther 6: 1305–1312.

31. Riddell SR, Elliott M, Lewinsohn DA, Gilbert MJ, Wilson L, et al. (1996) T-cell mediated rejection of gene-modified HIV-specific cytotoxic T lymphocytes in HIV-infected patients. Nat Med 2: 216–223.

32. Bonini C, Ferrari G, Verzeletti S, Servida P, Zappone E, et al. (1997) HSV-TK gene transfer into donor lymphocytes for control of allogeneic graft-versus-leukemia. Science 276: 1719–1724.

PERMISSIONS

LIST OF CONTRIBUTORS

Lamprini G. Kalampoki, Constantin N. Flytzanis
Department of Biology, University of Patras, Patras 26504, Greece

Michael H.-L. Lai, Catherine M. Lanagan, Linda E. O'Connor
Nathan R. Martinez and Christopher W. Schmidt
Cancer Immunotherapy Laboratory, QIMR Berghofer Medical Research Institute, Brisbane, Queensland, Australia

Antonia L. Pritchard
Oncogenomics Laboratory, QIMR Berghofer Medical Research Institute, Brisbane, Queensland, Australia

Michelle A. Neller
Cancer Immunotherapy Laboratory, QIMR Berghofer Medical Research Institute, Brisbane, Queensland, Australia
School of Medicine, The University of Queensland Mayne Medical School, Brisbane, Queensland, Australia

Carla M. R. Varanda, Marco Machado, Maria I. E. Clara and Maria R. Félix
Laboratório de Virologia Vegetal, Instituto de Ciências Agráriase Ambientais Mediterrânicas Universidade de Évora, Évora, Portugal

Paulo Martel
Departamento de Ciências Bioló gicase Bioengenharia, Faculdade de Ciênciase Tecnologia da Universidade do Algarve, Faro, Portugal

Gustavo Nolasco
Laboratório de Virologia Vegetal, Universidade do Algarve, Faro, Portugal

Matthias T. Buhmann, Nicole Poulsen and Jennifer Klemm
BCUBE Center for Molecular Bioengineering, Technische Universität Dresden, Dresden, Germany

Nils Kräger
BCUBE Center for Molecular Bioengineering, Technische Universität Dresden, Dresden, Germany
Department of Chemistry and Food Chemistry, Technische Universität Dresden, Dresden, Germany

Matthew R. Kennedy and C. David Sherrill
School of Chemistry and Biochemistry, Georgia Institute of Technology, Atlanta, Georgia, United States of America

Vijay K. Chaudhary, Nimisha Shrivastava, Vaishali Verma, Shilpi Das, Charanpreet Kaur and Payal Grover
Department of Biochemistry, University of Delhi South Campus, Benito Juarez Road, New Delhi 110021, India

Amita Gupta
Department of Microbiology, University of Delhi South Campus, Benito Juarez Road, New Delhi 110021, India

Juliana Caierão, Fernando Hayashi Sant'anna, Gabriela Rosa da Cunha, Pedro Alves d'Azevedo and Cícero Dias
Federal University of Health Science of Porto Alegre, Rio Grande do Sul, Brazil

Paulina Hawkins
Emory University, Atlanta, Georgia, United States of America

Lesley McGee
Emory University, Atlanta, Georgia, United States of America
Centers for Disease Control and Prevention, Atlanta, Georgia, United States of America

Cristiana Leite, N. Tatiana Silva, Tânia Lourenço and Má rio Grãos
Biocant - Technology Transfer Association, Biocant Park, Cantanhede, Portugal

Sandrine Mendes, Andreia Ribeiro and Artur Paiva
Blood and Transplantation Center of Coimbra, Portuguese Institute of the Blood and Transplantation, Coimbra, Portugal

Joana Paes de Faria and João B. Relvas
Instituto de Biologia Moleculare Celular, Porto, Portugal

Francisco dos Santos, Pedro Z. Andrade Cláudia L. da Silva and Joaquim M. S. Cabral
Institute for Biotechnology and Bioengineering and Department of Bioengineering, Instituto Superior Técnico, Universidade de Lisboa, Lisboa, Portugal

Carla M. P. Cardoso and Margarida Vieira
Crioestaminal Saúde e Tecnologia, S.A., Biocant Park, Cantanhede, Portugal

Patrick Chinestra, Cyril Inard and Jean-Charles Faye
Inserm, UMR 1037-CRCT, GTPases Rho dans la progression tumorale, Toulouse, France

Gilles Favre
Inserm, UMR 1037-CRCT, GTPases Rho dans la progression tumorale, Toulouse, France
Université Toulouse III-Paul Sabatier, Facultédes Sciences Pharmaceutiques,

Toulouse, France
Institut Claudius Regaud, Toulouse, France

Aurélien Olichon, Claire Medale-Giamarchi, Isabelle Lajoie-Mazenc and Rémi Gence
Université Toulouse III-Paul Sabatier, Facultédes Sciences Pharmaceutiques, Toulouse, France
Inserm, UMR 1037-CRCT, GTPases Rho dans la progression tumorale, Toulouse, France

Laetitia Ligat
CRCT, plateau de protéomique, Toulouse, France

Yung-Heng Chang and Yi Henry Sun
Graduate Institute of Life Sciences, National Defense Medical Center, Taipei, Taiwan, Republic of China
Institute of Molecular Biology, Academia Sinica, Taipei, Taiwan, Republic of China

Krishna L. Kanchi, Nathan D. Dees, Charles Lu, Christopher A. Miller,

Michael C. Wendl and Joshua F. McMichael
The Genome Institute, Washington University in St. Louis, St. Louis, Missouri, United States of America

Obi Griffith
The Genome Institute, Washington University in St. Louis, St. Louis, Missouri, United States of America
Department of Medicine, Washington University in St. Louis, St. Louis, Missouri, United States of America

Malachi Griffith
The Genome Institute, Washington University in St. Louis, St. Louis, Missouri, United States of America
Department of Genetics, Washington University in St. Louis, St. Louis, Missouri, United States of America

Richard K. Wilson
The Genome Institute, Washington University in St. Louis, St. Louis, Missouri, United States of America
Department of Genetics, Washington University in St. Louis, St. Louis, Missouri, United States of America
Siteman Cancer Center, Washington University in St. Louis, St. Louis, Missouri, United States of America

Vernon K. Sondak
The Genome Institute, Washington University in St. Louis, St. Louis, Missouri, United States of America
Donald A. Adam Comprehensive Melanoma Research Center, Moffitt Cancer Center, Tampa, Florida, United States of America

Li Ding
The Genome Institute, Washington University in St. Louis, St. Louis, Missouri, United States of America
Department of Medicine, Washington University in St. Louis, St.
Department of Genetics, Washington University in St. Louis, St. Louis, Missouri, United States of America
Siteman Cancer Center, Washington University in St. Louis, St. Louis, Missouri, United States of America

Timothy J. Ley
The Genome Institute, Washington University in St. Louis, St. Louis, Missouri, United States of America
Department of Medicine, Washington University in St. Louis, St. Louis, Missouri, United States of America
Siteman Cancer Center, Washington University in St. Louis, St. Louis, Missouri, United States of America

Gerald P. Linette
Department of Medicine, Washington University in St. Louis, St.
Louis, Missouri, United States of America
Siteman Cancer Center, Washington University in St. Louis, St. Louis, Missouri, United States of America

Lynn A. Cornelius
Department of Medicine, Washington University in St. Louis, St.
Louis, Missouri, United States of America
Department of Surgery, Washington University in St. Louis, St. Louis, Missouri, United States of America

Ryan C. Fields
Siteman Cancer Center, Washington University in St. Louis, St. Louis, Missouri, United States of America
Department of Surgery, Washington University in St. Louis, St. Louis, Missouri, United States of America

Minjung Kim, David Fenstermacher, Hyeran Sung, James J. Mulé and Jeffrey S. Weber
Donald A. Adam Comprehensive Melanoma Research Center, Moffitt Cancer Center, Tampa, Florida, United States of America

Brian Goetz
Department of Surgery, Washington University in St. Louis, St. Louis, Missouri, United States of America

Batzaya Davaadelger, Hong Shen, Carl G. Maki
Department of Anatomy and Cell Biology, Rush University Medical Center, Chicago, Illinois, United States of America

Andrá s Vida, Tímea Ocskó , Beata Tryniszewska, Tibor A. Rauch, Tibor T. Glant and Katalin Mikecz
Section of Molecular Medicine, Department of Orthopedic Surgery, Rush University Medical Center, Chicago, Illinois, United States of America

Jú lia Kurkó
Section of Molecular Medicine, Department of Orthopedic Surgery, Rush University Medical Center, Chicago, Illinois, United States of America
Department of Rheumatology, University of Debrecen, Faculty of Medicine, Debrecen, Hungary

Zoltán Szekanecz
Department of Rheumatology, University of Debrecen, Faculty of Medicine, Debrecen, Hungary

Kenneth Day, Jun Song, Devin Absher
HudsonAlpha Institute for Biotechnology, Huntsville, Alabama, United States of America

Shuo Yan, Weilong Zhu, Xinfang Zhang, Zhen Li, Xiaoxia Liu and Qingwen Zhang
Department of Entomology, China Agricultural University, Beijing, P.R. China

Jialin Zhu
Beijing Entry-Exit Inspection and Quarantine Bureau, Beijing, P.R. China

Yee Sun Tan
Center for Stem Cell Biology & Regenerative Medicine, University of Maryland School of Medicine, Baltimore, Maryland, United States of America

Tami J. Kingsbury
Center for Stem Cell Biology & Regenerative Medicine, University of Maryland School of Medicine, Baltimore, Maryland, United States of America
Greenebaum Cancer Center, University of Maryland School of Medicine, Baltimore, Maryland, United States of America
Department of Physiology, University of Maryland School of Medicine,
Baltimore, Maryland, United States of America

Curt I. Civin
Center for Stem Cell Biology & Regenerative Medicine, University of Maryland School of Medicine, Baltimore, Maryland, United States of America
Greenebaum Cancer Center, University of Maryland School of Medicine, Baltimore, Maryland, United States of America
Department of Physiology, University of Maryland School of Medicine,
Baltimore, Maryland, United States of America
Department of Pediatrics, University of Maryland School of Medicine, Baltimore, Maryland, United States of America

MinJung Kim
Center for Stem Cell Biology & Regenerative Medicine, University of Maryland School of Medicine, Baltimore, Maryland, United States of America
Department of Physiology, University of Maryland School of Medicine,
Baltimore, Maryland, United States of America
Department of Pediatrics, University of Maryland School of Medicine, Baltimore, Maryland, United States of America

Wen-Chih Cheng
Center for Stem Cell Biology & Regenerative Medicine, University of Maryland School of Medicine, Baltimore, Maryland, United States of America
Department of Pediatrics, University of Maryland School of Medicine, Baltimore, Maryland, United States of America

Cameron Fox, Qiuping Ma and Gabrielle S. Wong
Department of Medical Oncology, Dana-Farber Cancer Institute, Boston, Massachusetts, United States of America

Yong Sang Hong
Department of Medical Oncology, Dana-Farber Cancer Institute, Boston, Massachusetts, United States of America
Department of Oncology, Asan Medical Center, University of Ulsan College of Medicine, Seoul, Korea

Jihun Kim
Department of Medical Oncology, Dana-Farber Cancer Institute, Boston, Massachusetts, United States of America
Department of Pathology, Asan Medical Center, University of Ulsan College of Medicine, Seoul, Korea

Eirini Pectasides
Department of Medical Oncology, Dana-Farber Cancer Institute, Boston, Massachusetts, United States of America
Division of Hematology/Oncology, Beth Israel Deaconess Medical Center, Boston, Massachusetts, United States of America

Seung-Woo Hong
Innovative Cancer Research, Asan Institute for Life Science, Asan Medical Center, University of Ulsan College of Medicine, Seoul, Korea

Shouyong Peng
Department of Medical Oncology, Dana-Farber Cancer Institute, Boston, Massachusetts, United States of America

Cancer Program, The Broad Institute of MIT and Harvard, Cambridge, Massachusetts, United States of America

Matthew D. Stachler
Department of Medical Oncology, Dana-Farber Cancer Institute, Boston, Massachusetts, United States of America
Department of Pathology, Brigham and Women's Hospital, Boston, Massachusetts, United States of America

Aaron R. Thorner and Paul Van Hummelen
Center for Cancer Genome Discovery, Dana-Farber Cancer Institute, Boston, Massachusetts, United States of America

Adam J. Bass
Department of Medical Oncology, Dana-Farber Cancer Institute, Boston, Massachusetts, United States of America
Cancer Program, The Broad Institute of MIT and Harvard, Cambridge, Massachusetts, United States of America
Department of Medicine, Harvard Medical School, Boston, Massachusetts,
United States of America
Department of Medicine, Brigham and Women's Hospital, Boston, Massachusetts, United States of America

Tarmo Ketola and Johanna Mappes
Centre of Excellence in Biological Interactions, Department of Biological and Environmental Science, University of JyväskyläJyväskylä Finland

Ji Zhang and Jouni Laakso
Centre of Excellence in Biological Interactions, Department of Biological and Environmental Science, University of JyväskyläJyväskylä, Finland
Department of Biological and Environmental Science, University of Helsinki, Helsinki, Finland

Anni-Maria ŐrmäläOdegrip
Department of Biological and Environmental Science, University of Helsinki, Helsinki, Finland

Geon A Kim, Hyun Ju Oh, Min Jung Kim, Young Kwang Jo, Jin Choi, Jung Eun Park, Eun Jung Park, Goo Jang and Byeong Chun Lee
Department of Theriogenology & Biotechnology, College of Veterinary Medicine, Seoul National University, Seoul, Republic of Korea

Sang Hyun Lim and Sung Keun Kang
Central Research Institutes, Kstem
cell, Seoul, Republic of Korea

Byung Il Yoon
Laboratory of Histology and Molecular Pathogenesis, College of Veterinary Medicine, Kangwon National University, Chuncheon, Gangwon-do, Republic of Korea

Index